Space Nuclear Propulsion and Power

Principles, Systems, and Applications

Bahram Nassersharif, Ph.D.
Distinguished University Professor
Department of Mechanical, Industrial, and Systems Engineering
University of Rhode Island, United States of America

Table of Contents

List of Figures

List of Tables

Preface

Welcome to *Space Nuclear Propulsion and Power: Principles, Systems, and Applications*, a comprehensive examination of the principles and technologies underpinning space nuclear propulsion and power systems. This volume serves as an essential resource for students, educators, and professionals interested in the fundamental and applied aspects of space nuclear engineering.

The domain of space nuclear propulsion and power encompasses a wide range of technologies and applications, from the theoretical foundations of nuclear reactions to the practical considerations of reactor design and safety management. As humanity ventures further into space, the demand for efficient and reliable propulsion and power systems becomes increasingly critical. This text not only delves into the complexities of these technologies but also highlights their potential to revolutionize space exploration and expand our capabilities beyond Earth, inspiring hope and excitement for the future.

This book is structured to provide a thorough grounding in the subject, beginning with an introduction to the basic concepts of space propulsion and power systems. It covers chemical propulsion and nuclear thermal propulsion, highlighting their respective roles in various phases of space missions. Subsequent chapters delve into more advanced topics, including the design and operation of nuclear reactors for space applications, safety considerations, and the latest innovations in the field.

Throughout the text, we emphasize the integration of theory and practice, providing detailed explanations, illustrative examples, and problem-solving exercises. The book aims to not only impart technical knowledge but also to inspire curiosity and innovation among its readers. It challenges them to think critically about the future of space exploration and the role of nuclear technology in that future.

Space Nuclear Propulsion and Power is more than a textbook; it is a call to explore the frontiers of space technology. Whether you are a student embarking on your journey into the field or a seasoned professional seeking to expand your knowledge, this book offers valuable insights and guidance. We invite you to engage with the material, challenge the concepts presented, and contribute to the ongoing research and development that propels this fascinating field forward. Your role is integral to the future of space technology, and we value your contributions.

May this book serve as a guide and inspiration as you navigate the challenges and opportunities of space nuclear engineering. Welcome to an exploration of the technologies that will shape the future of space exploration and our understanding of the universe.

Bahram Nassersharif, Ph.D.
August 2024

1 Introduction

Propulsion and power systems form the backbone of space exploration, enabling humanity to venture beyond the confines of Earth and explore the wonders of our solar system and beyond. This book, "Space Nuclear Propulsion and Power Systems," is designed to provide a comprehensive overview of the principles, mechanisms, and applications of various nuclear propulsion and power systems for space exploration.

As we are positioned at the cusp of a new era in space exploration, with ambitious plans for returning to the Moon, establishing a human presence on Mars, and venturing even further into the solar system, the need for advanced propulsion and power technologies has never been more critical. This book aims to equip the reader with an understanding of principles, concepts, and systems preparing them for the challenges and opportunities that lie ahead in the realm of space exploration.

The content of this book is based on lecture notes developed for a course on Space Nuclear Propulsion and Power, designed at the level of a professional senior engineering elective course and also suitable for graduate students. While the focus is on nuclear propulsion and power systems, the book provides a holistic view of space propulsion technologies, including chemical propulsion and touching on emerging concepts like aero nuclear propulsion. This comprehensive approach allows readers to understand the strengths, limitations, and potential applications of each technology in the context of various space missions.

Our journey through the realm of space propulsion begins with an exploration of chemical propulsion systems, which have been the workhorses of space exploration since its inception. We delve into the fundamentals of chemical propulsion, examining the principles of thrust generation through exothermic reactions and the expulsion of high-temperature gases.

A cornerstone of our discussion on chemical propulsion is the Tsiolkovsky rocket equation, often referred to as the ideal rocket equation. This fundamental principle describes the relationship between a rocket's velocity change and the effective exhaust velocity of its propellant. We explore the derivation of this equation, its implications for spacecraft design and limitations in real-world applications.

As we move beyond chemical propulsion, we enter the realm of nuclear propulsion systems, which offer the promise of higher specific impulse and greater efficiency for long-duration missions. We provide an in-depth examination of Nuclear Thermal Propulsion (NTP) systems, exploring their fundamental principles, types (such as solid, liquid, and gas core nuclear thermal rockets), and potential applications.

The mechanics of nuclear thermal rockets are discussed in detail, including the process of heat transfer from the nuclear reactor to the propellant and the factors affecting efficiency. We examine the challenges and opportunities presented by NTP systems, including their potential for enabling faster transit times to distant planets and the technical hurdles that must be overcome for their practical implementation.

Complementing our discussion of propulsion systems is an exploration of space nuclear power systems. We delve into the principles and applications of Radioisotope Power Systems (RPS), including Radioisotope Thermoelectric Generators (RTGs) and Radioisotope Heater Units (RHUs). These technologies have been crucial in powering spacecraft on long-duration missions far from the Sun, where solar power is impractical. We examine their design, operation, and the missions where they have been employed, such as the Voyager probes and the New Horizons mission to Pluto.

The book also covers the potential of nuclear reactors for space applications, both for power generation and as part of a Nuclear Electric Propulsion (NEP) systems. We explore how these technologies could enable sustained human presence on the Moon and Mars, powering habitats, life support systems, and resource extraction activities in environments where traditional power sources are insufficient or unreliable.

As we look to the future of space propulsion, we examine emerging concepts such as aero nuclear propulsion. This technology, which involves using atmospheric air as part of the propulsion process during the ascent phase of space launches, presents intriguing possibilities for reducing launch costs and increasing payload capacities. We explore various aero nuclear propulsion systems, including turbojets, ramjets, and the cutting-edge scramjet technology, discussing their principles of operation and potential applications in space launch systems.

Throughout the book, we emphasize the practical applications of these propulsion and power technologies through many example problems. We examine Earth surface launch concepts, discussing the use of multi-stage rockets and the integration of aeropropulsion systems to enhance launch efficiency. We also explore orbital maneuvers and the specific propulsion requirements for different types of missions, from satellite deployment to interplanetary travel.

A key theme running through our discussions is the interplay between different propulsion and power technologies. We examine how various systems can be combined or used in complementary ways to meet the diverse requirements of space missions. For instance, we explore how chemical propulsion might be used for initial launch and orbital insertion, while nuclear thermal or electric propulsion could be employed for interplanetary transit.

As we delve into these topics, we also address the broader implications of advanced space propulsion and power technologies. We consider the potential impact on future space exploration missions, the possibilities for human expansion into the solar system, and the challenges that must be overcome to realize these ambitious goals. This includes discussions on the technical, economic, and regulatory aspects of developing and deploying these technologies.

The book also touches upon the environmental and safety considerations associated with nuclear propulsion and power systems, both in terms of their use in space and the potential risks during launch or re-entry. We examine the measures taken to ensure the safe use of these technologies and the ongoing research aimed at minimizing potential risks.

In addition to covering established technologies, we look ahead to the future of space propulsion and power. We explore emerging concepts and technologies that are currently in the research and development phase, such as advanced electric propulsion systems, fusion propulsion, and even more speculative ideas like antimatter propulsion. While some of these concepts may seem like science fiction today, they represent the cutting edge of propulsion research and could potentially revolutionize space travel in the coming decades.

Throughout the book, we strive to balance theoretical understanding with practical applications. Each chapter includes examples and case studies, illustrating how the theories discussed can be applied in practice. This approach helps to ground the theoretical concepts in practical situations and provides valuable insights into the engineering challenges involved.

The mathematical and physical principles underlying space propulsion and power systems are presented in a clear and accessible manner, with derivations and explanations provided to help readers develop an understanding of the fundamental concepts. At the same time, we recognize that the field of space propulsion is rapidly evolving, and we encourage readers to think critically and creatively about the potential for innovation and improvement in these technologies.

As we embark on this journey of learning about space nuclear systems, we invite readers to embrace the spirit of exploration and innovation that has driven humanity's ventures into space.

Whether you are a student embarking on a career in engineering, a professional seeking to deepen your understanding of space technologies, or simply an enthusiast fascinated by the possibilities of space exploration, this book aims to provide you with a comprehensive and engaging exploration of space nuclear systems.

So, let us begin our exploration of the nuclear technologies that will propel humanity into the future of space exploration. The journey ahead is challenging, but the potential rewards are limitless.

1.1   Chemical Propulsion

Chemical propulsion is based on exothermic chemical reactions that produce high-temperature and high-pressure gases, which are expelled through a nozzle to generate thrust. This system exploits Newton's third law of motion, *i.e.,* for every action, there is an equal and opposite reaction.

1.1.1   Types of Chemical Propellants

Chemical propellants are categorized into liquid propellants, solid propellants, and hybrid propellants.

1. **Liquid Propellants**: These use separate fuel and oxidizer components stored in liquid form. Examples include RP-1 (a refined form of kerosene) and liquid oxygen (LOX). Liquid propellant engines can be throttled and restarted to provide a high specific impulse. However, they require complex storage and handling infrastructure.
2. **Solid Propellants**: These combine fuel and oxidizer in a solid matrix. Common materials include ammonium perchlorate and powdered aluminum. Solid rockets are more straightforward, more reliable, and easier to store than liquid rockets but cannot be throttled or shut down once ignited.
3. **Hybrid Propellants**: These systems use a liquid oxidizer and a solid fuel. An example is nitrous oxide with hydroxyl-terminated polybutadiene (HTPB). Hybrids offer a compromise between the simplicity of solid rockets and the controllability of liquid rockets.

1.1.2   The Tsiolkovsky Rocket Equation

The Tsiolkovsky rocket equation (the ideal rocket equation) is a fundamental principle in astronautics that describes the relationship between a rocket's velocity change and the effective exhaust velocity of its propellant. This equation is central in understanding the dynamics of rocket propulsion and is widely used in both theoretical and practical applications in space exploration.

The Tsiolkovsky rocket equation is mathematically expressed as:

$$\Delta v = v_e \ln \left(\frac{m_0}{m_f} \right) \tag{1-1}$$

where:
- Δv is the change in velocity (delta-v) of the rocket.
- v_e is the exhaust velocity of the propellant relative to the rocket.
- m_0 is the initial mass of the rocket (including propellant).
- m_f is the final mass of the rocket (after expelling the propellant).

This equation derives from the conservation of momentum. It assumes that the exhaust velocity is constant and the mass of the expelled propellant is infinitesimally small at any given moment, thus allowing for a continuous application of thrust.

The derivation of the Tsiolkovsky rocket equation begins with the conservation of linear momentum. Consider a rocket of mass. m moving with velocity v. As the rocket expels a small mass of propellant dm with velocity v_e, relative to the rocket, the conservation of momentum dictates:

$$m\,dv = -v_e\,dm \tag{1-2}$$

This relationship can be rearranged and integrated to yield:

$$\int_{v_0}^{v_f} dv = -v_e \int_{m_0}^{m_f} \frac{dm}{m} \tag{1-3}$$

$$v_f - v_0 = v_e \ln\left(\frac{m_0}{m_f}\right) \tag{1-4}$$

where v_f and v_0 are the final and initial velocities of the rocket, respectively. In many cases, v_0 is taken as zero, simplifying the equation to the commonly used form:

$$\Delta v = v_e \ln\left(\frac{m_0}{m_f}\right) \tag{1-5}$$

1.1.2.1 Key Parameters and Their Implications

- Effective Exhaust Velocity (v_e) : This parameter is crucial as it directly affects the delta-v capability of the rocket. A higher v_e indicates a more efficient propellant system, allowing for more significant changes in velocity for the same amount of propellant.
- Mass Ratio (m_0/m_f) : This ratio, also known as the mass fraction, significantly influences the achievable delta-v. A higher initial-to-final mass ratio implies a more significant amount of propellant, which can result in a higher delta-v, essential for reaching higher velocities or altering trajectories.

1.1.2.2 Practical Considerations and Limitations

While the Tsiolkovsky rocket equation is fundamental, it operates under idealized assumptions, such as constant exhaust velocity and instantaneous propellant burn. Real-world deviations include varying. v_e due to changes in exhaust temperature or pressure, non-instantaneous burns leading to gravity losses, and aerodynamic drag in atmospheric phases of flight. Furthermore, the structural mass of the rocket and payload must be minimized to optimize the mass ratio, presenting engineering challenges in materials and design.

1.1.2.3   Applications and Extensions

The Tsiolkovsky rocket equation is foundational in mission planning, particularly in determining the required amount of propellant for specific maneuvers such as orbital insertion, interplanetary transfer, or landing on celestial bodies. It also serves as a basis for more complex models that account for variable specific impulse, multi-stage rockets, and non-ideal conditions.

In modern space missions, this equation is a starting point for delta-v budgeting, which dictates the feasibility of mission objectives given the limitations of current propulsion technologies. Advanced propulsion systems, such as ion thrusters, also leverage this equation, albeit with modifications to account for low thrust but high efficiency over extended periods.

The Tsiolkovsky rocket equation remains a cornerstone in rocket science, providing critical insights into the physics of space travel. Despite its simplicity, it encapsulates the essence of propulsion dynamics and underscores the importance of efficient propellant use and mass optimization. As space exploration advances, this equation continues to be a vital tool in the design and execution of increasingly ambitious missions.

1.2   Nuclear Thermal Propulsion

Nuclear thermal propulsion (NTP) utilizes nuclear reactions to heat a propellant, typically hydrogen, to very high temperatures. The heated propellant is expelled through a nozzle to generate thrust, offering a higher specific impulse compared to chemical propulsion.

1.2.1   Types of Nuclear Thermal Propulsion

1. **Solid Core Nuclear Thermal Rockets**: These use a nuclear reactor with solid fuel elements. Hydrogen propellant flows through the reactor, absorbing heat from the nuclear reactions and achieving temperatures up to 3000 K, resulting in a specific impulse of around 900 seconds.
2. **Gas Core Nuclear Thermal Rockets**: In these systems, the nuclear fuel remains in a gaseous state. They can achieve even higher temperatures and specific impulses exceeding 5000 seconds, but they pose significant engineering challenges, particularly in containing the gaseous nuclear fuel.

1.2.2 Mechanics of Nuclear Thermal Rockets

The efficiency of NTP systems is determined by the reactor's ability to transfer heat to the propellant and the effective exhaust velocity of the heated hydrogen. The specific impulse (I_{sp}) is given by:

$$I_{sp} = \frac{v_e}{g_0} \tag{1-6}$$

where g sub 0 is the standard gravitational acceleration. The effective exhaust velocity, v_e, is proportional to the square root of the propellant's temperature and inversely proportional to its molecular weight.

1.3 Aero Nuclear Propulsion

Aero nuclear propulsion is a significant advancement in propulsion technology, blending principles of aerodynamics with nuclear energy to achieve unprecedented performance levels for atmospheric applications. This innovative approach leverages the immense energy of nuclear reactions to power aircraft and spacecraft, offering distinct advantages over conventional chemical propulsion systems. The integration of nuclear power in the area of aerospace not only promises enhanced efficiency and range but also opens new possibilities for long-duration flights, higher speeds, and sustained operations in environments where traditional propulsion systems face severe limitations.

1.3.1 Historical Context and Development

The concept of utilizing nuclear energy for propulsion dates back to the mid-20th century, during the height of the Cold War, when the United States and the Soviet Union were exploring advanced propulsion technologies for military and space applications. The development of aero nuclear propulsion was driven by the need for aircraft and missiles capable of extended operational ranges and higher speeds without the limitations imposed by chemical fuels. The most notable projects during this era include the U.S. Aircraft Nuclear Propulsion (ANP) program and the Soviet Tupolev Tu-119 project, both of which sought to develop nuclear-powered aircraft capable of long-range missions with no refueling requirements.

The U.S. ANP program, initiated in the late 1940s, aimed to develop a nuclear-powered bomber that could remain airborne for extended periods, potentially weeks or months, without the need for refueling. The program explored several reactor designs, including the Direct Air Cycle and Indirect Air Cycle concepts, where the reactor either directly or indirectly heated the air passing through the engine. Although the ANP program ultimately did not produce an operational nuclear-powered aircraft, it laid the groundwork for future research into nuclear propulsion systems and highlighted the potential benefits and challenges of integrating nuclear technology into aerospace applications.

In parallel, the Soviet Union pursued similar objectives with its Tupolev Tu-119 project, a modified version of the Tu-95 bomber, designed to test the feasibility of nuclear-powered flight. While both the U.S. and Soviet programs encountered significant technical and safety challenges, the research conducted during this period provided valuable insights into the complexities of aero nuclear propulsion, particularly in areas such as reactor design, radiation shielding, and thermal management.

1.3.2   Fundamental Principles of Aero Nuclear Propulsion

At the core of aero nuclear propulsion is the use of a nuclear reactor as the primary energy source, replacing the combustion of chemical fuels. In a nuclear propulsion system, the reactor generates heat through nuclear fission—a process in which heavy atomic nuclei, such as uranium-235 or plutonium-239, are split into smaller fragments, releasing a tremendous amount of energy in the form of heat. This heat is then transferred to a working fluid, typically air, which is subsequently expanded and accelerated through a nozzle to produce thrust.

There are several configurations for aero nuclear propulsion systems, each with distinct operational characteristics and design considerations. Two of the most prominent designs were the Direct Air Cycle and Indirect Air Cycle systems:

1. **Direct Air Cycle**: In this configuration, air from the atmosphere is compressed and directed into the reactor core, where it absorbs heat directly from the nuclear reactions. The heated air then expands through a nozzle to generate thrust. The Direct Air Cycle system is conceptually simple and efficient, as it eliminates the need for intermediate heat exchangers. However, it presents significant challenges in managing radiation contamination of the exhaust stream and ensuring adequate shielding for the reactor.
2. **Indirect Air Cycle**: The Indirect Air Cycle system involves using a secondary working fluid, such as helium or liquid metal, which circulates through the reactor core to absorb heat. This fluid then transfers the heat to the atmospheric air via a heat exchanger before the air is expanded to produce thrust. While this configuration adds complexity and reduces efficiency caused by the additional heat exchange step, it offers better control over radiation exposure and contamination, making it a safer option for manned missions.

1.3.3   Advantages and Challenges

The potential advantages of aero nuclear propulsion are numerous and compelling, particularly for missions that require extended operational endurance, high-speed travel, or sustained operations in harsh environments. Some of the key benefits include:

- **Extended Range and Endurance**: Nuclear-powered systems can operate for extended periods without the need for refueling, significantly increasing the range and endurance of aircraft and spacecraft. This capability is particularly valuable for strategic military applications and long-duration space missions, where fuel logistics pose a significant challenge.

- **Higher Speeds**: The high energy density of nuclear fuel allows for greater acceleration and higher speeds compared to chemical propulsion systems. This advantage could enable faster transit times for space missions and more rapid response capabilities for military aircraft.

- **Operational Flexibility**: Nuclear propulsion systems are less dependent on external environmental conditions, such as the availability of oxygen or solar energy, making them suitable for operations in environments where traditional propulsion systems are ineffective. This flexibility is crucial for missions in the outer solar system, where solar power becomes impractical, or for sustained operations in the upper atmosphere or near-space environments.

Despite these advantages, aero nuclear propulsion also presents significant technical, safety, and regulatory challenges that must be addressed before these systems can be widely adopted:

- **Radiation Management**: The operation of a nuclear reactor generates radiation, which poses a hazard to both the crew and the environment. Effective shielding is required to protect the crew from radiation exposure, but this adds weight and complexity to the system. Additionally, managing the radioactive exhaust and preventing environmental contamination are critical concerns, particularly for air-breathing designs.

- **Thermal Management**: The heat generated by the reactor must be effectively managed to prevent overheating and ensure the safe and efficient operation of the propulsion system. This requires advanced materials and cooling techniques, particularly in high-temperature environments encountered during atmospheric flight or re-entry.

- **Safety and Reliability**: The potential consequences of a nuclear reactor failure in flight are severe, necessitating rigorous safety protocols and fail-safe designs. Ensuring the reliability of the reactor and associated systems under the extreme conditions of flight is a significant engineering challenge.

- **Regulatory and Environmental Considerations**: The use of nuclear technology in aerospace applications is subject to strict regulatory oversight, particularly concerning the launch and operation of nuclear-powered systems in Earth's atmosphere or near-Earth space. International treaties and agreements, such as the Outer Space Treaty and the Nuclear Test Ban Treaty, impose limitations on the deployment of nuclear technology in space, and compliance with these regulations is essential for the development of aero nuclear propulsion systems.

1.3.4   Future Prospects and Applications

The future of aero nuclear propulsion is closely tied to advancements in nuclear reactor technology, materials science, and thermal sciences. Continued research and development in these areas are essential to overcoming the challenges associated with nuclear propulsion and realizing its full potential. Emerging technologies, such as compact modular reactors, advanced materials for radiation shielding, and innovative cooling systems, hold promise for making nuclear propulsion systems more practical and safer for a wide range of applications.

For atmospheric and near-space applications, aero nuclear propulsion could revolutionize military and strategic operations, providing aircraft and missiles with unparalleled range, speed, and endurance. The ability to deploy nuclear-powered drones, reconnaissance aircraft, or strategic bombers that can remain airborne for extended periods without refueling would offer a significant tactical advantage. Furthermore, the development of nuclear-powered high-altitude platforms or spaceplanes could facilitate rapid global response capabilities and enhance space commercialization.

As the aerospace industry continues to explore the potential of nuclear propulsion, collaboration between government agencies, research institutions, and private industry will be essential to address the technical, safety, and regulatory challenges involved. The successful development and deployment of aero nuclear propulsion systems could mark a new era in aerospace technology, unlocking capabilities that were previously beyond reach and opening new frontiers for exploration and defense.

1.3.5   Russia's Aero Nuclear Hypersonic Weapons

In recent years, Russia has made significant strides in the development and public announcement of advanced hypersonic weapons, several of which reportedly incorporate nuclear propulsion or nuclear payload capabilities. These developments have been highlighted as part of Russia's broader military modernization efforts, aimed at reinforcing its strategic deterrence and global military influence. The introduction of such weapons marks a potential shift in the global military balance, raising critical questions regarding their technical feasibility, strategic implications, and the resulting impact on international security dynamics.

1.3.5.1   Technical Feasibility and Capabilities

The most prominent of Russia's hypersonic weapons announcements include the

Avangard hypersonic glide vehicle and the *Burevestnik* nuclear-powered cruise missile. These systems purportedly leverage advancements in both hypersonic technology and nuclear propulsion to achieve unprecedented speed, range, and evasive capabilities.

Avangard Hypersonic Glide Vehicle: The *Avangard* system is a hypersonic glide vehicle capable of carrying a nuclear payload. It is launched atop an intercontinental ballistic missile (ICBM) and is designed to travel at speeds exceeding Mach 20 (approximately 24,500 km/h) while maneuvering to avoid missile defense systems. The combination of extreme speed and unpredictable flight paths significantly complicates interception, potentially rendering current missile defense architectures obsolete.

Burevestnik Nuclear-Powered Cruise Missile: The *Burevestnik*, or "Petrel," is described as a nuclear-powered, nuclear-armed cruise missile with virtually unlimited range. The missile is believed to use a nuclear thermal propulsion system, allowing it to stay airborne for extended periods and circumvent traditional missile defense systems by taking complex, unpredictable flight paths. The idea behind such a missile is to maintain a credible second-strike capability, ensuring the ability to retaliate even after a first-strike scenario.

While Russia's claims about these weapons' capabilities have been met with skepticism, particularly regarding the feasibility and safety of a nuclear-powered cruise missile, there is little doubt that significant advancements have been made in hypersonic technology. However, the exact operational status of these systems, especially the *Burevestnik*, remains unclear. Testing and deployment of nuclear propulsion in a cruise missile raise substantial technical challenges, including reactor miniaturization, reliable operation over extended periods, and ensuring safe recovery or disposal in the event of a failure. The environmental and safety risks associated with the crash of a nuclear-powered missile also present significant concerns.

1.4   Launch Concepts

Nuclear propulsion offers the potential for higher efficiency, greater thrust, and the ability to conduct missions that are currently beyond the reach of traditional chemical rockets. As the space industry advances, the debate over how best to integrate nuclear propulsion into space missions has increasingly focused on two primary concepts: surface launch and orbital launch. Each approach presents distinct advantages and challenges, driven by the unique demands of launching nuclear-powered systems from Earth's surface or deploying them once already in orbit.

1.4.1 Surface Launch of Nuclear Propulsion Systems

Surface launch involves deploying nuclear propulsion systems directly from Earth's surface, integrated into a launch vehicle that carries the reactor and its associated propulsion system into space. This concept has been the subject of extensive discussion and analysis, given the significant engineering and safety challenges involved.

The main advantage of surface launch is that it allows for the direct use of nuclear propulsion systems from the start of the mission, potentially providing higher thrust and specific impulse right from the initial launch phase. This could enable more massive payloads or faster transit times to distant destinations, as the nuclear propulsion system can be activated immediately upon reaching space.

However, surface launching nuclear propulsion systems raises substantial safety concerns. The launch of a nuclear reactor from the Earth's surface involves the risk of reactor damage or containment failure during ascent, which could potentially lead to the release of radioactive materials. This necessitates rigorous safety measures, including robust containment systems and emergency response plans, to minimize the risk of contamination in the event of a launch failure. Additionally, stringent international regulations and public concern about the safety of launching nuclear material from Earth must be navigated, complicating the approval and execution of such missions.

1.4.2   Orbital Launch of Nuclear Propulsion Systems

Orbital launch, by contrast, refers to the deployment of nuclear propulsion systems only after the spacecraft has been delivered to orbit by a conventional, non-nuclear rocket. In this concept, the nuclear propulsion system is not activated until the spacecraft has reached a stable orbit, minimizing the risks associated with launching nuclear material from the Earth's surface.

One of the primary advantages of the orbital launch approach is enhanced safety. By keeping the nuclear propulsion system dormant during the initial launch and ascent, the risk of radioactive contamination in the event of a launch failure is significantly reduced. The reactor can be activated only once the spacecraft is safely in orbit, where it can then provide sustained propulsion for deep-space missions.

This approach also allows for greater flexibility in mission design. For instance, the spacecraft could be assembled in orbit, with the nuclear propulsion module delivered separately and integrated with the payload in space. This could enable larger, more complex missions that would be difficult to launch directly from the surface. Additionally, this method aligns with existing space treaties and regulations, which often impose stricter controls on the launch of nuclear materials from Earth compared to their use in space.

However, the orbital launch of nuclear propulsion systems introduces challenges related to the need for dual launch systems—one for placing the spacecraft in orbit and another for operating the nuclear propulsion system. This can increase the complexity, cost, and duration of the mission, as multiple launches and on-orbit assembly operations may be required. Furthermore, the spacecraft must carry sufficient conventional propellant or rely on other means to reach its initial orbit, before the nuclear propulsion system can take over.

1.4.3 Nuclear Thermal Propulsion (NTP) in Surface vs. Orbital Launch

In Nuclear Thermal Propulsion (NTP) systems, a nuclear reactor heats a propellant, typically hydrogen, which is then expelled through a nozzle to produce thrust. NTP offers a specific impulse roughly twice that of chemical rockets, making it an attractive option for missions to Mars and beyond.

For a surface launch, the ability to activate the NTP system immediately after reaching space could significantly reduce travel times to planetary destinations, allowing for more direct trajectories and less reliance on gravity assists. However, the challenges of launching a nuclear reactor from Earth's surface, particularly the need to ensure reactor safety during launch, cannot be understated.

In contrast, using NTP in an orbital launch scenario would involve placing the reactor in orbit first, then activating it for deep-space propulsion. While this approach reduces the risks associated with launching nuclear material from Earth, it requires additional mission stages and careful planning to ensure that the reactor can be safely and effectively brought online once in orbit.

1.4.4 Nuclear Electric Propulsion (NEP) in Surface vs. Orbital Launch

Nuclear Electric Propulsion (NEP) systems, which generate electricity via a nuclear reactor to power electric thrusters, also highlight the considerations between surface and orbital launches. NEP systems are ideal for long-duration missions where continuous, low-thrust propulsion is needed. Because of the low thrust in NEP systems, it is not feasible to consider a surface launch. The only option for an NEP system would be orbital launch.

1.4.5 Hybrid Nuclear Propulsion Systems

Hybrid nuclear propulsion systems, which combine aspects of both NTP and NEP, further illustrate the trade-offs between surface and orbital launch. These systems could leverage the high thrust of NTP for rapid departure from Earth orbit, followed by the efficient, continuous propulsion of NEP for deep-space travel. The decision to launch such systems from the surface or to assemble and activate them in orbit depends on mission requirements, safety considerations, and the logistical complexities of each approach.

For hybrid systems, a surface launch might be desirable for missions where time is critical, allowing the spacecraft to take full advantage of the high thrust provided by NTP. In contrast, an orbital launch might be preferred for missions requiring a more gradual approach, where safety and regulatory concerns take precedence.

1.4.6   Considerations for Future Missions

As space exploration goals become more ambitious, the choice between surface and orbital launch for nuclear propulsion systems will be a critical decision in mission planning. Each approach has its own set of challenges and benefits, influencing mission design, safety protocols, and regulatory compliance. Whether launching directly from Earth or deploying in orbit, the integration of nuclear propulsion systems represents a significant step forward in our ability to explore and utilize space more effectively.

1.4.7   Earth Surface Launch

Launching from the Earth's surface involves overcoming gravitational and atmospheric drag forces. Chemical rockets, with their high thrust-to-weight ratios, are typically used for surface launches.

1. **Stages**: Multi-stage rockets are used to shed excess weight as fuel is consumed, optimizing the mass ratio and improving the efficiency of reaching orbit.
2. **Launch Profile**: The launch begins with a vertical ascent followed by a gravity turn maneuver to achieve a horizontal trajectory for orbital insertion. Aero-nuclear propulsion systems can be integrated into the initial ascent phase to enhance efficiency.

1.4.8   Orbital Launch

In space, propulsion requirements shift to velocity adjustments and trajectory control. Depending on mission requirements, both chemical and nuclear propulsion systems can be utilized for orbital maneuvers.
1. Orbital Maneuvers: Chemical propulsion is commonly used for short burns to change velocity (Δv), such as orbital insertion, plane changes, and deorbit burns.
2. Deep Space Missions: Nuclear thermal propulsion is advantageous for long-duration missions due to its high specific impulse, reducing the amount of propellant needed for interplanetary travel.

1.5   Space Nuclear Power

Radioisotope power systems (RPS), including radioisotope thermoelectric generators (RTGs) and radioisotope heater units (RHUs), have been instrumental in powering spacecraft for over six decades. RTGs convert the heat released by the natural decay of radioisotopes, typically plutonium-238, into electricity. This technology has been employed in missions such as Voyager, Cassini, and New Horizons, enabling these spacecrafts to operate far from the Sun, where solar energy is scarce. RHUs, on the other hand, provide localized heating, protecting spacecraft instruments from extreme cold.

Nuclear reactors, on the other hand, offer a different set of capabilities. Unlike radioisotopes, nuclear reactors can provide high power levels, making them suitable for propulsion as well as power generation. Nuclear Electric Propulsion (NEP) and Nuclear Thermal Propulsion (NTP) are two primary applications of nuclear reactors in space. NTP systems use a nuclear reactor to heat a propellant, e.g., hydrogen, which is then expelled through a nozzle to produce thrust. NEP systems, conversely, generate electricity to power electric thrusters, offering high efficiency for deep space missions.

The application of these nuclear technologies is not limited to space travel. On the Moon and Mars, nuclear reactors can serve as robust power sources for habitats, scientific experiments, and resource extraction activities. The Moon's prolonged night and Mars' distance from the Sun make solar power less viable, especially for long-duration missions. Furthermore, nuclear power can support life support systems, communication, and mobility, providing a sustainable energy solution in environments where sunlight is unreliable or insufficient.

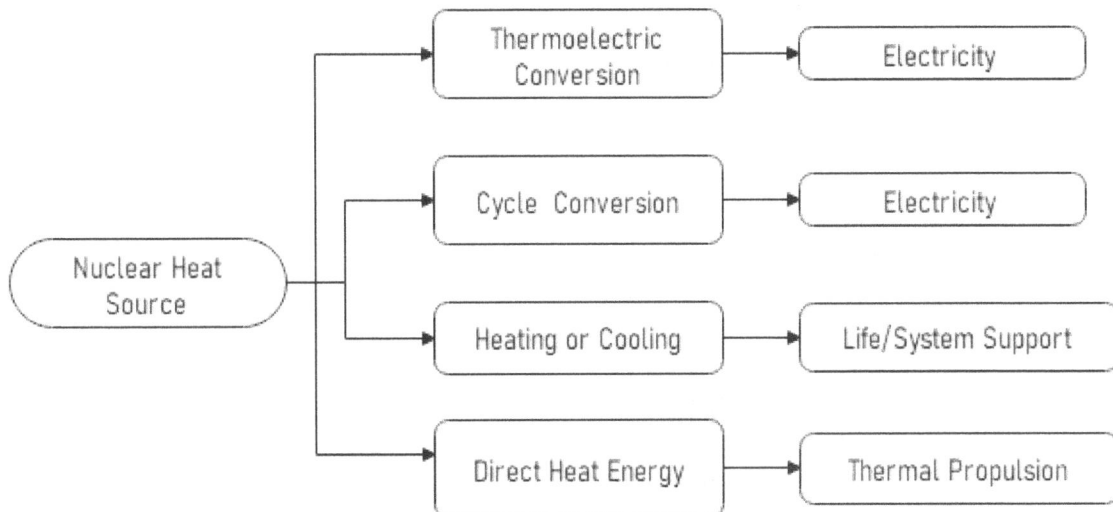

Figure 1-1. Applications of Nuclear Energy Systems in Space.

1.6   Problems

Example 1-1:

A single-stage rocket has an initial mass (including propellant) of 5000 kg and a final mass (after propellant is expended) of 2000 kg. The exhaust velocity of the propellant is 3000 m/s. Calculate the change in velocity (delta-v) that the rocket can achieve using the Tsiolkovsky rocket equation.

Solution:

Given:

- Initial mass, $m_0 = 5000$ kg
- Final mass, $m_f = 2000$ kg
- Exhaust velocity, $v_e = 3000$ m/s

The Tsiolkovsky rocket equation is given by:

$$\Delta v = v_e \ln \left(\frac{m_0}{m_f} \right)$$

Substituting the given values:

$$\Delta v = 3000 \ln \left(\frac{5000}{2000} \right)$$

Calculating the natural logarithm:

$$\Delta v = 3000 \times 0.9163 \approx 2748.9 \text{ m/s}$$

Answer:

The change in velocity (delta-v) that the rocket can achieve is approximately 2748.9 m/s.

Example 1-2:

An interplanetary spacecraft requires a delta-v of 8000 m/s to complete its mission. The spacecraft's initial total mass, including propellant, is 12000 kg , and the exhaust velocity of its propulsion system is 3500 m/s. Determine the final mass of the spacecraft after propellant expenditure and the amount of propellant used.

Solution:

Given:

- Required delta- v, $\Delta v = 8000$ m/s
- Initial mass, $m_0 = 12000$ kg
- Exhaust velocity, $v_e = 3500$ m/s

Using the Tsiolkovsky rocket equation:

$$\Delta v = v_e \ln \left(\frac{m_0}{m_f} \right)$$

Solving for the final mass m_f :

$$8000 = 3500 \ln \left(\frac{12000}{m_f} \right)$$

$$\frac{8000}{3500} = \ln \left(\frac{12000}{m_f} \right)$$

$$2.2857 = \ln \left(\frac{12000}{m_f} \right)$$

Exponentiating both sides:

$$e^{2.2857} = \frac{12000}{m_f}$$

Calculating the exponential:

$$9.8325666 \approx \frac{12000}{m_f}$$

Thus:

$$m_f = \frac{12000}{9.8325666} \approx 1220.43 \text{ kg}$$

To find the propellant mass used:

$$\text{Propellant mass} = m_0 - m_f$$

$$\text{Propellant mass} = 12000 \text{ kg} - 1220.43 \text{ kg} \approx 10779.6 \text{ kg}$$

Answer:
The final mass of the spacecraft after propellant expenditure is approximately 1220.4 kg, and the amount of propellant used is approximately 10779.6 kg.

Example 1-3:

A two-stage rocket consists of a first stage with an initial mass of 6000 kg (including propellant) and a final mass of 2500 kg after expending its propellant. The second stage has an initial mass of 3000 kg (including propellant) and a final mass of 1000 kg. The exhaust velocity for both stages is 2900 m/s. Calculate the total delta-v achieved by the rocket.

Solution:
First Stage Calculation:

Given:
- Initial mass of the first stage, $m_{0,1} = 6000$ kg
- Final mass of the first stage, $m_{f,1} = 2500$ kg
- Exhaust velocity, $v_e = 2900$ m/s

Using the Tsiolkovsky rocket equation for the first stage:

$$\Delta v_1 = v_e \ln \left(\frac{m_{0,1}}{m_{f,1}} \right)$$

$$\Delta v_1 = 2900 \ln \left(\frac{6000}{2500} \right)$$

$$\Delta v_1 = 2900 \ln (2.4) \approx 2538.86 \text{ m/s}$$

Second Stage Calculation:

After the first stage burns out, the total initial mass for the second stage is the final mass of the first stage plus the initial mass of the second stage minus the empty mass of the first stage:
- Initial mass of the second stage, $m_{0,2} = 2500 + 3000 = 5500$ kg
- Final mass of the second stage, $m_{f,2} = 2500 + 1000 = 3500$ kg

Using the Tsiolkovsky rocket equation for the second stage:

$$\Delta v_2 = v_e \ln \left(\frac{m_{0,2}}{m_{f,2}} \right)$$

$$\Delta v_2 = 2900 \ln \left(\frac{5500}{3500} \right)$$

$$\Delta v_2 = 2900 \ln(1.5714) \approx 1310.76 \text{ m/s}$$

Total Delta-v:

$$\Delta v_{total} = \Delta v_1 + \Delta v_2 \approx 2538.86 + 1310.76 = 3849.62 \text{ m/s}$$

Answer:
The total delta-v achieved by the two-stage rocket is approximately 3849 m/s.

Example 1-4:
A space probe uses a single-stage rocket motor to achieve a delta-v of 9000 m/s. The initial mass of the probe, including the propellant, is 8000 kg, and the final mass after burning all the propellant is 2000 kg. Determine the specific impulse (I_{sp}) of the rocket engine. The specific impulse is related to the exhaust velocity by the relation $I_{sp} = \frac{v_e}{g_0}$, where $g_0 = 9.81 \text{ m/s}^2$.

Solution:
Given:
- Delta- v, $\Delta v = 9000$ m/s
- Initial mass, $m_0 = 8000$ kg
- Final mass, $m_f = 2000$ kg
- Standard gravitational acceleration, $g_0 = 9.81 \text{ m/s}^2$

Using the Tsiolkovsky rocket equation to solve for v_e :

$$\Delta v = v_e \ln\left(\frac{m_0}{m_f}\right)$$

Using the Tsiolkovsky rocket equation to solve for v_e :

$$\Delta v = v_e \ln\left(\frac{m_0}{m_f}\right)$$

$$9000 = v_e \ln\left(\frac{8000}{2000}\right)$$

$$9000 = v_e \ln(4)$$

Calculating the natural logarithm:
$$\ln(4) \approx 1.386294$$

$$9000 = v_e \times 1.386294$$

Solving for v_e :

$$v_e \approx \frac{9000}{1.386294} \approx 6492.13 \text{ m/s}$$

Now, calculate the specific impulse I_{sp} :

$$I_{sp} = \frac{v_e}{g_0}$$

$$I_{sp} = \frac{6492.13}{9.81} \approx 661.8 \text{ s}$$

Answer:
The specific impulse of the rocket engine is approximately 661.8 seconds.

Example 1-5:
   A spacecraft needs to achieve a delta-v of 4500 m/s using a chemical rocket engine with a specific impulse (I_{sp}) of 300 seconds. The spacecraft's initial mass, including propellant, is 15000 kg. Calculate the amount of propellant required for this maneuver.

Solution:
Given:
   - Delta- v, $\Delta v = 4500$ m/s
   - Specific impulse, $I_{sp} = 300$ s
   - Initial mass, $m_0 = 15000$ kg
   - Standard gravitational acceleration, $g_0 = 9.81$ m/s^2

First, calculate the exhaust velocity v_e using the specific impulse:
$$v_e = I_{sp} \times g_0$$

$$v_e = 300 \times 9.81 \approx 2943 \text{ m/s}$$

Now, use the Tsiolkovsky rocket equation to find the final mass m_f :

$$\Delta v = v_e \ln \left(\frac{m_0}{m_f} \right)$$

$$4500 = 2943 \ln \left(\frac{15000}{m_f} \right)$$

$$\frac{4500}{2943} = \ln \left(\frac{15000}{m_f} \right)$$

$$1.529052 = \ln \left(\frac{15000}{m_f} \right)$$

Exponentiating both sides:

$$e^{1.529052} = \frac{15000}{m_f}$$

$$4.61380081 \approx \frac{15000}{m_f}$$

Solving for m_f :

$$m_f \approx \frac{15000}{4.61380081} \approx 3251.12 \text{ kg}$$

The amount of propellant used is:

$$\text{Propellant mass} = m_0 - m_f$$

$$\text{Propellant mass} = 15000 \text{ kg} - 3251.12 \text{ kg} \approx 11748.88 \text{ kg}$$

Answer:
The spacecraft requires approximately 11748.88 kg of propellant to achieve the desired delta-v.

Example 1-6:
A deep-space probe has a specific impulse (I_{sp}) of 400 seconds and uses 1000 kg of propellant. The initial total mass of the probe, including propellant, is 5000 kg. Calculate the delta-v the probe can achieve.

Solution:

Given:

- Specific impulse, $I_{sp} = 400$ s
- Propellant mass, $m_p = 1000$ kg
- Initial mass, $m_0 = 5000$ kg
- standard gravitational acceleration, $g_0 = 9.80665$ m/s^2

First, calculate the exhaust velocity v_e :

$$v_e = I_{sp} \times g_0$$

$$v_e = 400 \times 9.81 \approx 3924 \text{ m/s}$$

Determine the final mass m_f after burning the propellant:

$$m_f = m_0 - m_p$$

$$m_f = 5000 - 1000 = 4000 \text{ kg}$$

Now, use the Tsiolkovsky rocket equation to calculate delta-v:

$$\Delta v = v_e \ln\left(\frac{m_0}{m_f}\right)$$

$$\Delta v = 3924 \ln\left(\frac{5000}{4000}\right)$$

$$\Delta v = 3924 \ln(1.25)$$

$$\Delta v \approx 3924 \times 0.223144 \approx 875.6 \text{ m/s}$$

Answer:
The deep-space probe can achieve a delta-v of approximately 875.6 m/s.

Example 1-7:

A rocket engine expels exhaust gases at a velocity of 2500 m/s relative to the rocket. Calculate the specific impulse (I_{sp}) of the rocket engine.

Solution:

Given:

- Exhaust velocity, $v_e = 2500$ m/s
- Standard gravitational acceleration, $g_0 = 9.81$ m/s^2

The specific impulse $\left(I_{sp}\right)$ is related to the exhaust velocity by the formula:

$$I_{sp} = \frac{v_e}{g_0}$$

Substitute the given values:

$$I_{sp} = \frac{2500}{9.81}$$

$$I_{sp} \approx 254.8 \text{ s}$$

Answer:
The specific impulse of the rocket engine is 254.8 seconds.

Example 1-8:
   A spacecraft achieves a Δv of 6000 m/s. The initial mass of it, including propellant, is 18000 kg, and the final mass after propellant expenditure is 4500 kg. Determine the specific impulse (I_{sp}) of the propulsion system.

Solution:

Given:

- $\Delta v = 6000$ m/s
- Initial mass, $m_0 = 18000$ kg
- Final mass, $m_f = 4500$ kg
- Standard gravitational acceleration, $g_0 = 9.81$ m/s^2

First, use the Tsiolkovsky rocket equation to find the exhaust velocity v_e:

$$\Delta v = v_e \ln\left(\frac{m_0}{m_f}\right)$$

Solving for v_e :

$$v_e = \frac{\Delta v}{\ln \left(\frac{m_0}{m_f} \right)}$$

$$v_e = \frac{6000}{\ln \left(\frac{18000}{4500} \right)}$$

$$v_e = \frac{6000}{\ln (4)}$$

$$v_e = \frac{6000}{1.386294} \approx 4328.1 \text{ m/s}$$

Now, calculate the specific impulse I_{sp} :

$$I_{sp} = \frac{v_e}{g_0}$$

$$I_{sp} = \frac{4328.1}{9.81} \approx 441.2 \text{ s}$$

Answer:

The specific impulse of the propulsion system is 441.2 seconds.

2 Rocket Propulsion

2.1 Thrust in Rocket Propulsion

Thrust is the driving force behind the motion of rockets and spacecraft, crucial for overcoming gravitational forces and achieving flight. It is typically measured in pounds (lb), kilograms (kg), or Newtons (N). Thrust is fundamentally the result of pressure exerted on the walls of the combustion chamber within a rocket engine.

2.1.1 Combustion Chamber and Nozzle Dynamics

Combustion chamber exhaust gases are expelled through a nozzle. Inside the chamber, the pressure distribution is largely uniform, but it decreases significantly near the nozzle. This pressure differential is critical in generating thrust.

The internal pressure within the combustion chamber exerts a force on its walls. On the bottom of the chamber, this force is not counterbalanced by an external force, resulting in a net force (F) caused by the difference in pressures between the internal and external environments. This net force, or thrust, acts in the direction opposite to the gas jet, propelling the chamber (and the rocket) upwards.

2.1.2 Generation of High-Speed Exhaust Gases

To produce high-speed exhaust gases necessary for substantial thrust, two primary conditions must be met:

1. High Temperatures and Pressures: Achieved by using highly energetic fuels.
2. Low Molecular Weight of Exhaust Gases: Ensures that the exhaust gases can reach higher velocities.

Additionally, the design of the nozzle plays a crucial role. By maximizing the nozzle's expansion ratio, i.e., the ratio of the exit area (A_e) to the throat area (A_t), the pressure of the exhaust gases is minimized as they exit the nozzle, enhancing thrust.

2.1.3   Thrust Calculation

The resultant thrust (F) is derived from the forces due to pressure differences on the inner and outer walls of the combustion chamber and nozzle. This can be quantified using the principle of conservation of momentum. The equation for thrust is:

$$F = \dot{m} V_e + (P_e - P_a)A_e \qquad\qquad (2\text{-}1)$$

where:

- $\dot{m}$ is the mass flow rate of the exhaust gases,
- V_e is the velocity of the exhaust gases as they exit the nozzle,
- P_e is the nozzle exit pressure,
- P_a is the ambient atmospheric pressure,
- A_e is the nozzle exit cross-sectional area.

Equation (2-1) highlights two critical components of thrust:

1. Momentum Thrust ($\dot{m}V_e$): Resulting from the ejection of mass at high velocity.
2. Pressure Thrust $(P_e - P_a)A_e$: Arising from the pressure difference between the exhaust gases and the ambient atmosphere.

Example 2-1:

The exhaust gas mass flow rate from the engine of a spacecraft's engine is 35 kg/s and the effective exhaust velocity is 3,500 m/s. The pressure at the nozzle exit is 4 kPa, and the exit area is 0.8 m². Calculate the engine thrust in a vacuum?

Solution:

Given:

- Mass flow rate, $\dot{m}$ = 35 kg/s
- Exhaust velocity, V_e = 3,500 m/s
- Exit area, A_e = 0.8 m²
- Pressure at the nozzle exit, P_e = 4kPa = 4,000 N/m²
- Ambient pressure in a vacuum, P_a = 0

Calculate using:

$$F = \dot{m} \times V_e + (P_e - P_a) \times A_e$$

Substitute the given values:

$$F = 35 \times 3500 + (4000 - 0) \times 0.8$$

$$F = 122{,}500 + 3{,}200$$

$$F = 125{,}700 \text{ N}$$

Thrust of the engine: $F = 125{,}700$ N

2.1.4 Conditions of Thrust Measurement

Thrust is specified under two primary conditions:

- Sea Level Thrust: Measured in the presence of Earth's atmospheric pressure.
- Vacuum Thrust: Measured in the absence of atmospheric pressure, typically in space.

The performance of a rocket engine varies between these two conditions due to changes in P_a. In a vacuum, P_a is zero, simplifying the thrust equation and often resulting in higher thrust values compared to sea level conditions.

2.1.5 Practical Considerations

1. **Rocket Design**: To optimize thrust, the combustion chamber and nozzle must be designed to ensure efficient combustion and gas expansion.
2. **Fuel Selection**: The choice of fuel impacts the temperatures and pressures achievable within the combustion chamber, directly affecting thrust.
3. **Nozzle Expansion Ratio**: The nozzle's design, particularly its expansion ratio, is critical for maximizing the conversion of thermal energy into kinetic energy, thereby optimizing thrust.

2.2 Conservation of Momentum

2.2.1 Definition and Newton's Second Law

Momentum (p) is the product of a particle's mass (m) and its velocity (v):

$$p = m\,v \tag{2-2}$$

Newton's second law of motion can be expressed as:

$$\sum \vec{F} = \frac{d}{dt}(m\,\vec{v}) \tag{2-3}$$

For a system of particles, the total momentum (P) is the sum of the individual momenta. If the resultant external force acting on the system is zero, the total momentum of the system remains constant. This principle is known as the conservation of linear momentum.

2.2.2   Application to Rocket Mechanics

Consider a rocket in gravity-free space. When the rocket's engine fires for a time Δt, it ejects gases at a constant rate and constant exhaust velocity relative to the rocket. Assuming no external forces, such as gravity or air resistance, the situation is as follows:

1. Initial State (Figure 2-1): At time t, the rocket (with its fuel) has a total mass M and velocity v.
2. Final State (Figure 2-1): After a time Δt, a mass ΔM is ejected, moving with velocity u. The rocket's mass decreases to $M - \Delta M$ and its velocity changes to $v + \Delta v$.

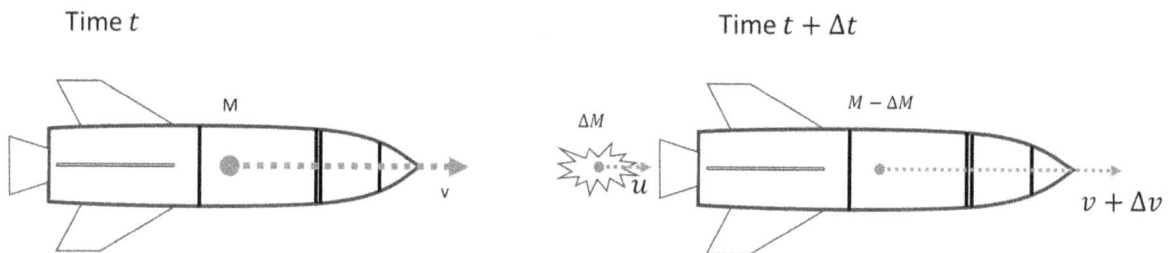

Figure 2-1. Rocket propulsion model.

Since there are no external forces:

$$\frac{dP}{dt} = 0 \tag{2-4}$$

For the time interval Δt :

$$P_2 = P_1 \tag{2-5}$$

Where P_2 is the final momentum and P_1 is the initial momentum. We write:

$$(M - \Delta M)(v + \Delta v) + (\Delta M)u = Mv \tag{2-6}$$

As Δt approaches zero, $\frac{\Delta v}{\Delta t}$ approaches $\frac{dv}{dt}$, the acceleration. The mass ejected in Δt, ΔM, leads to a decrease in the rocket's mass. Since $\frac{dM}{dt}$ is negative, we replace $\frac{\Delta M}{\Delta t}$ with $-\frac{dM}{dt}$. The relative velocity of the mass ejected with respect to the rocket is $V_{rel} = u - (v + \Delta v)$. Thus, the equation becomes:

$$M\frac{dv}{dt} = V_{rel}\frac{dM}{dt} \tag{2-7}$$

2.3 Thrust in Rocket Propulsion

The right-hand side of the equation represents the thrust force (F) :

$$F = V_{rel}\frac{dM}{dt} \tag{2-8}$$

Thrust is the reaction force exerted on the rocket by the ejected mass. To maximize thrust, a rocket should eject mass rapidly ($\frac{dM}{dt}$ large) and at high relative speed (V_{rel} large).

2.3.1 Basic Thrust Equation

In rocketry, the primary thrust equation is:

$$F = \dot{m}V_e + (P_e - P_a)A_e \tag{2-9}$$

where:
- $\dot{m}$ is the rate of the ejected mass flow,
- V_e is the exhaust gas ejection speed,
- P_e is the nozzle exit pressure,
- P_a is the ambient atmospheric pressure,
- A_e is the area of the nozzle exit.

The term $\dot{m}V_e$ is the momentum or velocity thrust, and $(P_e - P_a)A_e$ is the pressure thrust. Maximum thrust occurs when $P_e = P_a$.

2.3.2 Simplification Using Effective Exhaust Velocity

The effective exhaust gas velocity (v_{ee}) is defined as:

$$v_{ee} = V_e + \frac{(P_e - P_a)A_e}{\dot{m}} \qquad (2\text{-}10)$$

Thus, the thrust equation simplifies to:

$$F = \dot{m}\, v_{ee} \qquad (2\text{-}11)$$

This simplification helps analyze and design rocket propulsion systems for optimal performance.

2.4   Impulse and Momentum

In the previous section, we explored how Newton's second law can be expressed as:

$$F = \frac{dp}{dt} \qquad (2\text{-}12)$$

By multiplying both sides by dt and integrating over a time interval from t_1 to t_2, we obtain:

$$\int_{t_1}^{t_2} F dt = \int_{t_1}^{t_2} \frac{dp}{dt} dt \qquad (2\text{-}13)$$

The integral on the left side represents the linear impulse (or simply impulse) of the force F over the time interval. This equation indicates that when a particle is subjected to a force F over a given time interval, the final momentum p_2 of the particle can be found by adding the initial momentum p_1 and the impulse of the force F during that time. Symbolically, this is:

$$p_2 = p_1 + \int_{t_1}^{t_2} F dt \qquad (2\text{-}14)$$

When multiple forces act on a particle, the impulse of each force must be considered. For a system of particles, we sum the momenta of all particles and the impulses of all forces involved, leading to:

$$\sum p_2 = \sum p_1 + \sum \int_{t_1}^{t_2} F \, dt \qquad (2\text{-}15)$$

For a short time interval, Δt, this can be written as:

$$\sum p_2 = \sum p_1 + \sum F \Delta t \qquad (2\text{-}16)$$

2.4.1 Application to Rocket Mechanics

Let's apply the principle of impulse and momentum to rocket mechanics. Consider a rocket of initial mass M launched vertically at time $t = 0$. The fuel is consumed at a constant rate q and expelled at a constant speed V_e relative to the rocket. At time t, the mass of the rocket shell and remaining fuel is $M - qt$, and its velocity is v. During the interval Δt, a mass of fuel $q\Delta t$ is expelled. Denote u as the absolute velocity of the expelled fuel. Applying the principle of impulse and momentum from time t to $t + \Delta t$:

$$(M - q\Delta t)(v + \Delta v) + (q\Delta t)u = Mv \qquad (2\text{-}17)$$

Dividing by Δt and replacing $u - (v + \Delta v)$ with V_e, the relative velocity of the expelled mass, as Δt approaches zero, we get:

$$M \frac{dv}{dt} = -V_e \frac{dM}{dt} \qquad (2\text{-}18)$$

2.4.2 Integration and Rocket Equation

Separating variables and integrating from $t = 0, v = 0$ to $t = t, v = v$:

$$\int_0^v dv = -V_e \int_M^{M-gt} \frac{dM}{M} \qquad (2\text{-}19)$$

This simplifies to:

$$v = -V_e \ln \left(\frac{M - gt}{M} \right) \qquad (2\text{-}20)$$

Including the effect of Earth's gravity, the term $-gt$ accounts for gravitational pull:

$$v = -V_e \ln \left(\frac{M - gt}{M} \right) - gt \qquad (2\text{-}21)$$

For a rocket drifting in space, the $-gt$ term is omitted. We can express the resulting velocity as a change in velocity (ΔV) :

$$\Delta V = V_e \ln \left(\frac{M}{M - gt} \right) \qquad (2\text{-}22)$$

Example 2-2:

The spacecraft from the previous problem has an initial mass of 28,000 kg. Calculate the change in velocity if the burn time is 70 seconds.

Solution:

Given:

- Initial mass, M = 28,000 kg
- Mass flow rate, $\dot{m}$ = 35 kg/s
- Exhaust velocity, V_e = 3,500 m/s
- Burn time, t = 70 s

Calculate using

$$\Delta V = V_e \times \ln \left(\frac{M}{M - \dot{m}\, t} \right)$$

Substitute the given values:

$$\Delta V = 3,500 \times \ln \left(\frac{28,000}{28,000 - (35 \times 70)} \right)$$

$$\Delta V = 3,500 \times \ln \left(\frac{28,000}{25,550} \right)$$

$$\Delta V \approx 320.485 \text{ m/s}$$

2.4.3   Tsiolkovsky's Rocket Equation

Using the initial mass (M) and final mass ($M - qt$), the equation is often written as:

$$\Delta V = V_e \ln \left(\frac{m_o}{m_f} \right) \tag{2-23}$$

Where $\frac{m_o}{m_f}$ is the mass ratio. This is known as Tsiolkovsky's rocket equation, named after Konstantin E. Tsiolkovsky, who first derived it.

In practical applications, V_e is replaced by the effective exhaust gas velocity (C) :

$$\Delta V = v_{ee} \ln \left(\frac{m_0}{m_f} \right) \tag{2-24}$$

Alternatively, we can express it as:

$$\Delta V = v_{ee} \ln \left(\frac{m_o}{m_f} \right) = v_{ee} \ln \left(e^{\frac{\Delta V}{C}} \right) \tag{2-25}$$

Example 2-3:

A spacecraft's dry mass is 80,000 kg. The main engine effective exhaust gas velocity is 3,200 m/s. Calculate the amount of propellant required for the propulsion system is to achieve a total Δv of 750 m/s ?

Solution:

Given:
- Dry mass, $M_f = 80,000$ kg
- Effective exhaust velocity, $v_{ee} = 3,200$ m/s
- Desired Δv, $\Delta V = 750$ m/s

Calculation using

$$M_0 = M_f \times e^{(\Delta V / v_{ee})}$$

Substitute the given values:

$$M_0 = 80,000 \times e^{(750/3,200)}$$

$$M_0 \approx 101129 \text{ kg}$$

Propellant mass:

$$M_p = M_0 - M_f$$

$$M_p = 101129 - 80000$$

$$M_p = 21129 \text{ kg}$$

Propellant mass required: $M_p = 21129 \text{ kg}$

2.4.4   Burn Duration

To calculate engine burn time, rearrange the variables, giving:

$$t = \frac{M}{q}\left(1 - e^{-\frac{\Delta V}{v_{ee}}}\right) \tag{2-26}$$

Example 2-4:

A 4,500 kg spacecraft is in Earth orbit traveling at a velocity of 8,000 m/s. An engine burn accelerates the spacecraft to a velocity of 13,000 m/s. At this speed the spacecraft is placed on a escape trajectory. The engine exhaust mass flow rate is 12 kg/s. The effective exhaust velocity is 3,200 m/s. Calculate the duration of the burn.

Solution:
Given:
- Initial mass, $M = 4,500 \text{ kg}$
- Mass flow rate, $\dot{m} = 12 \text{ kg/s}$
- Effective exhaust velocity, $v_{ee} = 3,200 \text{ m/s}$
- Desired change in velocity, $\Delta V = 13,000 - 8,000 = 5,000 \text{ m/s}$

Calculation using:

$$t = \frac{M}{\dot{m}}\left[1 - \frac{1}{e^{(\Delta V/v_{ee})}}\right]$$

Substitute the given values:

$$t = \frac{4,500}{12}\left[1 - \frac{1}{e^{(5,000/3,200)}}\right]$$

$$t = 375\left[1 - \frac{1}{4.770733}\right]$$

$$t = 375 \times 0.7903886$$

$$t = 296.396$$

2.5 Combustion and Exhaust Velocity

The combustion process involves the oxidation of fuel constituents capable of being oxidized, represented by a chemical equation. During combustion, the mass of each element remains constant, following the law of conservation of mass. For instance, consider the combustion of methane with oxygen:

$$CH_4 + 2O_2 \rightarrow CO_2 + 2H_2O \tag{2-27}$$

Two moles of oxygen and one mole of methane react producing two moles of water and one mole of carbon dioxide. In combustion terminology, the initial substances (methane and oxygen) are the reactants, and the resulting substances (carbon dioxide and water) are the products.

2.5.1 Stoichiometric Mixture and Mixture Ratio

The given reaction is a stoichiometric mixture, meaning there is exactly enough oxygen to react with all the fuel. This mixture achieves the highest flame temperature. However, rocket engines often operate at a "fuel-rich" mixture ratio to optimize performance and prevent overheating.

The mass flow of the oxidizer divided by the mass flow of the fuel is defined as as the mixture ratio (O/F). Consider the combustion of kerosene (approximated as dodecane, $C_{12}H_{26}$) with oxygen:

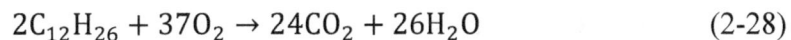

$$2C_{12}H_{26} + 37O_2 \rightarrow 24CO_2 + 26H_2O \tag{2-28}$$

Given the molecular weights ($C_{12}H_{26}$ is 170 and O_2 is 32), the mixture ratio is:

$$O/F = \frac{37 \times 32}{2 \times 170} \approx 3.5 \tag{2-29}$$

This ratio is typical for many rocket engines using kerosene or RP-1 fuel.

2.5.2 Optimizing Mixture Ratio

The optimum mixture ratio delivers the highest engine performance, measured by specific impulse (I_{sp}). However, in some cases, a different. $\frac{O}{F}$ the ratio is preferable. For instance, in a volume-constrained vehicle using low-density fuels like liquid hydrogen, a higher O/F This ratio significantly reduces vehicle size, compensating for performance losses with reduced fuel tank requirements. Additionally, bipropellant systems like NTO/MMH often use a mixture ratio of 1.67 to simplify tank manufacturing and integration by equalizing the fuel and oxidizer tank sizes.

2.5.3   Impulse Thrust and Exhaust Velocity

Impulse thrust is calculated by multiplying the propellant mass flow ($\dot{m}$) by the exhaust velocity (V_e). The ideal exhaust velocity is given by:

$$V_e = \sqrt{\frac{2kR^*T_c}{M(k-1)}\left(1-\left(\frac{P_e}{P_c}\right)^{\frac{k-1}{k}}\right)} \qquad (2\text{-}30)$$

Where:
- k is the specific heat ratio,
- R^* is the universal gas constant (8,314.4621 J/kmol-K or 49,720ft $-$ lb/(slug $\cdot$ mol) $\cdot$ °R),
- T_c is the combustion temperature,
- M is the exhaust gas average molecular weight,
- P_c is the combustion chamber pressure,
- P_e is the pressure at the nozzle exit.

Example 2-5:

A rocket engine burning liquid hydrogen and liquid oxygen operates at a mixture ratio of 6 and a combustion chamber pressure of 60 atm. Assume the nozzle exit pressure is 1 atm. Calculate the exhaust gas velocity.

Solution:
Given:

- Mixture ratio, $O/F = 6$
- Combustion chamber pressure, $P_c = 60$ atm
- Exit pressure at sea level, $P_e = P_a = 1$ atm

From LOX/LH $_2$ charts we estimate:
- Combustion temperature, $T_c = 3,200$ K

- Molar mass of exhaust, $M = 14.50$
- Specific heat ratio, $k = 1.240$

Calculation using

$$V_e = \sqrt{\left(\frac{2k}{k-1}\right)\left(\frac{R^* T_c}{M}\right)\left[1 - \left(\frac{P_e}{P_c}\right)^{\frac{k-1}{k}}\right]}$$

Substitute the given values:

$$V_e = \sqrt{\left(\frac{2 \times 1.240}{1.240 - 1}\right)\left(\frac{8,314.46 \times 3,200}{14.50}\right)\left[1 - \left(\frac{1}{60}\right)^{\frac{1.240-1}{1.240}}\right]}$$

$$V_e \approx 3221.28 \text{ m/s}$$

Exhaust gas velocity relative to the rocket: $V_e \approx 3221.3$ m/s

2.5.4 Specific Heat Ratio and Combustion Temperatures

The specific heat ratio (k) varies with the composition and temperature of the exhaust gases, typically around 1.2. Calculating combustion temperatures involves complex thermodynamics, with flame temperatures ranging from 2,500 to $3,600°C(4,500 - 6,500°F)$. Chamber pressures can range from 7 to 250 atmospheres. For optimal performance, P_e should match the ambient pressure where the engine operates.

2.5.5 Achieving High Thrust

From the exhaust velocity equation, high thrust is achieved through:
- High combustion temperature (T_c),
- High chamber pressure (P_c),
- Low molecular weight of exhaust gases (M).

These factors explain the desirability of liquid hydrogen as a rocket fuel due to its low molecular weight and high specific energy.

2.5.6 Dissociation in Combustion

During combustion, high heat causes molecular dissociation, resulting in a mixture of atoms and simpler molecules capable of recombining. For example, the combustion of kerosene with oxygen produces an equilibrium mixture of $C, CO, CO_2, H, H_2, H_2O, OH, O$, and O_2. Dissociation significantly affects flame temperature.

2.5.7   Practical Considerations

For further understanding, refer to Rocket Thermodynamics or Propellant Combustion Charts, which provide data on optimum mixture ratios, adiabatic flame temperatures, gas molecular weights, and specific heat ratios for common rocket propellants.

2.5.8   Simplified Example

In practical applications, the exhaust gas velocity (V_e) is often replaced by the effective exhaust gas velocity (C), simplifying the thrust equation:

$$F = \dot{m}C$$

This equation is fundamental for calculating the necessary parameters for efficient rocket engine design and operation.

2.6   Specific Impulse

The specific impulse (I_{sp}) indicates the efficiency of the engine. It is defined as:

$$I_{sp} = \frac{F}{\dot{m}g_0} \tag{2-31}$$

where:
- F is the thrust,
- $\dot{m}$ is the rate of mass flow,
- g_0 is the standard gravitational acceleration (9.80665 m/s^2).

Specific impulse is expressed in seconds and it gives a measure of how long one unit of propellant can produce thrust equal to its own weight. A higher specific impulse indicates a more efficient rocket engine.

Example 2-6:

A rocket engine produces a thrust of 1200 kN at sea level with a propellant flow rate of 450 kg/s. Calculate the specific impulse.

Solution:

Given:
- Thrust, $F = 1{,}200{,}000$ N
- Propellant flow rate, $\dot{m} = 450$ kg/s

Calculate using

$$I_{sp} = \frac{F}{\dot{m} \times g}$$

Where $g = 9.80665$ m/s^2 (standard gravitational acceleration).

$$I_{sp} = \frac{1200000}{450 \times 9.80665}$$

$$I_{sp} = \frac{1200000}{4,412.9925}$$

$$I_{sp} \approx 272 \text{ s(sea level)}$$

Specific impulse: $I_{sp} \approx 272$ s

2.6.1   Specific Impulse at Different Conditions

The specific impulse of a rocket engine varies depending on whether it is measured at sea level or in the vacuum of space. This variation is due to the influence of ambient pressure on the engine's thrust. Thus, it is crucial to specify the conditions under which the specific impulse is measured.

1. Sea Level Specific Impulse: This takes into account the atmospheric pressure, which reduces the effective thrust.
2. Vacuum-Specific Impulse: Measured in the absence of atmospheric pressure, vacuum-specific impulse typically results in higher specific impulse values.

2.6.2   Factors Affecting Specific Impulse

Several factors can reduce the efficiency of a rocket engine, thereby affecting its specific impulse:

1. Combustion Efficiency: Inefficiencies in the chemical reaction can result in incomplete combustion of the fuel, reducing thrust.
2. Nozzle Losses: Imperfect nozzle design can cause flow separation and suboptimal expansion of exhaust gases.
3. Pump Losses: Pumps feed propellant into the combustion chamber in liquid rocket engines. Inefficiencies in these pumps can reduce the overall specific impulse.

The real specific impulse ($I_{sp,real}$) is always less than the theoretical specific impulse (

$I_{sp,\,theoretical}$), which assumes ideal conditions and perfect efficiency. The efficiency of the specific impulse is the ratio of the real specific impulse to the theoretical specific impulse, typically a few percent lower than the theoretical value.

2.6.3   Practical Use and Calculations

From the thrust equation, we can substitute $\dot{m}C$ for F in the specific impulse formula:

$$I_{sp} = \frac{\dot{m}\,v_e}{\dot{m}\,g_0} = \frac{v_e}{g_0} \tag{2-32}$$

This relationship is useful for solving various rocket performance equations. The effective exhaust velocity (v_e) is calculated from the specific impulse:

$$v_e = I_{sp} \cdot g_0 \tag{2-33}$$

2.7   Rocket Engines

A typical rocket engine consists of three main components: the nozzle, the combustion chamber, and the injector. Each part plays a crucial role in the propulsion process, converting chemical energy into kinetic energy to generate thrust.

2.7.1   Combustion Chamber

The combustion chamber is where the propellants (fuel and oxidizer) burn at high pressure. This chamber must be robust enough to withstand the high pressure and high temperature resulting from the combustion process. Due to the extreme temperatures and significant heat transfer, the combustion chamber and nozzle are typically cooled to prevent damage. The chamber must also be sufficiently long to ensure complete combustion of the propellants before the gases enter the nozzle.

2.7.2   Nozzle

The nozzle's primary function is to convert the high-pressure, high-temperature gases from the combustion chamber into high-velocity gases of lower pressure and temperature, thus generating thrust. This process converts the chemical-thermal energy into kinetic energy, where high gas velocity is crucial because thrust is the product of mass flow rate and velocity.

A nozzle consists of a convergent section and a divergent section, with the minimum flow area between these sections known as the throat of the nozzle. The nozzle exit area is the flow area at the end of the divergent section. The nozzle is designed so that the pressure in the combustion chamber is reduced to the pressure outside the nozzle at the exit, achieving maximum thrust when the exit pressure. (P_e) equals the ambient pressure (P_a). This condition is known as optimum or correct expansion.

- Under-Extended Nozzle: When $P_e > P_a$, resulting in suboptimal thrust.
- Over-Extended Nozzle: When $P_e < P_a$, also resulting in suboptimal thrust.

2.7.2.1   Nozzle Design for Different Conditions

Rocket nozzles must be tailored for the altitude at which they operate. At sea level, atmospheric pressure limits the discharge of exhaust gases due to jet separation from the nozzle wall. In the vacuum of space, this limitation does not exist, allowing for different designs for atmospheric and vacuum operation. Thus, launch vehicles typically have different engines and nozzles for their first stage (operating through the atmosphere) and subsequent stages (operating in a vacuum). Figure 2-2 shows the nozzle parameters that we will use in the following sections.

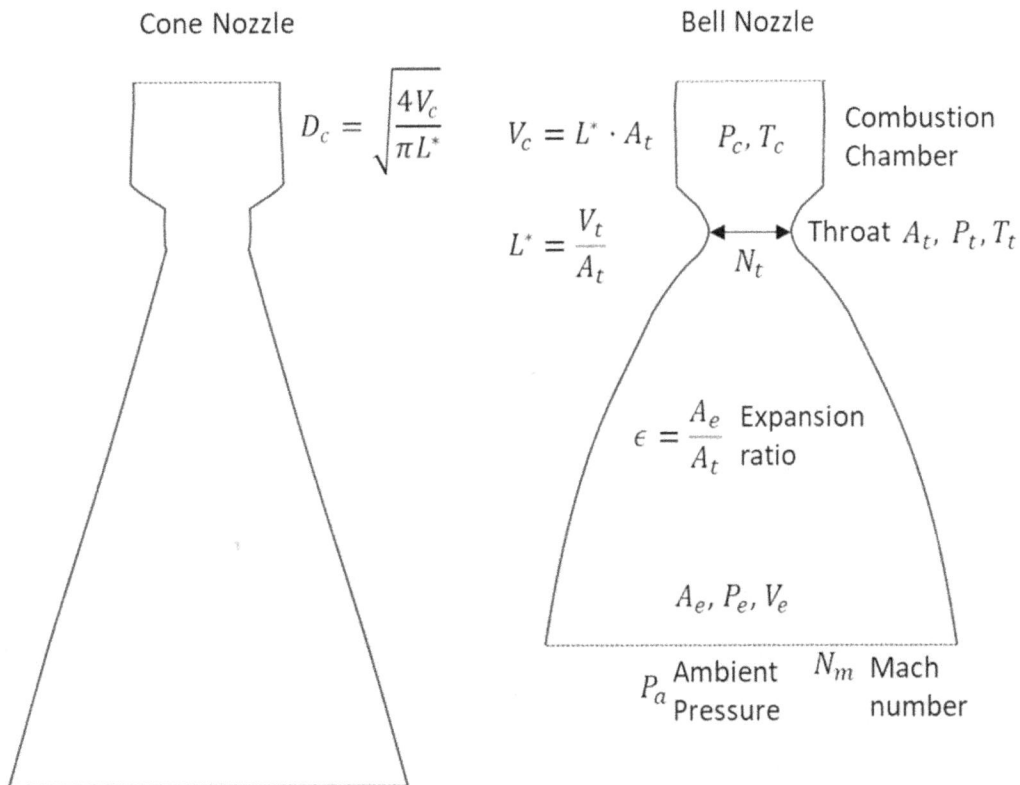

Cone Nozzle

$$D_c = \sqrt{\frac{4V_c}{\pi L^*}}$$

Bell Nozzle

$$V_c = L^* \cdot A_t$$

$$L^* = \frac{V_t}{A_t}$$

P_c, T_c Combustion Chamber

Throat A_t, P_t, T_t

N_t

$$\epsilon = \frac{A_e}{A_t}$$ Expansion ratio

A_e, P_e, V_e

P_a Ambient Pressure N_m Mach number

Figure 2-2. Nozzle parameters and shapes.

2.7.2.2   Calculating Nozzle Parameters

To design a nozzle, the total propellant flow rate and operating conditions must be known. Assuming ideal gas law conditions, the throat area (A_t) can be determined using:

$$\dot{m} = \frac{P_t A_t}{\sqrt{T_t}} \sqrt{\frac{k}{R^*}} \qquad (2\text{-}34)$$

where $\dot{m}$ is the propellant mass flow rate, P_t is the gas pressure at the nozzle throat, T_t is the gas temperature at the throat, R^* is the universal gas constant, and k is the specific heat ratio.

The pressures and temperatures at the nozzle throat (P_t and T_t) are related to the combustion chamber conditions:

$$P_t = P_c \left(\frac{2}{k+1}\right)^{\frac{k}{k-1}} \qquad (2\text{-}35)$$

$$T_t = T_c \left(\frac{2}{k+1}\right) \qquad (2\text{-}36)$$

where P_c is the combustion chamber pressure and T_c is the combustion chamber temperature.

Example 2-7:

A rocket engine burning liquid hydrogen and liquid oxygen operates at a mixture ratio of 6 and a combustion chamber pressure of 60 atm. If the propellant flow rate is 600 kg/s, calculate the area of the nozzle throat.

Solution:

Given:

- Combustion chamber pressure, $P_c = 60 \times 0.101325 = 6.0795 \text{MPa}$
- Combustion temperature, $T_c = 3{,}200 \text{ K}$
- Molar mass of exhaust gases, $M = 14.50$
- Specific heat ratio, $k = 1.240$
- Propellant flow rate, $\dot{m} = 600 \text{ kg/s}$

Calculation of Throat Pressure using:

$$P_t = P_c \left[1 + \frac{k-1}{2}\right]^{-\frac{k}{k-1}}$$

$$P_t = 6.0795 \left[1 + \frac{1.240 - 1}{2} \right]^{-\frac{1.240}{1.240-1}}$$

$$P_t \approx 3.3851 \text{ MPa} = 3.3851 \times 10^6 \text{ N/m}^2$$

Calculate Throat Temperature using:

$$T_t = \frac{T_c}{1 + \frac{k - 1}{2}}$$

$$T_t = \frac{3200}{1 + \frac{1.240 - 1}{2}}$$

$$T_t = \frac{3200}{1.120}$$

$$T_t \approx 2857 \text{ K}$$

Calculate Throat Area:

$$A_t = \left(\frac{\dot{m}}{P_t} \right) \sqrt{\left(\frac{R^* T_t}{Mk} \right)}$$

$$A_t = \left(\frac{600}{3.3851 \times 10^6} \right) \sqrt{\left(\frac{8314.46 \times 2857}{14.50 \times 1.240} \right)}$$

$$A_t = (0.0001772473487)\sqrt{(1321157.52)}$$

$$A_t \approx 0.20373 \text{ m}^2$$

Example 2-8:

Consider the previous example's rocket engine but with a different configuration. The convergent cone half-angle is modified to 25 degrees. Calculate the dimensions of the modified combustion chamber.

Solution:

Given:

- Throat area, $A_t = 0.20373 \text{ m}^2 = 2037.3 \text{ cm}^2$
- Throat diameter, $D_t = 2 \times \sqrt{\frac{2037.3}{\pi}} \approx 50.93 \text{ cm}$
- Convergent cone half-angle, $\theta = 25°$
- Characteristic Length, $L^* = 115 \text{ cm}$

Calculate Combustion Chamber Volume using:

$$V_c = A_t \times L^*$$

$$V_c = 2037.3 \times 115 = 234290 \text{ cm}^3$$

Combustion Chamber Length L_c :

Would have to be computed iteratively, assume $L_c = 70 \text{ cm}$

Calculate Combustion Chamber Diameter D_c using

$$D_c = \sqrt{\left(\frac{D_t^3 + \frac{24}{\pi} \tan \theta \times V_c}{D_c + 6 \tan \theta \times L_c} \right)}$$

Substitute the given values:

$$D_c = \sqrt{\left(\frac{53.7^3 + \frac{24}{\pi} \tan(25°) \times 234290}{D_c + 6 \tan(25°) \times 70} \right)}$$

First Iteration:

$$D_c = \sqrt{\left(\frac{154854 + \frac{24}{\pi} \times 0.4663 \times 234290}{D_c + 6 \times 0.4663 \times 70} \right)}$$

$$D_c = \sqrt{\left(\frac{989458}{D_c + 195.846} \right)}$$

Solve as a quadratic equation by squaring both sides, or iteratively

$$D_c \approx 61.95 \text{ cm}$$

Summary:
- Combustion Chamber Volume: $V_c \approx 260,245 \text{ cm}^3$
- Combustion Chamber Length: $L_c \approx 70 \text{ cm}$
- Combustion Chamber Diameter: $D_c \approx 61.95 \text{ cm}$

2.7.2.3 Expanding Hot Gases

For maximum thrust, hot gases must be expanded in the nozzle's diverging section. The gas pressure decreases as energy is used to accelerate the gases. The nozzle exit area (A_e) is determined by finding the area where the gas pressure equals the ambient pressure. The exit Mach number (N_m) is given by:

$$N_m = \sqrt{\frac{2}{k-1}\left(\left(\frac{P_a}{P_a}\right)^{\frac{k-1}{k}} - 1\right)} \tag{2-37}$$

The nozzle exit area corresponding to the exit Mach number is:

$$A_e = A_t\left(\frac{N_m}{N_t}\right) \tag{2-38}$$

where N_t is the Mach number at the throat.

$$A_e = \frac{A_t}{N_m}\left[\frac{1 + \frac{k-1}{2}N_m^2}{\frac{k+1}{2}}\right]^{\frac{k+1}{2(k-1)}} \tag{2-39}$$

Example 2-9:

Consider the same rocket problem as the previous example. The engine is optimized to operate at an elevation of 3000 m. Determine (a) the nozzle exit area (b) the ratio of the nozzle exit to the throat area.

Solution:

Given:

- Combustion chamber pressure, $P_c = 6.0795 \text{MPa}$
- Throat area, $A_t = 0.2263 \text{ m}^2$
- Specific heat ratio, $k = 1.240$
- Ambient pressure at 3000m, $P_a = 0.0701 \text{MPa}$

Calculation of Exit Mach Number:

$$N_m^2 = \left(\frac{2}{k-1}\right)\left[\left(\frac{P_c}{P_a}\right)^{\frac{k-1}{k}} - 1\right]$$

$$N_m^2 = \left(\frac{2}{1.240-1}\right)\left[\left(\frac{6.0795}{0.0701}\right)^{\frac{1.240-1}{1.240}} - 1\right]$$

$$N_m^2 = \left(\frac{2}{0.240}\right)\left[\left(\frac{6.0795}{0.0701}\right)^{0.193548387} - 1\right]$$

$$N_m^2 = 8.333333[86.72610556^{0.193548387} - 1]$$

$$N_m^2 \approx 11.43384$$

Calculation of Exit Area:

$$A_e = \frac{A_t}{N_m}\left[\frac{1 + \frac{k-1}{2}N_m^2}{\frac{k+1}{2}}\right]^{\frac{k+1}{2(k-1)}}$$

$$A_e = \frac{0.2263}{3.3813958}\left[\frac{1 + \frac{1.240-1}{2} \times 11.43384}{\frac{1.240+1}{2}}\right]^{\frac{1.240+1}{2(1.240-1)}}$$

$$A_e = \frac{0.2263}{3.3813958}\left[\frac{1 + 0.120 \times 11.43384}{1.120}\right]^{4.66667}$$

$$A_e = \frac{0.2263}{3.3813958}\left[\frac{1 + 1.3720608}{1.120}\right]^{4.66667}$$

$$A_e \approx 0.066925025 \times 33.18212$$

$$A_e \approx 2.22071 \text{ m}^2$$

$$\text{Section Ratio } = \frac{A_e}{A_t} = \frac{2.22071}{0.2263} \approx 9.813$$

2.7.2.4 Nozzle Expansion Ratio

The expansion ratio (ϵ) is defined as:

$$\epsilon = \frac{A_e}{A_t} \qquad (2\text{-}40)$$

Trajectory computations determine the optimum exit pressure for launch vehicles, particularly the first stage. However, constraints like the maximum allowable nozzle exit diameter often limit this. For space engines, the ambient pressure is zero, and thrust increases with a higher nozzle expansion ratio until additional weight offsets the performance gains.

2.7.2.5 Nozzle Shape and Design

The shape of the nozzle, particularly the diverging section, significantly impacts performance. Essential design goals include:
1. Parallel, uniform, and axial gas flow at the nozzle exit.
2. Minimum separation and turbulence losses.
3. The nozzle length should be the shortest possible to minimize space, weight, wall friction losses, and cooling requirements.

2.7.3 Ease of manufacturing.

Conical Nozzle: This traditional nozzle design is easy to manufacture and can be adjusted for different expansion ratios. The typical design includes a throat section with a circular arc contour and specific half-angles for convergent (20-45 degrees) and divergent (12-18 degrees) sections. However, conical nozzles suffer from performance losses due to non-axial exhaust gas components.

Bell Nozzle: Developed for higher performance and shorter length, bell-shaped nozzles feature a fast-expansion section leading to uniform axial flow at the exit. The bell nozzle's length is often 80 100% of an equivalent conical nozzle's length, with optimized contours to maximize performance without excessive weight.

2.7.4   Practical Considerations

Designing an optimum nozzle requires careful consideration of throat diameter, axial length, expansion ratio, and wall angles. Although different propellant combinations have minimal impact on contour, selecting the proper inputs can yield near-optimal nozzle designs for various applications.

Conical Nozzles

In early rocket engine applications, conical nozzles were used almost exclusively due to their satisfactory performance and ease of manufacturing. A conical nozzle allows flexibility in converting existing designs to higher or lower expansion ratios without significant redesign. Typically, the nozzle throat section has a circular arc contour with a radius ranging from 0.25 to 0.75 times the throat diameter (D_t). The half-angle of the convergent cone section can range from 20 to 45 degrees. The divergent cone half-angle can approximately vary 12-18 degrees. The 15-degree divergent half-angle conical nozzle has become a standard due to its balance of weight, length, and performance.

Performance losses in a conical nozzle arise from the non-axial component of the exhaust gas velocity. A correction factor, thrust efficiency (η), is applied in the exit-gas momentum calculation. It is the ratio between the exit-gas velocity of the conical nozzle and exhaust velocity for an ideal nozzle.

Bell Nozzles

To achieve higher performance and reduced nozzle lengths, engineers developed the bell-shaped nozzle. The bell design features a rapidly expanding section in the initial divergent region. The bell design is characterized by a uniform axial flow at the nozzle exit. The wall contour is gradually adjusted to prevent oblique shocks, with a standard half-angle of 15 degrees often used to specify bell nozzles. For example, an 80% bell nozzle (the distance between the throat and exit plane) is 80% the length of a 15-degree half-angle conical nozzle with the same throat area, radius below the throat, and area expansion ratio. Beyond approximately 80% length, further reductions in bell nozzle length offer minimal performance gains, especially when considering weight penalties. However, lengths up to 100% are optimal for applications requiring exceptionally high performance.

Designing a specific nozzle requires key data points: throat diameter, axial length from the throat to the exit plane, expansion ratio, initial wall angle of the parabola, and nozzle exit wall angle. The wall angles are determined by the expansion ratio. By selecting the proper inputs, optimum nozzle contours can be accurately approximated, and variations in propellant combinations typically have minimal impact on the contour.

Rocket Nozzle Design: Optimizing Expansion for Maximum Thrust

A rocket engine operates by burning propellants in a combustion chamber, generating high-pressure gases that are then expanded through a specially shaped nozzle to produce thrust. The nozzle's function is to convert the thermal energy transferred to the gas propellant into kinetic energy. This process transforms the slow-moving, high-pressure, high-temperature gases within the combustion chamber or in the reactor into high-velocity gases with lower pressure and temperature. Rocket nozzles are capable of achieving gas velocities ranging from 1.5 to 5 kilometers per second. The nozzles that accomplish this, known as De Laval nozzles, feature convergent and divergent sections. The narrowest point of the nozzle, located between the convergent and divergent sections, is called the throat, while the end of the divergent section is referred to as the exit area.

Expansion in the Diverging Section

In the diverging section of the nozzle, the hot exhaust gases expand, decreasing in pressure as their energy is used to accelerate the gas to high velocities. The nozzle is typically designed to be long enough or have a sufficiently large exit area so that the combustion chamber pressure is reduced to match the ambient pressure at the nozzle exit. The optimum or correct expansion condition, where $P_e = P_a$ (exit pressure equals ambient pressure), results in maximum thrust. Under this condition, the nozzle is said to be adapted.

Basic Thrust Equation

The basic thrust equation can be expressed as:

$$F = \dot{m}V_e + (P_e - P_a)A_e \qquad (2\text{-}41)$$

where:
- F = Thrust
- $\dot{m}$ = Propellant mass flow rate
- V_e = Velocity of exhaust gases
- P_e = Pressure at nozzle exit
- P_a = Ambient pressure
- A_e = Area of nozzle exit

In this equation, $\dot{m}V_e$ is called the momentum thrust, and $(P_e - P_a)A_e$ is called the pressure thrust. As V_e and P_e are inversely proportional, optimizing nozzle performance requires balancing these two components to achieve maximum thrust at $P_e = P_a$.

Example Calculation

Consider a rocket engine equipped with an extendible nozzle, tested in an environment with constant ambient pressure. During the burn, the nozzle is extended from fully retracted to fully extended. At some intermediate point, $P_e = P_a$.

- Given Conditions:
- Propellant mass flow rate $(\dot{m})$: 100 kg/s
- Specific heat ratio (k): 1.20
- Exhaust gas molecular weight (M): 24
- Combustion chamber temperature (T_c): 3600 K
- Combustion chamber pressure (P_c): 5 MPa
- Ambient pressure (P_a): 0.05 MPa

First, calculate the pressure (P_t) and temperature (I_t') at the nozzle throat:

$$P_t = P_c \left[1 + \frac{k-1}{2}\right]^{-\frac{k}{k-1}}$$

$$P_t = 5 \times \left[1 + \frac{1.20 - 1}{2}\right]^{-\frac{1.20}{1.20-1}}$$

$$P_t = 2.82 \, \text{MPa}$$

$$T_t = T_c \left[1 + \frac{k-1}{2}\right]^{-1}$$

$$T_t = 3600 \times \left[1 + \frac{1.20 - 1}{2}\right]^{-1}$$

$$T_t = 3273 \, \text{K}$$

Next, calculate the throat area (A_t) :

$$A_t = \frac{\dot{m}}{P_t} \sqrt{\frac{R^* T_t}{Mk}}$$

$$A_t = \frac{100}{2.82 \times 10^5} \sqrt{\frac{8314 \times 3273}{24 \times 1.20}}$$

$$A_t = 0.0345 \, \text{m}^2$$

The Mach number (N_m) at the nozzle exit is:

$$N_m^2 = \frac{2}{k-1}\left[\left(\frac{P_c}{P_a}\right)^{\frac{k-1}{k}} - 1\right]$$

$$N_m^2 = \frac{2}{1.20-1}\left[\left(\frac{5}{0.05}\right)^{\frac{1.20-1}{1.20}} - 1\right]$$

$$N_m^2 = 11.54$$

$$N_m = \sqrt{11.54} = 3.40$$

The nozzle exit area (A_e) is:

$$A_e = \frac{A_t}{N_m}\left[\frac{1 + \frac{k-1}{2}N_m^2}{\frac{k+1}{2}}\right]^{\frac{k+1}{2(k-1)}}$$

$$A_e = \frac{0.0345}{3.40}\left[\frac{1 + \frac{1.20-1}{2} \times 11.54}{\frac{1.20+1}{2}}\right]^{\frac{1.20+1}{2(1.20-1)}}$$

$$A_e = 0.409 \text{ m}^2$$

The velocity (V_e) of the exhaust gases at the nozzle exit is:

$$V_e = \sqrt{\frac{2k}{k-1}\frac{R^*T_c}{M}\left[1 - \left(\frac{P_e}{P_c}\right)^{\frac{k-1}{k}}\right]}$$

$$V_e = \sqrt{\frac{2 \times 1.20}{1.20-1}\frac{8314 \times 3600}{24}\left[1 - \left(\frac{0.05}{5}\right)^{\frac{1.20-1}{1.20}}\right]}$$

$$V_e = 2832 \text{ m/s}$$

Finally, calculate the thrust (F) :

$$F = \dot{m}V_e + (P_e - P_a)A_e$$

$$F = 100 \times 2832 + (0.05 \times 10^6 - 0.05 \times 10^6) \times 0.409$$

$$F = 283{,}200 \text{ N}$$

Under-Extended Nozzle ($P_e > P_a$)

Assume $P_e = 2P_a = 0.10 \text{MPa}$:

$$N_m^2 = \frac{2}{1.20 - 1}\left[\left(\frac{5}{0.10}\right)^{\frac{1.20-1}{1.20}} - 1\right]$$

$$N_m^2 = 9.19$$

$$N_m = \sqrt{9.19} = 3.03$$

$$A_e = \frac{0.0345}{3.03}\left[\frac{1 + \frac{1.20 - 1}{2} \times 9.19}{\frac{120 + 1}{2}}\right]^{\frac{1.20+1}{2(1.20-1)}}$$

$A_e = 0.243 \text{ m}^2$

$$V_e = \sqrt{\frac{2 \times 1.20}{1.20 - 1}\frac{8314 \times 3600}{24}\left[1 - \left(\frac{0.10}{5}\right)^{\frac{1.20-1}{1.20}}\right]}$$

$$V_e = 2677 \text{ m/s}$$

$$F = 100 \times 2677 + (0.10 \times 10^6 - 0.05 \times 10^6) \times 0.243$$

$$F = 279{,}850 \text{ N}$$

Over-Extended Nozzle ($P_e < P_a$)

Assume $P_e = \frac{P_a}{2} = 0.025 \text{MPa}$:

$$N_m^2 = \frac{2}{1.20 - 1}\left[\left(\frac{5}{0.025}\right)^{\frac{1.20-1}{1.20}} - 1\right]$$

$$N_m^2 = 14.18$$

$$N_m = \sqrt{14.18} = 3.77$$

$$A_e = \frac{0.0345}{3.77}\left(\frac{1 + \frac{1.20-1}{2} \times 14.18}{\frac{1.20+1}{2}}\right)^{\frac{1.20+1}{2(1.20-1)}}$$

$$A_e = 0.696 \text{ m}^2$$

$$V_e = \sqrt{\frac{2 \times 1.20}{1.20-1}\frac{8314 \times 3600}{24}\left[1 - \left(\frac{0.025}{5}\right)^{\frac{1.20-1}{1.20}}\right]}$$

$$V_e = 2963 \text{ m/s}$$

$$F = 100 \times 2963 + (0.025 \times 10^6 - 0.05 \times 10^6) \times 0.696$$

$$F = 278{,}900 \text{ N}$$

Analysis

Both under-extended, and over-extended nozzles produce less thrust than when $P_e = P_a$. The analysis confirms that maximum thrust occurs when the nozzle is correctly adapted, with $P_e = P_a$.

2.7.4.1 Combustion Chamber

The rocket engine combustion chamber ensures the complete mixing and combustion of propellants. Its primary function is to provide an enclosure where propellants can reside for a sufficient period to achieve efficient combustion. The required residence time for combustion depends on various parameters, and the theoretically required combustion chamber volume is a function of the propellant mass flow, the average density of the combustion products, and the necessary stay time for complete combustion. This relationship can be mathematically expressed as:

$$V_c = \dot{m} \cdot V \cdot t_s \qquad (2\text{-}42)$$

where:

- V_c is the chamber volume,
- $\dot{m}$ is the propellant mass flow rate,
- V is the average specific volume of the combustion products,
- t_s is the propellant residence time.

2.7.4.2   Characteristic Length (L^*)

A useful parameter related to the chamber volume and residence time is the characteristic length, L^* (pronounced "L star"). It is defined as the chamber volume divided by the nozzle's sonic throat area:

$$L^* = \frac{V_t}{A_t} \qquad (2\text{-}43)$$

The L^* concept is more intuitive than "combustion residence time," which is typically expressed in small fractions of a second. Since the throat area (A_t) is proportional to the product of the mass flow rate ($\dot{m}$) and the average specific volume (V), L^* essentially becomes a function of the residence time (t_s).

Table 2-1: Chamber Characteristic Length (L^*)

Propellant Combination	L^* (cm)
Ntric acid/hydrazine–base fuel	76 − 89
Ntrogen tetroxide/hydrazine–base fuel	76 − 89
Hydrogen peroxide/RP-1 (including catalyst bed)	152 − 178
Liquid oxygen/RP-1	102 − 127
Liquid oxygen/ammonia	76 − 102
Liquid oxygen/liquid hydrogen (GH2 injection)	56 − 71
Liquid oxygen/liquid hydrogen (LH2 injection)	76 − 102
Liquid fluorine/liquid hydrogen (GH2 injection)	56 − 66
Liquid fluorine/liquid hydrogen (LH2 injection)	64 − 76
Liquid fluorine/hydrazine	61 − 71
Chlorine trifluoride/hydrazine–base fuel	51 − 89

The customary method of establishing the L^* of a new thrust chamber design relies heavily on past experience with similar propellants and engine sizes. The minimum required L^* can only be accurately evaluated through experimental firings of test chambers. Typical L^* values for various propellants are shown in the table below. Once the throat area and minimum required L^* are established, the chamber volume can be calculated using the relationship:

$$V_c = L^* \cdot A_t \qquad (2\text{-}44)$$

2.7.4.3 Combustion Chamber Geometries

Three geometrical shapes are commonly used in combustion chamber design: spherical, near-spherical, and cylindrical. The cylindrical chamber is most frequently employed in the United States. A spherical chamber offers less cooling surface and weight per the same volume as other geometries. However, spherical chambers are more challenging to manufacture and have provided poorer performance in other respects.

The total combustion process, from the injection of reactants to the completion of chemical reactions and the conversion of products into hot gases, requires finite amounts of time and volume, as represented by the characteristic length L^*. This value is significantly greater than the linear length between the injector face and throat plane. The contraction ratio, defined as the major cross-sectional area of the combustor divided by the throat area, also plays a role. Typically, smaller chambers use a large contraction ratio with a shorter length, still providing sufficient L^* for adequate vaporization and combustion dwell time. Larger engines are designed with a low contraction ratio and a comparatively long length.

2.7.4.4 Sizing a New Combustion Chamber

When sizing a new combustion chamber, a useful starting point is to examine the dimensions of similarly sized previous designs. It makes sense to plot historical data in relation to throat diameter to design a throat size for a new engine.

The volume of the combustion chamber is given by:

$$V_c = L^* \cdot A_t \qquad (2\text{-}45)$$

Rearranging this equation allows for solving the chamber diameter through iteration:

$$D_c = \sqrt{\frac{4V_c}{\pi L^*}} \qquad (2\text{-}46)$$

2.7.4.5   Injector

The injector is responsible for delivering the propellants into the combustion chamber in the correct proportions and conditions to achieve efficient and stable combustion. It is located at the forward or upper end of the combustor. The injector also serves a structural role. It creates a seal against the high temperatures and pressures in the combustion chamber.

2.7.4.5.1   Functions of the Injector

The injector's primary function is akin to that of a carburetor in an automobile engine, as it supplies the fuel and oxidizer at the proper rates and proportions. However, the injector's role in a rocket engine is far more complex and critical. Positioned directly above the high-pressure combustion zone, the injector must handle multiple tasks related to both combustion and cooling processes, making it a vital component for the rocket engine's overall performance.

2.7.4.5.2   Impact on Engine Performance

No other component has as significant an impact on rocket engine performance as the injector. In different applications, well-designed injectors can vary widely in combustion efficiency. It's not uncommon for an injector with a characteristic exhaust velocity (C^*) efficiency as low as 92% to be considered acceptable. For small engines designed for specialized purposes, such as attitude control, the design may prioritize response time and weight over combustion efficiency, leading to satisfactory performance even if efficiency falls below 90%.

In general, however, modern injection systems have achieved efficiencies so close to 100% of theoretical values that the accuracy of measurement tools becomes the limiting factor. High levels of combustion efficiency are achieved through the uniform distribution of the desired mixture ratio and fine atomization of liquid propellants.

2.7.4.5.3   Combustion Stability

A satisfactory injector design must also ensure combustion stability. Under certain conditions, local disturbances in the chamber, such as fluctuations in mixing or propellant flow, can generate shock and detonation waves. These disturbances may trigger pressure oscillations that are amplified and sustained by the combustion process, leading to a phenomenon known as combustion instability.

High-amplitude waves produced by this instability result in significant vibration and heat flux, which can be highly destructive.

Designing for stable combustion is a major focus in the development of injectors. Many parameters that enhance performance can reduce the stability margin, making the balance between efficiency and stability a critical aspect of injector design.

2.7.4.5.4   Key Parameters for Injector Design

Several key parameters influence injector design to achieve both high performance and stability:

1. Uniform Distribution: Ensuring that the propellants are evenly distributed in the correct proportions.
2. Fine Atomization: Achieving a fine spray of propellants to facilitate efficient mixing and combustion.

2.7.4.5.5   Key Parameters for Injector Design

Several key parameters influence injector design to achieve both high performance and stability:

1. Uniform Distribution: Ensuring that the propellants are evenly distributed in the correct proportions.
2. Fine Atomization: Achieving a fine spray of propellants to facilitate efficient mixing and combustion.
3. Stable Combustion: Designing the injector to minimize the risk of combustion instability through careful control of flow dynamics and mixing processes.

The injector is indispensable for the efficient and stable operation of a rocket engine. Its design directly affects the engine's performance and reliability, requiring meticulous attention to detail in achieving uniform distribution, fine atomization, and combustion stability. The challenges associated with injector design highlight its importance and complexity, making it a focal point in the development of advanced propulsion systems.

2.7.5   Power Cycles in Liquid Bipropellant Rocket Engines

Liquid bipropellant rocket engines are categorized based on their power cycles. The method of feeding propellants to the combustion chamber is determined based on the power cycle. Below are descriptions of some common power cycles.

2.7.5.1   Gas-Generator Cycle

The gas-generator cycle, also known as the open cycle, taps off a small portion (typically 2 to 7 percent) of fuel and oxidizer from the main flow to feed a separate burner called a gas generator. The hot gas produced by this generator passes through a turbine, generating power for the pumps that feed propellants into the main combustion chamber. After passing through the turbine, the hot gas is either vented overboard or directed into the main nozzle downstream. Increasing the propellant flow into the gas generator speeds up the turbine, increasing the propellant flow into the main combustion chamber and thus increasing thrust. The gas generator burns propellants at a suboptimal mixture ratio to keep turbine blade temperatures low, making this cycle suitable for moderate power requirements but less efficient for high-power systems, which would require diverting a significant portion of the main flow to the less efficient gas-generator flow.

Some propellant in the gas-generator cycle is used to cool the nozzle and combustion chamber to enhance efficiency and allow higher engine temperatures.

2.7.5.2   Staged Combustion Cycle

In the staged combustion cycle, also known as the closed cycle, propellants are burned in stages. Similar to the gas-generator cycle, it uses a burner to generate gas for the turbine. The burner burns a small amount of one propellant and a large amount of the other, producing either an oxidizer-rich or fuel-rich hot gas mixture that is primarily unburned-vaporized propellant. This hot gas is passed through the turbine, injected into the chamber, and combusted.

In the staged combustion cycle, also known as the closed cycle, propellants are burned in stages. Similar to the gas-generator cycle, it uses a preburner to generate gas for the turbine. The preburner burns a small amount of one propellant and a large amount of the other, producing either an oxidizer-rich or fuel-rich hot gas mixture that is primarily unburned vaporized propellant. This hot gas is passed through the turbine, injected into the chamber, and combusted again with the remaining propellants. Unlike the gas-generator cycle, the staged combustion cycle burns all propellants at the optimal mixture ratio in the main chamber without dumping any flow overboard, making it suitable for high-power applications. Higher chamber pressures achieved in this cycle allow for smaller and lighter engines to produce the same thrust. However, the development cost is higher due to the complexities of handling high pressures, harsh turbine conditions, high-temperature piping, and intricate feedback and control systems.

Staged combustion was first developed by Soviet engineers and appeared in 1960. The first Western laboratory-staged combustion test engine was built in Germany in 1963.

2.7.5.3   Expander Power Cycle

The expander power cycle is similar to the staged combustion power cycle but does not use a preburner. Instead, the heat from the cooling jacket of the main combustion chamber vaporizes the fuel. The fuel vapor then passes through the turbine before being injected into the main chamber to burn with the oxidizer. This cycle is effective with fuels like hydrogen or methane, which have low boiling points and can be easily vaporized. As in the staged combustion cycle, propellants are combusted at an optimal mixture ratio, typically without any overboard flow. However, the power available to the turbine is limited by the heat transfer to the fuel, making this cycle suitable for small to mid-sized engines.

2.7.5.4 Pressure-Fed Cycle

The pressure-fed cycle is the simplest system, lacking pumps or turbines, and relies on tank pressure to feed propellants into the main chamber. This cycle is typically limited to low chamber pressures since higher pressures would require excessively heavy vehicle tanks. Despite its limitations, the pressure-fed cycle can be very reliable due to its reduced part count and complexity compared to other systems.

2.7.6 Engine Cooling

Combustion in a rocket engine generates significant heat contained within the exhaust gases. While most of this heat is expelled with the exhaust gases, sufficient heat is transferred to the thrust chamber walls, necessitating effective cooling.

Thrust chamber designs are categorized by their hot gas wall cooling methods or the configuration of coolant passages. The coolant pressure inside these passages can reach up to 500 atmospheres.

The high combustion temperatures (2500 to 3600 K) and high heat transfer rates (up to 16 kJ/cm^2-s) present significant challenges for designers. Several chamber cooling techniques have been successfully utilized, with the optimal cooling method selected depending on factors such as propellant type, chamber pressure, available coolant pressure, combustion chamber configuration, and material.

Effective cooling techniques ensure the combustion chamber's structural integrity and performance, allowing for sustained high-temperature operations and maximizing the rocket engine's efficiency and reliability.

2.7.6.1 Regenerative Cooling and Other Cooling Methods in Rocket Engines

The most common method used for cooling a rocket engine thrust chamber is regenerative cooling.

This technique involves flowing high-velocity coolant over the backside of the combustion chamber's hot gas wall. The hot gas liner is convectively cooled by the cold propellants, effectively cooling the chamber. After absorbing heat, the coolant is discharged into the injector and utilized as a propellant.

2.7.6.2 Historical and Modern Thrust Chamber Designs

Early thrust chamber designs, such as those used in the V − 2 and Redstone rockets featured low chamber pressure, heat flux, and coolant pressure requirements. These could be satisfied with a simplified "double wall chamber" design that combined regenerative and film cooling. However, as rocket engine technology advanced, chamber pressures increased, making the cooling requirements more challenging. This necessitated the development of new coolant configurations that were more structurally efficient and had improved heat transfer characteristics.

2.7.6.3   Tubular Wall Thrust Chambers

This need led to the design of "tubular wall" thrust chambers, which became the most widely used approach for large rocket engine applications. Tubular wall designs have been successfully implemented in engines for the Thor, Jupiter, Atlas, H-1, J-2, F-1, and RS-27, among others used by the Air Force and NASA. The primary advantages of tubular wall designs include their lightweight structure and the extensive experience base that has accumulated over time. However, as chamber pressures and hot gas wall heat fluxes have continued to increase, exceeding 100 atmospheres, even more effective cooling methods have been required.

2.7.6.4   Channel Wall Thrust Chambers

One advanced solution is the "channel wall" thrust chamber. In this design, coolant flows through rectangular channels that are machined or formed into a hot gas liner made from high-conductivity materials such as copper or copper alloys. An exemplary implementation of this design is the Space Shuttle Main Engine (SSME), which operates at a nominal chamber pressure of 204 atmospheres and a temperature of 3,600 K for durations up to 520 seconds. Channel wall thrust chambers offer excellent heat transfer and structural characteristics.

2.7.6.5   Alternative Cooling Methods

In addition to regeneratively cooled designs, other cooling methods have been employed in rocket engines. These include dump cooling, film cooling, transpiration cooling, ablative liners, and radiation cooling. Each method has unique applications and advantages.

2.7.6.6   Dump Cooling

Similar to regenerative cooling, dump cooling involves flowing coolant through small passages over the backside of the thrust chamber wall. However, the coolant passing by the wall is discharged through openings at the aft end of the divergent nozzle. This method results in performance loss from dumping the coolant overboard and is not used in actual applications to date.

2.7.6.7   Film Cooling

Film cooling mitigates excessive heating by introducing a thin layer of coolant or propellant through orifices around the injector perimeter or via manifolded openings in the chamber wall near the injector or throat area. This technique is commonly employed in regions with high heat flux and is frequently used in conjunction with regenerative cooling.

2.7.6.8   Transpiration Cooling

Transpiration cooling involves delivering coolant, whether in gaseous or liquid propellant form, through a porous chamber wall at a rate that keeps the hot gas wall within the desired temperature range. This method is a specialized variant of film cooling.

2.7.6.9   Ablative Cooling

Ablative cooling involves sacrificing the combustion gas-side wall material through melting, vaporization, and chemical changes to dissipate heat. The resultant relatively cool gases flow over the wall surface, lowering the boundary-layer temperature and assisting in the cooling process.

2.7.6.10 Radiation Cooling

Radiation cooling disperses heat by emitting it from the external surface of the combustion chamber or nozzle extension. This method is frequently used in small thrust chambers constructed with high-temperature refractory materials and in low heat flux areas, like nozzle extensions.

For large liquid rocket engines, regeneratively cooled combustion chambers have emerged as the most efficient cooling solution. While various cooling methods have been successfully applied, regenerative cooling offers distinct benefits, particularly when tailored to the specific requirements of the rocket engine, such as the type of propellant, chamber pressure, and structural considerations. By carefully selecting and optimizing these cooling techniques, engineers can preserve the structural integrity and improve the performance of rocket engines under extreme conditions, ensuring dependable operation.

2.7.7   Solid Rocket Motors

Solid rocket motors are recognized for their versatility, using propellants stored in a solid state. These propellants typically consist of powdered aluminum as the fuel and ammonium perchlorate as the oxidizer. A synthetic rubber binder, like polybutadiene, binds the fuel and oxidizer powders, creating a cohesive propellant grain. Although solid rocket motors usually deliver lower performance than liquid propellant rockets, their simplicity and operational flexibility often make them the preferred option for various propulsion applications, offering confidence in their adaptability.

2.7.7.1   Solid Fuel Geometry

The geometry of the solid fuel dictates the area and shape of its exposed surfaces, which in turn influences the burn pattern and thrust profile. In the space industry, two primary types of solid fuel blocks are utilized: cylindrical blocks that burn at the surface (end burners) and cylindrical blocks with internal combustion channels.

2.7.7.2   End Burners

In end burners, the front of the flame travels in layers from the nozzle end toward the top of the casing. This configuration produces constant thrust throughout the burn.

2.7.7.3   Internal Combustion Channels

In motors with internal combustion channels, the combustion surface develops along the length of a central channel. This channel can have various geometries, such as a star shape, to moderate the growth of the combustion surface and control the thrust profile.

2.7.7.4   Thrust Profiles Based on Fuel Geometry

The shape of the fuel block is selected based on the mission requirements and desired thrust profile. As the combustion progresses from the fuel block's free surface, the changing surface area affects whether the thrust increases, decreases, or remains constant.

1. Cylindrical Channel:
   - These fuel blocks develop thrust progressively as the combustion surface increases over time.
2. Channel with Central Cylinder of Fuel:
   - This configuration produces a relatively constant thrust, which drops to zero quickly once the fuel is exhausted.
3. Five-Pointed Star Profile:
   - This profile provides relatively constant thrust, which decreases slowly to zero as the fuel is consumed.
4. Cruciform Profile:
   - This geometry results in progressively decreasing thrust throughout the burn.
5. Double Anchor Profile:
   - Similar to the cruciform profile, this design produces decreasing thrust, which rapidly drops off near the end of the burn.

2.7.7.5   Cog Profile:

This profile generates a strong initial thrust, followed by an almost constant lower thrust for the remainder of the burn.

2.7.8   Burn Rate

The burn rate of a rocket propellant grain refers to the regression rate of its burning surface, which recedes perpendicularly to the surface as combustion progresses. This rate, typically measured in millimeters per second (mm/s) or inches per second (in/s), varies significantly depending on the propellant formulation and operating conditions. Understanding and quantifying the burn rate under various conditions is crucial for the successful design of a solid rocket motor.

2.7.8.1 Factors Influencing Burn Rate

Several factors greatly impact propellant burn rates, including combustion chamber pressure, initial propellant temperature, gas velocity along the burning surface, local static pressure, and the motor's acceleration and spin.

2.7.8.2 Chamber Pressure

Chamber pressure profoundly affects the burn rate, often represented by the Saint-Robert's Law:

$$r = aP_c^n \tag{2-47}$$

where:
- r is the burn rate,
- a is the burn rate coefficient,
- n is the pressure exponent,
- P_c is the combustion chamber pressure.

The coefficients a and n are empirically determined for specific propellant formulations. It's essential to note that these coefficients are typically valid over a specific pressure range, and multiple sets may be required to cover the entire pressure regime of interest. For example, the Space Shuttle SRBs have $a = 0.0445289$ (pressure in pascals, burn rate in mm/s) and $n = 0.35$, resulting in a burn rate of 9.34 mm/s at an average chamber pressure of 4.3MPa.

2.7.8.3 Initial Temperature

The initial temperature of the propellant grain affects the burn rate by influencing the speed of chemical reactions. Propellants that are sensitive to grain temperature may show variations in their time-thrust profiles under extreme temperatures, such as during winter launches when grain temperatures might be lower than normal.

2.7.8.4 Combustion Gas Velocity

The velocity of combustion gases flowing parallel to the burning surface can augment the burn rate, a phenomenon known as erosive burning. This effect varies with the propellant type and chamber pressure. Many propellants exhibit a threshold flow velocity; below this level, there may be no augmentation or even a decrease in burn rate (negative erosive burning). Minimizing erosive burning effects involves designing the motor with a sufficient port-to-throat area ratio (A_{port}/A_t). A ratio of at least 2 is recommended for a grain with an L/D ratio of 6, with higher ratios advisable for grains with larger L/D ratios.

2.7.8.5 Local Static Pressure

There is a pressure drop along the axis of an operating rocket motor's combustion chamber, which is necessary to accelerate the increasing mass flow of combustion products toward the nozzle. The highest static pressure is at the front of the motor, where gas flow is zero, leading to the highest burn rate at this location. However, this effect is usually minor and is often offset by the countereffect of erosive burning.

2.7.8.6 Motor Acceleration

The burn rate increases with motor acceleration, whether due to longitudinal force (thrust) or spin.

Burning surfaces forming an angle of about 60-90 degrees with the acceleration vector are particularly prone to increased burn rates.

2.7.8.7 Modifying Burn Rate

Modifying the burn rate to suit specific grain configurations and mission requirements can be achieved through several methods:

2.7.8.8 Oxidizer Particle Size

The particle size of the oxidizer, particularly ammonium perchlorate (AP), significantly affects the burn rate. Smaller AP particle sizes generally lead to higher burn rates, as the decomposition of AP is the rate-determining step in the combustion process.

2.7.8.9 Oxidizer/Fuel Ratio

The oxidizer/fuel (O/F) ratio strongly influences the burn rate. Adjusting this ratio modifies the burn rate, but it also affects the propellant's performance and mechanical properties, making this method somewhat restrictive.

2.7.8.10 Catalysts and Suppressants

Adding catalysts to the propellant mixture is the most effective means of increasing the burn rate.

Catalysts are chemical compounds added in small quantities to tailor the burn rate. Conversely, burn rate suppressants are additives used to decrease the burn rate.

2.7.8.11 Chamber Pressure Adjustment

For propellants following Saint-Robert's burn rate law, designing the rocket motor to operate at a lower chamber pressure can reduce the burn rate. However, due to the nonlinear pressure-burn rate relationship, a significant reduction in operating pressure may be necessary to achieve the desired burn rate. This approach has the drawback of reducing motor performance, as specific impulse decreases with lower chamber pressure.

2.7.8.12 Example Calculation

For the Space Shuttle SRBs, assuming $a = 0.0445289$ (pressure in pascals, burn rate in mm/s) and $n = 0.35$:

$$P_c = 4.3 \text{ MPa} = 4.3 \times 10^6 \text{ Pa}$$

$$r = 0.0445289 \times (4.3 \times 10^6)^{0.35}$$

$$r \approx 9.34 \text{ mm/s}$$

The rate at which combustion products are produced in a solid rocket motor is dictated by the regression speed of the propellant grain. This product generation rate, when integrated over the surface area of the burning propellant, can be represented as:

$$q = \rho_p \cdot A_b \cdot r \tag{2-48}$$

where:

- q is the combustion product generation rate,
- ρ_p is the solid propellant density,
- A_b is the area of the burning surface,
- r is the burn rate.

The propellant density can be calculated from the mass fraction and density of the individual constituents using the following formula:

$$\rho_p = \frac{1}{\sum_i \frac{1}{\rho_i}} \tag{2-49}$$

where:

- w_i is the mass fraction of constituent i,
- ρ_i is the density of constituent i.

This formula gives the ideal density. The actual density is typically $94\% - 97\%$ of the ideal density due to small voids in the grain, depending on the manufacturing technique.

Example 2-10:

Determine the density for a solid rocket propellant given a composition of 65% ammonium perchlorate (AP), 20% aluminum (Al), and 15% hydroxyl-terminated polybutadiene (HTPB) by mass.

Solution:

Given:

- $w_{AP} = 0.65$
- $w_{Al} = 0.20$
- $w_{HTPB} = 0.15$

From the Properties of Rocket Propellants, we have:

- $\rho_{AP} = 1.95$ g/ml
- $\rho_{Al} = 2.70$ g/ml
- $\rho_{HTPB} \approx 0.93$ g/ml

Calculation

$$\rho = \frac{1}{\Sigma_i \left(\frac{w_i}{\rho_i}\right)}$$

$$\rho = \frac{1}{\left(\frac{0.65}{1.95} + \frac{0.20}{2.70} + \frac{0.15}{0.93}\right)}$$

$$\rho \approx 1.7584 \text{ g/ml}$$

2.7.8.13 Condensed-Phase Mass

Combustion products may include both gaseous and condensed-phase mass (solid or liquid particles, often seen as smoke). While only the gaseous products contribute to pressure development, the condensed-phase particles, due to their mass and velocity, also impact the thrust of the rocket motor.

However, the presence of condensed-phase particles can reduce performance for several reasons:

- Expansion Work: Condensed-phase mass does not perform any expansion work, reducing the efficiency of converting thermal energy into kinetic energy.
- Molecular Weight: The higher effective molecular weight of the products lowers the characteristic exhaust velocity, C^*.

- Thermal Lag: The thermal inertia of the condensed species results in some heat being ejected before transferring to the surrounding gas, known as particle thermal lag.
- Velocity Lag: Because of their greater mass, particles cannot accelerate as quickly as the surrounding gases, particularly in regions of high acceleration, such as the nozzle throat. This difference in velocity, referred to as particle velocity lag, results in particles exiting the nozzle at slower speeds compared to the gases.

2.7.8.14 Chamber Pressure

The pressure profile of a rocket motor exhibits both transient and steady-state behavior. Transient phases include the ignition and start-up phase and the tail-off phase after nearly complete grain consumption when pressure drops to ambient levels. During steady-state burning, variations in chamber pressure are mainly due to changes in grain geometry and burn rate. However, factors like nozzle throat erosion and erosive burning can also play a role.

2.7.8.15 Monopropellant Engines

Monopropellant engines are extensively used for spacecraft attitude and velocity control, with hydrazine being the most common propellant due to its favorable handling properties, relative stability, and clean decomposition products. The operational sequence of a hydrazine thruster involves the following steps:
- **Activation:** When the attitude control system signals the need for thruster operation, an electric solenoid valve opens, allowing hydrazine to flow.
- **Injection:** The pressure from the tank forces liquid hydrazine into the injector, where it is sprayed into the chamber contacting the catalyst.
- **Catalyst Reaction:** The catalyst bed, typically composed of alumina pellets impregnated with iridium, heats the hydrazine to its vaporization point, initiating its decomposition. The chemical reactions then become self-sustaining as the hydrazine breaks down.
- **Flow Control:** By adjusting flow variables and the geometry of the catalyst chamber, designers can fine-tune the chemical products, exhaust temperature, molecular weight, and enthalpy. For high-specific-impulse applications, it is common to achieve 30-40% ammonia dissociation. For applications requiring lower-temperature gases, higher levels of ammonia dissociation are targeted.
- **Thrust Generation:** The decomposition products of the hydrazine are expelled through a high-expansion-ratio nozzle, generating thrust.

Monopropellant hydrazine thrusters typically achieve a specific impulse of about 230 to 240 seconds. Other propellants, such as hydrogen peroxide (H_2O_2) and nitrous oxide (N_2O), are also used, though they offer lower performance, with specific impulses of around 150 seconds and 170 seconds, respectively.

2.7.9   Advanced Propulsion Systems

Monopropellant systems are adequate for orbit maintenance and attitude control but lack the efficiency for large delta-v maneuvers required for orbit insertion. Bipropellant systems, which use separate fuel and oxidizer, offer higher performance and can perform all three functions (orbit insertion, maintenance, and attitude control) but are more complex.

2.7.9.1   Dual Mode Systems

Dual mode systems are hybrid designs using hydrazine as both a monopropellant for low-thrust catalytic thrusters and as a fuel for high-performance bi-propellant engines. This configuration allows the same fuel tank to supply both monopropellant and bipropellant thrusters, simplifying the propulsion system.

2.7.9.2   Cold Gas Propulsion

Cold gas propulsion systems, the simplest form of rocket engines, consist of a pressurized gas source and a nozzle. These systems are used where simplicity and the need to avoid hot gases are critical. As an example, the Manned Maneuvering Unit (MMU) used by astronauts is a cold gas propulsion system.

Understanding the product generation rate, the impact of condensed-phase mass, and the nuances of different propulsion systems is crucial for optimizing rocket engine performance. Each propulsion method offers unique advantages and trade-offs, influencing the design choices for specific mission requirements.

2.7.10 Staging

Multistage rockets are essential for improving payload capabilities in vehicles with high delta-V (ΔV) requirements, such as launch vehicles and interplanetary spacecraft. In a multistage rocket, propellants are stored in smaller, separate tanks rather than a single large tank as in a single-stage rocket. Each tank, along with its associated engines, is discarded when empty. This approach ensures that energy is not wasted on accelerating empty tanks, thereby achieving a higher total ΔV or enabling a larger payload mass to be accelerated to the same total ΔV. Each of these discardable units is referred to as a stage.

2.7.10.1 Performance Calculation

The performance of a multistage rocket is described by the same rocket equation used for single-stage rockets, but calculations must be performed on a stage-by-stage basis. The velocity increment (ΔV_i) for each stage is calculated using:

$$\Delta V_i = I_{sp,i} \cdot g_0 \cdot \ln \left(\frac{m_{o,i}}{m_{f,i}} \right) \tag{2-50}$$

where:

- ΔV_i is the velocity increment for the stage i,
- $I_{sp,i}$ is the specific impulse of the stage i,
- g_0 is the standard gravitational acceleration,
- $m_{o,i}$ is the total vehicle mass at the ignition of stage i,
- $m_{f,i}$ is the total vehicle mass at the burnout of stage i but before discarding the stage.

It's important to note that the payload mass for any stage includes the mass of all subsequent stages plus the ultimate payload itself. The total velocity increment (ΔV) for the vehicle is the sum of the velocity increments for the individual stages:

$$\Delta V = \sum_{i=1}^{n} \Delta V_i \tag{2-51}$$

where n is the total number of stages.

Example 2-11:

A two-stage rocket has the following masses: 1st-stage propellant mass 110,000 kg, 1st-stage dry mass 8,500 kg, 2nd-stage propellant mass 28,000 kg, 2nd-stage dry mass 2,500 kg, and payload mass 3,500 kg. The specific impulses of the 1 st and 2 nd stages are 270 s and 330 s , respectively. Calculate the rocket's total ΔV.

Solution:

Given:

- 1st Stage:
- Propellant mass: $M_{p1} = 110{,}000$ kg
- Dry mass: $M_{d1} = 8{,}500$ kg
- 2nd Stage:
- Propellant mass: $M_{p2} = 28{,}000$ kg
- Dry mass: $M_{d2} = 2{,}500$ kg
- Payload mass: $M_{pl} = 3{,}500$ kg
- Specific Impulse:
- 1st stage: $I_{sp1} = 270$ s
- 2nd stage: $I_{sp2} = 330$ s

Initial and Final Mass Calculations:

- 1st Stage:

  Initial mass: $M_{o1} = M_{p1} + M_{d1} + M_{p2} + M_{d2} + M_{pl}$

  Final mass: $M_{f1} = M_{d1} + M_{p2} + M_{d2} + M_{pl}$

$$M_{o1} = 110{,}000 + 8{,}500 + 28{,}000 + 2{,}500 + 3{,}500 = 152{,}500 \text{ kg}$$

$$M_{f1} = 8{,}500 + 28{,}000 + 2{,}500 + 3{,}500 = 42{,}500 \text{ kg}$$

- 2nd Stage:

  Initial mass: $M_{o2} = M_{p2} + M_{d2} + M_{pl}$

  Final mass: $M_{f2} = M_{d2} + M_{pl}$

$$M_{o2} = 28{,}000 + 2{,}500 + 3{,}500 = 34{,}000 \text{ kg}$$

$$M_{f2} = 2{,}500 + 3{,}500 = 6{,}000 \text{ kg}$$

Calculation of Effective Exhaust Velocities:

$$C_1 = I_{sp1}g = 270 \times 9.80665 = 2{,}648 \text{ m/s}$$

$$C_2 = I_{sp2}g = 330 \times 9.80665 = 3{,}236 \text{ m/s}$$

Calculation of ΔV for Each Stage:

$$\Delta V_1 = C_1 \ln\left(\frac{M_{o1}}{M_{f1}}\right)$$

$$\Delta V_1 = 2{,}648\ln\left(\frac{152{,}500}{42{,}500}\right)$$

$$\Delta V_1 = 2{,}648\ln(3.5882)$$

$$\Delta V_1 \approx 3{,}477 \text{ m/s}$$

$$\Delta V_2 = C_2 \ln\left(\frac{M_{o2}}{M_{f2}}\right)$$

$$\Delta V_2 = 3{,}236\ln \left(\frac{34{,}000}{6{,}000} \right)$$

$$\Delta V_2 \approx 5{,}725 \text{ m/s}$$

Total ΔV Calculation:

$$\Delta V_{\text{Total}} = \Delta V_1 + \Delta V_2$$

$$\Delta V_{\text{Total}} = 3{,}477 + 5{,}725$$

$$\Delta V_{\text{Total}} = 9{,}202 \text{ m/s}$$

2.7.10.2 Payload Fraction

The payload fraction is given by:

$$\text{Payload Fraction} = \frac{m_{pl}}{m_o} \tag{2-52}$$

where,

m_{pl} is payload mass, and
m_o is initial mass.

For dissimilar stages in a multistage vehicle, the overall payload fraction depends on how the ΔV requirement is partitioned among the stages. If ΔV is partitioned suboptimally, payload fractions will be reduced. The optimal distribution can be determined through trial and error. By postulating different ΔV distributions and calculating the resulting payload fractions, the optimal distribution is found when the payload fraction is maximized.

2.7.10.3 Sizing the Vehicle

Once the distribution is optimized, vehicle sizing begins with the uppermost or final stage, which carries the actual deliverable payload. The initial mass of this stage is calculated first. This stage then serves as the payload for the preceding stage, and the process is repeated until all stages are sized. To maximize the payload fraction for a given requirement, the following guidelines are typically followed:

1. Stages with higher specific impulse (I_{sp}) should be positioned above stages with lower I_{sp}.

2. A greater portion of the ΔV should be provided by stages with higher I_{sp}.
3. Each succeeding stage should be smaller than its predecessor.
4. Similar stages should provide the same ΔV.

3 Orbital Mechanics

Orbital mechanics, also known as flight mechanics, is the field of study that focuses on the motions of artificial satellites and space vehicles as they travel through space under the influence of various forces. These forces include gravity, atmospheric drag, and thrust, among others. The principles of orbital mechanics are essential for the design and operation of space missions, allowing engineers to predict and control the paths of spacecraft.

The origins of orbital mechanics stem from the broader field of celestial mechanics, which focuses on the movements of natural celestial bodies like the moon, planets, and stars. This discipline dates back to the 17th century, when the English mathematician and physicist Sir Isaac Newton (1642-1727) established his three laws of motion and the law of universal gravitation. Newton's groundbreaking work laid the groundwork for understanding the motion of objects under the influence of gravity, forming the foundation for both celestial and orbital mechanics.

Newton's Contributions

- **Laws of Motion:** Newton's three laws of motion describe how an object's motion is influenced by the forces acting on it:
    - **First Law (Law of Inertia):** An object at rest remains at rest, and an object in motion continues to move at a constant speed and in the same direction unless acted upon by an external force.
    - **Second Law (Law of Acceleration):** The net force acting on an object is equal to the product of its mass and acceleration.
    - **Third Law (Action and Reaction):** For every action, there is an equal and opposite reaction.

- **Law of Universal Gravitation:** Every mass attracts other masses with a force that is directly proportional to the product of their masses and inversely proportional to the square of the distance between their centers.

3.1   Evolution of Orbital Mechanics

The study of orbital mechanics has undergone profound evolution over the centuries, tracing its origins to the early observations and mathematical descriptions of celestial motions. The foundations of this field were laid in the 17th century, beginning with Johannes Kepler's pioneering work on planetary orbits. Kepler's laws of planetary motion provided empirical relationships that described how planets move in elliptical orbits around the Sun. These laws, based on meticulous astronomical observations, were groundbreaking in that they deviated from the previously held belief in circular orbits, establishing a more accurate model of planetary movement.

Kepler's First Law, often referred to as the Law of Ellipses, described the shape of planetary orbits as ellipses, with the Sun occupying one of the two foci. His Second Law, the Law of Equal Areas, demonstrated that planets sweep out equal areas in equal times, indicating that their speed varies depending on their distance from the Sun. Finally, the Third Law, the Law of Harmonies, established a proportional relationship between the square of a planet's orbital period and the cube of the semi-major axis of its orbit. These laws were purely empirical, derived from observations rather than a theoretical understanding of the forces involved.

The theoretical framework that explained why Kepler's laws worked came from Sir Isaac Newton, who formulated the law of universal gravitation. Newton's law provided a fundamental explanation of the forces acting on celestial bodies, showing that the gravitational attraction between two masses was the underlying cause of orbital motion. This theoretical development not only solidified the understanding of elliptical orbits but also extended to explain a wide range of phenomena in celestial mechanics, from the motions of comets to the orbits of moons around planets.

With the advent of space exploration in the 20th century, orbital mechanics transitioned from a purely theoretical discipline to a critical field of engineering and applied science. The launch of artificial satellites and manned space missions necessitated precise control over spacecraft trajectories, requiring engineers to develop detailed mathematical models to predict and manage the orbits of these objects. The successful execution of missions such as the Apollo moon landings and the deployment of interplanetary probes like Voyager relied heavily on advances in orbital mechanics.

3.1.1   Orbital Elements:

Orbital elements are the fundamental parameters that uniquely define the size, shape, and orientation of an orbit, as well as the position of a spacecraft along that orbit. These elements include:

- o   **Semi-major axis (a):** Describes the size of the orbit.
- o   **Eccentricity (e):** Indicates the shape of the orbit, with values ranging from 0 (circular) to 1 (parabolic).

- o **Inclination (i):** The tilt of the orbit relative to the plane of reference, usually the equatorial plane.
- o **Right Ascension of the Ascending Node (Ω):** Specifies the horizontal orientation of the orbit in the reference plane.
- o **Argument of Periapsis (ω):** Describes the orientation of the orbit within its plane.
- o **True Anomaly (v):** Indicates the spacecraft's position along the orbit at a specific time.

3.1.2 Kepler's Laws of Planetary Motion:

- o **First Law (Law of Ellipses):** The orbit of a planet around the Sun is an ellipse, with the Sun located at one of the foci. This law applies universally to all celestial bodies, including artificial satellites.
- o **Second Law (Law of Equal Areas):** A line segment joining a planet and the Sun sweeps out equal areas in equal time intervals, implying that a planet moves faster when it is closer to the Sun and slower when it is farther away.
- o **Third Law (Law of Harmonies):** The square of the orbital period (T) of a planet is directly proportional to the cube of the semi-major axis (a) of its orbit, mathematically expressed as $T2\propto a3 T^2 \propto a^3 T2\propto a3$. This relationship is crucial for calculating orbital periods and distances in space missions.

3.1.3 Hohmann Transfer Orbit:

The Hohmann transfer orbit is an energy-efficient way to move a spacecraft between two circular orbits of different radii. This maneuver involves two engine burns: the first to move the spacecraft into an elliptical transfer orbit and the second to circularize the orbit at the target altitude. The Hohmann transfer is particularly useful for missions like moving from low Earth orbit (LEO) to geostationary orbit (GEO).

3.1.4 Gravity Assist (Slingshot Maneuver):

Gravity assist is a technique that uses the gravitational pull of a planet or other celestial body to alter the speed and trajectory of a spacecraft. By carefully navigating through a planet's gravity well, a spacecraft can gain or lose energy, effectively "slingshotting" to a higher or lower orbit without expending additional fuel. This maneuver is critical for deep space missions, such as the Voyager probes, which used gravity assists to reach the outer planets.

3.1.5 Engineering Applications

Orbital mechanics plays a vital role in the engineering and execution of space missions, with applications that span the entire lifecycle of a mission:

- **Ascent Trajectories**: The ascent trajectory is the path a rocket follows from the point of launch to its insertion into orbit. Engineers must carefully plan this trajectory to optimize fuel usage, minimize aerodynamic drag, and ensure that the spacecraft reaches its intended orbit. This process involves calculating the thrust needed to overcome gravity, atmospheric resistance, and the timing of stage separations.

- **Reentry and Landing**: Designing the trajectory for reentry and landing is a complex task that requires careful consideration of aerodynamic forces, thermal loads, and precision targeting. The spacecraft must follow a controlled descent path to safely reenter the Earth's atmosphere or land on another celestial body, such as the Moon or Mars. This involves managing the spacecraft's speed and angle of entry to avoid excessive heating or skipping off the atmosphere.

- **Rendezvous Computations**: Rendezvous operations, such as docking with another spacecraft or space station, require precise calculations to ensure that the two objects meet in orbit at the same time and place. This involves adjusting the spacecraft's orbit through a series of burns to match the velocity and trajectory of the target, a process that requires accurate predictions of orbital dynamics.

- **Lunar and Interplanetary Trajectories**: Missions to the Moon, Mars, and other celestial bodies involve complex trajectory planning that includes determining launch windows, transfer orbits, and gravitational assists. Engineers must account for the positions and movements of celestial bodies, the influence of multiple gravitational fields, and the spacecraft's propulsion capabilities to chart a successful course.

3.1.6 Modern Developments

The field of orbital mechanics has been revolutionized by advancements in computational power and algorithms, enabling engineers to simulate and optimize increasingly complex missions with unprecedented accuracy. Modern software tools can model the effects of non-uniform gravitational fields, atmospheric drag, solar radiation pressure, and perturbations from other celestial bodies. These tools are essential for planning and executing contemporary space missions, which often involve intricate maneuvers and long-duration flights.

For example, the deployment of satellite constellations, such as those used for global communications or Earth observation, requires precise coordination of multiple satellites in various orbits. Similarly, space station resupply missions, which involve docking with a moving target in low Earth orbit, demand high precision in orbital calculations. Interplanetary exploration missions, which may involve landing on distant moons or deploying rovers on Mars, rely on detailed simulations to ensure mission success.

In addition to traditional missions, modern developments in orbital mechanics also support emerging technologies, such as reusable launch vehicles and space tourism. The ability to accurately predict and control the trajectories of these vehicles is critical for their safe and efficient operation. As the field continues to evolve, orbital mechanics will remain a cornerstone of space exploration, enabling humanity to reach farther and achieve more in the cosmos.

3.2   Conics

Conic Sections, often called conics, are the curves obtained by intersecting a plane with a right circular cone. The shape of the conic section depends on the angle at which the plane intersects the cone. This chapter will explore the different types of conic sections, their properties, and their applications, particularly in orbital mechanics. Figure 3-1. Conic sections. shows the four types of conic sections.

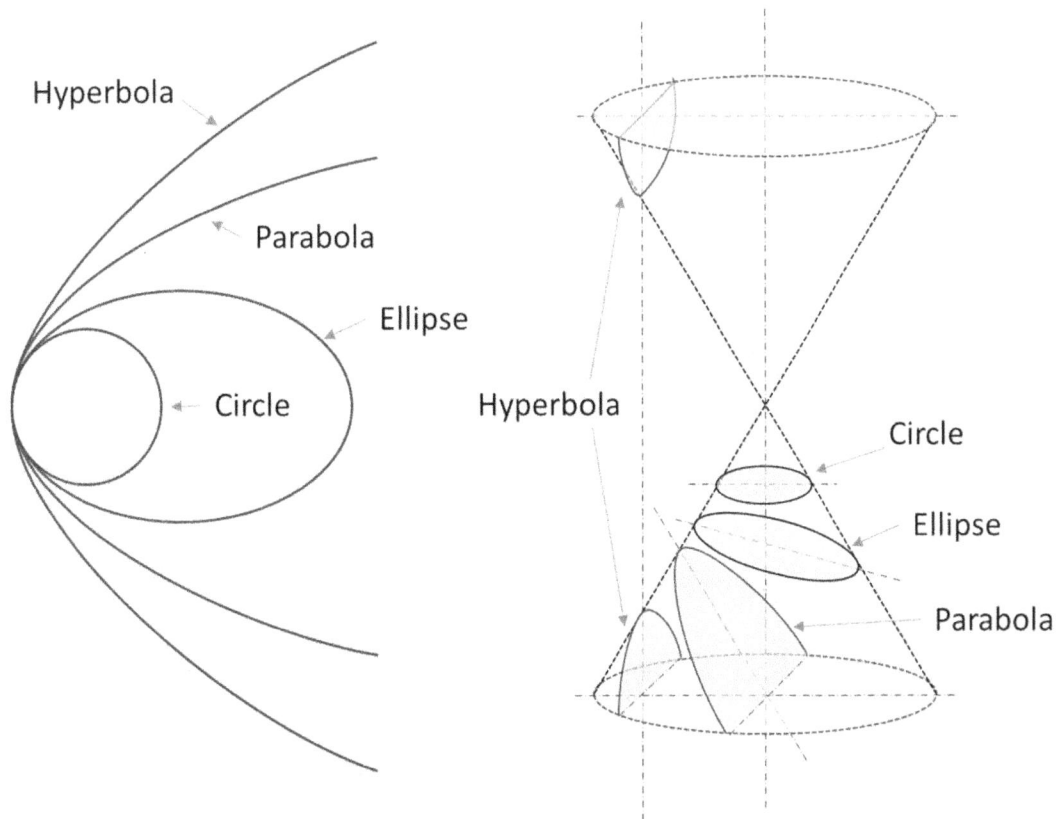

Figure 3-1. Conic sections.

3.2.1   Types of Conic Sections

1. **Circle**
2. **Ellipse**
3. **Parabola**
4. **Hyperbola**

Each conic section can be defined and distinguished based on the angle and position of the intersecting plane relative to the cone.

3.2.1.1   Circle

A circle is a special case of an ellipse in which the plane is perpendicular to the cone's axis, resulting in a perfectly round shape. Mathematically, a circle is defined as the set of points in a plane that are equidistant from a fixed point, the center.

- **Eccentricity (ε):** 0
- **Semi-major Axis (a):** Equal to the radius (r)

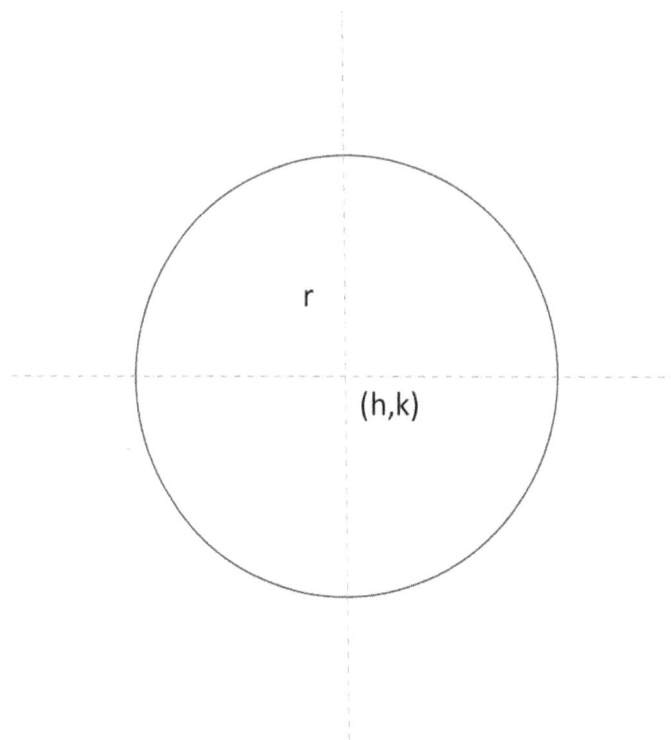

Figure 3-2. Circle.

3.2.1.2 Ellipse

An ellipse is formed when the intersecting plane is angled such that it cuts through one nappe of the cone, resulting in a closed, oval-shaped curve. An ellipse can be defined as a set of points where the sum of the distances from two fixed points (foci) is constant.

- **Eccentricity (ε):** Between 0 and 1
- **Semi-major Axis (a):** Greater than 0

$$(x - h)^2 + (y - k)^2 = r^2$$

$$\varepsilon = 0$$

$$\frac{(x - h)^2}{a^2} + \frac{(y - k)^2}{b^2} = 1 \tag{3-1}$$

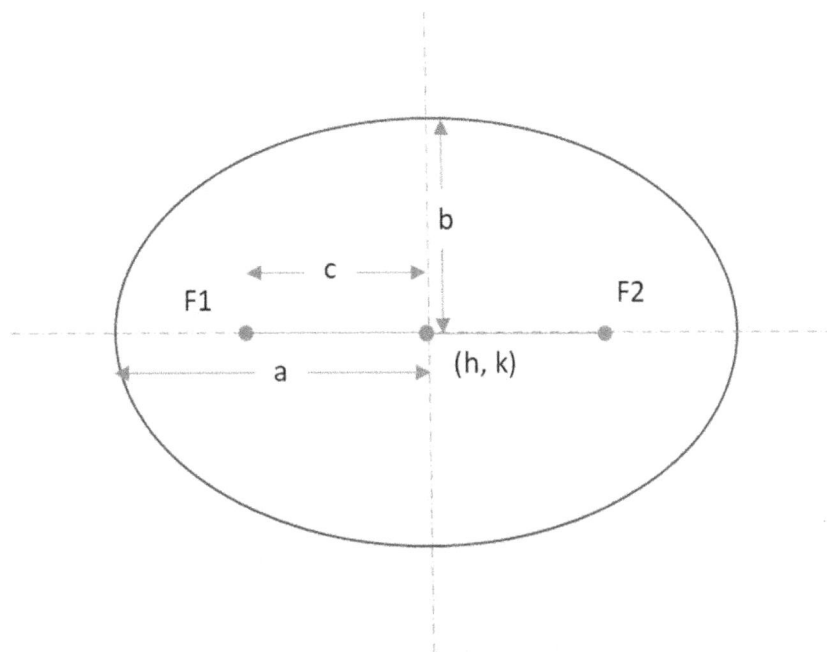

Figure 3-3. Ellipse.

If $a > b$

$$the\ the\ \varepsilon = \frac{\sqrt{a^2 - b^2}}{a} \tag{3-2}$$

If $b > a$

$$\varepsilon = \frac{\sqrt{b^2 - a^2}}{b} \qquad (3\text{-}3)$$

3.2.1.3   Parabola

A parabola is formed when the plane is parallel to a generator line of the cone, resulting in an open curve. A parabola can be defined as a set of points equidistant from a fixed point (focus) and a fixed line (directrix).

- **Eccentricity (e):** 1
- **Semi-major Axis (a):** Infinity

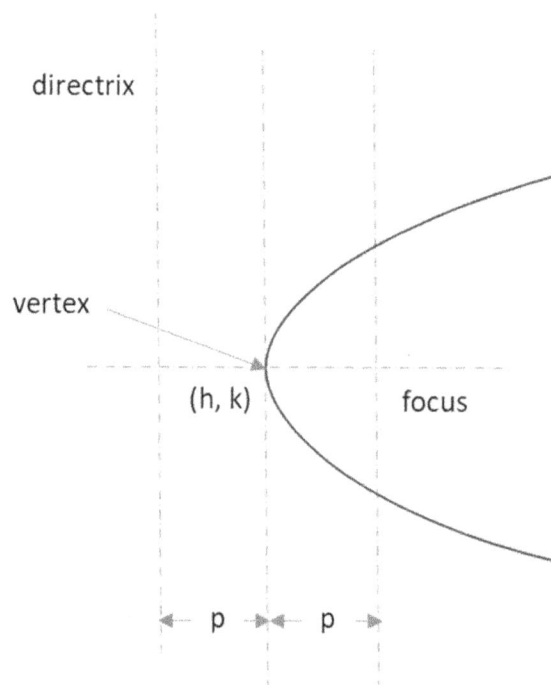

Figure 3-4. Parabola.

$$(x - h)^2 = 4p(y - k) \qquad (3\text{-}4)$$

$$\varepsilon = 1 \qquad (3\text{-}5)$$

3.2.1.4   Hyperbola

A hyperbola is formed when the intersecting plane cuts through both nappes of the cone, resulting in two separate open curves. A hyperbola can be defined as a set of points where the absolute difference of the distances from two fixed points (foci) is constant.

- **Eccentricity (ε)**: Greater than 1
- **Semi-major Axis (a)**: Less than 0

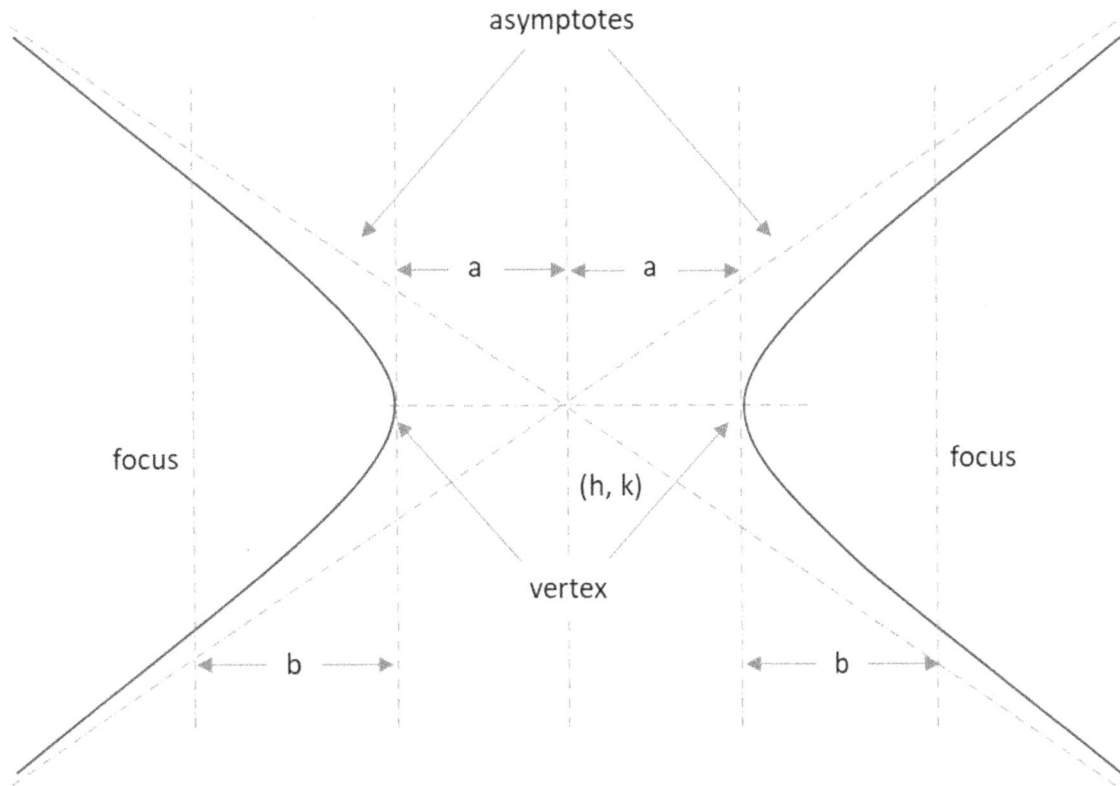

Figure 3-5. Hyperbola.

$$\frac{(x-h)^2}{a^2} - \frac{(y-k)^2}{b^2} = 1 \qquad (3\text{-}6)$$

$$the\ \varepsilon = \frac{\sqrt{a^2 + b^2}}{a} \qquad (3\text{-}7)$$

3.2.2   Orbital Elements

To describe the orbit of a satellite or celestial body mathematically, six orbital elements are required:

1. **Semi-Major Axis (a)**: Represents the satellite's average distance from the primary focus. It is half the longest diameter of the ellipse.
2. **Eccentricity (ε)**: Describes the shape of the orbit, measuring the deviation from a perfect circle.
3. **Inclination (i)**: The angle between the orbital plane and the equatorial plane of the primary body.
4. **Argument of Periapsis (ω)**: The angle from the ascending node to the periapsis, measured in the direction of the orbit.
5. **Time of Periapsis Passage (T)**: The time at which the satellite passes through the point of periapsis.
6. **Longitude of the Ascending Node (Ω)**: The angle from a reference direction (usually the vernal equinox) to the ascending node.

3.2.3   Properties of Elliptical Orbits

Elliptical orbits are commonly encountered in celestial mechanics and are defined by several properties:

- **Foci**: Two fixed points inside the ellipse. The primary body is located at one of the foci.
- **Major Axis**: The longest diameter of the ellipse, passing through both foci.
- **Minor Axis**: The shortest diameter, perpendicular to the major axis.
- **Semi-Major Axis**: Half of the major axis length.
- **Semi-Minor Axis**: Half of the minor axis length.
- **Eccentricity**: The ratio of the distance between the foci to the length of the major axis.

3.2.4   Inclination

Inclination (iii) measures the tilt of the orbital plane relative to the equatorial plane of the primary body:

- **0 degrees**: The orbit lies on the equator, moving in the same direction as the primary's rotation (prograde).
- **90 degrees**: Polar orbit, passing over the poles.
- **180 degrees**: Retrograde equatorial orbit, moving opposite to the primary's rotation.

3.2.5   Periapsis and Apoapsis

- **Periapsis**: Closest point in the orbit to the primary body.

- **Apoapsis**: Farthest point in the orbit from the primary body.

3.2.5.1   Nodes and Longitude

- **Ascending Node**: Point where the satellite crosses the equatorial plane from south to north.
- **Descending Node**: Point where it crosses from north to south.
- **Longitude of the Ascending Node (Ω\Omega$\Omega$)**: Measured from the vernal equinox to the ascending node.

3.2.5.2   Period and True Anomaly

- **Period (P)**: Time taken for one complete orbit.
- **True Anomaly (v)**: Angle between the periapsis and the current position of the satellite, measured at the primary focus.

3.2.5.3   Energy of Orbits

The total mechanical energy (E) of an orbit determines its type:
- **Negative Energy**: Indicates bound orbits (circle, ellipse).
- **Zero Energy**: Represents parabolic orbits.
- **Positive Energy**: Represents hyperbolic orbits.

3.2.6   Applications of Conic Sections in Orbital Mechanics

- **Satellite Orbits**: Satellites follow elliptical orbits around Earth, with specific parameters determining their path and behavior.
- **Interplanetary Trajectories**: Missions to other planets often use hyperbolic trajectories for departure and arrival.
- **Reentry and Landing**: The understanding of parabolic and hyperbolic paths helps design safe reentry trajectories.

Conic sections provide the mathematical foundation for understanding and predicting the motion of objects in space. By mastering these concepts, one can design, analyze, and optimize various aspects of space missions, ensuring accurate and efficient navigation through the cosmos

3.3   Types of Orbits

Achieving orbit around Earth requires a spacecraft to reach a specific altitude and velocity. The most energy-efficient orbit is a direct low-inclination orbit, which minimizes propellant use. This orbit is achieved by launching the spacecraft eastward from a location near the Earth's equator, capitalizing on the rotational speed of the Earth to boost the spacecraft's final orbital speed. For instance, a launch from Cape Canaveral (located at 28.5 degrees north latitude) gains a significant speed advantage of approximately 1,471 km/h (914 mph) due to Earth's rotation.

Launching in other directions or from locations far from the equator results in higher inclination orbits. These orbits cannot fully utilize the Earth's rotational speed, requiring more energy from the launch vehicle to reach orbital velocity. Despite being less energy-efficient, high-inclination orbits offer specific advantages for various applications. Below are detailed descriptions of several types of orbits and their unique benefits:

3.3.1   Geosynchronous Orbits (GEO)

Geosynchronous orbits are circular orbits around Earth with an orbital period of 24 hours. When the inclination is zero degrees, this orbit is called a geostationary orbit. A spacecraft in a geostationary orbit remains fixed relative to a single point on Earth's equator, appearing stationary in the sky. This property makes geostationary orbits ideal for communication and meteorological satellites, which require a constant position relative to Earth.

To achieve a geosynchronous orbit, a spacecraft is typically first placed into a geosynchronous transfer orbit (GTO) with an apogee of 35,786 km (22,236 miles). The spacecraft's engine is then fired at apogee to circularize the orbit. This maneuver ensures that the spacecraft attains a stable orbit, providing continuous coverage over a specific area on Earth.

3.3.1.1   Applications and Benefits:

- **Communication Satellites**: Provide consistent signal strength for television, radio, and internet services.
- **Weather Satellites**: Monitor weather patterns and provide real-time data for forecasting.
- **Surveillance**: Offer constant observation of specific regions for security and defense purposes.

3.3.2   Polar Orbits (PO)

Polar orbits have an inclination of 90 degrees, allowing satellites to pass over Earth's poles. These orbits are particularly useful for global mapping and surveillance operations. As the Earth rotates beneath the satellite, the satellite can cover nearly every part of the Earth's surface over time.

In a polar orbit, a satellite travels from the North Pole to the South Pole on each orbit, covering a different longitudinal strip of Earth with each pass. This comprehensive coverage is essential for Earth observation satellites used in environmental monitoring, military reconnaissance, and global mapping projects.

3.3.2.1 Applications and Benefits:

- **Earth Observation**: Detailed mapping of the Earth's surface, including land use, vegetation, and urban development.
- **Environmental Monitoring**: Track changes in climate, deforestation, and pollution levels.
- **Surveillance and Reconnaissance**: Provide high-resolution images and data for military and security purposes.

3.3.3 Walking Orbits (Precessing Orbits)

Walking orbits, also known as precessing orbits, are influenced by gravitational perturbations from the Earth's uneven mass distribution and other celestial bodies like the Sun and Moon. These gravitational forces cause the orbital plane of the satellite to gradually rotate, or precess, over time.

Engineers can design walking orbits to take advantage of these natural precessions, which can be beneficial for certain mission objectives. For instance, a satellite in a walking orbit can maintain a consistent ground track, allowing it to repeatedly pass over the same areas on Earth, which is useful for continuous monitoring and data collection.

3.3.3.1 Applications and Benefits:

- **Consistent Ground Track**: Enables repeated observations of specific regions for time-series data analysis.
- **Environmental Monitoring**: Track changes over time in areas such as deforestation, urban growth, and glacier movement.
- **Scientific Research**: Monitor phenomena such as ocean currents, weather patterns, and atmospheric conditions.

3.3.4 Sun-Synchronous Orbits (SSO)

Sun-synchronous orbits are a type of walking orbit where the orbital plane precesses at the same rate as Earth's orbit around the Sun. This synchronization allows a satellite to cross the equator at the same local solar time on each pass, providing consistent lighting conditions for imaging and observational instruments.

Sun-synchronous orbits are especially valuable for satellites conducting Earth observation, environmental monitoring, and scientific research. The consistent lighting conditions enable accurate comparisons of images taken at different times, facilitating the analysis of changes in the environment, urban development, and agricultural patterns.

3.3.4.1   Applications and Benefits:

- **Earth Observation**: High-resolution imaging for mapping, agriculture, and urban planning.
- **Climate Studies**: Monitor long-term changes in climate variables such as temperature, ice cover, and sea level.
- **Scientific Research**: Provide consistent data for studying Earth's processes and changes over time.

3.3.5   Molniya Orbits

Molniya orbits are highly eccentric with a period of approximately 12 hours (two revolutions per day). These orbits have inclinations of around 63.4 degrees or 116.6 degrees, ensuring that the argument of perigee remains stable, which means the satellite spends most of its time over the northern hemisphere.

This extended coverage makes Molniya orbits ideal for communication and observation in high-latitude regions, such as Russia and Canada, where geostationary satellites are less effective due to the high latitude. The high apogee over the northern hemisphere provides nearly continuous coverage of these regions.

3.3.5.1   Applications and Benefits:

- **Communication**: Provide reliable communication services to high-latitude regions.
- **Observation**: Extended monitoring of high-latitude areas for weather forecasting and environmental monitoring.
- **Military Applications**: Continuous surveillance and reconnaissance over critical regions.

3.3.6   Hohmann Transfer Orbits

Hohmann transfer orbits are used for interplanetary missions due to their fuel efficiency. These orbits allow a spacecraft to transfer between two orbits with minimal propellant use.

To travel to an outer planet, such as Mars, a spacecraft is launched and accelerated in the direction of Earth's orbit around the Sun until it escapes Earth's gravity and reaches a velocity that places it in an elliptical orbit around the Sun, with an aphelion at the orbit of the target planet. Upon arrival, the spacecraft must decelerate for the target planet's gravity to capture it into orbit.

For missions to inner planets, such as Venus, the spacecraft is decelerated (launched in the direction opposite Earth's orbit) to achieve an orbit around the Sun with a perihelion at the target planet's orbit. The spacecraft's speed decreases relative to Earth, even though it continues to move in the same direction as Earth.

3.3.6.1   Applications and Benefits:

- **Interplanetary Missions**: Efficient transfers to both inner and outer planets.
- **Fuel Efficiency**: Minimizes the amount of propellant required, reducing mission costs.
- **Precise Timing**: Essential for planning successful interplanetary missions.

3.3.7   Launch Windows

Timing is crucial in interplanetary missions. The spacecraft must be launched at a specific time to ensure that it reaches the target planet's orbit when the planet is at the desired position. This precise timing is known as the launch window, similar to how a quarterback must "lead" a receiver to ensure the football and the receiver converge at the same point simultaneously.

3.3.7.1   Importance of Launch Windows:

- **Successful Rendezvous**: Ensures the spacecraft meets the target planet at the correct time.
- **Efficiency**: Minimizes travel time and fuel consumption.
- **Mission Planning**: Critical for coordinating launch schedules and mission timelines.

By understanding and utilizing these various types of orbits, space missions can be designed to achieve specific goals, whether it's for communication, observation, or exploration. Each type of orbit offers unique advantages and is selected based on the mission's requirements. This knowledge is fundamental to the field of aerospace engineering and plays a vital role in the success of space missions.

3.4   Newton's Laws of Motion and Universal Gravitation

Newton's laws of motion and his law of universal gravitation form the cornerstone of classical mechanics. These principles explain the relationship between the forces acting upon objects their motion. Newton's laws provide a comprehensive framework for understanding the behavior of physical systems ranging from terrestrial objects to celestial bodies.

3.4.1   Newton's Laws of Motion

3.4.1.1   First Law: Law of Inertia

Newton's first law states that a particle will remain at rest or in uniform motion in a straight line unless acted upon by a net external force. This law implies that the natural state of a body is to maintain its velocity (both speed and direction) if no forces are present. Mathematically, it can be expressed as:

$$\sum \mathbf{F} = 0 \Rightarrow \mathbf{v} = \text{constant} \tag{3-8}$$

where $\sum \mathbf{F}$ is the sum of all forces acting on the body, and $\mathbf{v}$ is its velocity.

3.4.1.2   Second Law: Law of Acceleration

Newton's second law quantifies the effect of forces on the motion of a body. It states that the acceleration of a particle is directly proportional to the net external force acting on it and inversely proportional to its mass. This relationship is given by the equation:

$$\mathbf{F} = m\mathbf{a} \tag{3-9}$$

where $\mathbf{F}$ is the net external force, m is the mass of the particle, and $\mathbf{a}$ is the acceleration. This law is fundamental in dynamics, as it provides a means to calculate the motion of an object when the forces acting upon it are known.

3.4.1.3   Third Law: Action and Reaction

Newton's third law states that for every action, there is an equal and opposite reaction. If a body A exerts a force $\mathbf{F}_{A \to B}$ on body B, then body B exerts a force $\mathbf{F}_{B \to A}$ on body A that is equal in magnitude and opposite in direction:

$$\mathbf{F}_{A \to B} = -\mathbf{F}_{B \to A} \tag{3-10}$$

This law ensures the conservation of momentum in isolated systems, as the forces between interacting bodies are equal and opposite.

3.4.2   Universal Gravitation

Newton's law of universal gravitation describes the attractive force between two masses. It states that every point mass attracts every other point mass with a force that is proportional to the product of their masses and inversely proportional to the square of the distance between them. The gravitational force $\mathbf{F}$ between two masses m_1 and m_2 separated by a distance r is given by:

$$\mathbf{F} = G\frac{m_1 m_2}{r^2}\hat{\mathbf{r}} \tag{3-11}$$

where:

- G is the gravitational constant (6.67430×10^{-11} N m² kg⁻²).
- $\hat{\mathbf{r}}$ is the unit vector pointing from m_1 to m_2.

This force is mutual and acts along the line joining the two masses.

3.4.2.1 Gravitational Force Exerted by Earth

Consider an object of mass m at a distance r from the center of the Earth, which has mass M. The gravitational force exerted by the Earth on the object is:

$$\mathbf{F} = G\frac{mM}{r^2} \tag{3-12}$$

If the object is near the surface of the Earth, where $r \approx R_E$ (the radius of the Earth), this force causes the object to accelerate towards the center of the Earth. By Newton's second law, the acceleration g is:

$$g = \frac{GM}{R_E^2} \tag{3-13}$$

At the Earth's surface, this acceleration is approximately 9.80665 m/s².

In many gravitational computations, the product GM (where M is the mass of the Earth and G is the gravitational constant) is often treated as a constant denoted by μ. For Earth, this constant is:

$$\mu = GM = 3.986004418 \times 10^{14} \text{ m}^3/\text{s}^2 \tag{3-14}$$

Using μ, the equations of motion and gravitational interactions can be simplified, making it easier to perform calculations related to orbital mechanics and satellite dynamics.

3.4.3 Practical Applications

The combination of Newton's laws of motion and universal gravitation allows for the precise prediction of celestial and terrestrial dynamics. These principles are fundamental in the fields of astrophysics, aerospace engineering, and many other disciplines that involve the motion of objects under the influence of forces. Below are some specific applications:

- Orbital Mechanics: Determining the trajectories of satellites, planets, and other celestial bodies.
- Astrodynamics: Planning space missions, including launch windows and transfer

orbits.

- Engineering: Designing structures and systems that account for gravitational forces and accelerations.
- Physics: Understanding fundamental interactions and the behavior of matter under various forces.

3.4.3.1   Uniform Circular Motion

Uniform circular motion refers to the motion of a particle traveling along a circular path with a constant speed. Unlike linear motion, where a particle may move in a straight line, uniform circular motion involves continuous change in the direction of the particle's velocity, though its magnitude remains constant. This change in direction implies the presence of an acceleration, even though the speed does not change.

3.4.3.1.1   Characteristics of Uniform Circular Motion

1. Constant Speed: The magnitude of the velocity vector remains unchanged as the particle moves along the circular path.
2. Changing Velocity Direction: The direction of the velocity vector is constantly changing, which means the particle is undergoing acceleration.
3. Centripetal Acceleration: The acceleration that changes the direction of the velocity is directed towards the center of the circle. This inward acceleration is known as centripetal acceleration.

3.4.3.2   Centripetal Acceleration

Centripetal acceleration (a_c) is the acceleration that points towards the center of the circle and keeps the particle in its circular path. It can be derived from the particle's speed (v) and the radius of the circle (r) using the following formula:

$$a_c = \frac{v^2}{r} \tag{3-15}$$

This acceleration is essential for maintaining circular motion, as it constantly redirects the particle's velocity towards the center of the circle.

3.4.3.3   Centripetal Force

According to Newton's second law ($F = ma$), any accelerating particle must be acted upon by a force. For a particle in uniform circular motion, this force is called the centripetal force. The magnitude of the centripetal force (F_c) required to keep a particle moving in a circular path is given by:

$$F_c = ma_c = m\frac{v^2}{r} \tag{3-16}$$

where m is the mass of the particle. The direction of this force is always radially inward, towards the center of the circle.

3.4.3.4 Gravitational Force as Centripetal Force in Orbits

In the context of orbital mechanics, a satellite in orbit around a planet is an example of a body in uniform circular motion. The only significant force acting on the satellite is the gravitational pull of the planet. This gravitational force provides the necessary centripetal force to keep the satellite in its circular orbit.

The centripetal acceleration (a_c) of a satellite in a circular orbit is provided by the gravitational acceleration (g) due to the planet. Therefore, we can equate the centripetal acceleration to the gravitational acceleration:

$$g = \frac{v^2}{r} \tag{3-17}$$

From Newton's law of universal gravitation, the gravitational acceleration (g) at a distance r from the center of a planet with mass M is given by:

$$g = \frac{GM}{r^2} \tag{3-18}$$

where G is the gravitational constant.

By equating the expressions for centripetal acceleration and gravitational acceleration, we get:

$$\frac{v^2}{r} = \frac{GM}{r^2} \tag{3-19}$$

Solving for the orbital speed (v) of the satellite, we find:

$$v = \sqrt{\frac{GM}{r}} \tag{3-20}$$

This equation shows that the orbital speed of a satellite in uniform circular motion depends on the gravitational parameter (GM) of the planet and the radius (r) of the orbit.

Example 3-1:

Determine the speed of an artificial satellite orbiting the Earth in a circular orbit. The altitude of the satellite is 300 km above the Earth's surface.

Solution:

From Basic Constants:
- Radius of Earth: $R_E = 6,378.14$ km
- Gravitational parameter GM of Earth: 3.986005×10^{14} m^3/s^2

Given:
- Altitude above Earth's surface: 300 km

Calculation:

Total orbital radius, r :

$$r = (6,378.14 + 300) \times 1,000$$

$$r = 6,678,140 \text{ m}$$

Orbital velocity using Equation (4.6):

$$v = \sqrt{\frac{GM}{r}}$$

$$v = \sqrt{\frac{3.986005 \times 10^{14}}{6,678,140}}$$

$$v \approx \sqrt{5.9721 \times 10^7}$$

$$v \approx 7,725 \text{ m/s}$$

Velocity of the satellite: $v \approx 7,725$ m/s

3.4.3.5   Implications of Uniform Circular Motion in Orbital Mechanics

Understanding uniform circular motion is crucial for various applications in orbital mechanics:

1. Satellite Orbits: Calculating the orbital speed and period of satellites in circular orbits.
2. Artificial Gravity: Designing rotating space habitats to create artificial gravity through centripetal force.
3. Stability of Orbits: Analyzing the stability of orbits and the effects of perturbations on satellites.

Orbital Period: The time taken for a satellite to complete one full orbit around the planet is called the orbital period (T). For a circular orbit, the orbital period can be derived using the circumference of the orbit and the orbital speed:

$$T = \frac{2\pi r}{v} \tag{3-21}$$

Substituting $v = \sqrt{\frac{GM}{r}}$ into the equation, we get:

$$T = 2\pi r \sqrt{\frac{r}{GM}} = 2\pi \sqrt{\frac{r^3}{GM}} \tag{3-22}$$

This equation indicates that the orbital period depends on the radius of the orbit and the gravitational parameter of the planet.

Example Calculation

Consider a satellite orbiting Earth at an altitude where the radius of the orbit is 7,000 km(7,000,000 meters). Given $GM_{\text{Earth}} = 3.986 \times 10^{14}$ m^3/s^2, the orbital speed (v) and orbital period (T) can be calculated as follows:

1. Orbital Speed:

$$v = \sqrt{\frac{GM}{r}} = \sqrt{\frac{3.986 \times 10^{14}}{7 \times 10^6}} \approx 7.12 \times 10^3 \text{ m/s}$$

2. Orbital Period:

$$T = 2\pi \sqrt{\frac{r^3}{GM}} = 2\pi \sqrt{\frac{(7 \times 10^6)^3}{3.986 \times 10^{14}}} \approx 5{,}830 \text{ s} \approx 1.62 \text{ hours}$$

These calculations demonstrate the application of uniform circular motion principles in determining key parameters of satellite orbits.

Uniform circular motion provides a foundational understanding of the dynamics of particles and objects moving in circular paths with constant speed. By analyzing the centripetal acceleration and force, as well as their implications in orbital mechanics, we gain valuable insights into the behavior of satellites and other celestial bodies. This knowledge is essential for designing and operating space missions, ensuring the stability and efficiency of orbits, and exploring the mechanics of rotational systems in space.

3.5   Motions of Planets and Satellites

The motions of planets and satellites are governed by the principles of celestial mechanics, which were significantly advanced by Johannes Kepler's three laws of planetary motion. Kepler's work, based on meticulous observations and data collected by Tycho Brahe, provided a foundation that was later explained by Newton's laws of motion and his law of universal gravitation.

3.5.1   Kepler's Laws of Planetary Motion

1. Elliptical Orbits: All planets move in elliptical orbits with the sun at one focus.
2. Equal Areas: A line joining any planet to the sun sweeps out equal areas in equal times.
3. Harmonic Law: The square of the orbital period of any planet is proportional to the cube of the planet's mean distance from the sun.

These laws describe the kinematic properties of planetary orbits without necessarily explaining the underlying dynamics, which Newton later provided through his laws of motion and gravitation.

3.5.2   Newtonian Mechanics and Planetary Motion

To analyze planetary and satellite motion, we consider the gravitational forces acting between two bodies. We start with the simpler case of circular orbits before extending the principles to elliptical orbits.

3.5.2.1   Circular Orbits

Consider two bodies of masses M and m moving in circular orbits around their common center of mass C. The position of the center of mass is determined by the relationship $mr = MR$, where r and R are the distances of the smaller and larger masses from C, respectively.

Both bodies move with the same angular velocity ω. The gravitational force between the two bodies provides the necessary centripetal force to keep them in their circular orbits:

$$F_g = \frac{GMm}{(r+R)^2} \tag{3-23}$$

This gravitational force must equal the centripetal force needed to keep each body in orbit:

$F_c = m\omega^2 r$ and $F_c = M\omega^2 R$

Since these forces are equal and opposite, we have:

$$m\omega^2 r = M\omega^2 R \tag{3-24}$$

$$\frac{GMm}{(r+R)^2} = m\omega^2 r \tag{3-25}$$

Assuming $M \gg m$ (e.g., a planet and a satellite), we can approximate $R \approx 0$ and the distance between the bodies as r. Thus, the equation simplifies to:

$$\frac{GMm}{r^2} = m\omega^2 r \tag{3-26}$$

$$\omega^2 = \frac{GM}{r^3} \tag{3-27}$$

Expressing angular velocity ω in terms of the orbital period P, where $\omega = \frac{2\pi}{P}$:

$$\left(\frac{2\pi}{P}\right)^2 = \frac{GM}{r^3} \tag{3-28}$$

$$P^2 = \frac{4\pi^2 r^3}{GM} \tag{3-29}$$

This result is a specific case of Kepler's third law, showing that the square of the orbital period P is proportional to the cube of the orbit's radius r.

3.5.2.2   Elliptical Orbits

In elliptical orbits, the semi-major axis a replaces the radius r used in circular orbits. Kepler's first law states that planets move in ellipses with the Sun at one focus. The gravitational force between the planet and the Sun provides the centripetal force necessary to maintain this elliptical motion.

For an elliptical orbit, the relationship between the orbital period P and the semi-major axis a is:

$$P^2 = \frac{4\pi^2 a^3}{GM} \qquad (3\text{-}30)$$

This equation holds for both circular and elliptical orbits when a is used as the radius of the circular orbit.

Example 3-2:

Calculate the period of revolution for the satellite orbiting the Earth at an altitude of 300 km above the Earth's surface.

Solution:

Given:

- Orbital radius, $r = 6{,}678{,}140$ m

$$P^2 = \frac{4\pi^2 r^3}{GM}$$

$$P = \sqrt{\frac{4\pi^2 r^3}{GM}}$$

Substitute the known values:

$$P = \sqrt{\frac{4\pi^2 (6{,}678{,}140)^3}{3.986005 \times 10^{14}}}$$

$$P = \sqrt{\frac{4\pi^2(2.9755 \times 10^{20})}{3.986005 \times 10^{14}}}$$

$$P = \sqrt{\frac{3.7292 \times 10^{22}}{3.986005 \times 10^{14}}}$$

$$P = \sqrt{9.3588 \times 10^7}$$

$$P \approx 9{,}672 \text{ s}$$

Period of revolution: $P \approx 9{,}672$ s

Example 3-3:

Calculate the radius of orbit for an Earth satellite in a geosynchronous orbit, where the Earth's rotational period is 86,400 seconds.

Solution:

Given:

- Orbital period, $P = 86{,}400$ s

Calculation

$$P^2 = \frac{4\pi^2 r^3}{GM}$$

Rearranging for r :

$$r = \left[\frac{P^2 \times GM}{4\pi^2}\right]^{1/3}$$

Substitute the known values:

$$r = \left[\frac{(86{,}400)^2 \times 3.986005 \times 10^{14}}{4\pi^2}\right]^{1/3}$$

$$r = \left[\frac{7.46496 \times 10^9 \times 3.986005 \times 10^{14}}{39.4784}\right]^{1/3}$$

$$r = \left[\frac{2.9785 \times 10^{24}}{39.4784}\right]^{1/3}$$

$$r = [7.5487 \times 10^{22}]^{1/3}$$

$$r \approx 42{,}164{,}020 \text{ m}$$

Radius of the geosynchronous orbit: $r \approx 42{,}164{,}020$ m

3.5.3   Angular Momentum and Kepler's Second Law

Kepler's second law, the law of equal areas, can be derived from the conservation of angular momentum. For a body moving under a central force, the specific angular momentum h (angular momentum per unit mass) is conserved:

$$h = rv\sin\theta \tag{3-31}$$

In a circular orbit, $\theta = 90°$ (since velocity is perpendicular to radius), so:

$$h = rv \tag{3-32}$$

For elliptical orbits, the areal velocity (rate at which area is swept out) is constant:

$$\frac{dA}{dt} = \frac{1}{2}rv\sin\theta = \frac{h}{2} \tag{3-33}$$

This constancy of areal velocity is a direct consequence of the conservation of angular momentum and implies that a planet moves faster when closer to the Sun and slower when farther away, sweeping out equal areas in equal times.

3.5.4   Energy in Orbits

The total mechanical energy E of a satellite in orbit is the sum of its kinetic energy T and potential energy V :

$$E = T + V \tag{3-34}$$

$$T = \frac{1}{2}mv^2 \tag{3-35}$$

$$V = -\frac{GMm}{r} \tag{3-36}$$

For circular orbits, using $v = \sqrt{\frac{GM}{r}}$:

$$T = \frac{1}{2}m\left(\sqrt{\frac{GM}{r}}\right)^2 = \frac{1}{2}\frac{GMm}{r} \tag{3-37}$$

$$V = -\frac{GMm}{r} \tag{3-38}$$

Thus, the total energy is:

$$E = \frac{1}{2}\frac{GMm}{r} - \frac{GMm}{r} = -\frac{1}{2}\frac{GMm}{r} \tag{3-39}$$

This negative energy indicates a bound system. For elliptical orbits, the total energy remains:

$$E = -\frac{GMm}{2a} \tag{3-40}$$

where a is the semi-major axis.

3.5.5 Application to Planetary and Satellite Motion

1. Two-Body Problem: In celestial mechanics, when considering two bodies of comparable mass (e.g., Earth and Moon), the analysis must account for the motion of both bodies about their common center of mass (barycenter). The reduced mass μ is used in such cases:

$$\mu = \frac{Mm}{M + m} \tag{3-41}$$

2. Orbital Parameters: Understanding the parameters that describe an orbit (semi-major axis, eccentricity, inclination, etc.) is crucial for mission planning and satellite operations.
3. Transfer Orbits: Hohmann transfer orbits and other techniques leverage these principles to move satellites between orbits efficiently.

The motions of planets and satellites are deeply rooted in the principles established by Kepler and Newton. Their laws provide a comprehensive framework for understanding and predicting the behavior of celestial bodies in various orbital configurations. These principles are not only fundamental to astrophysics but also essential for the practical application of satellite technology and space exploration.

3.6   Space Vehicle Launch

The launch of a satellite or space vehicle involves a powered flight phase. The vehicle is powered by a rocket or a launch vehicle and is lifted above the Earth's atmosphere and accelerated to orbital velocity. This powered flight ends when the final stage of the rocket burns out, at which point the vehicle enters free flight. During this phase, the space vehicle is influenced only by Earth's gravitational pull. However, if the vehicle travels far from Earth, its trajectory may also be affected by the gravitational forces of the Sun, Moon, or other planets.

3.6.1   Determining the Orbit

The orbit of a space vehicle can be determined from its position and velocity at the beginning of its free flight. These parameters are described by the variables r, v, and ϕ, where:
- r is the distance from the center of the Earth,
- v is the velocity of the vehicle,
- ϕ is the angle between the position and velocity vectors, also known as the flight-path angle.

Let's denote the initial values at the beginning of free flight as r_1, v_1, and ϕ_1. These initial conditions determine the vehicle's orbit.

3.6.2   Orbital Parameters

To find the orbital parameters, we consider the specific energy ϵ and specific angular momentum h. The specific energy is given by:

$$\epsilon = \frac{v^2}{2} - \frac{GM}{r} \qquad (3\text{-}42)$$

where G is the gravitational constant and M is the mass of the Earth. The specific angular momentum is:

$$h = rv\cos\phi \qquad (3\text{-}43)$$

For an elliptical orbit, the semi-major axis a is related to the specific energy by:

$$\epsilon = -\frac{GM}{2a} \qquad (3\text{-}44)$$

$$a = -\frac{GM}{2\epsilon} \qquad (3\text{-}45)$$

3.6.3   Perigee and Apogee

The perigee (R_p) and apogee (R_a) distances can be derived using the semi-major axis and eccentricity (e). The distances are given by:

$$R_p = a(1 - e) \qquad (3\text{-}46)$$

$$R_a = a(1 + e) \qquad (3\text{-}47)$$

The eccentricity e can be determined from the initial conditions:

$$e = \sqrt{1 + \frac{2\epsilon h^2}{(GM)^2}} \qquad (3\text{-}48)$$

3.6.3.1   Solving for Perigee and Apogee

The specific energy and specific angular momentum can be computed from the initial conditions r_1, v_1, and ϕ_1 :

$$\epsilon = \frac{v_1^2}{2} - \frac{GM}{r_1} \qquad (3\text{-}49)$$

$$h = r_1 v_1 \cos \phi_1 \tag{3-50}$$

Substituting these into the equation for eccentricity:

$$e = \sqrt{1 + \frac{2\left(\frac{v_1^2}{2} - \frac{GM}{r_1}\right)(r_1 v_1 \cos \phi_1)^2}{(GM)^2}} \tag{3-51}$$

The quadratic equation derived from the balance of forces can also provide the perigee and apogee:

$$\left(\frac{R_p}{r_1}\right) = \frac{2GM - r_1 v_1^2 \cos^2 \phi_1 \pm \sqrt{(2GM - r_1 v_1^2 \cos^2 \phi_1)^2 - 4G^2 M^2}}{2G^2 M^2} \tag{3-52}$$

The smaller root corresponds to R_p, and the larger root corresponds to R_a.

3.6.4   True Anomaly and Flight-Path Angle

To precisely determine a satellite's orbit, the true anomaly v (the angle between the periapsis and the satellite's current position) is needed. This angle can be found using the relationship between the semi-major axis, eccentricity, and the radial distance r :

$$r = \frac{a(1 - e^2)}{1 + e\cos v} \tag{3-53}$$

Given the initial conditions r_1 and ϕ_1, the true anomaly can be solved:

$$\cos v = \frac{a(1 - e^2)/r_1 - 1}{e} \tag{3-54}$$

In practice, it is often useful to work with the flight-path angle γ, which is the complement of the zenith angle ϕ. The flight-path angle is positive when the velocity vector is directed away from the primary body (Earth):

$$\sin \gamma = e \sin v \tag{3-55}$$

3.6.5   Semi-Major Axis Calculation

To determine the semi-major axis directly, we use the energy equation:

$$a = -\frac{GM}{2\epsilon} = \frac{GM}{r_1 v_1^2 - 2GM} \tag{3-56}$$

3.6.6 Application to Space Missions

In space missions, efficient use of propellant is critical. Typically, spacecraft launches are optimized to minimize propellant use by terminating the powered flight at perigee or apogee where the zenith angle ϕ is 90°. At these points, the velocity is tangential to the orbit, simplifying the energy calculations and ensuring maximum efficiency.

Example Problems

Problem 1: Calculating Perigee and Apogee

Given initial conditions:
- $r_1 = 7000$ km
- $v_1 = 7.8$ km/s
- $\phi_1 = 0°$

Calculate the specific energy:

$$\epsilon = \frac{(7.8 \times 10^3)^2}{2} - \frac{GM}{7000 \times 10^3}$$

Using $GM = 3.986 \times 10^{14}$ m³/s² :

$$\epsilon = 30.42 \times 10^6 - 5.69 \times 10^4 = 2.9731 \times 10^7 \text{ J/kg}$$

Using $GM = 3.986 \times 10^{14}$ m³/s² :

$$\epsilon = 30.42 \times 10^6 - 5.69 \times 10^4 = 2.9731 \times 10^7 \text{ J/kg}$$

Calculate the specific angular momentum:

$$h = 7000 \times 10^3 \times 7.8 \times 10^3 \cos 0° = 5.46 \times 10^7 \text{ m}^2/\text{s}$$

Calculate eccentricity:

$$e = \sqrt{1 + \frac{2 \times 2.9731 \times 10^7 \times (5.46 \times 10^7)^2}{(3.986 \times 10^{14})^2}} \approx 0.716$$

Calculate semi-major axis:

$$a = \frac{7000 \times 10^3}{2(1 - 0.716)} \approx 6.175 \times 10^6 \text{ m}$$

Calculate perigee and apogee:

$$R_p = a(1 - e) = 6.175 \times 10^6 \times (1 - 0.716) \approx 1.76 \times 10^6 \text{ m}$$

$$R_a = a(1 + e) = 6.175 \times 10^6 \times (1 + 0.716) \approx 10.59 \times 10^6 \text{ m}$$

These detailed calculations ensure precise determination of orbital parameters, critical for mission planning and satellite operations. Understanding these principles allows engineers to design efficient trajectories and optimize the use of propellant, ensuring successful space missions.

3.6.7   Orbit Tilt, Rotation, and Orientation

Understanding the size and shape of an orbit is crucial, but determining its orientation in space is equally important. The orientation of an orbit is defined by several parameters that describe the tilt, rotation, and spatial positioning of the orbital plane. These parameters are essential for accurate navigation, mission planning, and ensuring that the spacecraft reaches its intended destination.

3.6.8   Orbital Elements and Orientation

The orientation of an orbit in space is defined by six orbital elements:
1. Inclination (i): The tilt of the orbit relative to the equatorial plane of the Earth.
2. Longitude of the Ascending Node (Ω) : The angle measured from the vernal equinox to the ascending node, where the orbit crosses the equatorial plane moving from south to north.
3. Argument of Periapsis (ω) : The angle from the ascending node to the periapsis, the point in the orbit closest to Earth.
4. True Anomaly (v): The angle from the periapsis to the current position of the spacecraft.
5. Semi-major Axis (a): Describes the size of the orbit.
6. Eccentricity (e): Describes the shape of the orbit.

3.6.9   Orbital Orientation

To determine the orientation of the orbit in space, we need to know the latitude, longitude, and heading of the space vehicle at the moment of engine burnout or orbit insertion. The key parameters are:

- Azimuth Heading (θ): Measured clockwise from north.
- Geocentric Latitude (δ): The declination of the burnout point.
- Ascending Node Distance (η) : The angular distance (measured in the equatorial plane) between the ascending node and the burnout point.
- Orbital Plane Distance (ξ): The angular distance (measured in the orbital plane) between the ascending node and the burnout point.
- Longitude of Ascending Node (λ_1): The geographical longitude of the ascending node.
- Longitude at Burnout Point (λ_2): The geographical longitude of the burnout point at the instant of engine burnout.

3.6.10 Calculating Orbital Elements

If the azimuth heading (θ), geocentric latitude (δ), and longitude at burnout (λ_2) are given, the other values can be calculated using the following relationships:

3.6.10.1 Inclination (i):

$$i = \arccos (\sin\,\delta\sin\,\theta) \tag{3-57}$$

3.6.10.2 Longitude of the Ascending Node (Ω):

$$\Omega = \lambda_1 + \eta \tag{3-58}$$

3.6.10.3 Argument of Periapsis (ω):

$$\omega = \eta - \xi \tag{3-59}$$

where:

$$\eta = \arctan \left(\frac{\cos\,\delta\sin\,\theta}{\cos\,\theta}\right) \tag{3-60}$$

$$\xi = \arctan \left(\frac{\cos\ i \sin\ \theta}{\cos\ \theta} \right) \qquad (3\text{-}61)$$

3.6.11 Practical Application

Understanding the orientation of an orbit is crucial for mission planning, especially for interplanetary missions, Earth observation satellites, and communication satellites. Knowing the inclination, longitude of the ascending node, and argument of periapsis allows for precise trajectory planning and ensures the spacecraft achieves its intended orbit.

3.6.11.1 Geodetic Latitude, Geocentric Latitude, and Declination

Latitude is the angular distance of a point on Earth's surface north or south of Earth's equator, measured in degrees. It is positive north of the equator and negative south. Understanding the different types of latitude-geodetic, geocentric, and declination-is essential for precise calculations in geodesy, astronomy, and orbital mechanics.

3.6.11.2 Geodetic Latitude (φ)

Geodetic latitude, also known as geographical latitude, is the angle between the equatorial plane and the normal to the reference ellipsoid that intersects the point of interest. The reference ellipsoid is a mathematically defined surface that approximates the shape of the Earth.

The geodetic latitude ϕ is crucial for cartography and navigation, as it accounts for the Earth's oblateness (flattening at the poles and bulging at the equator).

3.6.11.3 Geocentric Latitude (φ')

Geocentric latitude is the angle between the equatorial plane and the radius vector extending from the Earth's center to a point on the reference ellipsoid. Unlike geodetic latitude, geocentric latitude does not account for the ellipsoidal shape of the Earth.

The relationship between geodetic latitude ϕ and geocentric latitude ϕ' is given by:

$$\tan\ \phi' = \left(\frac{b^2}{a^2} \right) \tan\ \phi$$

Where:
- a is the equatorial radius of the Earth,
- b is the polar radius of the Earth.

For the Earth:
- $a = 6{,}378{,}137$ meters,
- $b = 6{,}356{,}752$ meters.

3.6.11.4 Declination (δ)

Declination is the angular distance of a celestial object north or south of the celestial equator, analogous to latitude but projected onto the celestial sphere. It is the angle between the geocentric radius vector to the object and the equatorial plane. Declination is crucial for locating celestial objects in the sky.

3.6.11.5 Radius of the Reference Ellipsoid

The radius R of the reference ellipsoid at a given geodetic latitude ϕ is calculated as:

$$R(\phi) = \sqrt{\frac{(a^2\cos^2 \phi + b^2\sin^2 \phi)}{(a\cos \phi + b\sin \phi)^2}} \qquad (3\text{-}62)$$

3.6.11.6 Altitude and Radius Vector

To solve for the radius vector r from altitude h, or vice versa, use:

$$r = R + h \qquad (3\text{-}63)$$

$$h = r - R \qquad (3\text{-}64)$$

For high precision, particularly in low Earth orbit calculations, consider the slight differences between geodetic and geocentric latitudes. However, the differences are generally small and often negligible:

- In low Earth orbit, the difference is typically not more than 0.00001 degrees.
- Even at the Moon's distance, the difference is about 0.01 degrees.
  Thus, for many practical applications near Earth, we can assume:

$$\phi \approx \phi' \qquad (3\text{-}65)$$

$$r \approx R + h \qquad (3\text{-}66)$$

3.6.12 Practical Application

3.6.13 Mean Anomaly (M)

The mean anomaly is defined as the fraction of an orbital period that has elapsed since perigee passage, scaled to a full 2π radians. For a circular orbit, the mean anomaly equals the true anomaly. By definition:

$$M = M_0 + nt \tag{3-67}$$

where:

- M_0 is the mean anomaly at the reference time t_0,
- n is the mean motion, the average angular velocity of the satellite, defined as:

$$n = \sqrt{\frac{GM}{a^3}} \tag{3-68}$$

Here, G is the gravitational constant, M is the mass of the primary body, and a is the semi-major axis of the orbit. The mean anomaly provides the average position of the satellite, assuming it moves with a constant angular speed.

3.6.14 Eccentric Anomaly (E)

Since satellite orbits are elliptical, the distance from the primary body and the velocity of the satellite vary continuously. To account for this, Kepler introduced the eccentric anomaly (E), an angular parameter that relates the mean anomaly to the true anomaly (v) in elliptical orbits. The eccentric anomaly is defined by the equation:

$$M = E - e\sin E \tag{3-69}$$

where e is the eccentricity of the orbit. This relationship, known as Kepler's Equation, must often be solved iteratively due to its transcendental nature.

3.6.15 True Anomaly (v)

The true anomaly is the angle between the direction of the periapsis and the current position of the satellite, as seen from the primary focus of the ellipse. It is related to the eccentric anomaly by:

$$\tan\left(\frac{v}{2}\right) = \sqrt{\frac{1+e}{1-e}}\tan\left(\frac{E}{2}\right) \tag{3-70}$$

For small eccentricities, a good approximation of the true anomaly can be obtained using:

$$\nu \approx E + 2e\sin E \qquad (3\text{-}71)$$

This approximation is accurate to the order of e^3.

3.6.16 Practical Calculations

To determine the position of a satellite in its orbit at a given time or to find the time it takes to move from one point in an orbit to another, we use the following steps:

1. Compute the Mean Anomaly (M):

$$M = M_0 + nt \qquad (3\text{-}72)$$

2. Solve Kepler's Equation for Eccentric Anomaly (E):

$$M = E - e\sin E \qquad (3\text{-}73)$$

This step often requires numerical methods such as Newton-Raphson iteration.

3. Calculate the True Anomaly (v):

$$\tan\left(\frac{\nu}{2}\right) = \sqrt{\frac{1+e}{1-e}}\tan\left(\frac{E}{2}\right) \qquad (3\text{-}74)$$

4. Determine the Radius (r) and Position Vector (r, v):

The magnitude of the radius vector r (distance from the primary body) is given by:

$$r = a(1 - e\cos E) \qquad (3\text{-}75)$$

The position in polar coordinates can be expressed as (r, v).

5. Compute the Velocity (v):

The spacecraft's velocity in an elliptical orbit can be determined by:

$$v = \sqrt{GM\left(\frac{2}{r} - \frac{1}{a}\right)} \qquad (3\text{-}76)$$

6. Flight-Path Angle (γ):

The flight-path angle, which is the angle between the spacecraft's velocity vector and the local horizontal, can be calculated by:

$$\tan \gamma = \frac{e\sin \nu}{1 + e\cos \nu} \tag{3-77}$$

Example Problem
Given:

- Semi-major axis $a = 10,000$ km
- Eccentricity $e = 0.1$
- Mean anomaly at time $t_0 : M_0 = 0$
- Time $t = 5,000$ seconds

Steps:

1. Compute Mean Motion (n):

$$n = \sqrt{\frac{GM}{a^3}} \tag{3-78}$$

Given $GM = 3.986 \times 10^{14}$ m^3/s^2 :

$$n = \sqrt{\frac{GM}{a^3}} \tag{3-79}$$

$$n = \sqrt{\frac{3.986 \times 10^{14}}{(10,000 \times 10^3)^3}} \approx 0.000631 \text{rad/s}$$

2. Compute Mean Anomaly (M):

$$M = M_0 + nt = 0 + 0.000631 \times 5000 \approx 3.155 \text{rad}$$

3. Solve Kepler's Equation for E :

Using iterative methods (e.g., Newton-Raphson):

$$M = E - 0.1 \sin E$$

After several iterations, $E \approx 3.13 \text{rad}$

4. Calculate True Anomaly (v):

$$\tan\left(\frac{v}{2}\right) = \sqrt{\frac{1 + 0.1}{1 - 0.1}} \tan\left(\frac{3.13}{2}\right)$$

$$v \approx 3.11 \text{rad}$$

5. Determine Radius (**r**):

$$r = a(1 - e\cos E) = 10{,}000 \times (1 - 0.1\cos 3.13) \approx 9{,}997 \text{ km}$$

6. Compute Velocity (v):

$$v = \sqrt{3.986 \times 10^{14} \left(\frac{2}{9{,}997 \times 10^3} - \frac{1}{10{,}000 \times 10^3}\right)}$$

$$v \approx 7.12 \text{ km/s}$$

7. Calculate Flight-Path Angle (γ) :

$$\tan \gamma = \frac{0.1 \sin 3.11}{1 + 0.1 \cos 3.11} \approx 0.028$$

$$\gamma \approx 1.6°$$

3.7   Example Problems

Example 3-4:

A satellite in Earth orbit reaches its perigee at an altitude of 250 km above the Earth's surface, traveling at a speed of 7,900 m/s. Determine the altitude at apogee for the satellite, and calculate the orbit's eccentricity..

Solution:

Given:

- Perigee Radius: $R_p = (6{,}378.14 + 250) \times 1{,}000 = 6{,}628{,}140$ m
- Velocity at Perigee: $V_p = 7{,}900$ m/s

Calculation of Apogee Radius

$$R_a = \frac{R_p}{\left[\dfrac{2 \times GM}{R_p \times V_p^2} - 1\right]}$$

$$R_a = \frac{6628140}{\left(\dfrac{2 \times 3.986005 \times 10^{14}}{6628140 \times 7900^2} - 1\right)}$$

$$R_a \approx 7148719.3 \text{ m}$$

Calculation of Apogee Altitude:

$$\text{Altitude at apogee} = \frac{R_a}{1{,}000} - 6378.14$$

$$\text{Altitude at apogee} = \frac{7148719.3}{1000} - 6378.14$$

$$\text{Altitude at apogee} \approx 770.6 \text{ km}$$

Calculation of Eccentricity

$$e = \frac{R_p \times V_p^2}{GM} - 1$$

$$e = \frac{6628140 \times 7900^2}{3.986005 \times 10^{14}} - 1$$

$$e \approx 0.037786$$

Example 3-5:

A satellite is launched into Earth orbit, with its launch vehicle burning out at an altitude of 300 km. At burnout, the satellite's speed is 8,000 m/s, with a zenith angle of 88 degrees. Determine the satellite's altitude at both perigee and apogee.

Solution:

Given:

- Burnout Radius: $r_1 = (6{,}378.14 + 300) \times 1{,}000 = 6{,}678{,}140$ m
- Velocity at Burnout: $v_1 = 8{,}000$ m/s
- Zenith Angle: $\theta = 88°$

Calculation

$$\left(\frac{R_p}{r_1}\right)_{1,2} = \frac{-C \pm \sqrt{C^2 - 4 \times (1 - C) \times (-\sin^2 \theta)}}{2 \times (1 - C)}$$

Where:

$$C = \frac{2 \times GM}{r_1 \times v_1^2}$$

$$C = \frac{2 \times 3.986005 \times 10^{14}}{6678140 \times 8000^2}$$

$$C = \frac{7.97201 \times 10^{14}}{4.27088 \times 10^{15}}$$

$$C \approx 1.86523$$

Solving for $\left(\frac{R_p}{r_1}\right)_{1,2}$:

$$\left(\frac{R_p}{r_1}\right)_{1,2} = \frac{-1.86523 \pm \sqrt{1.86523^2 - 4 \times (1 - 1.86523) \times (-\sin^2 88°)}}{2 \times (1 - 1.86523)}$$

$$\left(\frac{R_p}{r_1}\right)_1 \approx 0.9914337, \quad \left(\frac{R_p}{r_1}\right)_2 \approx 1.1643284$$

Perigee Radius, $R_p = R_{p1} = r_1 \times \left(\frac{R_p}{r_1}\right)_1$

$$R_p = 6678140 \times 0.9914337$$

$$R_p \approx 6620933 \text{ m}$$

Altitude at Perigee:

$$\text{Altitude at perigee} = \frac{R_p}{1{,}000} - 6{,}378.14$$

$$\text{Altitude at perigee} = \frac{6620933}{1000} - 6378.14$$

$$\text{Altitude at perigee} \approx 242.79 \text{ km}$$

Apogee Radius, $R_a = R_{p2} = r_1 \times \left(\frac{R_p}{r_1}\right)_2$

$$R_a = 6678140 \times 1.1643284$$

$$R_a \approx 7775548 \text{ m}$$

Altitude at Apogee:

$$\text{Altitude at apogee} = \frac{R_a}{1{,}000} - 6{,}378.14$$

$$\text{Altitude at apogee} = \frac{7775548}{1000} - 6378.14$$

$$\text{Altitude at apogee} \approx 1397.41 \text{ km}$$

4 Nozzle Design

4.1 Optimizing Expansion for Maximum Thrust

A rocket engine is a device where propellants are burned in a combustion chamber, producing high-pressure gases that are then expanded through a specially shaped nozzle to generate thrust. The function of the nozzle is to convert the thermal energy produced in the combustion chamber or the reactor into kinetic energy. This process transforms slow-moving, high-pressure, high-temperature gas into high-velocity gas with lower pressure and temperature. Gas velocities from 2 to 4.5 kilometers per second can be achieved in rocket nozzles. The nozzles that perform this feat are called De Laval nozzles, consisting of a convergent section, a throat (the minimum flow area), and a divergent section.

4.2 Nozzle Design and Function

In the divergent section of the nozzle, hot exhaust gases expand, causing their pressure to decrease while accelerating to high velocities. Ideally, the nozzle is designed such that the pressure at the nozzle exit is equal to the ambient pressure outside the nozzle. This condition, known as optimal expansion or correct expansion, results in maximum thrust.

4.3 Basic Thrust Equation

To understand the relationship between nozzle design and thrust, we must examine the basic thrust equation:

$$F = \dot{m} \cdot V_e + (P_e - P_a) \cdot A_e \qquad (4\text{-}1)$$

where:
- F = Thrust
- $\dot{m}$ = Propellant mass flow rate
- V_e = Velocity of exhaust gases at the nozzle exit

- P_e = Pressure at the nozzle exit
- P_a = Ambient pressure
- A_e = Area of the nozzle exit

The term $\dot{m} \cdot V_e$ is called the momentum thrust, and the term $(P_e - P_a) \cdot A_e$ is called the pressure thrust. For maximum thrust, the nozzle must be designed so that $P_e = P_a$.

4.4   Nozzle Performance: Underextended vs. Overextended

- Underextended Nozzle: When the nozzle exit pressure P_e is greater than the ambient pressure P_a, the velocity V_e is relatively low, resulting in lower momentum thrust. However, the pressure thrust is higher because of the larger pressure differential.
- Overextended Nozzle: When the nozzle exit pressure P_e is less than the ambient pressure P_a the velocity V_e is higher, leading to higher momentum thrust. However, the pressure thrust decreases because the pressure differential is negative.

4.4.1   An Example: Nozzle Design:

Consider a rocket engine with the following conditions:
- Propellant mass flow rate $(\dot{m})$ = 100 kg/s
- Specific heat ratio (k) = 1.20
- Molecular weight of exhaust gas (M) = 24
- Combustion chamber temperature (T_c) = 3600 K
- Combustion chamber pressure (P_c) = 5MPa
- Ambient pressure (P_a) = 0.05MPa

For optimal expansion $(P_e = P_a)$:

1. Throat Conditions:
- Throat pressure (P_t) :

$$P_t = P_c \left[1 + \frac{k-1}{2} \right]^{-\frac{k}{k-1}}$$

- Throat pressure (P_t) :

$$P_t = P_c \left[1 + \frac{k-1}{2}\right]^{-\frac{k}{k-1}}$$

$$P_t = 5 \times \left[1 + \frac{1.20 - 1}{2}\right]^{-\frac{1.20}{1.20-1}}$$

$$P_t = 2.8224 \text{ MPa}$$

- Throat temperature (T_t) :

$$T_t = T_c \left[\frac{1}{1 + \frac{k-1}{2}}\right]$$

$$T_t = 3600 \times \left[\frac{1}{1 + \frac{1.20 - 1}{2}}\right]$$

$$T_t = 3272.73 \text{ K}$$

- Throat area (A_t) :

$$A_t = \frac{\dot{m}}{P_t} \sqrt{\frac{R' \times T_t}{M \times k}}$$

$$A_t = \frac{100}{2.8224 \times 10^6} \sqrt{\frac{8314 \times 3272.73}{24 \times 1.20}}$$

$$A_t = 0.0344386 \text{ m}^2$$

2. Exit Conditions:
- Exit Mach number (M_e) :

$$M_e^2 = \frac{2}{k-1}\left[\left(\frac{P_c}{P_a}\right)^{\frac{k-1}{k}} - 1\right]$$

$$M_e^2 = \frac{2}{1.20 - 1}\left[\left(\frac{5}{0.05}\right)^{\frac{1.20-1}{1.20}} - 1\right]$$

$$M_e^2 = 11.544347$$

$$M_e = \sqrt{11.544347} = 3.3977$$

- Exit area (A_e) :

$$A_e = \frac{A_t}{M_e}\left[\frac{1 + \frac{k-1}{2}M_e^2}{\left(\frac{k+1}{2}\right)}\right]^{\frac{k+1}{2(k-1)}}$$

$$A_e = \frac{0.0344386}{3.3977}\left[\frac{1 + \frac{1.20-1}{2}\times 11.544347}{\left(\frac{1.20+1}{2}\right)}\right]^{\frac{1.201+1}{2(1.201-1)}}$$

$$A_e = 0.40204 \text{ m}^2$$

- Exhaust velocity (V_e) :

$$V_e = \sqrt{\frac{2k}{k-1}\frac{R' \times T_c}{M}\left(1 - \left(\frac{P_e}{P_c}\right)^{\frac{k-1}{k}}\right)}$$

$$V_e = \sqrt{\frac{2 \times 1.20}{1.20 - 1}\frac{8314 \times 3600}{24}\left(1 - \left(\frac{0.05}{5}\right)^{\frac{1.20-1}{1.20}}\right)}$$

$$V_e = 2831.78 \text{ m/s}$$

- Thrust (F) :

$$F = \dot{m} \times V_e + (P_e - P_a) \times A_e$$

$$F = 100 \times 2831.78 + (0.05 \times 10^6 - 0.05 \times 10^6) \times 0.409$$

$$F = 283178 \, \text{N}$$

Under-extended and Over-extended Nozzle Conditions

Under-extended Nozzle $(P_e > P_a)$:
1. Assume $P_e = 0.10\text{MPa}$:
2. Calculate new exit parameters:

- Exit Mach number (M_e) :

$$M_e^2 = \frac{2}{1.20 - 1}\left[\left(\frac{5}{0.10}\right)^{\frac{1.20-1}{1.20}} - 1\right]$$

$$M_e^2 = 9.19383$$

$$M_e = \sqrt{9.19383} = 3.03213$$

- Exit area (A_e) :

$$A_e = \frac{0.0344386}{3.03213}\left[\frac{1 + \frac{1.20 - 1}{2} \times 9.19383}{\left(\frac{1.20 + 1}{2}\right)}\right]^{\frac{1.20+1}{2(1.20-1)}}$$

$$A_e = 0.242676 \, \text{m}^2$$

- Exhaust velocity (V_e) :

$$V_e = \sqrt{\frac{2 \times 1.20}{1.20 - 1}\frac{8314 \times 3600}{24}\left(1 - \left(\frac{0.10}{5}\right)^{\frac{1.20-1}{1.20}}\right)}$$

$$V_e = 2677.37 \, \text{m/s}$$

- Thrust (F) :

$$F = 100 \times 2677.37 + (0.10 \times 10^6 - 0.05 \times 10^6) \times 0.243$$

$$F = 267737 \text{ N}$$

Over-extended Nozzle $(P_e < P_a)$:
1. Assume $P_e = 0.025$MPa :
2. Calculate new exit parameters:

- Exit Mach number (M_e) :

$$M_e^2 = \frac{2}{1.20 - 1} \left[\left(\frac{5}{0.025}\right)^{\frac{1.20-1}{1.20}} - 1 \right]$$

$$M_e^2 = 14.1827$$

$$M_e = \sqrt{14.1827} = 3.76599$$

- Exit area (A_e) :

$$A_e = \frac{0.0344386}{3.76599} \left[\frac{1 + \frac{1.20 - 1}{2} \times 14.1827}{\left(\frac{1.20 + 1}{2}\right)} \right]^{\frac{1.20+1}{2(1.20-1)}}$$

$$A_e = 0.696278 \text{ m}^2$$

- Exhaust velocity (V_e) :

$$V_e = \sqrt{\frac{2 \times 1.20}{1.20 - 1} \frac{8314 \times 3600}{24} \left(1 - \left(\frac{0.025}{5}\right)^{\frac{1.20-1}{1.20}}\right)}$$

$$V_e = 2962.57 \text{ m/s}$$

- Thrust (F) :

$$F = 100 \times 2963 + (0.025 \times 10^6 - 0.05 \times 10^6) \times 0.696$$

$$F = 296257 \text{ N}$$

By analyzing the thrust equation and various nozzle conditions, we observe that the optimal nozzle design is achieved when the exit pressure equals the ambient pressure ($P_e = P_a$). Both underextended and over-extended nozzles produce less thrust compared to the optimally expanded nozzle. This optimal design maximizes the conversion of thermal energy into kinetic energy, thus achieving maximum thrust.

4.5   Example Problems:

Example 4-1:
Optimized Rocket Nozzle Design Problem: Achieving Optimal Thrust Under Varying Ambient Pressures

Objective:

To design an adaptable rocket nozzle capable of optimizing thrust by adjusting its expansion ratio in real-time to match varying ambient pressure conditions.

Background:
In a rocket engine, the DeLaval nozzle converts thermal energy into kinetic energy, producing thrust. The performance of the nozzle depends on the match between the nozzle exit pressure (P_e) and the ambient pressure (P_a). An optimal nozzle design should achieve maximum thrust by adjusting to maintain the condition $P_e = P_a$, despite changes in external conditions.

Key Parameters:
- Propellant mass flow rate ($\dot{m}$): 100 kg/s
- Specific heat ratio (k): 1.20
- Exhaust gas molecular weight (M): 24 kg/kmol
- Combustion chamber temperature (T_c): 3600 K
- Combustion chamber pressure (P_c): 5MPa
- Ambient pressure (P_a) : Variable (ranging from 0.025 MPa to 0.1 MPa)

Equations:

1. Throat Pressure and Temperature:

$$P_t = P_c \left[1 + \frac{k-1}{2}\right]^{-\frac{k}{k-1}}$$

$$T_t = T_c \left[1 + \frac{k-1}{2}\right]^{-1}$$

2. Throat Area (A_t) :

$$A_t = \frac{\dot{m}}{P_t}\sqrt{\frac{R'T_t}{Mk}}$$

Where $R' = 8{,}314$ J/(kmol $\cdot$ K).

3. Mach Number at Exit (M_e) :

$$M_e^2 = \frac{2}{k-1}\left[\left(\frac{P_c}{P_e}\right)^{\frac{k-1}{k}} - 1\right]$$

4. Exit Area $(\mathbf{A}_e)$:

$$A_e = \frac{A_t}{M_e}\left[\frac{1 + \frac{k-1}{2}M_e^2}{\frac{k+1}{2}}\right]^{\frac{k+1}{2(k-1)}}$$

5. Exhaust Velocity (V_e) :

$$V_e = \sqrt{\frac{2k}{k-1}\frac{R'T_c}{M}\left[1 - \left(\frac{P_e}{P_c}\right)^{\frac{k-1}{k}}\right]}$$

6. Thrust (F):

$$F = \dot{m}V_e + (P_e - P_a)A_e$$

Problem Statement:
1. Determine the optimal nozzle throat area (A_t) based on the combustion chamber conditions.
2. Calculate the necessary nozzle exit area (A_e) for maintaining optimal expansion (i.e., $P_e = P_a$) across a range of ambient pressures (from 0.025 MPa to 0.1 MPa).
3. Calculate the exhaust velocity (V_e) and the corresponding thrust (F) for each

ambient pressure.

4. Develop a mechanism or propose a design that allows the nozzle to dynamically adjust its exit area to maintain optimal expansion conditions.

Scenario Parameters:

- Ambient Pressures: P_a ranging from 0.025 MPa to 0.1 MPa .

Requirements:

1. Perform the necessary calculations using the provided equations.
2. Analyze the effect of varying ambient pressures on the nozzle's performance and determine the range of exit areas required.
3. Suggest a design mechanism for an adjustable nozzle capable of real-time adaptation to optimize thrust.

Expected Outcome:

A comprehensive analysis of the nozzle performance across a range of ambient pressures, detailing how the nozzle exit area needs to adjust to achieve optimal thrust. The problem aims to highlight the significance of adaptable nozzle designs in maximizing rocket efficiency under varying environmental conditions.

give a detailed solution

Solution:

Given Parameters:

- Propellant mass flow rate $(\dot{m})$: 100 kg/s
- Specific heat ratio (k): 1.20
- Exhaust gas molecular weight (M): 24 kg/kmol
- Combustion chamber temperature (T_c): 3600 K
- Combustion chamber pressure (P_c): 5MPa
- Universal gas constant (R'): 8,314 J/(kmol-K)
- Ambient pressures (P_a) : Varying from 0.025 MPa to 0.1 MPa

Throat Conditions

Throat Pressure (P_t) Calculation:

$$P_t = P_c \left[1 + \frac{k-1}{2} \right]^{-\frac{k}{k-1}}$$

$$P_t = 5\left[1 + \frac{1.20 - 1}{2}\right]^{-\frac{1.20}{1.20-1}}$$

$$P_t \approx 2.82 \text{MPa}$$

Throat Temperature (T_t) Calculation:

$$T_t = T_c\left[1 + \frac{k-1}{2}\right]^{-1}$$

$$T_t = 3600\left[1 + \frac{1.20 - 1}{2}\right]^{-1}$$

$$T_t \approx 3273 \text{ K}$$

Throat Area (A_t) Calculation:

$$A_t = \frac{\dot{m}}{P_t}\sqrt{\frac{R'T_t}{Mk}}$$

$$A_t = \frac{100}{2.82 \times 10^6}\sqrt{\frac{8{,}314 \times 3273}{24 \times 1.20}}$$

$$A_t \approx 0.03447 \text{ m}^2$$

Calculations for Different Ambient Pressures
We will calculate the exit area (A_e), exhaust velocity (V_e), and thrust (F) for different values of ambient pressure (P_a).

For $P_a = 0.025$MPa (Over-Extended Nozzle Condition)

Exit Mach Number (M_e) Calculation:
50

$$M_e^2 = \frac{2}{k-1}\left[\left(\frac{P_c}{P_e}\right)^{\frac{k-1}{k}} - 1\right]$$

$$M_e^2 = \frac{2}{0.20}\left[\left(\frac{5}{0.025}\right)^{\frac{0.20}{1.20}} - 1\right]$$

$$M_e^2 = 14.1827$$

$$M_e \approx 3.76599$$

Exit Area (A_e) Calculation:

$$A_e = \frac{A_t}{M_e}\left[\frac{1 + \frac{k-1}{2}M_e^2}{\frac{k+1}{2}}\right]^{\frac{k+1}{2(k-1)}}$$

$$A_e = \frac{0.03447}{3.76599}\left[\frac{1 + \frac{1.2-1}{2} \times 14.1827}{\frac{1.2+1}{2}}\right]^{\frac{1.2+1}{2(1.2-1)}}$$

$$A_e \approx 0.696913 \text{ m}^2$$

Exhaust Velocity (V_e) Calculation:

$$V_e = \sqrt{\frac{2k}{k-1}\frac{R'T_c}{M}\left[1 - \left(\frac{P_e}{P_c}\right)^{\frac{k-1}{k}}\right]}$$

$$V_e = \sqrt{\frac{2 \times 1.2}{1.2-1}\frac{8314 \times 3273}{24}\left[1 - \left(\frac{0.025}{5}\right)^{\frac{1.2-1}{1.2}}\right]}$$

$$V_e \approx 2824.82 \text{ m/s}$$

Thrust (F) Calculation:

$$F = \dot{m}V_e + (P_e - P_a)A_e$$

$$F = 100 \times 2824.82 + (0.025 \times 10^6 - 0.025 \times 10^6) \times 0.695$$

$$F = 282482 \ \text{N}$$

For $P_a = 0.05\text{MPa}$ (Optimal Expansion)

Exit Mach Number ($\mathbf{M}_e$) Calculation:

$$M_e^2 = \frac{2}{0.20}\left[\left(\frac{5}{0.05}\right)^{\frac{0.20}{1.20}} - 1\right]$$

$$M_e^2 = 11.5443$$

$$M_e \approx 3.3977$$

Exit Area (A_e) Calculation:

$$A_e = \frac{0.03447}{3.3977}\left[\frac{1 + \frac{0.20}{2} \times 11.5443}{1.10}\right]^{5.5}$$

$$A_e \approx 0.40919 \ \text{m}^2$$

Exhaust Velocity (V_e) Calculation:

$$V_e = \sqrt{12\frac{29930400}{24}\left[1 - \left(\frac{0.05}{5}\right)^{0.1667}\right]}$$

$$V_e \approx 2832 \ \text{m/s}$$

Thrust (F) Calculation:

$$F = \dot{m}V_e + (P_e - P_a)A_e$$

$$F = 100 \times 2832 + (0.05 \times 10^6 - 0.05 \times 10^6) \times 0.409$$

$$F = 283200 \ \text{N}$$

For $P_a = 0.1\text{MPa}$ (Under-Extended Nozzle Condition)

Exit Mach Number (M_e) Calculation:

$$M_e^2 = \frac{2}{0.20}\left[\left(\frac{5}{0.10}\right)^{\frac{0.20}{1.20}} - 1\right]$$

$$M_e^2 = 9.19383$$

$$M_e^{\square} = 3.03213$$

Exit Area (A_e) Calculation:

$$A_e = \frac{0.03447}{3.03213}\left(\frac{1 + \frac{0.20}{2} \times 9.19383}{1.10}\right)^{5.5}$$

$$A_e \approx 0.242897 \text{ m}^2$$

Exhaust Velocity (V_e) Calculation:

$$V_e = \sqrt{12\frac{29930400}{24}\left[1 - \left(\frac{0.10}{5}\right)^{0.1667}\right]}$$

$$V_e = \sqrt{149652000 \times 0.582}$$

$$V_e \approx 2677.56 \text{ m/s}$$

Thrust (F) Calculation:

$$F = \dot{m}V_e + (P_e - P_a)A_e$$

$$F = 100 \times 2677.56 + (0.10 \times 10^6 - 0.10 \times 10^6) \times 0.243$$

$$F = 267756 \text{ N}$$

Summary of Results

P_a(MPa)	M_e	A_e(m^2)	V_e(m/s)	F(N)
0.025	3.76599	0.696913	2824.82	282482
0.05	3.3977	0.40919	2832.00	283200
0.10	3.03213	0.242897	2677.56	267756

Design Mechanism for an Adjustable Nozzle

To achieve optimal thrust across varying ambient pressures, the nozzle design must include an adjustable exit area mechanism. This can be achieved through:

1. Extendible Nozzle Sections: Mechanically extendible and retractable sections can change the nozzle length and exit diameter, allowing adjustment of A_e.
2. Variable Geometry Nozzle (VGN): Incorporating movable petals or flaps around the nozzle exit can dynamically change the area ratio in response to real-time pressure sensors.
3. Active Control Systems: An active feedback system utilizing sensors to measure ambient pressure and engine performance can control actuators that adjust the nozzle geometry for optimal expansion.

The adaptability of the nozzle design ensures that the rocket engine maintains maximum efficiency and thrust, regardless of the ambient pressure changes during ascent or in different atmospheric conditions. This optimization is crucial for missions requiring precise control over thrust and fuel efficiency.

5 Mission to Mars

The human exploration of Mars represents a formidable challenge requiring significant technological advancements. One critical area is propulsion systems, which are essential for ensuring the safe transport of astronauts to and from Mars. Advanced nuclear propulsion systems, whether used alone or in conjunction with chemical propulsion systems, offer the potential to significantly reduce the duration of interplanetary travel. Shorter trip times are crucial as they mitigate risks associated with prolonged space radiation exposure, the effects of zero gravity, and the complexities of launch and orbital assembly.

This report evaluates the primary technical and programmatic challenges, advantages, and risks associated with developing nuclear thermal propulsion (NTP) and nuclear electric propulsion (NEP) systems, potentially augmented with chemical propulsion, for human missions to Mars. It also provides development roadmaps for both NTP and NEP technologies, outlining key milestones.

Numerous NASA studies have explored the feasibility of utilizing NTP and NEP technologies for human Mars exploration. These studies have considered a variety of mission scenarios involving nuclear, solar, and chemical propulsion systems, with varying mission parameters. The specific requirements for these missions are influenced by the relative positions of Earth and Mars, which vary significantly over a synodic period of approximately 26 months. This variation affects the distance between the two planets, ranging from 55 to 400 million kilometers, and consequently impacts the propulsion system requirements, particularly the total velocity increment (ΔV) necessary for a round-trip mission.

NASA has outlined a baseline mission for its report, "Baseline Mission to Mars: Crewed Opposition-Class Missions," focusing on an opposition-class crewed mission to Mars, scheduled for launch in 2039. Prior to the crewed mission, cargo missions are planned to deploy essential infrastructure and supplies. The selected propulsion system must be capable of supporting both opposition and conjunction-class missions. The baseline mission parameters include:

- **Launch Date**: Crew mission launch in the 2039 opportunity.

- **Total Crew Trip Duration**: Not exceeding 750 days.
- **Mission Structure**: Split mission with separate crew and cargo vehicles, utilizing the same propulsion systems. Cargo missions are scheduled to arrive at Mars before the crew departs from Earth.
- **Mars Surface Stay Duration**: 30 days.
- **Crew Composition**: Four astronauts, with two landing on Mars.
- **Vehicle and Propellant Launch**: Multiple launch vehicles to deliver components and propellant to an assembly orbit, either in low Earth orbit or cislunar space.

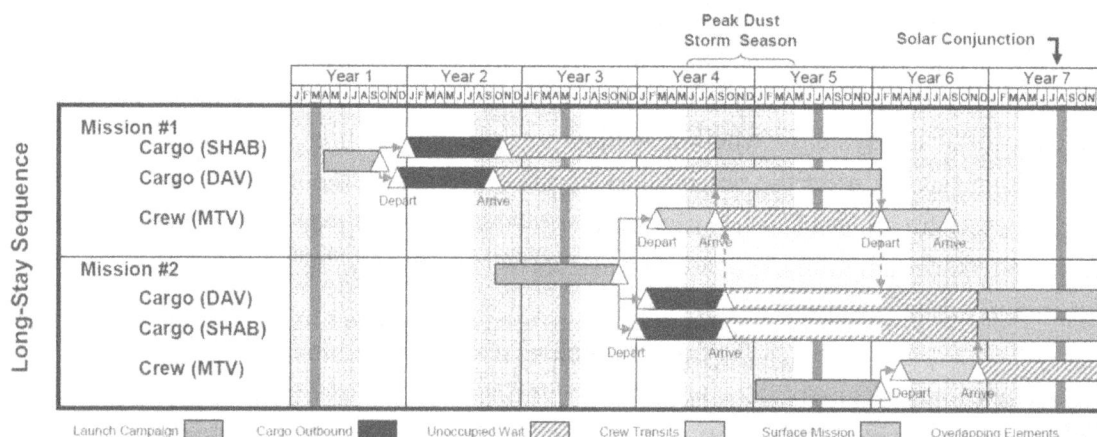

Figure 5-1. Mars mission timeline (source: NASA-SP-2009-566).

For the NEP system, the mission architecture includes additional in-space chemical propulsion using liquid methane and liquid oxygen to assist with Earth departure, Mars capture, and departure maneuvers. This system will handle acceleration and deceleration in interplanetary space. In contrast, the NTP system provides propulsion for all critical maneuvers.

The dynamic nature of the Earth-Mars relationship necessitates varying propulsion requirements for different launch windows, influencing mission planning and design. The flexibility and high specific impulse (Isp) of nuclear propulsion systems are particularly advantageous, offering the capability to meet mission requirements with greater consistency across different launch opportunities. This characteristic is critical for maintaining mission feasibility and minimizing launch mass, especially during less optimal planetary alignments.

Overall, the development of robust NTP and NEP systems is vital for enabling flexible and efficient Mars exploration missions. This report presents an in-depth analysis of the necessary technological advancements, challenges, and strategic considerations for these systems to achieve the ambitious goal of human Mars exploration.

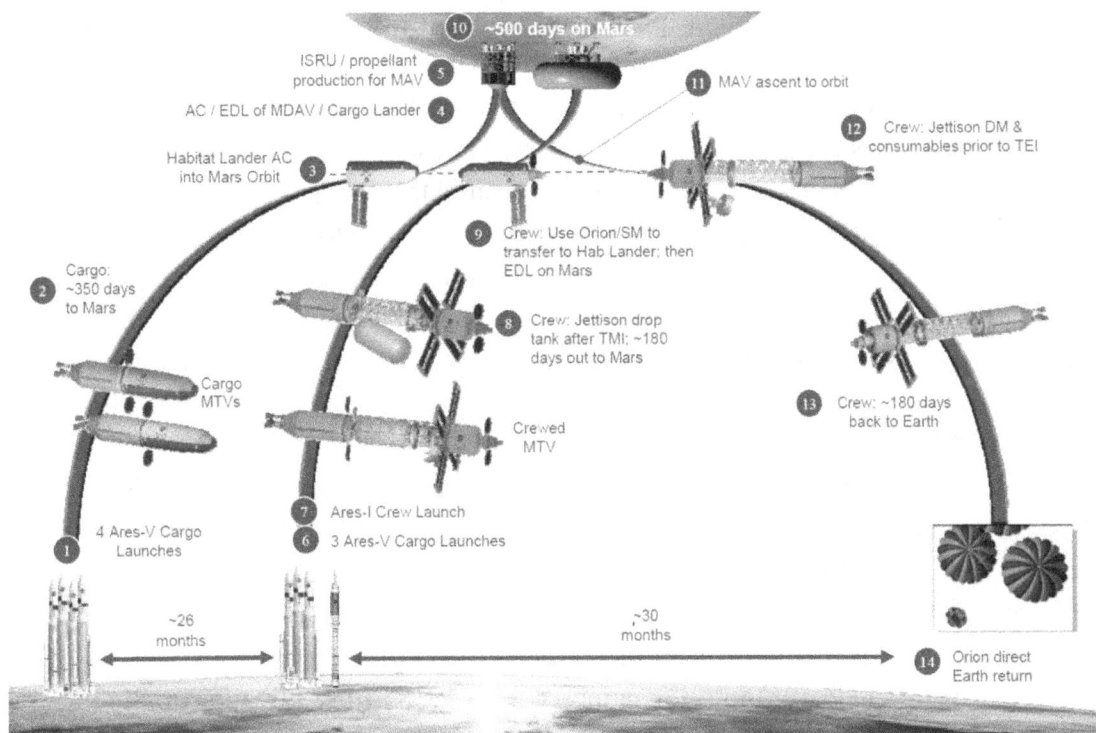

Figure 5-2. Mars mission reference architecture (source: NASA-SP-2009-566).

5.1   Nuclear Thermal Propulsion Concept

A Nuclear Thermal Propulsion (NTP) system operates on a similar principle to chemical propulsion systems but replaces the combustion chamber with a nuclear reactor to heat the propellant. The core components of an NTP system include a nuclear reactor, a rocket engine, and a propellant storage and management subsystem. These elements are highly integrated to ensure efficient operation. Specifically, the nuclear reactor subsystem comprises the core, control drums with actuators, reflector, shield, and pressure shell. The engine subsystem includes turbomachinery, associated valves and piping, and a nozzle. Additionally, the propellant storage and management subsystem consists of the liquid hydrogen (LH_2) tank and helium pressurization tanks.

In NTP systems, as well as in nuclear electric propulsion systems and terrestrial nuclear power plants, the reactor generates heat through nuclear fission. This process also produces high levels of radiation, necessitating robust shielding to protect both personnel and materials. In an NTP system, the LH_2 propellant is pumped from the cryogenic storage tank through turbopumps and associated management components into the reactor, where it is heated directly. The heated hydrogen then expands through the rocket nozzle, creating thrust. This heating mechanism differs from chemical rockets, where combustion provides the energy.

The reactor's control drums, positioned around the reactor core's outer annulus within the reflector, play a critical role in regulating the reactor's power output. These drums absorb neutrons and are used to start or shut down the reactor and adjust its power level. Hydrogen turbopumps control the mass flow rate and pressure of the hydrogen propellant, which is crucial for maintaining the desired thrust.

The design of NTP reactors draws heavily from the historical Rover and Nuclear Engine for Rocket Vehicle Applications (NERVA) programs. The core design features tightly packed hexagonal (prismatic) fuel elements. Surrounding the core are control drums partially covered with reflector material, which serves to reflect neutrons back into the core, sustaining the fission process. The drums are rotated to control the reactor's power output.

The reactor core is encased within inner and outer pressure vessels, with reflector materials surrounding the control drums. These components are critical for maintaining structural integrity and controlling the neutron flux within the reactor. The choice of uranium-235 (U-235) as the fuel for these reactors is standard across U.S. space nuclear propulsion designs, owing to its favorable fission characteristics.

The historical Rover project, conducted from 1955 to 1973 by the Atomic Energy Commission, aimed to develop suitable nuclear reactors for NTP systems. Following this, NASA's NERVA program, running from 1961 to 1973, focused on creating a complete NTP system. Both initiatives were jointly managed by the Space Nuclear Propulsion Office, laying the groundwork for modern NTP technology.

5.2   History of Nuclear Thermal Rocket (NTR) Development

The development of Nuclear Thermal Propulsion (NTP) in the United States began in earnest with Project Rover, which commenced in 1955 under the auspices of the Atomic Energy Commission. This initiative aimed to develop nuclear reactors capable of serving as propulsion systems for space missions. By 1961, NASA, in collaboration with the AEC, launched the Nuclear Engine for Rocket Vehicle Applications (NERVA) program, which focused on the engineering and testing of complete NTP systems.

These programs represented a significant technological leap, as they sought to replace conventional chemical propulsion systems with nuclear reactors capable of heating hydrogen propellant to extremely high temperatures. The early designs involved reactors using highly enriched uranium (HEU) graphite fuel elements, incorporating materials such as uranium dioxide (UO_2), uranium dicarbide (UC_2), and later, coated UC_2 particles. The primary objective was to harness the energy released from nuclear fission to achieve specific impulse (Isp) values significantly higher than those achievable by chemical rockets.

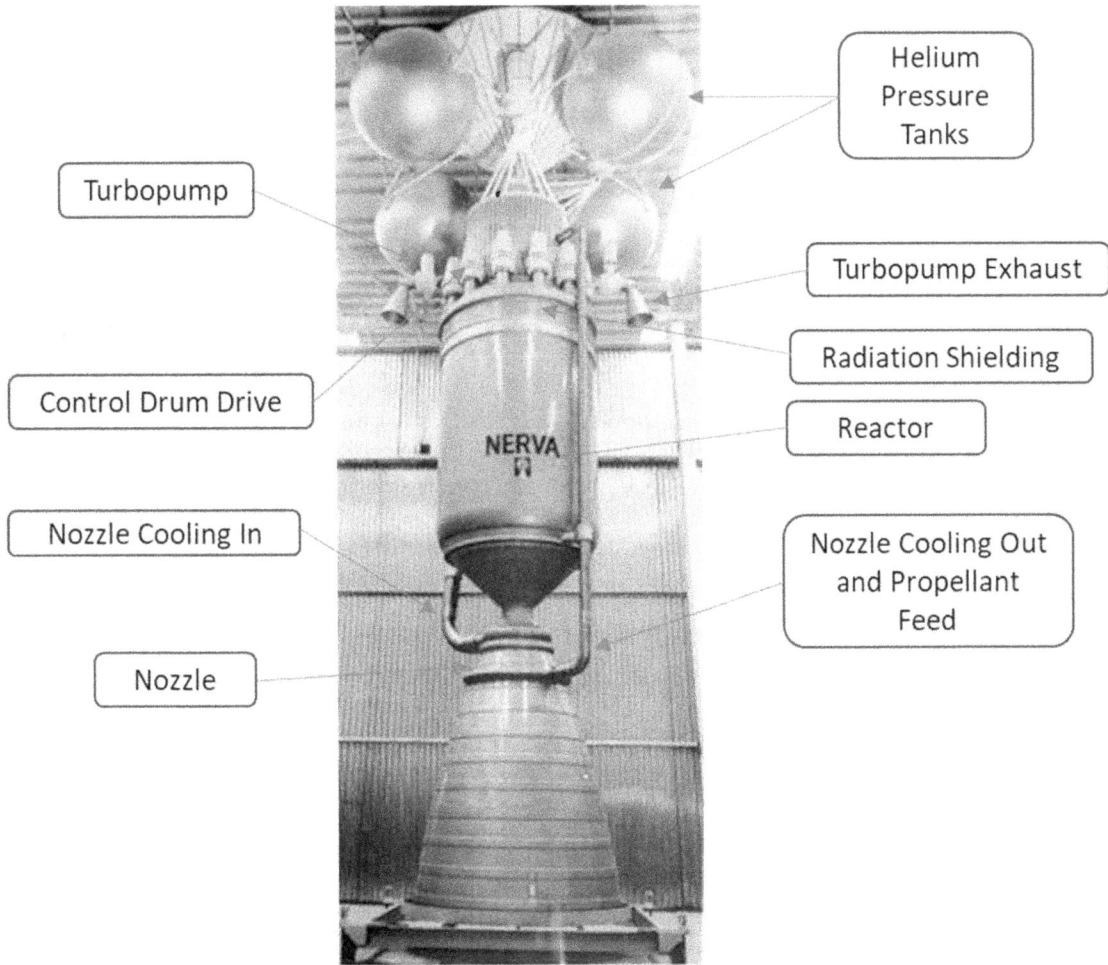

Figure 5-3. Photo of the Rover/NERVA.
(source: ntrs.nasa.gov/citations/20180006514).

5.2.1 Ground Testing and Reactor Development

The Rover/NERVA programs undertook an extensive ground testing campaign, constructing and testing 22 reactors. These tests were crucial for demonstrating controlled reactor operations and validating the NTP concept. The reactors, such as the Kiwi series, faced numerous technical challenges, including structural issues caused by flow-induced vibrations. For example, the Kiwi B4A reactor experienced significant vibrations that necessitated the destruction of several reactor units to diagnose and resolve the problem.

A critical component of the NTP design was neutron moderation, which was primarily achieved using graphite in the fuel elements. The graphite served a dual purpose: it acted as a heat transfer element and as the primary neutron moderator, slowing down neutrons to sustain the nuclear reaction. The Pewee reactor, a standout in the NERVA series, incorporated additional separately cooled zirconium hydride (ZrH) moderator material in its tie tubes, allowing it to reach a peak fuel temperature of 2750 K and a propellant temperature of 2550 K at the reactor exit. This configuration achieved an Isp of approximately 875 seconds in a vacuum, a significant improvement over earlier designs.

5.2.2   Technological Innovations and Challenges

Throughout the testing phase, the NERVA program explored various fuel compositions and configurations. The use of coatings such as niobium carbide (NbC) and zirconium carbide (ZrC) on graphite surfaces exposed to hot hydrogen was tested to improve material durability under high temperatures. NbC-coated fuel elements often experienced pronounced cracking, whereas ZrC coatings demonstrated better durability, leading to their consideration for future designs.

One of the most significant advancements was the XE-Prime reactor, which achieved a record number of starts, shutdowns, and restarts, far exceeding the operational requirements for potential Mars missions. XE-Prime demonstrated the feasibility of using NTP systems for complex mission profiles, including multiple engine firings and prolonged operations.

5.2.3   Key Reactor Designs and Achievements

The Rover/NERVA programs produced several notable reactor designs, including the NRX series and the Phoebus reactors. These reactors varied in terms of power output, propellant exit temperatures, and thrust. For instance, the Phoebus 2A reactor reached a power output of 4000 MWth and produced a thrust of 200,000 lbf. The Pewee reactor, designed as a compact and high-performance unit, demonstrated the highest fuel and propellant exit temperatures among the series, underscoring its potential for future applications.

5.2.4   Operational Considerations and Performance Optimization

The NTP systems' performance was directly tied to their operating temperatures, which influenced the specific impulse. A critical operational goal was to minimize the startup and shutdown times of the reactors to maintain high Isp. The faster the reactor could reach full operational temperature, the less time was spent at lower efficiency during each engine firing. This requirement posed design challenges, particularly in ensuring the integrity of materials and components under rapid thermal transients.

The NTP systems developed under Rover/NERVA also faced unique operational challenges, such as managing the hydrogen propellant's flow and pressure. The hydrogen turbopumps played a crucial role in maintaining the desired mass flow rate and pressure, which were essential for achieving the necessary thrust levels.

5.2.5 Program Termination and Legacy

Despite the technical successes and innovations demonstrated by the Rover/NERVA programs, the projects were ultimately canceled in 1973 due to shifting political and financial priorities. The cancellation occurred before any NTP system could be fully integrated and tested in a flight configuration or deployed in space. Nevertheless, the programs provided invaluable data and experience, establishing a foundation for future NTP development.

5.2.6 Post-NERVA Developments and International Efforts

Following the termination of Rover/NERVA, several efforts aimed to build on the knowledge gained. Notable among these was the Space Nuclear Thermal Propulsion (SNTP) program, which focused primarily on fuel development, including testing of coated HEU particles for potential use in NTP systems. The SNTP program also explored the use of polyethylene moderator materials and produced hardware for non-nuclear component testing. However, it did not progress to the stage of ground testing complete reactors.

In parallel, international efforts, particularly by the Soviet Union, pursued NTP development. The Soviet RD-410 reactor, for example, employed a unique fuel design using twisted ribbon carbide HEU fuel and a ZrH moderator. These efforts underscored the global interest in NTP as a viable technology for deep-space exploration.

5.2.7 Fuel Innovations and Material Testing

The evolution of NTP technology also involved significant advances in fuel materials. Various fuel forms, including ceramic-metal (cermet) composites and refractory metal composites, were tested to withstand the extreme conditions within an NTP reactor. These materials were evaluated for their maximum operating temperatures and their ability to maintain structural integrity under high radiation fields and hydrogen exposure. While most of these fuel forms were not tested in fully integrated NTP systems, they demonstrated the potential to achieve high-temperature operations necessary for efficient propulsion.

The history of NTP development in the United States, spearheaded by the Rover and NERVA programs, represents a remarkable period of technological innovation and exploration. Despite the premature termination of these programs, the groundwork laid by these early efforts continues to inform contemporary research and development in space nuclear propulsion. The lessons learned and the technological advancements achieved remain a cornerstone for ongoing and future endeavors in harnessing nuclear technology for space exploration. The potential of NTP systems to significantly reduce travel times and enable more ambitious missions remains a compelling objective for space agencies worldwide.

5.3   Modeling and Simulation Capabilities in Nuclear Thermal Propulsion

In the development of Nuclear Thermal Propulsion (NTP) systems, modeling and simulation (M&S) play a pivotal role. Given the limited experimental data available and the significant challenges associated with full-scale testing, these computational tools have become indispensable in advancing NTP technology. The field of M&S for NTP encompasses several key areas, each crucial to understanding and optimizing these complex systems.

At the heart of NTP M&S lies reactor physics. Neutronics simulations, often utilizing advanced codes such as the Monte Carlo N-Particle Transport Code (MCNP), allow researchers to model the behavior of neutrons within the reactor core. These simulations are essential for predicting critical parameters such as power distribution, fuel burnup, and reactivity changes over time. Modern neutronics models can handle both steady-state and transient reactor behavior, providing crucial insights into reactor performance under various operating conditions.

Complementing neutronics simulations is the field of thermal-hydraulics modeling. Here, Computational Fluid Dynamics (CFD) tools come to the forefront, enabling detailed simulations of propellant flow and heat transfer within the reactor core. Advanced software packages like ANSYS Fluent and Siemens STAR-CCM+ have been adapted to handle the extreme conditions present in NTP systems. These tools allow engineers to model complex geometries within the reactor, predict hot spots, and optimize coolant channel designs for maximum heat transfer efficiency.

The extreme temperatures and radiation environments inherent to NTP systems necessitate robust structural and mechanical analysis capabilities. Finite Element Analysis (FEA) tools are employed to model thermal stresses and mechanical behavior of reactor components under these harsh conditions. However, current capabilities in this area are somewhat limited, particularly when it comes to modeling dynamic phenomena such as flow-induced vibrations. This remains an area ripe for further development, as understanding and mitigating these effects is crucial for ensuring long-term reactor integrity.

System-level modeling represents another critical aspect of NTP M&S. By integrating various subsystem models, researchers can predict overall NTP system performance and behavior. Tools like NASA's Numerical Propulsion System Simulation (NPSS), originally developed for conventional propulsion systems, have been adapted for NTP applications. These integrated models allow engineers to optimize system-wide performance, identify potential issues at the interfaces between subsystems, and predict how changes in one component might affect the entire NTP system.

Beyond the propulsion system itself, mission analysis tools incorporate NTP performance characteristics into broader spacecraft and mission designs. These tools enable trajectory optimization, mission planning, and assessment of how NTP capabilities might impact overall mission architecture and requirements. By providing a clear picture of the potential benefits and challenges of NTP in various mission scenarios, these models play a crucial role in guiding technology development and mission planning efforts.

The state of the art in NTP M&S has advanced significantly in recent years, driven by increases in computational power and the development of sophisticated multi-physics simulation environments. Platforms like the Multi-physics Object-Oriented Simulation Environment (MOOSE) now enable the coupling of various physical phenomena, allowing for more comprehensive and realistic simulations of NTP systems. This ability to model the complex interactions between neutronics, thermal-hydraulics, and structural mechanics provides invaluable insights that would be difficult or impossible to obtain through physical testing alone.

Despite these advancements, significant challenges remain in the field of NTP M&S. Perhaps the most pressing is the lack of recent experimental data, particularly for systems using High-Assay Low-Enriched Uranium (HALEU). This shortage of validation data constrains the accuracy and reliability of many models, particularly when it comes to predicting behavior under realistic operating conditions. The complexity of multi-physics coupling in NTP systems also presents ongoing challenges, as accurately representing the interactions between various physical phenomena requires sophisticated models and significant computational resources.

Modeling transient behavior remains another area of difficulty. The dynamic nature of reactor operation, coupled with the complexities of propellant flow and changing thermal conditions, presents a formidable challenge to simulation efforts. Advances in this area will be crucial for predicting and mitigating potential issues during reactor startup, shutdown, and power transients.

Looking to the future, the continued development and refinement of M&S capabilities will be essential for the advancement of NTP technology. As computational power continues to increase and modeling techniques become more sophisticated, researchers will be able to create increasingly accurate and comprehensive simulations of NTP systems. However, these advancements in M&S must go hand-in-hand with experimental efforts. Only through a combination of advanced modeling and carefully designed experiments can the full potential of NTP be realized, paving the way for this revolutionary propulsion technology to enable ambitious deep space exploration missions.

5.4   Reactor and Systems Technologies

Nuclear Thermal Propulsion (NTP) represents a promising technology for future deep space exploration missions, particularly for crewed missions to Mars. We explore the key technology requirements, risks, and options associated with NTP systems, focusing on the reactor subsystem, engine subsystem, and propellant storage and management.

NTP system performance is primarily driven by heat transfer efficiency, which depends on several factors including temperature profiles during operations, time at maximum operating temperature, the number of planned operating cycles, and rates of temperature change across the system. The primary risks associated with NTP systems stem from four main challenges:

1. The high operating power density and temperature of the reactor, which must heat the propellant to approximately 2700 K at the reactor exit to meet the required specific impulse (Isp) of 900 seconds.
2. The need for long-term storage and management of cryogenic liquid hydrogen (LH2) propellant.
3. The relatively short NTP reactor startup times compared to other space or terrestrial power reactors.
4. The longer startup and shutdown transients of an NTP system relative to chemical engines.

The reactor subsystem presents some of the most significant challenges in NTP development. Operating at temperatures around 2700 K creates an extreme environment in terms of both temperature and hydrogen corrosion for the materials in the reactor core. This high operating temperature severely limits the number of viable fuel architectures. The fuel element, which includes the fuel and cladding, along with fuel assemblies, moderator, support structures, and the reactor pressure vessel, must maintain physical integrity while cycling through the thermomechanical stress induced during repeated cycles of reactor startup, operation at power, shutdown, and restart.

NASA is currently considering at least three new NTP fuel architectures:

1. Cercer-coated fuel particles in a refractory ceramic matrix
2. Cercer solid solutions of mixed (U, Zr, Nb)C fuel with multiple potential particles (UN, UC, UCZr, etc.)
3. Cermet-coated fuel particles in a refractory metal matrix

Each of these fuel options presents its own set of advantages and challenges. For instance, the reference cermet fuel architecture uses uranium nitride (UN) particle fuel at 40 to 70 volume percent packing density, with a molybdenum (Mo)-30% tungsten (W) metal matrix. This composition balances the need to limit parasitic thermal neutron absorption (by reducing tungsten content) with maximizing the alloy melting temperature (by increasing tungsten content).

Composite nuclear fuels such as cercer with coated fuel particles can potentially increase safety margins. However, cercer fuels are at a lower level of technological and fabrication maturity. Similarly, cercer solid solution fuels offer the potential for higher performance and safety margins but are also at a lower level of technological maturity.

The development of suitable moderator materials is another critical aspect of NTP reactor design. Potential moderator materials include zirconium hydride (ZrH), yttrium hydride (YH), beryllium (Be), beryllium oxide (BeO), and beryllium carbide (Be2C). Each of these materials has its own set of challenges, particularly in terms of maximum use temperatures and susceptibility to hydrogen embrittlement and dissociation.

The reactor structure serves as the primary interface to the LH2 propellant at the upper plenum inlet as well as the interface to the nozzle. While progress has been made in developing reactor structural materials and initial designs, significant work remains to optimize these systems for mass, propellant flow, pressure drop, flow rates and stability, and stress analysis. Additionally, shielding design for both gamma and neutron radiation must be carefully considered as part of the overall reactor system design.

Reactivity control is another crucial aspect of NTP reactor design. Many existing and current designs employ reactivity control drums located radially within a reflector assembly. These control drums, typically constructed of a neutron reflector material with a section of a thermal neutron absorber like boron carbide (B4C), can be rotated to control reactor operation. The control mechanism for these drums must manage the flow rates for LH2 and gaseous hydrogen through various parts of the reactor system.

Moving to the engine subsystem, while it has significant heritage from chemical rocket engines, including the use of gaseous H2 and LH2 as fuel, additional testing will be necessary to demonstrate integrated operability, lifetime, and reliability. However, compared to the reactor subsystem, ensuring the performance of the engine subsystem is a relatively low-risk element of developing an NTP system for Mars missions.

The propellant storage and management subsystem presents another significant challenge in NTP development. The long duration of a Mars mission, including time for in-space vehicle assembly and the round trip to Mars, necessitates the storage of metric tons of LH2 at cryogenic temperatures as low as 20 K, with minimal losses. Current expectations for a Mars mission suggest that at least six NTP system starts will be needed, with a total LH2 propellant requirement ranging from 7 to 21 10,000-kg tanks of LH2, depending on the launch vehicles used and the mission departure year.

Minimizing the boiloff of LH2 from the storage tanks is crucial to reduce the amount of LH2 that must be launched and the number of storage tanks that must be integrated into the Mars exploration spacecraft. While development of refrigeration technology is proceeding, existing cryocooling systems cannot reliably meet propellant tank requirements over a mission of this duration. Additionally, propellant mass must be accurately measured before and after each firing of the propulsion system to appropriately balance flow rate to the reactor startup and reactivity control operations.

While NTP technology offers significant potential benefits for deep space exploration, particularly for crewed Mars missions, it also presents numerous technical challenges that must be overcome. The development of suitable fuel architectures, moderator materials, and reactor structural components capable of withstanding the extreme conditions of NTP operation is critical. Additionally, the long-term storage and management of cryogenic LH2 propellant remains a significant hurdle.

To address these challenges, NASA and its partners must prioritize research and development in several key areas. First, the selection and validation of a fuel architecture capable of achieving the required propellant reactor exit temperature without significant deterioration during the mission lifetime is crucial. This process should consider the availability of appropriate fuel feedstock production capabilities.

Second, the development of high-capacity tank systems capable of storing LH2 at 20 K with minimal boiloff for the duration of the mission is essential. This will likely require advancements in cryocooling technology and insulation materials.

Third, continued refinement of reactor design, including moderator materials, structural components, and reactivity control systems, is necessary to ensure the reliability and safety of NTP systems for long-duration missions.

Finally, extensive testing and validation of integrated NTP systems will be required to demonstrate their readiness for use in crewed Mars missions. This will likely involve a combination of ground-based testing and potentially in-space demonstrations of key technologies.

As NASA continues to pursue its goals of deep space exploration and crewed missions to Mars, the development of NTP technology will play a crucial role. By addressing the technical challenges and risks associated with NTP systems, NASA can pave the way for more efficient and capable propulsion systems that could significantly enhance our ability to explore the solar system.

5.5   Testing, Modeling, and Simulation

Nuclear Thermal Propulsion (NTP) represents a promising technology for future deep space exploration, particularly for crewed missions to Mars. However, the development and validation of NTP systems present unique challenges that require a comprehensive approach to testing, modeling, and simulation. Here, we explore the critical aspects of NTP system development, focusing on the necessary steps to ensure the technology's readiness for the ambitious goal of human Mars exploration.

5.5.1   The Importance of Testing in NTP Development:

Testing is a fundamental aspect of any propulsion system development, but it takes on even greater significance in the context of NTP due to the system's complexity and the high stakes involved in human spaceflight. The primary goals of testing in NTP development are:

    a. Verifying material characteristics: NTP systems operate under extreme conditions of temperature and radiation, necessitating thorough testing of all materials used in the system.

    b. Evaluating operational performance: Testing must demonstrate that the NTP system can achieve the required specific impulse and thrust levels consistently and reliably.

c. Assessing functionality: This includes evaluating the system's operability, controllability, and thermal management capabilities over its intended operational lifetime, including during transients and with appropriate safety margins.

d. Validating models: Test data is crucial for validating the computational models used in system design and performance prediction.

5.5.2 The Testing Progression for NTP Systems:

A traditional progression of tests for NTP systems includes:

a. Separate effects testing: This involves characterizing materials properties and behaviors under controlled conditions that simulate aspects of the NTP environment.

b. Component, subassembly, and assembly testing: Individual parts and subsystems of the NTP are tested to ensure they meet performance requirements.

c. Scaled system testing: Smaller-scale versions of the NTP system may be tested to gather data on system behavior while minimizing costs and risks associated with full-scale testing.

d. Integrated system tests: These involve testing the complete NTP system, including the reactor, engine, and propellant management subsystems.

For the reactor subsystem specifically, testing may include:

- Testing of components in environment chambers or test reactors
- Subcritical tests to evaluate neutronic behavior
- Critical tests at zero-power and full-power conditions

5.5.3 Unique Challenges in NTP Testing:

NTP systems present several unique challenges that complicate the testing process:

a. High power density and temperature: NTP reactors operate at much higher power densities and temperatures than standard nuclear reactors, requiring specialized testing facilities and procedures.

b. Open-cycle operation: Unlike terrestrial nuclear reactors, NTP systems expel the reactor coolant (hydrogen) through the nozzle, necessitating specialized containment and hydrogen management systems for ground testing.

c. Limited material database: The extreme operating conditions of NTP systems mean that there is a limited existing database of material properties under relevant conditions, increasing the importance of comprehensive testing.

d. Coupled multi-physics phenomena: The complex interactions between nuclear, thermal-hydraulic, and structural phenomena in NTP systems require sophisticated testing approaches to capture all relevant effects.

5.5.4   Proposed Testing Phases for NTP Development:

Based on the unique requirements and challenges of NTP systems, a three-phase testing approach has been proposed:

5.5.4.1   Phase 1: Zero-power critical (ZPC) and low-power tests

- These tests verify the neutronic characteristics of the reactor without generating significant power or radiation.
- ZPC tests are crucial for validating reactivity control systems and basic reactor physics models.
- While essential, these tests do not demonstrate the full thermomechanical behavior of the system or the effects of hydrogen propellant on reactivity and performance.

5.5.4.2   Phase 2: Reactor operational tests

- These tests involve operating the complete reactor subsystem at full power, with liquid hydrogen pumped through the system as in actual operation.
- Such tests demonstrate reactor operability, performance, reliability, and controllability through all phases of operation, including startup, steady-state operation, shutdown, and multiple restarts.
- These tests provide crucial data for validating computational models and identifying potential failure mechanisms.
- However, they require specialized facilities capable of managing high-temperature hydrogen effluent and meeting stringent safety and environmental standards.

5.5.4.3   Phase 3: Integrated system tests

- These tests incorporate the full NTP system, including the reactor, engine, and propellant management subsystems.
- They provide the most comprehensive validation of system performance and behavior.
- However, they also require the most extensive and expensive testing facilities, which do not currently exist and would require significant investment to construct.

5.5.5   The Role of Modeling and Simulation in NTP Development:

Modeling and simulation (M&S) play a crucial role in NTP development, complementing and guiding physical testing efforts. Key aspects of M&S in NTP development include:

a. Reactor physics modeling: Advanced codes like MCNP (Monte Carlo N-Particle Transport Code) are used to simulate neutron transport and nuclear reactions within the reactor core.

b. Thermal-hydraulic modeling: Computational Fluid Dynamics (CFD) tools are employed to model the complex flow and heat transfer phenomena within the NTP system.

c. Structural and mechanical analysis: Finite Element Analysis (FEA) is used to predict the mechanical behavior of NTP components under extreme thermal and mechanical loads.

d. System-level modeling: Integrated models that couple various physical phenomena are crucial for predicting overall NTP system performance and behavior.

e. Mission analysis: M&S tools are used to optimize NTP system design for specific mission profiles and to assess the impact of NTP performance on overall mission architecture.

The importance of M&S in NTP development has grown significantly since earlier NTP programs like Rover/NERVA, due to advances in computational capabilities and the desire to reduce the extent and cost of physical testing. However, the validity of M&S results depends critically on the quality of input data and the accuracy of the underlying physical models, highlighting the continued importance of experimental validation.

5.5.6   Challenges and Limitations in NTP Modeling and Simulation:

While M&S capabilities have advanced significantly, several challenges remain in accurately modeling NTP systems:

a. Multi-physics coupling: The complex interactions between nuclear, thermal-hydraulic, and structural phenomena in NTP systems are challenging to capture accurately in integrated models.

b. Transient behavior: Modeling the dynamic behavior of NTP systems, particularly during startup and shutdown transients, remains a significant challenge.

c. Material property data: The lack of comprehensive material property data under NTP-relevant conditions limits the accuracy of many models.

d. Scaling effects: Many phenomena in NTP systems are scale-dependent, making it difficult to extrapolate results from subscale tests or models to full-scale system behavior.

e. Validation data: The limited availability of experimental data from full-scale

NTP systems complicates the validation of computational models.

5.5.7   Strategies for NTP Development and Validation:

Given the challenges and limitations in both testing and modeling of NTP systems, a balanced approach leveraging both physical testing and advanced M&S is necessary. Key strategies include:

a. Leveraging historical data: Thoroughly reviewing and analyzing data from previous NTP programs like Rover/NERVA can provide valuable insights and help benchmark modern M&S tools.

b. Phased testing approach: Implementing a progressive testing program that moves from separate effects tests through component and subsystem tests to integrated system tests can help manage risks and costs while providing crucial validation data.

c. Advanced M&S development: Continued investment in improving multi-physics modeling capabilities and expanding material property databases is essential for enhancing the predictive power of NTP simulations.

d. Subscale testing: Where possible, conducting subscale tests can provide valuable data for model validation while minimizing costs and risks associated with full-scale testing.

e. Non-nuclear testing: Utilizing non-nuclear heated tests (e.g., with electrical heating) can provide valuable data on thermal-hydraulic and structural behavior without the complexities of nuclear testing.

f. In-space testing: Incorporating NTP testing into planned cargo missions can provide crucial full-scale, operational data to validate system performance before crewed missions.

5.5.8   The Path Forward: Balancing Ground Testing, Modeling, and Flight Testing:

The development of NTP systems for Mars missions requires a carefully balanced approach that leverages ground testing, advanced M&S, and in-space testing. Key considerations include:

a. Ground testing: Extensive ground testing, including integrated system tests at full scale and thrust, is crucial for fully characterizing NTP system behavior and identifying potential failure modes. However, the costs and regulatory challenges associated with such testing are significant.

b. Modeling and simulation: Advanced M&S capabilities are essential for optimizing system design, predicting performance, and reducing the extent of physical testing required. However, M&S results must be validated against experimental data to ensure their reliability.

c. Flight testing: While subscale in-space testing cannot address all risks associated with full-scale NTP systems, using cargo missions as a means of flight qualification for the NTP system intended for crewed missions can provide valuable operational data.

d. Timeframe considerations: Sufficient time must be allowed between initial flight tests (e.g., on cargo missions) and the first crewed mission to allow for analysis of flight data and any necessary design updates or additional validation.

Developing Nuclear Thermal Propulsion systems for Mars missions presents unique challenges that require a comprehensive approach to testing, modeling, and simulation. While advanced M&S capabilities offer the potential to reduce the extent of physical testing required, the complexity and scale-dependence of many phenomena in NTP systems necessitate a robust ground testing program, including integrated system tests at full scale and thrust.

The most effective path forward likely involves a combination of extensive investments in M&S, a comprehensive ground testing program, and the use of cargo missions for flight qualification for the NTP system intended for crewed missions. This approach can balance the need for thorough system characterization and risk mitigation with the practical constraints of cost, schedule, and regulatory considerations.

Ultimately, the success of NTP development for Mars missions will depend on the ability to effectively integrate insights from ground testing, advanced M&S, and in-space operations, ensuring that the technology is fully validated and reliable before its use in crewed missions. This integrated approach, while challenging, offers the best path to realizing the potential of NTP for enabling human exploration of Mars and beyond.

5.6   NTP Development and Demonstration for Mars Missions

The development of Nuclear Thermal Propulsion (NTP) systems for Mars missions represents an ambitious and complex undertaking that requires careful planning and execution. NASA's roadmap for NTP development outlines a series of key milestones and phases aimed at achieving the goal of launching a crewed mission to Mars in 2039, preceded by an initial cargo mission in 2033. We will explore the critical aspects of this roadmap, discussing the major phases, challenges, and timelines involved in bringing NTP technology to fruition for Mars exploration.

The NTP development program is structured around several key phases, each building upon the previous to create a comprehensive approach to system development and validation. These phases include:

1. Technology and Modeling & Simulation (M&S) Development
2. Ground Testing of Subsystems and Components
3. Facility Development and Integrated System Testing
4. Cargo Mission Development and Launch

5.   Crewed Mission Development and Launch

One of the most critical aspects of the roadmap is the parallel development of multiple technologies and capabilities. To meet the ambitious timeline for prototype demonstration, several activities must run concurrently. These include fuel architecture technology development, reactor core design, cryogenic fluid management, integrated propulsion system design, and engine component technology development and testing.

The first major milestone in the roadmap, set for the end of 2021, is the decision between using highly enriched uranium (HEU) or high-assay low-enriched uranium (HALEU) fuel. This decision is crucial as it sets the stage for subsequent fuel technology development efforts, including determining fuel chemistry (e.g., uranium nitride, uranium carbide-oxide, uranium dioxide) and maturing the fuel architecture technology (cercer, cermet, ceramic, etc.).

Following the fuel decision, the roadmap outlines a concentrated effort on fuel performance demonstration. This is a critical step that must be completed prior to the prototype final design review, as the chosen fuel architecture will significantly influence final design decisions, including the choice of moderator block configuration and reactor core materials.

In parallel with fuel development, the roadmap calls for technology development of reactor core structural and moderator materials to begin in 2022. This timing allows these efforts to support preliminary design activities, utilizing existing facilities within industry, NASA, and the Department of Energy (DOE) to test and validate nuclear and non-nuclear component performance and material characterization.

Another crucial technology milestone in the roadmap is the demonstration of long-term storage technologies for liquid hydrogen (LH2) propellant with near-zero boiloff. This capability is targeted for demonstration around 2025, addressing one of the major challenges in NTP system development for long-duration Mars missions.

The roadmap projects that a successful prototype demonstration could be completed between 2027 and 2029. This demonstration will be a critical milestone, characterizing engine performance considerations and the resulting reactor core operational and safety margins associated with hydrogen flow through the system. The success of this prototype demonstration is vital, as it sets the stage for the Mars flight system design, which must begin in 2029 or 2030 to maintain the timeline for a crewed mission in 2039.

An essential aspect of the roadmap is the recognition that new and upgraded facilities will be required to support the recommended testing regime. If not addressed early in the program, the development of these facilities could become a schedule-limiting factor. The roadmap allows for ground tests to continue into the cargo mission design phase, providing flexibility in the testing schedule.

The roadmap incorporates multiple cargo precursor missions as a key part of the

NTP development strategy. These missions, with the first planned for launch no later than 2033, serve a dual purpose. First, they deliver necessary supplies to Mars in preparation for the crewed mission. Second, and perhaps more critically for NTP development, they provide an opportunity for flight qualification of the integrated NTP engine system.

These cargo missions are envisioned to use a single NTP engine module and a lower total propellant load than required for the crewed mission. However, they would demonstrate the maximum propellant throughput for a single engine and include enough performance capabilities to validate adequate engine performance, lifetime, and reliability for the crewed mission. This approach allows for any emergent issues to be addressed before the 2039 crewed mission.

The roadmap highlights several major challenges that NASA must address to ensure the success of the NTP program. These include:

1. Developing an NTP system capable of heating its propellant to approximately 2700 K at the reactor exit for the duration of each burn.
2. Advancing technology for long-term storage of liquid hydrogen in space with minimal loss.
3. Developing adequate ground-based test facilities to support comprehensive system testing.
4. Achieving rapid system startup, bringing the NTP system to full operating temperature, preferably in 1 minute or less.

The NTP development and demonstration roadmap presents an aggressive but potentially achievable plan for developing this revolutionary propulsion technology for Mars exploration. It emphasizes the parallel development of critical technologies, early decision-making on key design choices, and a phased approach to testing and validation. The incorporation of cargo missions as part of the flight qualification strategy represents an innovative approach to risk mitigation.

However, the roadmap also underscores the significant challenges that must be overcome, particularly in areas such as fuel development, high-temperature materials, cryogenic propellant management, and testing infrastructure. Success will require sustained commitment, funding, and collaboration between NASA, DOE, and industry partners. If executed successfully, this roadmap could lead to the realization of NTP technology, potentially revolutionizing human exploration of Mars and beyond.

6 Nuclear Energy Fundamentals

6.1 Atomic and Molecular Weights

Understanding atomic and molecular weights is crucial in chemistry, physics, and various scientific disciplines where precise measurements of substances are required. These weights provide a way to compare different atoms and molecules on a relative scale.

6.1.1 Atomic Weight

The atomic weight (or relative atomic mass) of an element is a measure of the mass of one of its atoms relative to the mass of one atom of carbon-12(^{12}C), which is set at exactly 12. The atomic weight thus gives us a dimensionless number that provides a scale for comparing the mass of any atom to that of carbon-12.

6.1.1.1 Definition and Calculation

Mathematically, if $m(^{A}Z)$ is the mass of a neutral atom (where ^{A}Z represents the atomic number Z and the mass number A), and $m(^{12}C)$ is the mass of a neutral carbon-12 atom, then the atomic weight $M(^{A}Z)$ is defined as:

$$M(^{A}Z) = 1\frac{m(^{A}Z)}{m(^{12}C)} \qquad (6\text{-}1)$$

This formula scales the mass of any atom relative to that of ^{12}C. For example, if an atom has a mass precisely twice that of ^{12}C, its atomic weight would be $12 \times 2 = 24$.

Example 6-1. Hydrogen, the lightest element, has an atomic weight close to 1.008. This number means a hydrogen atom is about 1.008 times as heavy as $1/12$ the mass of a carbon- 12 atom.

6.1.2 Isotopic Contributions to Atomic Weight

Many elements occur naturally as a mixture of different isotopes, each with its own atomic weight.

The average atomic weight of an element found on the periodic table accounts for the relative abundance of each of these isotopes. This average is calculated as follows:

If γ_i represents the isotopic abundance (in atom percent) of the ith isotope with atomic weight M_i , then the average atomic weight M of the element is given by:

$$M = \frac{\sum_i \gamma_i M_i}{100} \tag{6-2}$$

This equation calculates the weighted mean of the atomic weights of the isotopes, factoring in their respective abundances.

Example 6-2. Chlorine has two stable isotopes: ^{35}Cl (with an abundance of about 75.77%) and ^{37}Cl (with an abundance of about 24.23%). The average atomic weight of chlorine would be calculated using their respective atomic weights and abundances.

6.1.3   Molecular Weight

The molecular weight of a molecule is the total mass relative to the mass of a neutral ^{12}C atom. When a molecule is composed of multiple atoms, its molecular weight is the sum of the atomic weights of the constituent atoms. If a molecule consists of n atoms of type A and m atoms of type B (with atomic weights M_A and M_B respectively), the molecular weight is:

$$M_{\text{molecule}} = n \times M_A + m \times M_B \tag{6-3}$$

Example 6-3. The molecular weight of water, which consists of two hydrogen atoms and one oxygen atom, would be:

$$M_{H_2O} = 2 \times 1.008 + 15.999 = 18.015$$

This means a water molecule is about 18.015 times as heavy as $1/12$ the mass of a carbon- 12 atom.

The concept of atomic and molecular weights is central to the study of chemistry and physics, as these weights provide a standardized way of discussing the masses of different atoms and molecules relative to carbon-12 (^{12}C), the standard reference.

6.1.4   Atomic Weight

Atomic weight (or relative atomic mass) of an element is the ratio of the average mass of atoms of the element to $1/12$ of the mass of an atom of carbon-12. This is a dimensionless number that allows us to compare the relative masses of different atoms on a scale where the atomic weight of ^{12}C is exactly 12.

Example 6-4. Calculating the Atomic Weight of Oxygen

Consider naturally occurring oxygen, which consists of three isotopes: ^{16}O, ^{17}O, and ^{18}O. Their abundances and atomic weights are provided as follows:

^{16}O : Abundance = 99.759%, Atomic weight = 15.99492
^{17}O : Abundance = 0.037%, Atomic weight = 16.99913
^{18}O : Abundance = 0.204%, Atomic weight = 17.99916

Using the formula for average atomic weight,

$$M = \frac{1}{100}\left(\gamma(^{16}O) \cdot M(^{16}O) + \gamma(^{17}O) \cdot M(^{17}O) + \gamma(^{18}O) \cdot M(^{18}O)\right)$$

we plug in the values:

$$M(O) = \frac{1}{100}(99.759 \times 15.99492 + 0.037 \times 16.99913 + 0.204 \times 17.99916)$$

$$M(O) = \frac{1}{100}(1596.9124828 + 0.62884781 + 3.67190704)$$

$$M(O) = \frac{1}{100} \times 1601.21323765 = 16.0121323765 \approx 15.99938$$

Hence, the atomic weight of naturally occurring oxygen is approximately 15.99938.

6.1.5   Molecular Weight
        Molecular weight is the sum of the atomic weights of the atoms in a molecule. It provides the mass of a molecule relative to $1/12$ of the mass of carbon- 12.

Example 6-5. Calculating the Molecular Weight of Water ($\mathbf{H_2O}$)

Water consists of two hydrogen atoms and one oxygen atom. The atomic weight of hydrogen is approximately 1.008, and from our previous calculation, the atomic weight of oxygen is approximately 15.99938.

$$M(\text{H}_2\text{O}) = 2 \times 1.008 + 15.99938 = 2.016 + 15.99938 = 18.01538$$

Thus, the molecular weight of water is approximately 18.01538.

6.1.6   Avogadro's Number and Moles

The mole is a unit that links atomic/molecular scale to macroscopic measurements. One mole is defined as the number of atoms in exactly 12 grams of pure carbon-12. This number is known as Avogadro's number, N_A, and is approximately 6.022×10^{23} entities per mole.

6.1.7   Using Avogadro's Number

Example 6-6. Mass of a Single Carbon Atom

Given that one mole of ^{12}C has a mass of 12 grams and contains N_A atoms, the mass of a single carbon atom is:

$$m(^{12}C) = \frac{12 \text{ g}}{6.022 \times 10^{23}} \approx 1.99265 \times 10^{-23} \text{ g}$$

$$m(^{12}C) = 1.99265 \times 10^{-26} \text{ kg}$$

6.1.8   Atomic and Nuclear Radii

Understanding the sizes of atoms and their nuclei is crucial in fields such as physics, materials science, and chemistry. This section explains how these sizes are defined and provides numerical examples to illustrate these concepts.

6.1.8.1   Atomic Radii

Definition: The size of an atom can be challenging to pinpoint due to the diffuse nature of the electron cloud surrounding the nucleus. Atoms don't have a sharp physical boundary as the density of the electron cloud decreases with distance from the nucleus. The atomic radius is typically considered to be the average distance from the nucleus to the outer boundary of the surrounding cloud of electrons.

Measurement: The atomic radius is approximately 2×10^{-10} meters (or 200 picometers) for most atoms. This measurement gives a good approximation of atomic size and helps in understanding many physical and chemical properties of elements.

Example: Consider the hydrogen atom, which has one electron. In its ground state, the electron occupies the first electron shell (1s orbital), which the Bohr model can approximate. The Bohr radius for hydrogen (the radius of the lowest energy level of the hydrogen atom, according to the Bohr model) is about 0.529×10^{-10} meters, or 52.9

picometers.

6.1.8.2   Nuclear Radii

Definition: Similar to the atomic radius, the nuclear radius lacks a precisely defined outer boundary. However, approximating nuclear radii is considerably more straightforward compared to atomic radii, as the nucleus is significantly less diffuse than the electron cloud surrounding an atom.

Formula: The radius of a nucleus can be estimated using the formula:

$$R = 1.25\text{fm} \times A^{1/3} \tag{6-4}$$

where R is the nuclear radius in femtometers (fm), and A is the atomic mass number (the total number of neutrons and protons in the nucleus).

Example 6-7. For a carbon-12 nucleus (^{12}C), which has an atomic mass number $A = 12$:

$$R = 1.25\text{fm} \times 12^{1/3} \approx 1.25\text{fm} \times 2.289 = 2.86\text{fm}$$

Example 6-8. For a uranium-238 nucleus (^{238}U), which has an atomic mass number $A = 238$:

$$R = 1.25\text{fm} \times 238^{1/3} \approx 1.25\text{fm} \times 6.203 = 7.75\text{fm}$$

6.1.1 Mass and Energy

One of the most revolutionary aspects of Albert Einstein's theory of relativity is the equivalence of mass and energy. This principle, encapsulated by the equation $E = mc^2$, asserts that mass can be converted into energy and vice versa, fundamentally linking these two concepts in ways that are crucial to both theoretical physics and practical applications, particularly in nuclear physics.

6.1.1.1   Einstein's Energy-Mass Equivalence

Formula:

$$E_{\text{rest}} = m_0 c^2 \tag{6-5}$$

Here, E_{rest} is the rest-mass energy, m_0 is the rest mass of the object, and c is the speed of light in a vacuum ($\approx 2.998 \times 10^8$ meters / second).

Physical Meaning: This equation tells us that even a small amount of mass contains a tremendous amount of energy, caused by the square of the speed of light in the formula.

Example 6-9. Annihilation of 1 gram of Matter

If 1 gram of matter is completely annihilated, the energy released can be calculated as follows:

$$E = 1\,\text{g} \times \left(2.9979 \times \frac{10^{10}\ \text{cm}}{\text{s}}\right)^2$$

$$E = 1\,\text{g} \times (8.9874 \times 10^{20}\,\text{erg})$$

$$E = 8.9874 \times 10^{13}\ \text{joules}$$

This amount of energy, when converted to more familiar units, equals about 25 million kilowatt-hours, illustrating the vast energy contained in even small amounts of mass.

6.1.1.2   Units of Energy in Nuclear Physics

Electron Volt (eV): An electron volt is the energy gained by an electron when it accelerates through an electric potential difference of one volt. It provides a convenient unit for energy at the atomic and subatomic levels.

$$1\text{eV} = 1.60219 \times 10^{-19}\ \text{joules} \tag{6-6}$$

Mega Electron Volt (MeV) and Kilo Electron Volt (keV):

$$1\text{MeV} = 10^6\text{eV}$$

$$1\text{keV} = 10^3\text{eV}$$

Example 6-10. Rest-Mass Energy of the Electron

The electron has a mass of 9.109×10^{-31} kg, or approximately 9.109×10^{-28} g. To find the rest-mass energy in MeV :

$$E = m_e c^2$$

$$E = 9.109 \times 10^{-28} \text{ g} \times (2.9979 \times 10^{10} \text{ cm/s})^2$$

$$E = 8.187 \times 10^{-14} \text{ joules}$$

Converting to MeV:

$$8.187 \times 10^{-14} \text{ joules} \div 1.60219 \times 10^{-13} \text{MeV/joule} = 0.511 \text{MeV}$$

Example 6-11. Energy Equivalent of the Atomic Mass Unit

Given that 1 atomic mass unit (amu) is equivalent to the mass of one-twelfth of a carbon-12 atom, or about 1.6606×10^{-24} g, the energy equivalent in MeV is:

$$E_{amu} = 1 \text{amu} \times c^2$$

Using the rest-mass energy of the electron from the previous example, we can calculate:

$$E_{amu} = \left(\frac{1.6606 \times 10^{-24} \text{ g}}{9.109 \times 10^{-28} \text{ g/electron}} \right) \times 0.511 \text{MeV/electron} = 931.5 \text{MeV}$$

The relationship between mass and energy has practical implications in nuclear power generation, medical imaging techniques, and understanding cosmic phenomena. These examples illustrate the enormous amount of energy bound within the mass of subatomic particles, highlighting the power and utility of Einstein's formula.

6.1.2   The Chart of Nuclides

The Chart of Nuclides, often referred to as the Segre chart, is an essential tool in nuclear physics that provides a comprehensive overview of all known isotopes of every element. It is a visual representation that plots nuclides according to their proton (Z) and neutron (N) numbers, offering insights into their stability, modes of decay, and other nuclear properties.

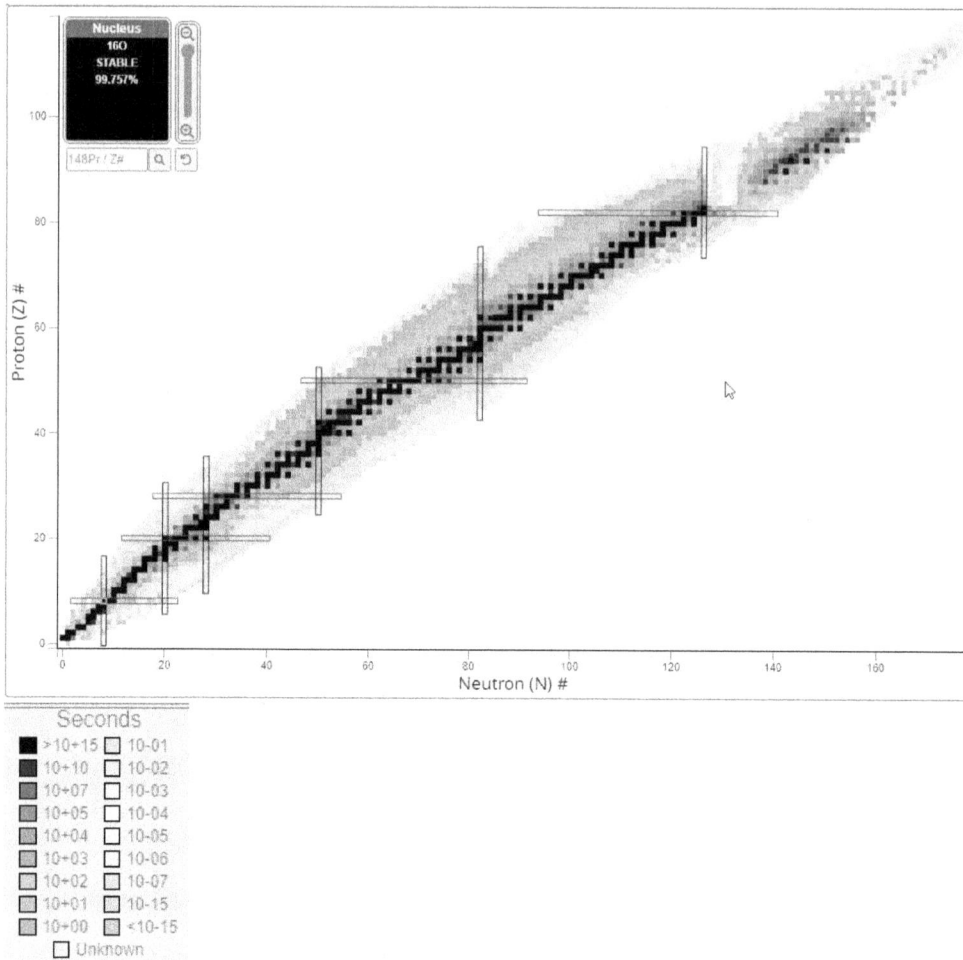

Figure 6-1. Chart of Nuclides (https://www.nndc.bnl.gov/nudat3/)**Structure of the Chart of Nuclides**

- Vertical Axis: Represents the number of protons (atomic number, Z), which defines the chemical element of the nuclide.
- Horizontal Axis: Represents the number of neutrons (neutron number, N).
- Stable Nuclides: Typically depicted in black or a dark color, these nuclides have a balanced ratio of protons to neutrons that allows them to exist indefinitely under normal conditions.
- Radioactive Nuclides: Shown in various colors based on their primary mode of decay. These nuclides do not have a stable proton-to-neutron ratio and, hence, are subject to radioactive decay to achieve stability.

6.1.2.2   Stability and Neutron-Proton Ratio

The chart reveals a "line of stability" where the nuclides are stable. This line indicates the ideal ratio of neutrons to protons, which generally increases with atomic number:

- For light elements ($Z < 20$), the stable ratio of neutrons to protons tends to approach 1:1.
- For heavier elements, the ratio gradually increases. This is because extra neutrons are necessary to mitigate the repulsive forces between the increasing number of protons within the nucleus. Beyond calcium ($Z > 20$), stable nuclides generally require more neutrons than protons.

Examples of Naturally Occurring Radioactive Nuclides

Isotopes of Uranium:

- Uranium-238: The most common isotope of uranium found in nature with 92 protons and 146 neutrons. It undergoes alpha decay, emitting an alpha particle ($He - 4$ nucleus) to form Thorium-234:

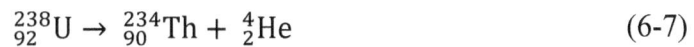

$$^{238}_{92}U \rightarrow ^{234}_{90}Th + ^{4}_{2}He \tag{6-7}$$

- Uranium-235: Another naturally occurring isotope, which is crucial for nuclear reactors and weapons due to its ability to sustain a nuclear chain reaction. It also decays via alpha emission:

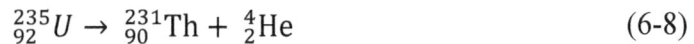

$$^{235}_{92}U \rightarrow ^{231}_{90}Th + ^{4}_{2}He \tag{6-8}$$

Isotopes of Potassium:
- Potassium-40: A notable example of a beta emitter, Potassium-40 has a significant role in geological dating. It undergoes beta-minus decay to form Calcium-40, or electron capture to form Argon-40:

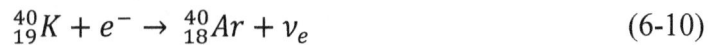

$$^{40}_{19}K \rightarrow ^{40}_{20}Ca + e^- + \bar{\nu}_e \tag{6-9}$$

$$^{40}_{19}K + e^- \rightarrow ^{40}_{18}Ar + \nu_e \tag{6-10}$$

Isotopes of Carbon:
- Carbon-14: Widely known for its use in radiocarbon dating of archaeological artifacts, Carbon-14 decays by beta-minus emission to Nitrogen-14:

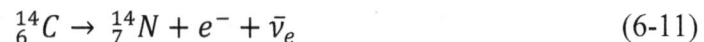

$$^{14}_{6}C \rightarrow ^{14}_{7}N + e^- + \bar{\nu}_e \tag{6-11}$$

6.1.2.3   Significance of the Chart of Nuclides

The Chart of Nuclides is invaluable for researchers and professionals in fields ranging from nuclear engineering and medicine to environmental science and archaeology. It aids in understanding:

- The stability of various isotopes and the factors influencing it.
- The types of radioactive decay and the resulting daughter nuclides.
- The selection of isotopes for specific applications, such as medical imaging or radiometric dating.

6.1.3   Radioactive Decay Types

Radioactive decay is a nuclear process where an unstable nucleus loses energy by emitting radiation, resulting in the transformation of its atomic structure. This phenomenon is pivotal for understanding elemental behavior, energy production, medical applications, and cosmic processes.

6.1.3.1   Alpha Decay (α-decay)

Alpha decay occurs predominantly in heavy nuclei (e.g., uranium, radium), where the nucleus emits an alpha particle (a helium- 4 nucleus, composed of two protons and two neutrons). This process reduces the atomic number by two and the mass number by four, leading to the formation of a new element.

Nuclear Reaction Example:

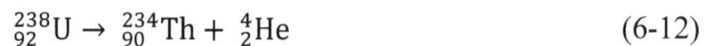

$$\ce{^{238}_{92}U} \rightarrow \ce{^{234}_{90}Th} + \ce{^{4}_{2}He} \tag{6-12}$$

Uranium-238 to Thorium-234: Uranium-238 decays to thorium-234, emitting an alpha particle. This is a common mode of decay for elements with atomic numbers higher than lead (Pb).

6.1.3.2   Beta Decay

Beta decay is categorized into two types based on the emission:

- Beta-minus (β^-)decay: A neutron is converted into a proton, an electron, and an antineutrino. This increases the atomic number by one. Involves the transformation of a down quark into an up quark within a neutron, turning it into a proton.
- Beta-plus (β^+)decay or Positron emission: A proton is converted into a neutron, a positron, and a neutrino, decreasing the atomic number by one. A proton's up quark is transformed into a down quark (converting the proton into a neutron).

Nuclear Reaction Examples:

- β^-Decay:

$$^{19}_{8}\text{O} \rightarrow \,^{19}_{9}\text{F} + e^- + \bar{\nu}_e \tag{6-13}$$

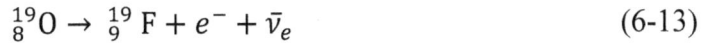

Oxygen-19 to Fluorine-19: The neutron-rich oxygen-19 emits an electron to become fluorine19.

- β^+Decay:

$$^{15}_{8}\text{O} \rightarrow \,^{15}_{7}\text{N} + e^+ + \nu_e \tag{6-14}$$

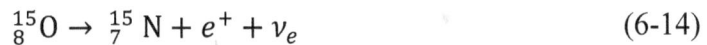

Oxygen-15 to Nitrogen-15: Oxygen-15 emits a positron, transforming into nitrogen-15.

6.1.3.3 Electron Capture

In electron capture, an inner orbital electron is absorbed by the nucleus. In this process a proton is converted into a neutron (and a neutrino is emitted). This process often competes with positron emission and is significant in proton-rich nuclides.

Example 6-12. Iron-55 to Manganese-55: Electron capture in iron-55 leads to the formation of manganese-55.

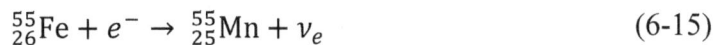

$$^{55}_{26}\text{Fe} + e^- \rightarrow \,^{55}_{25}\text{Mn} + \nu_e \tag{6-15}$$

6.1.3.4 Gamma Decay (γ-decay)

One way that an excited nucleus can release its excess energy is by emitting a gamma ray. In gamma decay the atomic number or mass number of the nucleus do not change. This process often follows other types of decay as the daughter nucleus transitions from an excited to a ground state.

Example 6-13. Cobalt-60 Excited State: An excited cobalt-60 nucleus emits a gamma ray to return to its ground state.

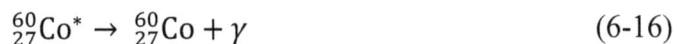

$$^{60}_{27}\text{Co}^* \rightarrow \,^{60}_{27}\text{Co} + \gamma \tag{6-16}$$

6.1.3.5   Decay Chains and Isomeric Transitions

Decay chains occur when successive decays lead from one radioactive nuclide to another until a stable nuclide is reached. Isomeric transitions involve nuclei in excited states (isomers) that emit gamma rays to reach their ground state or undergo further decay.

Example 6-14. Decay Chain for Neptunium-239 (Np-239)

1. Formation of Neptunium-239:

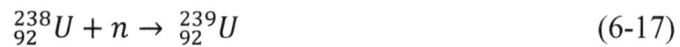

$$^{238}_{92}U + n \rightarrow {}^{239}_{92}U \tag{6-17}$$

Uranium-238 captures a neutron and becomes Uranium-239.

2. Beta-minus Decay of Uranium-239:

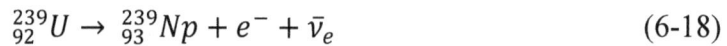

$$^{239}_{92}U \rightarrow {}^{239}_{93}Np + e^- + \bar{\nu}_e \tag{6-18}$$

Uranium-239 undergoes beta-minus decay to Neptunium-239. This process emits an electron and an antineutrino.

Beta-minus Decay of Neptunium-239 to Plutonium-239:

$$^{239}_{93}Np \rightarrow {}^{239}_{94}\text{Pu} + e^- + \bar{\nu}_e \tag{6-19}$$

Neptunium-239 further undergoes beta-minus decay to Plutonium-239, emitting another electron and antineutrino. The half-life of $Np - 239$ in this transition is about 2.356 days.

Alpha Decay of Plutonium-239:

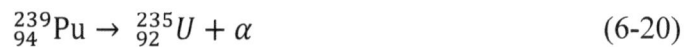

$$^{239}_{94}\text{Pu} \rightarrow {}^{235}_{92}U + \alpha \tag{6-20}$$

Plutonium-239 can undergo alpha decay to form Uranium-235. However, this is a very long-lived decay (the half-life of Plutonium-239 is about 24,110 years), and in a nuclear reactor context, Plutonium-239 is more likely to absorb a neutron or fission.

5. Further Decay of Uranium-235 (if formed):

Uranium-235 is a relatively stable isotope with a half-life of about 703.8 million years and is one of the main fissile isotopes used in nuclear reactors and weapons. It can absorb a neutron leading to fission or eventually undergo alpha decay:

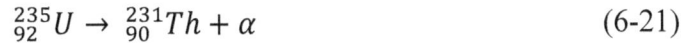

$$^{235}_{92}U \rightarrow ^{231}_{90}Th + \alpha \tag{6-21}$$

Thorium-231 then undergoes beta-minus decay to Protactinium-231, which itself has a long half-life and decays to actinium-227 through alpha decay.

This decay chain of Np-239, leading to the formation of Plutonium-239, is significant in the context of nuclear reactors and nuclear weapons. Plutonium-239 is a critical material for both energy production and nuclear weaponry, primarily due to its fissile nature.

The ability of Np-239 to rapidly decay into Plutonium-239, which has capabilities for sustaining a nuclear chain reaction, makes the Np-239 production and subsequent decay a point of interest for both energy generation and nuclear proliferation concerns.

Thus, the decay chain from Np-239 outlines not only the transformations from one isotope to another but also highlights the interconnections between nuclear physics and its applications in energy and defense.

6.1.4   Radioactivity Calculations

Radioactive decay is a spontaneous process by where an unstable nucleus discharges energy by emitting particles (alpha, beta, or conversion electrons) or electromagnetic radiation (gamma rays). Understanding the quantitative aspects of these processes is essential for applications ranging from nuclear power generation to medical diagnostics and treatment.

6.1.4.1   Fundamental Law of Radioactive Decay

The behavior of a radioactive decay process is governed by a simple yet profound law: the decay rate of a nucleus is proportional to the number of undecayed nuclei present at any given time. This decay rate is expressed mathematically by the differential equation:

$$-\frac{dn(t)}{dt} = \lambda n(t) \tag{6-22}$$

where:

- $n(t)$ is the number of undecayed nuclei at time t,

- λ is the decay constant, unique to each radioactive species.

This equation implies that the decay process is exponential in nature, leading to the exponential decay law.

6.1.4.2   Exponential Decay Law

Integrating the differential equation of decay gives the number of nuclei remaining undecayed at any time t :

$$n(t) = n_0 e^{-\lambda t} \tag{6-23}$$

where n_0 is the initial number of nuclei at $t = 0$.

6.1.4.3   Activity of a Radioactive Sample

The activity of a sample, denoted as $\alpha(t)$, is the rate of decay at time t and is given by:

$$\alpha(t) = \lambda n(t) \tag{6-24}$$

$$\alpha(t) = \lambda n_0 e^{-\lambda t} = \alpha_0 e^{-\lambda t} \tag{6-25}$$

where α_0 is the initial activity at $t = 0$. Activity measures the number of disintegrations per second, with units of becquerels (Bq), where 1 Bq = 1 disintegration per second.

6.1.4.4   Half-Life and Mean Life

6.1.4.4.1   Half-Life

The half-life $\left(T_{1/2}\right)$ of a radioactive substance is the time required for the activity of the substance to decrease to half of its initial value. From the exponential decay formula, the half-life can be determined:

$$\alpha\left(T_{1/2}\right) = \frac{\alpha_0}{2} = \alpha_0 e^{-\lambda T_{1/2}} \tag{6-26}$$

Solving for $T_{1/2}$:

$$\frac{1}{2} = e^{-\lambda T_{1/2}} \tag{6-27}$$

$$T_{1/2} = \frac{\ln 2}{\lambda} = \frac{0.693}{\lambda} \tag{6-28}$$

6.1.4.4.2   Mean Life

The mean life (t) is the average lifetime of a radioactive nucleus and is defined as the reciprocal of the decay constant:

$$t = \frac{1}{\lambda} \qquad (6\text{-}29)$$

This implies that in one mean life, the activity of the radioactive substance falls to $1/e$ of its initial value.

Example 6-15. Decay Calculation

Given: A sample starts with an activity of 1000 Bq and has a half-life of 10 years.
Find: The activity after 20 years.

Solution:

$$T_{1/2} = 10 \text{ years}$$

$$\lambda = \frac{0.693}{T_{1/2}} = \frac{0.693}{10} = 0.0693 \text{ year}^{-1}$$

$$\alpha(20) = 1000e^{-0.0693 \times 20}$$

$$\alpha(20) = 1000e^{-1.386} = 1000 \times 0.250 = 250 \text{ Bq}$$

6.1.5   Production and Decay Equations

In the study of radioactivity, it is often necessary to analyze situations where radioactive nuclides are not only decaying but are also being continuously produced, such as in nuclear reactors or when irradiating materials in accelerators. This chapter will derive the equations that describe the balance between the production and decay of a radioactive nuclide and extend these ideas to decay chains.

6.1.5.1   Production and Decay of a Single Nuclide

Assumptions and Setup

Consider a scenario where a radioactive nuclide is produced at a constant rate R (atoms per second) and decays with a decay constant λ. The rate of change of the number of radioactive atoms $n(t)$ in the system can be described by the differential equation:

$$\frac{dn}{dt} = R - \lambda n \tag{6-30}$$

Here:
- R is the rate of production of the nuclide,
- λn is the rate at which the nuclide decays (with λ being the decay constant).

Derivation of the Equation

The differential equation (2-52) can be solved to find $n(t)$, the number of nuclides at any time t. Rewriting the equation, we have:

$$\frac{dn}{dt} + \lambda n = R \tag{6-31}$$

This is a first-order linear differential equation and can be solved using an integrating factor. The integrating factor, $\mu(t)$, is given by:

$$\mu(t) = e^{\int \lambda dt} = e^{\lambda t} \tag{6-32}$$

Multiplying through by the integrating factor:

$$e^{\lambda t} \frac{dn}{dt} + \lambda e^{\lambda t} n = R e^{\lambda t} \tag{6-33}$$

$$\frac{d}{dt}\left(e^{\lambda t} n\right) = R e^{\lambda t} \tag{6-34}$$

Integrating both sides with respect to t :

$$e^{\lambda t} n = \int R e^{\lambda t} dt + C \tag{6-35}$$

$$n = \frac{R}{\lambda} + Ce^{-\lambda t} \qquad (6\text{-}36)$$

Where C is the integration constant, which can be determined using initial conditions. If $n(0) = n_0$ (initial number of nuclides at $t = 0$):

$$n_0 = \frac{R}{\lambda} + C \qquad (6\text{-}37)$$

$$C = n_0 - \frac{R}{\lambda} \qquad (6\text{-}38)$$

Thus, the solution for $n(t)$ becomes:

$$n(t) = n_0 e^{-\lambda t} + \frac{R}{\lambda}\left(1 - e^{-\lambda t}\right) \qquad (6\text{-}39)$$

Activity of the Nuclide

The activity $\alpha(t)$ of the nuclide, defined as the rate of disintegration, is given by:

$$\alpha(t) = \lambda n(t) \qquad (6\text{-}40)$$

$$\alpha(t) = \lambda\left(n_0 e^{-\lambda t} + \frac{R}{\lambda}\left(1 - e^{-\lambda t}\right)\right) \qquad (6\text{-}41)$$

$$\alpha(t) = \alpha_0 e^{-\lambda t} + R\left(1 - e^{-\lambda t}\right) \qquad (6\text{-}42)$$

where $\alpha_0 = \lambda n_0$ is the initial activity.

6.1.6   Production and Decay in Radioactive Decay Chains

The above framework can be extended to decay chains where one nuclide decays into another, which in turn may decay further. For a simple two-step decay chain, where a parent nuclide A decays to a daughter nuclide B, which then decays to C, and C decays to D, the equations become coupled.

To derive the solution for the activity as a function of time for a decay chain consisting of four isotopes $A \rightarrow B \rightarrow C \rightarrow D$, we must consider the radioactive decay of each nuclide and how each feeds the next in the chain. Each nuclide decays independently, and the production rate of each succeeding nuclide depends on the decay of its predecessor.

Setting Up the Differential Equations

We denote:

- $\lambda_A, \lambda_B, \lambda_C, \lambda_D$ as the decay constants for isotopes $A, B, C,$ and D respectively.
- $N_A(t), N_B(t), N_C(t), N_D(t)$ as the number of atoms of $A, B, C,$ and D at time t.

The decay process can be modeled by the following set of linear differential equations:

1. $\dfrac{dN_A}{dt} = -\lambda_A N_A$

2. $\dfrac{dN_B}{dt} = \lambda_A N_A - \lambda_B N_B$

3. $\dfrac{dN_C}{dt} = \lambda_B N_B - \lambda_C N_C$

4. $\dfrac{dN_D}{dt} = \lambda_C N_C - \lambda_D N_D$

Solving the Differential Equations

Solving for $N_A(t)$

For nuclide A, the solution to the differential equation is straightforward as it involves only a single decay mode:

$$N_A(t) = N_A(0)e^{-\lambda_A t} \tag{6-43}$$

where $N_A(0)$ is the initial number of atoms of A.

Solving for $N_B(t)$

To solve for $N_B(t)$, we need to integrate the source term from the decay of A :

$$\frac{dN_B}{dt} + \lambda_B N_B = \lambda_A N_A(t) \tag{6-44}$$

Substituting $N_A(t)$:

$$\frac{dN_B}{dt} + \lambda_B N_B = \lambda_A N_A(0) e^{-\lambda_A t} \qquad (6\text{-}45)$$

Using an integrating factor, $e^{\lambda_B t}$, we have:

$$\frac{d}{dt}\left(N_B e^{\lambda_B t}\right) = \lambda_A N_A(0) e^{(\lambda_B - \lambda_A)t} \qquad (6\text{-}46)$$

Integrating both sides:

$$N_B e^{\lambda_B t} = \frac{\lambda_A N_A(0)}{\lambda_B - \lambda_A} e^{(\lambda_B - \lambda_A)t} + C \qquad (6\text{-}47)$$

$$N_B(t) = \frac{\lambda_A N_A(0)}{\lambda_B - \lambda_A} e^{-\lambda_A t} + C e^{-\lambda_B t} \qquad (6\text{-}48)$$

Assuming $N_B(0) = 0$ (no B at $t = 0$):

$$N_B(t) = \frac{\lambda_A N_A(0)}{\lambda_B - \lambda_A} \left(e^{-\lambda_A t} - e^{-\lambda_B t}\right) \qquad (6\text{-}49)$$

Deriving $N_C(t)$

For C, using $N_B(t)$ as the source:

$$\frac{dN_C}{dt} = \lambda_B N_B - \lambda_C N_C \qquad (6\text{-}50)$$

$$\frac{dN_C}{dt} + \lambda_C N_C = \lambda_B N_B(t) \qquad (6\text{-}51)$$

$$N_B(t) = \frac{\lambda_A N_A(0)}{\lambda_B - \lambda_A} \left(e^{-\lambda_A t} - e^{-\lambda_B t}\right) \qquad (6\text{-}52)$$

Applying an integrating factor, $e^{\lambda_C t}$:

$$\frac{d}{dt}\left(N_C e^{\lambda_C t}\right) = \lambda_B \frac{\lambda_A N_A(0)}{\lambda_B - \lambda_A} \left(e^{(\lambda_C - \lambda_A)t} - e^{(\lambda_C - \lambda_B)t}\right) \qquad (6\text{-}53)$$

Integrating both sides:

$$N_C e^{\lambda_C t} = \frac{\lambda_B \lambda_A N_A(0)}{\lambda_B - \lambda_A}\left(\frac{e^{(\lambda_C - \lambda_A)t}}{\lambda_C - \lambda_A} - \frac{e^{(\lambda_C - \lambda_B)t}}{\lambda_C - \lambda_B}\right) + C \qquad (6\text{-}54)$$

Assuming $N_C(0) = 0$:

$$N_C(t) = \frac{\lambda_B \lambda_A N_A(0)}{\lambda_B - \lambda_A}\left(\frac{e^{-\lambda_A t}}{\lambda_C - \lambda_A} - \frac{e^{-\lambda_B t}}{\lambda_C - \lambda_B}\right) - \frac{e^{-\lambda_C t}}{\lambda_C - \lambda_A} + \frac{e^{-\lambda_C t}}{\lambda_C - \lambda_B} \qquad (6\text{-}55)$$

Deriving $N_D(t)$

To derive the explicit solution for $N_D(t)$ from the decay chain $A \to B \to C \to D$, we'll integrate the formula for $N_C(t)$ and apply it to the differential equation for $N_D(t)$. We already established that:

$$N_C(t) = \frac{\lambda_B \lambda_A N_A(0)}{\lambda_B - \lambda_A}\left(\frac{e^{-\lambda_A t}}{\lambda_C - \lambda_A} - \frac{e^{-\lambda_B t}}{\lambda_C - \lambda_B}\right) \qquad (6\text{-}56)$$

Step 1: Differential Equation for $N_D(t)$

The differential equation for $N_D(t)$ given that D is the product of C decaying, is:

$$\frac{dN_D}{dt} = \lambda_C N_C - \lambda_D N_D \qquad (6\text{-}57)$$

Step 2: Integrating Factor

Apply an integrating factor $e^{\lambda_D t}$ to both sides:

$$e^{\lambda_D t}\frac{dN_D}{dt} + \lambda_D e^{\lambda_D t} N_D = \lambda_C e^{\lambda_D t} N_C(t) \qquad (6\text{-}58)$$

$$\frac{d}{dt}\left(e^{\lambda_D t} N_D\right) = \lambda_C e^{\lambda_D t} N_C(t) \qquad (6\text{-}59)$$

Step 3: Integrate Both Sides

$$e^{\lambda_D t} N_D = \int \lambda_C e^{\lambda_D t} N_C(t)dt + C \qquad (6\text{-}60)$$

Substituting $N_C(t)$:

$$e^{\lambda_D t} N_D = \int \lambda_C e^{\lambda_D t} \left[\frac{\lambda_B \lambda_A N_A(0)}{\lambda_B - \lambda_A} \left(\frac{e^{-\lambda_A t}}{\lambda_C - \lambda_A} - \frac{e^{-\lambda_B t}}{\lambda_C - \lambda_B} \right) \right] dt + C \qquad (6\text{-}61)$$

Step 4: Integration

Since C should be zero (assuming $N_D(0) = 0$), we have:

$$e^{\lambda_D t} N_D = \lambda_C \frac{\lambda_B \lambda_A N_A(0)}{\lambda_B - \lambda_A} \left[\int e^{(\lambda_D - \lambda_A)t} \frac{dt}{\lambda_C - \lambda_A} \right. \qquad (6\text{-}62)$$
$$\left. - \int e^{(\lambda_D - \lambda_B)t} \frac{dt}{\lambda_C - \lambda_B} \right]$$

These integrals result in:

$$e^{\lambda_D t} N_D = \lambda_C \frac{\lambda_B \lambda_A N_A(0)}{\lambda_B - \lambda_A} \left[\frac{e^{(\lambda_D - \lambda_A)t}}{(\lambda_D - \lambda_A)(\lambda_C - \lambda_A)} - \frac{e^{(\lambda_D - \lambda_B)t}}{(\lambda_D - \lambda_B)(\lambda_C - \lambda_B)} \right] \qquad (6\text{-}63)$$

Step 5: Simplifying

$$N_D(t) = \lambda_C \frac{\lambda_B \lambda_A N_A(0)}{\lambda_B - \lambda_A} \left[\frac{e^{-\lambda_A t}}{(\lambda_D - \lambda_A)(\lambda_C - \lambda_A)} - \frac{e^{-\lambda_B t}}{(\lambda_D - \lambda_B)(\lambda_C - \lambda_B)} \right] \qquad (6\text{-}64)$$

This solution outlines $N_D(t)$ explicitly and shows the dependency of $N_D(t)$ on the decay constants and the exponential decays of its preceding nuclides.

Activity Equations

The activity of each nuclide $\alpha_X(t)$ is given by:

$$\alpha_X(t) = \lambda_X N_X(t) \qquad (6\text{-}65)$$

where X can be A, B, C, or D.

Example 6-16. Determine the rate of production of ^{51}V as a function of time when ^{50}Ti is irradiated with neutrons. Assume ^{51}V is produced through the neutron activation of ^{50}Ti and subsequent beta decay of ^{51}Ti. Given:

- Half-life of ^{51}Ti is 5.76 minutes.
- Neutron capture cross section for ^{50}Ti is 0.17 barns.
- Neutron flux is 10^{12} neutrons /cm^2/s.
- Sample mass of Titanium is 1 g with a natural abundance of ^{50}Ti at 5.41%.

Solution:

Step 1: Calculate the Decay Constant of ^{51}Ti

- Convert half-life to seconds: $T_{1/2} = 5.76$ minutes $\times$ 60 s/min $= 345.6$ s
- Decay constant (λ) formula: $\lambda = \frac{\ln(2)}{T_{1/2}}$

$$\lambda = \frac{0.693}{345.6} \approx 0.002005 \text{ s}^{-1}$$

Step 2: Calculate the Number of ^{50}Ti Atoms

- Density of Titanium: 4.506 g/cm^3
- Molar Mass of Titanium: 47.867 g/mol
- Volume for 1 g of Ti: $\frac{1 \text{ g}}{4.506 \text{ g/cm}^3} \approx 0.2219$ cm^3
- Number of ^{50}Ti atoms (N_0) :

$$N_0 = \left(\frac{\frac{4.506 \text{ g}}{\text{cm}^3} \times 0.2219 \text{ cm}^3}{\frac{47.867 \text{ g}}{\text{mol}}} \right) \times 0.0541 \times 6.022 \times 10^{23}$$

$$N_0 \approx 6.80535 \times 10^{20} \text{ atoms}$$

Step 3: Calculate the Activity $N(t)$ of ^{51}Ti

- Activity formula for ongoing production and decay under irradiation:

$$N(t) = N_0 \sigma \Phi \left(\frac{1 - e^{-\lambda t}}{\lambda} \right)$$

$$N(t) \approx 6.80535 \times 10^{20} \times 0.17 \times 10^{-24} \times 10^{12} \times \left(\frac{1 - e^{-0.002005t}}{0.002005} \right)$$

$$N(t) \approx 5.7701 \times 10^{10} \times (1 - e^{-0.002005t})$$

Using these parameters, the activity $A(t)$ of ^{51}Ti as a function of time under constant neutron irradiation is:

$$A(t) = 6.80535 \times 10^{20} \times 0.17 \times 10^{-24} \times 10^{12} \times (1 - e^{-0.002005 \times t})$$

$$A(t) = 1.1569095 \times 10^8 \times (1 - e^{-0.002005 \times t}) \text{ Bq}$$

Step 4: Calculate the Rate of Production of ^{51} V

- Rate of Production $R(t)$:

$$R(t) = \lambda N(t)$$

$$R(t) \approx 0.002005 \times 5.7701 \times 10^{10} \times (1 - e^{-0.002005t})$$

$$R(t) \approx 1.15691 \times 10^8 \times (1 - e^{-0.002005t}) \frac{\text{nuclei}}{\text{s}}$$

The rate of production of ^{51} V reaches a maximum and then stabilizes as the ^{51}Ti reaches a steady state under continuous neutron irradiation. The function

$$R(t) = 1.15691 \times 10^8 \times (1 - e^{-0.002005t}) \text{ nuclei/s}$$

describes how ^{51} V is produced over time from the activated ^{50}Ti. This model provides insights into the dynamics of nuclear transmutation under neutron bombardment in a laboratory setting.

6.2   Nuclear Reactions

Nuclear reactions involve interactions between two nuclear particles—such as two nuclei or a nucleus and a nucleon—that result in the production of two or more nuclear particles or gamma rays. These processes are fundamental in various fields, such as energy production, medical applications, and astrophysics.

6.2.1 Fundamental Concepts of Nuclear Reactions

1. Types of Nuclear Reactions

Several types of nuclear reactions can be broadly classified based on the type of the interaction and the particles involved:

- **Fusion:** Two light nuclei combine to form a heavier nucleus, releasing energy.
- **Fission:** A heavy nucleus splits into two lighter nuclei, typically triggered by the absorption of a neutron.
- **Radioactive Decay:** An unstable nucleus emits radiation to transform into a more stable configuration.
- **Transmutation:** One element is transformed into another through nuclear reactions, often involving neutron capture.

2. Conservation Laws

Several conservation laws govern the behavior of nuclear reactions:
- **Conservation of Nucleons:** The total number of nucleons (protons and neutrons) remains constant in a nuclear reaction.
- **Conservation of Charge:** The total electric charge is conserved in nuclear reactions.
- **Conservation of Momentum:** The total momentum before and after the reaction is conserved.
- **Conservation of Energy:** Includes the conservation of rest-mass energy, kinetic energy, and any resultant energy from the mass-energy equivalence principle.

3. Reaction Equation and Kinematics

A typical nuclear reaction can be represented as:

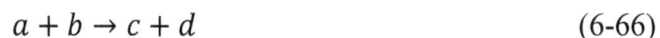

$$a + b \rightarrow c + d \tag{6-66}$$

Here, a and b are the reactant nuclei, and c and d are the products which may include nuclear particles or gamma rays. The kinematics of these reactions is determined by applying the conservation laws. The total energy, including kinetic and rest-mass energies of the reactants and products, must balance according to Einstein's equation:

$$E_{\text{total}} = \sum (K.E. + M.c^2) \tag{6-67}$$

6.2.1.1   The Q-Value of Nuclear Reactions

The Q-value is a critical parameter in nuclear reactions, indicating the net energy change resulting from the reaction. It is defined as:

$$Q = [(M_a + M_b) - (M_c + M_d)] \times c^2 \qquad (6\text{-}68)$$

- Exothermic Reactions: If $Q > 0$, the reaction releases energy, increasing the kinetic energy of the products.
- Endothermic Reactions: If $Q < 0$, the reaction absorbs energy, requiring an input of kinetic energy to proceed.

6.2.2   Experimental Arrangements

Nuclear reactions are commonly studied in controlled settings where one reactant (usually in a gaseous or plasma state) is accelerated towards a target containing the other reactant. Modern experimental setups involve particle accelerators and detectors to measure the energies and trajectories of the reaction products.

When neutrons are used as projectiles, they can induce fission in heavy nuclei (like Uranium-235) or transmute lighter elements (like turning Iron into Cobalt in steel under neutron exposure).

6.2.3   Nuclear Reactions and Calculating the Q Value

Nuclear reactions concern changes in the nucleus of an atom, where nucleons (neutrons and protons) rearrange to form new nuclei. These reactions are fundamental to nuclear physics. This section will focus on the techniques for balancing nuclear reactions and calculating their values, which indicate the net energy change associated with the reaction.

6.2.3.1   Fundamentals of Balancing Nuclear Reactions

Balancing a nuclear reaction requires adherence to the conservation laws of physics: conservation of nucleons, charge, and energy.

Conservation of Nucleons

Nuclear reactions must conserve the total number of nucleons. The sum of the mass numbers (total protons and neutrons) of the reactants and products must be equal.

Conservation of Charge

The total charge before and after the reaction must remain constant. This means the sum of the atomic numbers (protons) in the reactants must equal the sum in the products.

Steps to Balance a Nuclear Reaction

1. Identify Reactants and Products: Write down the initial and final components of

the reaction.

2. Determine Mass Numbers and Atomic Numbers: List the mass numbers (A) and atomic numbers (Z) for each component.

3. Adjust for Conservation Laws: Ensure that the sums of A and Z on both sides of the reaction equation are equal. If they are not, identify and add the required nucleons or nuclear particles (like protons, neutrons, alpha particles) to balance the equation.

Example 6-17. Balancing a Nuclear Reaction

Consider the alpha decay of Uranium-238:

$$^{238}U \rightarrow ? + {}^{4}He$$

- Reactants: $^{238}U(A = 238, Z = 92)$
- Products: $^{4}He(A = 4, Z = 2)$, and unknown product?
- Balancing:
- For A: $238 = 4 + A_? \rightarrow A_? = 234$
- For Z: $92 = 2 + Z_? \rightarrow Z_? = 90$
- Identified Product: $^{234}Th(A = 234, Z = 90)$
- Balanced Reaction: $^{238}U \rightarrow {}^{234}Th + {}^{4}He$

6.2.4   Calculating the Q Value of Nuclear Reactions

The Q value quantifies the net energy released or absorbed during a nuclear reaction, derived from the mass differences between the reactants and products.

Formula for Q Value

$$Q = \left(M_{reactants} - M_{products}\right) \times c^2 \tag{6-69}$$

Where $M_{reactants}$ and $M_{products}$ are the total rest masses of the reactants and products, respectively, and c is the speed of light.

Calculation Steps

1. Determine Rest Masses: Obtain the rest masses of all reactants and products from nuclear data tables.

2. Compute Mass Defect: Subtract the total mass of the products from that of the reactants.

3. Calculate Q Value: Multiply the mass defect by c^2 to convert the mass difference into energy.

Example 6-18. Example Calculation

For the reaction $^{238}\text{U} \rightarrow {}^{234}\text{Th} + {}^{4}\text{He}$:

- Masses: $M_{238_U} = 238.050784\text{u}, M_{234}\text{Th} = 234.043601^{-}\text{u}, M_4\text{He} = 4.002602\text{u}$
- Total Mass of Products: $234.043601\text{u} + 4.002602\text{u} = 238.046203\text{u}$
- Mass Defect: $238.050784\text{u} - 238.046203\text{u} = 0.004581\text{u}$
- Energy Equivalent (Q Value): $0.004581\text{u} \times 931.5\text{MeV/u} \approx 4.27\text{MeV}$

This positive Q value indicates that the reaction is exothermic, releasing approximately 4.27 MeV of energy.

Example 6-19. To illustrate the process of balancing a nuclear reaction and calculating the Q value, let's consider the example of a neutron being absorbed by Lithium- 6, ^{6}Li). This reaction is common in nuclear physics, particularly in applications involving neutron absorbers and moderators.

Step 1: Write the Unbalanced Reaction

The unbalanced reaction involves a neutron (n) being absorbed by a Lithium-6 nucleus (^{6}Li), forming a new nucleus. We can denote this as:

$$^{6}\text{Li} + n \rightarrow ?$$

Let's reconsider that the reaction might involve another alpha particle and a tritium (^{3}H) instead:

$$^{6}\text{Li} + n \rightarrow {}^{4}\text{He} + {}^{3}\text{H}$$

It balances:

- Mass Number (A):
  - Reactants: $6 + 1 = 7$
  - Products: $4 + 3 = 7$
- Atomic Number (Z):
  - Reactants: $3 + 0 = 3$
  - Products: $2 + 1 + 0 = 3$

Step 4: Calculate the Q Value

To calculate the Q value, we need the masses of the reactants and products:

- $M_{^6Li} \approx 6.0151223$ u
- $M_n \approx 1.008665$ u
- $M_{^4He} \approx 4.002602$ u
- $M_{^3H} \approx 3.016049$ u

Using the formula:

$$Q = \left[\left(M_{\text{reactants}} - M_{\text{products}}\right) \times c^2\right]$$

$$Q = \left[(6.0151223 \text{ u} + 1.008665 \text{ u}) - (4.0026032 \text{ u} + 3.016049 \text{ u})\right] \times \frac{931.5\text{MeV}}{1\text{ u}}$$

$$Q = \left[(7.0237873 \text{ u}) - (7.0186522 \text{ u})\right] \times \frac{931.5 \text{ MeV}}{1\text{ u}}$$

$$Q = 0.0051351 \text{ u} \times \frac{931.5\text{MeV}}{\text{u}}$$

$$Q \approx 4.783346 \text{ MeV}$$

This positive Q value indicates that the reaction is exothermic, releasing approximately 4.78 MeV of energy.

Example 6-20. To calculate the Q value for a Deuterium-Deuterium $(D - D)$ fusion reaction, we first need to understand the specifics of the reaction. Let's consider the reaction where two deuterium nuclei (each ^2H) fuse to form a helium-3 nucleus (^3He) and a neutron (n):

$$^2H + {}^2H \rightarrow {}^3He + n$$

The Q value of a reaction is defined as the difference in the total mass-energy of the reactants and the products, converted into energy via Einstein's equation $E = mc^2$. For nuclear reactions, this is typically given in units of mega electron volts (MeV).

To compute the Q value, you need the mass of each particle involved in the reaction:

- Mass of Deuterium (^2H): 2.014102 atomic mass units (amu)
- Mass of Helium-3 (^{3}He): 3.016029amu
- Mass of a neutron (n): 1.008665amu

The mass of the reactants is the sum of the masses of two deuterium nuclei:

$$M_{\text{reactants}} = 2.014102\text{amu} + 2.014102\text{amu} = 4.028204\text{amu}$$

The mass of the products is the sum of the masses of helium -3 and a neutron:

$$M_{\text{products}} = 3.016029\text{amu} + 1.008665\text{amu} = 4.024694\text{amu}$$

The mass defect (Δm) is:

$$\Delta m = M_{\text{reactants}} - M_{\text{products}} = 4.028204\text{amu} - 4.024694\text{amu} = 0.003510\text{amu}$$

To convert this mass defect into energy, we use the conversion factor where 1 amu corresponds to approximately 931.5MeV :

$$Q = \Delta m \times 931.5\text{MeV/amu} = 0.003510\text{amu} \times 931.5\text{MeV/amu} \approx 3.27\text{MeV}$$

6.3   Number Density in Pure Substances

Number density is a crucial concept in various scientific fields, particularly in physics and materials science. It describes the number of atoms or molecules per unit volume of a material. For pure substances, where the composition is consistent throughout, the number density provides insight into the structural arrangement and physical properties of the material.

For a pure substance, the number density N can be calculated using the formula:

$$N = \frac{\rho N_A}{M} \tag{6-70}$$

Where:

- ρ is the density of the substance in grams per cubic centimeter (g/cm^3),
- N_A is Avogadro's number (6.022×10^{23} per mole), representing the number of particles in a mole,

- M is the molar mass of the substance in grams per mole (g/mol).

This formula essentially converts the mass density to a count of individual particles by relating it to the molar mass and Avogadro's number.

Example 6-21. Number Density of Aluminum

Problem Statement: Calculate the number density of aluminum, which has a density of 2.70 g/cm³ and a molar mass of 26.98 g/mol.

Solution:

Using the formula for number density:

$$N = \frac{\rho N_A}{M} = \frac{2.70 \times 6.022 \times 10^{23}}{26.98}$$

$$N \approx \frac{2.70 \times 6.022 \times 10^{23}}{26.98} = 6.02 \times 10^{22} \text{ atoms /cm}^3$$

Interpretation: This result means that in every cubic centimeter of aluminum, there are approximately 6.02×10^{22} aluminum atoms. This high density of atoms is characteristic of metals, which contributes to their electrical conductivity and mechanical strength.

Example 6-22. Number Density of Gold

Problem Statement: Calculate the number density of gold, known for its high density of 19.32 g/cm³ and a molar mass of 197.0 g/mol.

Solution:

Applying the number density formula:

$$N = \frac{\rho N_A}{M} = \frac{19.32 \times 6.022 \times 10^{23}}{197.0}$$

$$N \approx \frac{19.32 \times 6.022 \times 10^{23}}{197.0} = 5.89 \times 10^{22} \text{ atoms /cm}^3$$

Interpretation:

Gold's number density is approximately 5.89×10^{22} atoms per cubic centimeter. Despite its higher mass density compared to aluminum, its number density is slightly lower due to its much larger atomic mass. This reflects how closely packed the atoms are, influencing properties like thermal conductivity and malleability.

The concept of number density provides valuable insights into the microscale arrangement of atoms within a material, influencing both its physical and chemical properties. These calculations are fundamental in areas ranging from materials science to nuclear physics, helping predict and explain the behavior of materials under various conditions.

6.4 Number Density of Atoms in Molecular and Crystalline Forms

Number density in molecular and crystalline forms extends beyond the straightforward calculation used for pure elements, considering the complexities of molecular and lattice structures. For molecular substances, the focus is on the molecules themselves as well as individual atom types within the molecule. For crystalline materials, understanding the lattice structure and unit cell composition is key.

6.4.1 Molecular Forms

Number density in molecular forms calculates not just the density of molecules but also considers the number of specific types of atoms within those molecules. The general formula is:

$$N = \frac{\rho N_A}{M}$$

Here, N can be calculated for molecules, and for individual atoms, it's multiplied by the number of each atom in the molecule.

Example 6-23. Number Density of Carbon Dioxide (CO_2)

Problem Statement: Calculate the number density of molecules and individual carbon and oxygen atoms in carbon dioxide, assuming its density at STP is 1.98×10^{-3} g/cm^3 and its molar mass is 44.01 g/mol.

Solution:

1. Calculate the molecule density of CO_2 :

$$N_{CO_2} = \frac{\rho N_A}{M} = \frac{1.98 \times 10^{-3} \times 6.022 \times 10^{23}}{44.01} \approx 2.71 \times 10^{19} \text{molecules/cm}^3$$

2. Calculate the atom densities:

- Carbon atoms per molecule: 1
- Oxygen atoms per molecule: 2

$$N_C = 2.71 \times 10^{19} \text{ atoms /cm}^3$$

$$N_O = 2 \times 2.71 \times 10^{19} = 5.42 \times 10^{19} \text{ atoms /cm}^3$$

Interpretation: There are approximately $2.71 \times 10^{19} CO_2$ molecules per cm^3, 2.71×10^{19} carbon atoms per cm^3, and 5.42×10^{19} oxygen atoms per cm^3 at STP.

6.4.2 Crystalline Forms

In crystalline forms, the unit cell structure determines the number density. Here, the molecular weight is calculated by summing of all the atomic weights in the unit cell.

Example 6-24. Number Density in Sodium Chloride (NaCl) Crystals

Problem Statement: Calculate the number density of Na^+ and Cl^- ions in a NaCl crystal with a density of 2.165 g/cm^3 and atomic weights of 22.990 g/mol for Na and 35.453 g/mol for Cl. The unit cell for NaCl contains one Na and one Cl atom.

Solution:

1. Calculate the molar mass of the NaCl unit cell:

$$M_{NaCl} = 22.990 + 35.453 = 58.443 \text{ g/mol}$$

2. Calculate the number density of the NaCl units (molecules):

$$N_{NaCl} = \frac{\rho N_A}{M_{NaCl}} = \frac{2.165 \times 6.022 \times 10^{23}}{58.443} \approx 2.23 \times 10^{22} \text{ units /cm}^3$$

Since each unit cell of NaCl contains one sodium and one chloride ion:

$$N_{Na} = N_{Cl} = N_{NaCl} = 2.23 \times 10^{22} \text{ ions /cm}^3$$

Interpretation: In each cubic centimeter of NaCl crystal, there are approximately 2.23×10^{22} sodium ions and 2.23×10^{22} chloride ions.

Example 6-25. Number Density in Uranium Dioxide (UO_2)

Uranium dioxide (UO_2) is a ceramic compound used as a nuclear fuel in reactors. It consists of uranium atoms and oxygen atoms in a 1:2 ratio, forming a crystalline structure. To calculate the number densities of uranium and oxygen in UO_2, we use the principles outlined for molecular and crystalline forms.

Problem Statement: Calculate the number densities of uranium (U) and oxygen (O) atoms in uranium dioxide, UO_2, given the following data:

- Density of UO_2: 10.97 g/cm^3
- Atomic mass of Uranium (U): 238.03 g/mol
- Atomic mass of Oxygen (O): 16.00 g/mol

Solution

Step 1: Calculate the Molar Mass of UO_2

Uranium dioxide (UO_2) consists of one uranium atom and two oxygen atoms per formula unit. The molar mass M of UO_2 can be calculated as follows:

$$M = M_U + 2 \times M_O$$

$$M = 238.03 + 2 \times 16.00 = 270.03 \text{ g/mol}$$

Step 2: Calculate the Molecule Density of UO_2

The molecule density N_{UO_2} of UO_2 is given by the formula:

$$N_{UO_2} = \frac{\rho N_A}{M}$$

Where ρ is the density of UO_2, N_A is Avogadro's number, and M is the molar mass of UO_2.

$$N_{UO_2} = \frac{10.97 \times 6.022 \times 10^{23}}{270.03}$$

$$N_{UO_2} = \frac{10.97 \times 6.022 \times 10^{23}}{270.03} \approx 2.45 \times 10^{22} \text{ molecules } /cm^3$$

Step 3: Calculate the Number Densities of U and O Atoms

Each molecule of UO_2 contains one uranium atom and two oxygen atoms. Therefore, the number density of uranium atoms N_U is the same as the molecule density of UO_2 :

$$N_U = N_{UO_2} = 2.45 \times 10^{22} \text{ atoms } /cm^3$$

The number density of oxygen atoms N_O is twice the molecule density of UO_2 because each molecule contains two oxygen atoms:

$$N_O = 2 \times N_{UO_2} = 2 \times 2.45 \times 10^{22}$$

$$N_O = 4.90 \times 10^{22} \text{ atoms } /cm^3$$

In uranium dioxide (UO_2) :

- The number density of uranium (U) atoms is approximately 2.45×10^{22} atoms $/cm^3$.
- The number density of oxygen (O) atoms is approximately 4.90×10^{22} atoms $/cm^3$.

6.4.3   Calculating Masses from Isotopic Compositions

The atomic mass of an element from the periodic table is the weighted average of the masses of its isotopes, taking into account their natural abundances. Molecular mass calculation further involves combining the atomic masses of the constituent elements in their stoichiometric ratios.

The average atomic mass M of an element from its isotopic composition is calculated using:

$$M = \sum (\gamma_i \times m_i) \tag{6-71}$$

Where:

- γ_i is the fractional isotopic abundance of the i-th isotope (expressed as a decimal),

- m_i is the atomic mass of the i-th isotope.

Steps to Calculate Average Atomic Mass:

1. Convert the isotopic abundance percentages to fractions (by dividing by 100 if given in percent).
2. Multiply the fractional abundance of each isotope by its atomic mass.
   3. Sum the products to obtain the average atomic mass.

Example 6-26. Calculating the Atomic Mass of Boron

Boron primarily exists as two isotopes: Boron-10 and Boron-11, with the following isotopic abundances and masses:

- Boron-10: $\gamma_{B-10} = 19.9\%$, Atomic mass $= 10.0129u$
- Boron-11: $\gamma_{B-11} = 80.1\%$, Atomic mass $= 11.0093u$

Solution:

1. Convert Abundances to Fractions:
   - $\gamma_{B-10} = 0.199$
   - $\gamma_{B-11} = 0.801$
2. Calculate Contributions:
   - Contribution of B-10: $0.199 \times 10.0129u = 1.9926u$
   - Contribution of B-11: $0.801 \times 11.0093u = 8.8184u$
Sum the Contributions:
   - $M_B = 1.9926u + 8.8184u = 10.811u$

Example 6-27. Calculating Molecular Mass from Isotopic Compositions
Calculate the molecular mass of water (H_2O) considering the isotopic compositions of hydrogen and oxygen:

- Hydrogen has two main isotopes: ^1H (99.985%) and ^2H (deuterium, 0.015%), with masses of $1.0078u$ and $2.014u$, respectively.
- Oxygen primarily consists of ^{16}O (99.762%), with a mass of $15.994u$.

Solution:

1. Calculate Average Atomic Mass for Hydrogen:

$$M_H = (0.99985 \times 1.0078) + (0.00015 \times 2.014)$$

$$M_H = 1.0076 + 0.0003 = 1.0079u$$

2. Average Atomic Mass for Oxygen (simplification assuming only ^{16}O):

$$M_O = 15.994u \text{ (neglecting minor contributions from } ^{17}O \text{ and } ^{18}O \text{)}$$

Calculate Molecular Mass for Water:

$$M_{H_2O} = 2 \times M_H + M_O$$

$$M_{H_2O} = 2 \times 1.0079 + 15.994 = 18.010u$$

Result: The molecular mass of water calculated from isotopic compositions is approximately 18.010u.

6.5   Isotopic Number Densities in Pure Substances

Isotopic number density describes the concentration of specific isotopes within a sample of a pure substance. This is especially relevant in fields like nuclear physics and geochemistry where isotopic composition significantly affects material properties and processes.

6.5.1   General Formula for Isotopic Number Density

The number density N_i of a specific isotope i can be calculated using the formula:

$$N_i = \frac{\gamma_i \rho N_A}{M} \tag{6-72}$$

Where:

- γ_i is the isotopic abundance of isotope i as a decimal,
- ρ is the density of the substance in g/cm^3,
- N_A is Avogadro's number (6.022×10^{23} atoms /mol),
- M is the average atomic mass of the substance in g/mol, calculated from isotopic masses and their abundances.

6.5.2   Calculating Average Atomic Mass from Isotopic Abundances

The average atomic mass M of an element based on its isotopic composition is calculated by:

$$M = \sum (\gamma_i \times m_i)$$

- m_i represents the mass of the i-th isotope,
- γ_i is the fractional abundance of the i-th isotope.

Example 6-28. Isotopic Number Densities for Copper Isotopes

We'll re-calculate the isotopic number densities for the isotopes of copper (Cu-63 and $Cu - 65$) using six significant figures for precision, ensuring a more accurate result for applications requiring detailed material properties.

Known Data:

- Density of copper (pure): 8.96000 g/cm^3
- Isotopic masses:
    - Copper-63 (Cu-63): 62.92960u
    - Copper-65 (Cu-65): 64.92780u
- Isotopic abundances:
    - Copper-63 (Cu-63): 69.170%
    - Copper-65 (Cu-65): 30.830%

Formula for Isotopic Number Density:

$$N_i = \frac{\gamma_i \rho N_A}{M}$$

Where:

γ_i is the isotopic abundance as a fraction,
ρ is the density of the substance,
N_A is Avogadro's number (6.022141×10^{23} atoms /mol),
M is the average atomic mass of the substance.

Steps to Solve:

1. Convert isotopic abundances to decimal form:

$$\gamma_{Cu-63} = 0.69170$$

$$\gamma_{Cu-65} = 0.30830$$

2. Calculate the average atomic mass of copper:

$$M = (0.69170 \times 62.92960) + (0.30830 \times 64.92780)$$

$$M = 43.5284 + 20.0172 = 63.54565 \text{ u}$$

$$M = \frac{63.54565 \text{ g}}{\text{mol}}$$

Calculate the number density for each isotope using six significant figures:
- For Cu-63:

$$N_{Cu-63} = \frac{0.69170 \times 8.96000 \times 6.022141 \times 10^{23}}{63.54565}$$

$$N_{Cu-63} = \frac{6.197632 \times 6.022141 \times 10^{23}}{63.54565}$$

$$N_{Cu-63} \approx 5.87342 \times 10^{22} \frac{\text{atoms}}{\text{cm}^3}$$

- For Cu-65:

$$N_{Cu-65} = \frac{0.30830 \times 8.96000 \times 6.022141 \times 10^{23}}{63.55785}$$

$$N_{Cu-65} = \frac{2.762368 \times 6.022141 \times 10^{23}}{63.55785}$$

$$N_{Cu-65} \approx 2.61736 \times 10^{22} \frac{\text{atoms}}{\text{cm}^3}$$

6.6   Weight Percent in Isotopic Calculations

Isotopic calculations often involve different measures of composition such as weight percent and atom percent. Understanding the distinction between these two is crucial for precise chemical and nuclear engineering applications.

6.6.1 Weight Percent (wt%)

Weight percent (or mass percent) of an isotope in a substance refers to the fraction of the total mass of the substance that is made up of that isotope, expressed as a percentage. It is a common measure used in specifying enrichment levels for nuclear materials such as uranium in the nuclear industry.

The weight percent ω_i of isotope i can be calculated by:

$$\omega_i = \frac{m_i \times n_i}{\sum (m_j \times n_j)} \times 100 \tag{6-73}$$

Where:

- m_i is the mass of isotope i,
- n_i is the number of moles of isotope i,
- j indexes all isotopes present.

6.6.2 Isotopic Number Density Using Weight Percent

The isotopic number density N_i using weight percent can be calculated by adjusting the general formula for number density to account for the weight percent:

$$N_i = \frac{\omega_i \rho N_A}{100 M_i} \tag{6-74}$$

Where:

ω_i is the weight percent of isotope i,
ρ is the density of the substance,
N_A is Avogadro's number (6.022×10^{23} atoms/mol),
M_i is the molar mass of isotope i.

6.6.3 Calculating Atomic Mass from Weight Percent of Each Isotope

The average atomic mass M of a pure substance, considering the weight percent of each isotope, can be derived from the weight fractions:

$$\frac{1}{M} = \sum \left(\frac{\omega_i}{m_i} \right) \tag{6-75}$$

Where:

ω_i is the weight percent of isotope i,

m_i is the atomic mass of isotope i.

The atomic mass M is then given by:

$$M = \frac{1}{\Sigma \left(\frac{\omega_i}{m_i}\right)} \tag{6-76}$$

This formulation accounts for the fact that the contribution of each isotope to the overall mass is inversely proportional to its individual atomic mass.

Example 6-29: Calculating Number Densities of U-235 and U-238 in 5% by Weight Enriched UO_2.

Given Data:
- Density of UO_2: 10.5000 g/cm^3
- Isotopic masses:
  - U-235: 235.043930u
  - U-238: 238.050788u
- Weight percent:
  - U-235: 5.00000%
  - U-238: 95.0000%

Step 1: Calculate Average Atomic Mass of Uranium

Using the formula for average atomic mass based on weight percentages:

$$\frac{1}{M_U} = \left(\frac{5.00000}{235.043930}\right) + \left(\frac{95.0000}{238.050788}\right)$$

$$\frac{1}{M_U} = 0.021272619 + 0.3990745 = 0.4203471$$

$$M_U = \frac{1}{0.4203471} = 237.89862 \text{ u}$$

Step 2: Calculate the Molar Mass of UO_2

$$M_{UO_2} = 237.89862 + 2 \times 15.994000 = 269.88662 \text{ g/mol}$$

Step 3: Calculate Isotopic Number Densities

Using Avogadro's number $N_A = 6.022141 \times 10^{23}$ atoms /mol :

- For U-235:

$$N_{U-235} = \frac{5.00000 \times 10.5000 \times 6.022141 \times 10^{23}}{100 \times 269.88662}$$

$$N_{U-235} \approx 1.171464 \times 10^{21} \frac{\text{atoms}}{\text{cm}^3}$$

- For U-238:

$$N_{U-238} = \frac{95.0000 \times 10.5000 \times 6.022141 \times 10^{23}}{100 \times 269.88662}$$

$$N_{U-238} \approx 2.2257812 \times 10^{22} \frac{\text{atoms}}{\text{cm}^3}$$

6.7   Fission

Nuclear fission, the process by which a heavy nucleus splits into two lighter nuclei along with the release of energy, neutrons, and other smaller particles, is a fundamental reaction that powers nuclear reactors and drives the explosive force of atomic bombs. The phenomenon of nuclear fission is intricately tied to concepts such as binding energy of the last neutron and critical energy, which are essential for understanding how fission is initiated and sustained.

6.7.1   Binding Energy of the Last Neutron

Binding Energy off the last neutron refers to the amount of energy required to remove a neutron from a nucleus. It effectively measures the stability of a nucleus: the greater the binding energy, the more energy is required to break the nucleus apart. When discussing the fission process, the binding energy of the last neutron is particularly significant.

When a neutron is absorbed by a fissionable nucleus, it temporarily creates a heavier, often unstable isotope. The energy state of this composite nucleus is crucial in determining whether it will undergo fission. The binding energy of the last neutron can be thought of as the energy gain when a neutron is captured by the nucleus. This energy is a result of the strong nuclear force acting between the newly added neutron and the existing nucleons (protons and neutrons of the nucleus).

For isotopes prone to fission, such as U-235 or Pu-239, the addition of a neutron makes them energetically unstable but does not immediately prompt them to split. Instead, the added neutron's binding energy facilitates forming a compound nucleus in an excited state. This excitation is because the binding energy of the absorbed neutron, combined with the neutron's kinetic energy if any, is converted into internal excitation energy of the nucleus.

6.7.2   Critical Energy

Critical Energy refers to the minimum amount of energy required for a compound nucleus to undergo fission. This energy is a threshold that must be exceeded for the nucleus to split into two or more smaller nuclei. The critical energy is a measure of the amount of energy required to destabilize a fissile or fissionable nucleus into fission. Table 3-1 shows the binding energy of the last neutron calculations and comparison to critical energy for fission.

The process of achieving or surpassing this critical energy can occur in several ways:

1. **Neutron Induced Fission**: Here, the critical energy is provided by the kinetic energy of an incoming neutron plus the binding energy of the neutron when captured by the nucleus. Neutrons released by other fission events, known as fast neutrons, can induce further fission if they retain sufficient kinetic energy upon collision with other fissionable atoms. This is the basis for a chain reaction in nuclear reactors.

2. **Spontaneous Fission**: Some isotopes can fission without external neutron capture if they possess internal energy greater than the critical energy. This type of fission is rare and usually occurs in very heavy isotopes.

6.7.3   Fission Process and Energy Release

When the compound nucleus's energy exceeds the critical energy, the nucleus becomes highly deformed. The electrostatic repulsion (due to the protons) overcomes the nuclear attraction (strong force) holding the nucleus together, and the nucleus elongates and constricts at the center, forming a dumbbell shape. Eventually, the nucleus splits into two fission fragments, which are typically isotopes of different elements. Compound nucleus formation for U-235 is shown below.

$$^{235}_{92}\text{U} + ^{1}_{0}\text{n} \rightarrow ^{236}_{92}\text{U}^{*}$$

The compound nucleus $^{236}_{92}\text{U}^{*}$ is highly unstable and has a binding energy of the last neutron of 6.54 MeV and a critical energy for fission of 5.3 MeV. Because the added energy by the absorption of a neutron is larger than the critical energy in U-236, the original nucleus, U-235 is fissile, i.e., it can fission by absorption of a neutron with almost zero energy. U-235 has a relatively high absorption cross section for thermal neutrons so absorption of any thermal neutron can cause a fission in U-235.

On the other hand, the absorption of a neutron in U-238 results in the following compound nucleus formation reaction:

$$^{238}_{92}U + {}^{1}_{0}n \rightarrow {}^{239}_{92}U^*$$

From Table 6-1, we see that the minimum neutron energy to cause a fission in this case is 0.69 MeV. So, fission in U-238 is possible if the neutron energy is greater than the threshold of 0.69 MeV. U-238 is therefore called fissionable but it is not fissile. This type of fission is called fast fission, referring to the energy of the incoming neutron, and it is a significant event in reactor criticality.

The actual fission event releases a considerable amount of energy, primarily due to the nuclear forces involved. The binding energy per nucleon in the fission fragments is generally higher than that in the original heavy nucleus. This difference in binding energy appears as the kinetic energy of the fission fragments, the kinetic energy of fission neutrons and beta particles, and gamma rays.

Typically, each fission event releases 2-3 neutrons. The fate of these neutrons - whether they will initiate further fissions or not - depends on their energy (whether it matches the critical energy of nearby nuclei) and how they are managed within the reactor environment (moderated or absorbed).

6.7.4   Characteristics of U-235 Fission Cross Section

For U-235, the fission cross section can be divided into three distinct regions based on neutron energy:

Low Energy Region (1/v Region):

At lower energies, σ_f typically follows a 1/v behavior, indicating that the fission cross section is inversely proportional to the neutron velocity. This region is characterized by a high probability of fission as the neutron energy decreases.

Resonance Region:

Following the low-energy region, there is a zone marked by numerous resonances. In this range, the cross section exhibits sharp peaks which correspond to specific energies where the nucleus has quasi-bound states that strongly absorb neutrons.

Table 6-1. Critical Energy for Fission and Binding Energy of the Last Neutron (BELN)

Isotope	Critical Energy (MeV)	Binding Energy (keV)	BELN (MeV)	Neutron Energy for Fission (MeV)	Designation
Th-232	5.9	1766691.409		1.71	fissionable
Th-233	6.5	1771477.760	4.79		
U-233	5.5	1771728.280		0	fissile
U-234	4.6	1778572.444	6.84		
U-235	5.75	1783870.285	5.30	0	fissile
U-236	5.3	1790415.042	6.54		
U-238	5.85	1801694.651		0.69	fissionable
U-239	5.5	1806500.907	4.81		
Pu-239	5.5	1806921.454		0	fissile
Pu-240	4.0	1813454.935	6.53		

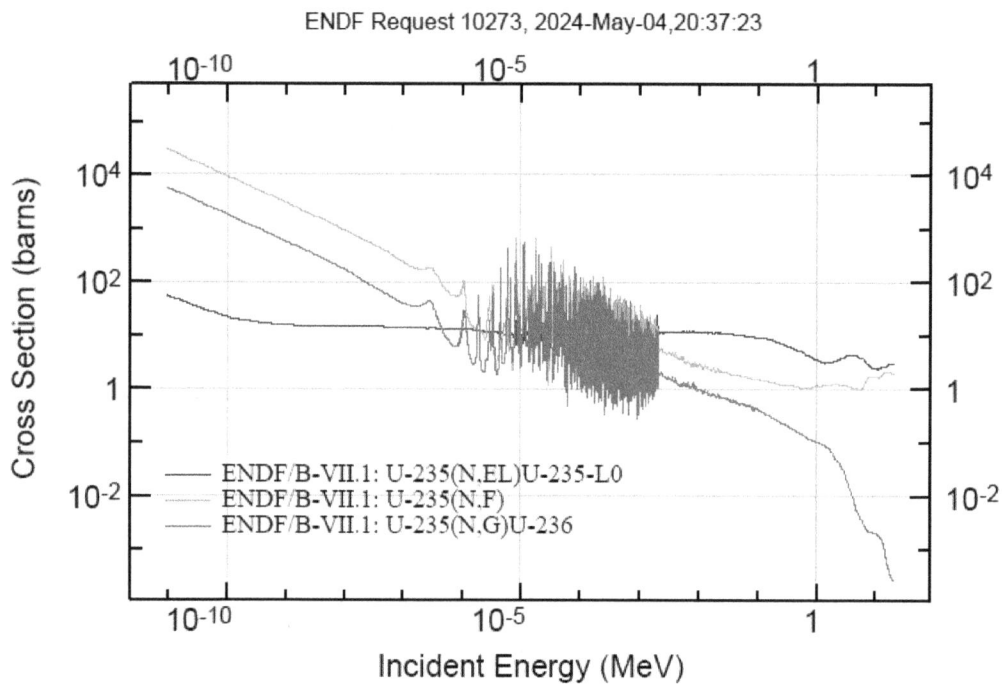

Figure 6-2. Neutron cross sections for fission, capture, and elastic scattering in U-235.

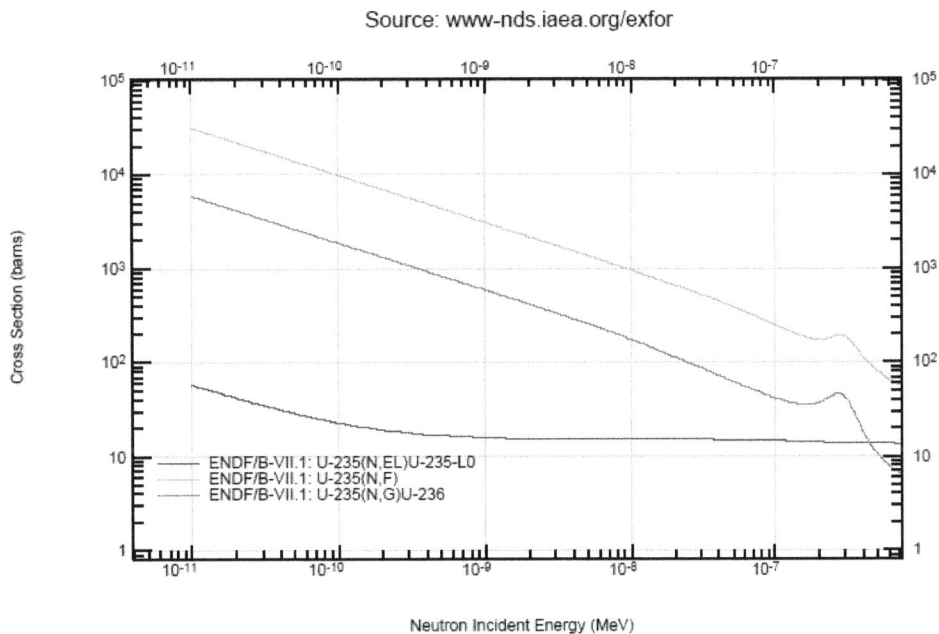

Figure 6-3. Thermal (1/v) cross sections for fission, capture, and elastic scattering in

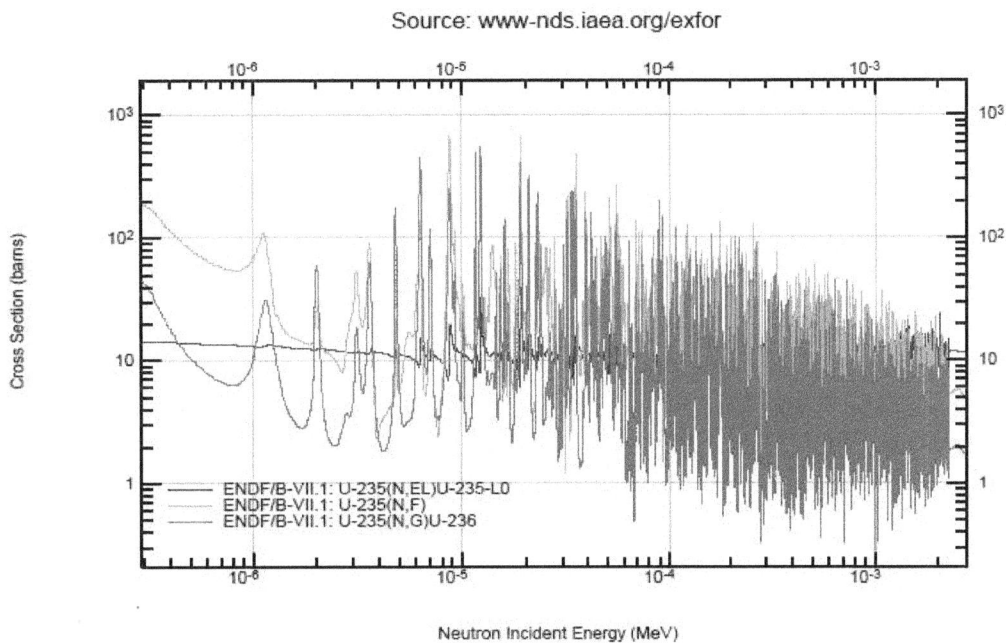

U-235.

Figure 6-4. Resonance cross sections for fission, capture, and elastic scattering

Source: www-nds.iaea.org/exfor

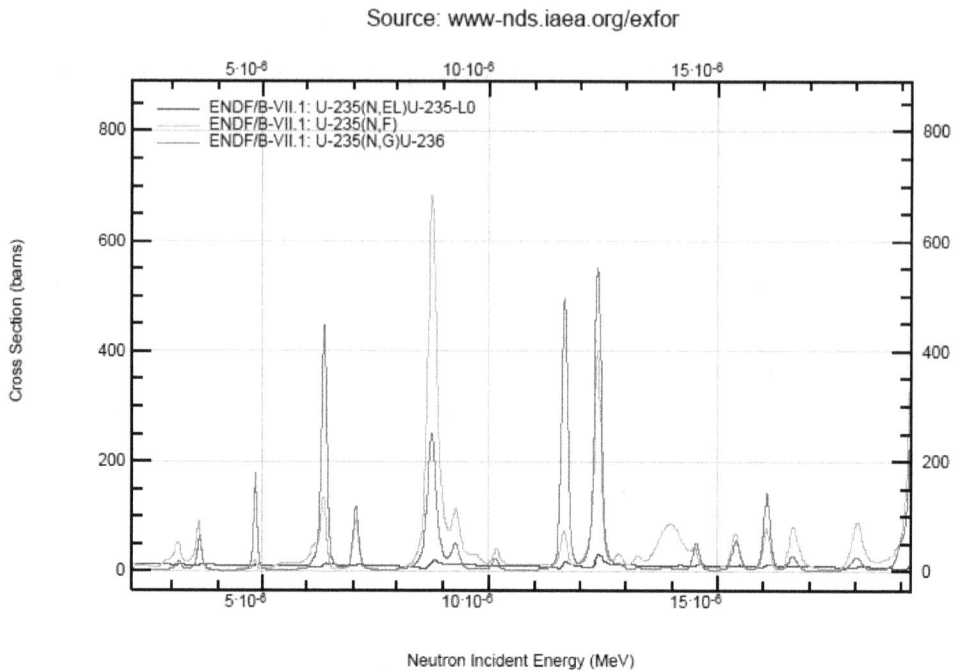

in U-235.

Figure 6-5. Resonance cross sections for fission, capture, and elastic scattering in U-235.

Source: www-nds.iaea.org/exfor

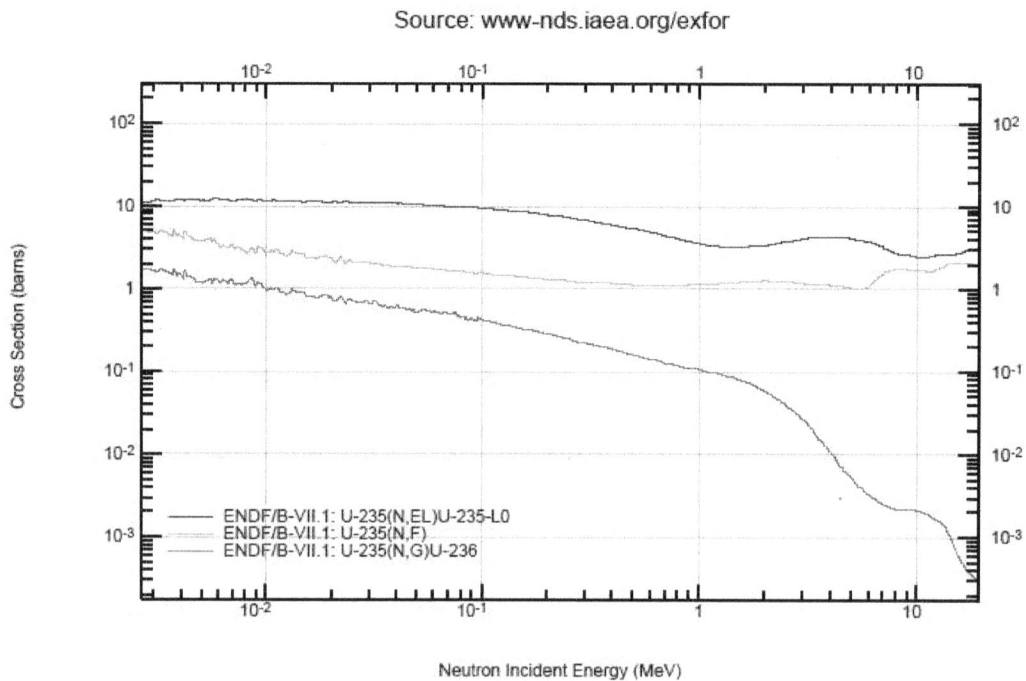

Figure 6-6. High-energy cross sections for fission, capture, and elastic scattering in U-235.**High Energy Region:**

Beyond the resonance peaks, the cross section becomes smoother and generally decreases, showing a gentle, rolling decline as the energy increases.

These regions are integral to understanding the behavior of fissile materials like U-235 under varying neutron energies.

Fission Cross Sections of Fissionable, Non-Fissile Nuclei

In contrast to fissile isotopes like U-235, fissionable but non-fissile isotopes, such as Uranium-238 (U-238), exhibit zero fission cross sections until a certain threshold energy is exceeded. This threshold lies above the resonance region, resulting in a comparatively smooth σ_f across all energy levels. This characteristic is demonstrated in Figure 6-7. Neutron cross sections for fission, capture, and elastic scattering in U-238.Figure 6-7, which also utilizes data from ENDF/B 6.

Additional Interactions and the Capture-to-Fission Ratio

It is important to note that the interaction of neutrons with fissile or fissionable nuclei does not invariably result in fission. Neutrons may also be scattered elastically or inelastically, or absorbed via radiative capture, among other interactions. While all these processes have been studied and documented in the ENDF/B 6 data, at low energies (especially in fissile nuclei), only three outcomes are predominantly observed: elastic scattering, radiative capture, and fission.

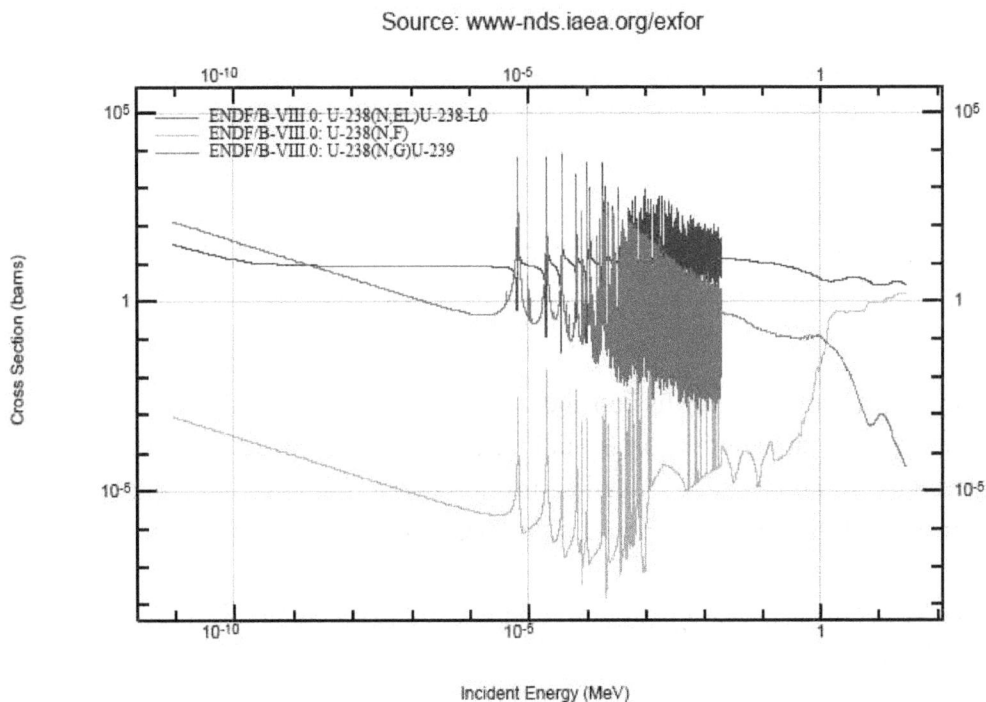

Figure 6-7. Neutron cross sections for fission, capture, and elastic scattering in U-238.In such low-energy scenarios, the cross section for elastic scattering (σ_s) is typically much smaller than those for radiative capture (σ_γ) and fission (σ_f), making the

latter two far more probable. The ratio of these two significant processes is known as the capture-to-fission ratio, denoted by

$$\alpha = \frac{\sigma_\gamma}{\sigma_f} \qquad (6\text{-}77)$$

This ratio, a critical parameter varying with neutron energy, plays a pivotal role in reactor design, influencing decisions on how reactors are configured and operated to optimize performance and safety.

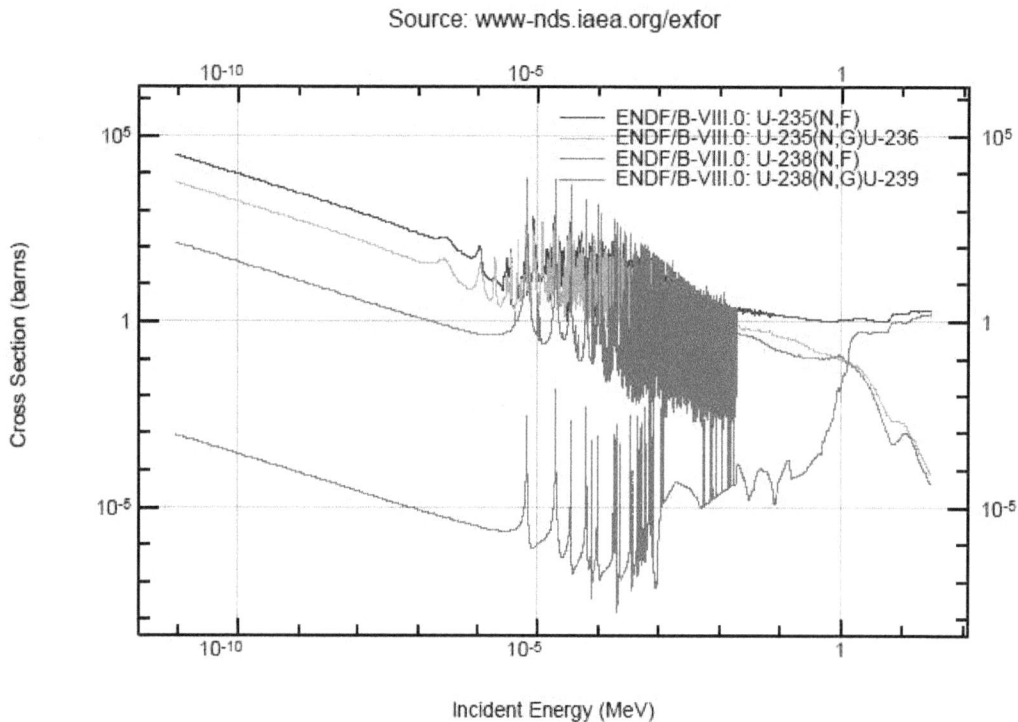

Figure 6-8. Comparison of fission and capture cross sections in U-235 and U-238.
Values of α, along with corresponding cross sections for fissile nuclei at a thermal neutron energy of 0.0253 eV, are detailed in subsequent reference materials, emphasizing the importance of these measurements in practical nuclear engineering applications.

Example 6-30. Calculate the fraction of time that a 0.0253 eV neutron absorbed in U-235 results in fission.

The fission cross-section (σ_f) and the capture cross-section (σ_γ) define the probabilities for these respective outcomes once a neutron is absorbed by the nucleus.

Given Values:

Fission Cross-Section (σ_f): 584.4 barns,

Capture Cross-Section (σ_γ): 98.81 barns.

The fraction of neutrons that result in fission upon absorption, often termed the fission probability, can be calculated using the formula:

$$f = \frac{\sigma_f}{\sigma_f + \sigma_\gamma} \qquad (6\text{-}78)$$

Calculation:

Insert the given values into the formula:

$$f = \frac{584.4}{584.4 + 98.81} = \frac{584.4}{683.21}$$

Performing the calculation:

$$f \approx \frac{584.4}{683.21} \approx 0.855$$

The calculation results in $f \approx 0.855$, meaning that about 85.5% of the neutrons absorbed by U-235 lead to fission, while the remaining 14.5% result in radiative capture, where the neutron is captured but does not cause the nucleus to split.

6.7.5   Fission Products

Fissioning of uranium-235 or plutonium-239, after capturing a neutron, results in splitting of the nucleus into two fission fragments. The fission process is highly asymmetric, meaning that the two resulting fragments have unequal masses.

Figure 6-9. illustrates this asymmetry by plotting the fission product yield (percentage of fissions producing a given fragment) as a function of fragment mass number for U-235 fission induced by thermal (0.0253 eV), epithermal (500 keV) and 14 MeV neutrons. The most probable heavy and light fission fragment masses are around 140 and 95 amu, respectively. The fission fragments are unstable because they contain too many neutrons relative to protons to be stable. They undergo a series of negative beta decays, where a neutron transforms into a proton while emitting an electron and an antineutrino, until reaching a stable nuclide. Each beta decay is typically accompanied by the emission of one or more gamma rays as the daughter nucleus de-excites.

A representative decay chain is shown in with arrows showing the decay chain from an example fission product for U-235.

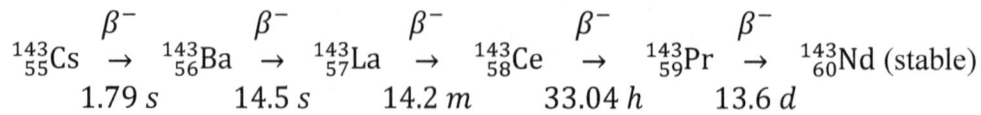

$$^{143}_{55}\text{Cs} \xrightarrow{\beta^-} {}^{143}_{56}\text{Ba} \xrightarrow{\beta^-} {}^{143}_{57}\text{La} \xrightarrow{\beta^-} {}^{143}_{58}\text{Ce} \xrightarrow{\beta^-} {}^{143}_{59}\text{Pr} \xrightarrow{\beta^-} {}^{143}_{60}\text{Nd} \text{ (stable)}$$
$$1.79\ s \qquad 14.5\ s \qquad 14.2\ m \qquad 33.04\ h \qquad 13.6\ d$$

In the chart of nuclides, the fission products are produced below the line of stability so they are mostly β^- emitters. The emission of a β^- on the chart is a diagonal move towards the line of stability. As evidenced by the above example the half lives increase through subsequent β^- emissions.

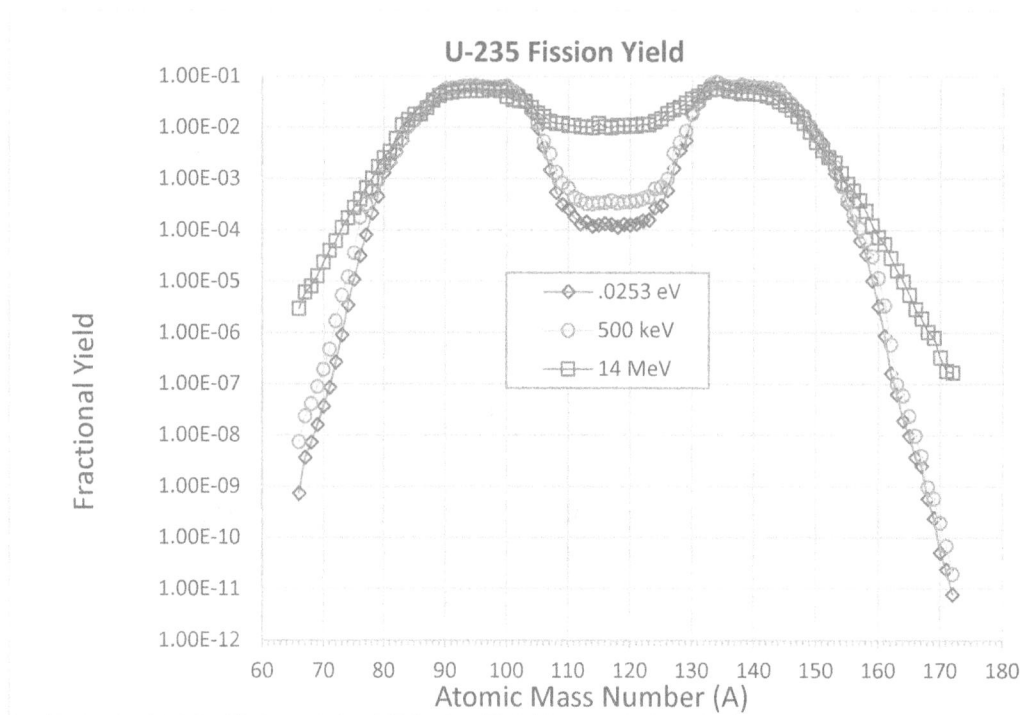

Figure 6-9. Fission product yield curves for U-235 at thermal, epithermal, and fast neutron energies.

6.7.6   Energy Release from Fission

The majority of the energy from fission, approximately 85%, manifests as kinetic energy of the fission fragments. These fragments quickly lose their kinetic energy within approximately 10^{-3} cm from the fission site, converting entirely to heat. The energies associated with the β-rays and γ-rays from fission products, along with both prompt and delayed neutrons, and prompt γ-rays, are also largely recoverable. This is because these forms of radiation seldom escape the confines of a nuclear power system. In contrast, the neutrinos produced during β-decay interact minimally with matter and typically escape the reactor, leading to a loss of about 12MeV per fission event.

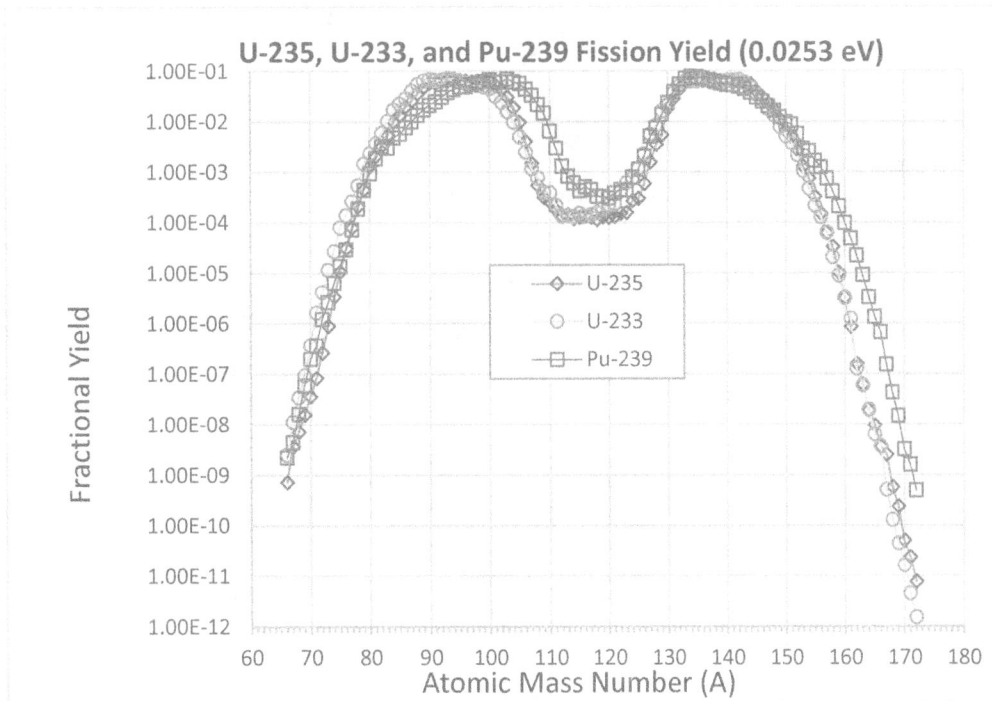

Figure 6-10. Fission product yield curves for U-235, U-233, and Pu-239 at thermal neutron energies.**Neutron Capture and Utility**

Within the reactor's environment, most fission neutrons are captured by other nuclei, contributing to the reactor's ongoing chain reaction and heat production. As detailed in the subsequent chapter, for a reactor to maintain its operational state, one of the emitted neutrons per fission must induce further fission in another fissionable nucleus. The remaining neutrons ($\nu - 1$) are absorbed in non-fission reactions, often leading to the emission of capture γ-rays, which vary in energy depending on the binding energy of the neutron with the compound nucleus. The number of neutrons per fission (ν) for $^{235}_{92}U$ averages 2.42, depending on the incident neutron energy, yielding approximately 3 to 12MeV of recoverable γ-ray energy per fission.

144Sm STABLE 3.07%	145Sm 340 d	146Sm 10.3E+7 y	147Sm 1.060E11 y 14.99%	148Sm 7E+15 y 11.24%	149Sm STABLE 13.82%	150Sm STABLE 7.38%	151Sm 90 y
	ε = 100.00%	α = 100.00%	α = 100.00%	α = 100.00%			β⁻ = 100.00%
0.0	0.0	0.0	0.0	1.63E-14	1.70E-12	1.21E-10	4.74E-9

143Pm 265 d	144Pm 363 d	145Pm 17.7 y	146Pm 5.53 y	147Pm 2.6234 y	148Pm 5.368 d	149Pm 53.08 h	150Pm 2.698 h
ε = 100.00%	ε = 100.00%	ε = 100.00% α 2 .8E-7%	ε = 65.70% β⁻ = 34.30%	β⁻ = 100.00%	β⁻ = 100.00%	β⁻ = 100.00%	β⁻ = 100.00%
0.0	0.0	0.0	4.49E-12	2.48E-11	8.09E-11	3.86E-8	2.99E-7

142Nd STABLE 27.152%	143Nd STABLE 12.174%	144Nd 2.29E+15 y 23.798%	145Nd STABLE 8.293%	146Nd STABLE 17.189%	147Nd 10.98 d	148Nd STABLE 5.756%	149Nd 1.728 h
		α = 100.00%			β⁻ = 100.00%		β⁻ = 100.00%
0.0	4.79E-14	9.56E-11	5.54E-10	3.18E-8	6.71E-7	9.92E-6	6.80E-5

Probability
- ■ ≥1.00E-1
- ■ 1.00E-2
- ■ 1.00E-3
- ■ 1.00E-4
- ☐ 1.00E-5
- ☐ 1.00E-6
- ☐ 1.00E-7
- ☐ 1.00E-8
- ☐ 1.00E-9
- ☐ 1.00E-11
- ■ 1.00E-13
- ■ ≤1.00E-15
- ☐ Unknown

	143Pr 13.57 d	144Pr 17.28 min	145Pr 5.984 h	146Pr 24.15 min	147Pr 13.4 min	148Pr 2.29 min
98% 2%	β⁻ = 100.00%	β⁻ = 100.00%	β⁻ = 100.00%	β⁻ = 100.00%	β⁻ = 100.00%	β⁻ = 100.00%
11	4.49E-9	1.27E-7	3.32E-6	3.62E-5	3.59E-3	3.88E-4

	142Ce > 5E+16 y 11.114%	143Ce 33.039 h	144Ce 284.91 d	145Ce 3.01 min	146Ce 13.49 min	147Ce 56.4 s
d .00%	2β⁻	β⁻ = 100.00%	β⁻ = 100.00%	β⁻ = 100.00%	β⁻ = 100.00%	β⁻ = 100.00%
-8	1.75E-6	3.13E-4	3.45E-4	8.54E-4	5.82E-3	9.97E-3

	141La 3.92 h	142La 91.1 min	143La 14.2 min	144La 40.8 s	145La 24.8 s	146La 6.27 s
d .00%	β⁻ = 100.00%	β⁻ = 100.00%	β⁻ = 100.00%	β⁻ = 100.00%	β⁻ = 100.00%	β⁻ = 100.00%
5	1.84E-4	9.65E-4	3.79E-3	1.06E-2	1.91E-2	7.44E-3

	140Ba 12.7527 d	141Ba 18.27 min	142Ba 10.6 min	143Ba 14.5 s	144Ba 11.5 s	145Ba 4.31 s
in .00%	β⁻ = 100.00%	β⁻ = 100.00%	β⁻ = 100.00%	β⁻ = 100.00%	β⁻ = 100.00%	β⁻ = 100.00%
4	4.88E-3	1.65E-2	3.01E-2	4.10E-2	3.97E-2	1.86E-2

30.08 y	33.41 min	139Cs 9.27 min	140Cs 63.7 s	141Cs 24.84 s	142Cs 1.684 s	143Cs 1.791 s	144Cs 0.994 s
β⁻ = 100.00%	β⁻ = 100.00%	β⁻ = 100.00%	β⁻ = 100.00%	β⁻ = 100.00% β⁻n = 0.04%	β⁻ = 100.00% β⁻n = 0.09%	β⁻ = 100.00% β⁻n = 1.64%	β⁻ = 100.00% β⁻n = 3.03%
5.99E-4	2.23E-3	1.30E-2	2.06E-2	2.91E-2	2.27E-2	1.40E-2	4.17E-3

136Xe > 2.4E+21 y 8.8573%	137Xe 3.818 min	138Xe 14.14 min	139Xe 39.68 s	140Xe 13.60 s	141Xe 1.73 s	142Xe 1.23 s	143Xe 0.511 s
2β⁻	β⁻ = 100.00%	β⁻ = 100.00%	β⁻ = 100.00%	β⁻ = 100.00%	β⁻ = 100.00% β⁻n = 0.04%	β⁻ = 100.00% β⁻n = 0.21%	β⁻ = 100.00% β⁻n = 1.00%
2.19E-2	3.19E-2	4.81E-2	4.32E-2	3.50E-2	1.21E-2	4.34E-3	2.64E-4

135I 6.58 h	136I 83.4 s	137I 24.5 s	138I 6.23 s	139I 2.280 s	140I 0.86 s	141I 0.43 s	142I 222 ms
β⁻ = 100.00%	β⁻ = 100.00%	β⁻ = 100.00% β⁻n = 7.14%	β⁻ = 100.00% β⁻n = 5.56%	β⁻ = 100.00% β⁻n = 10.00%	β⁻ = 100.00% β⁻n = 9.30%	β⁻ = 100.00% β⁻n = 21.20%	β⁻ = 100.00% β⁻n ?
2.92E-2	1.25E-2	2.62E-2	1.42E-2	7.71E-3	1.36E-3	4.06E-4	5.85E-5

134Te 41.8 min	135Te 19.0 s	136Te 17.63 s	137Te 2.49 s	138Te 1.4 s	139Te > 150 ns	140Te > 300 ns	141Te > 150 ns
β⁻ = 100.00%	β⁻ = 100.00%	β⁻ = 100.00% β⁻n = 1.31%	β⁻ = 100.00% β⁻n = 2.99%	β⁻ = 100.00% β⁻n = 6.30%	β⁻ = 100.00% β⁻n	β⁻ = 100.00% β⁻n	β⁻ = 100.00% β⁻n ?
6.21E-2	3.21E-2	1.31E-2	3.92E-3	6.61E-4	6.66E-5	1.69E-4	3.24E-7

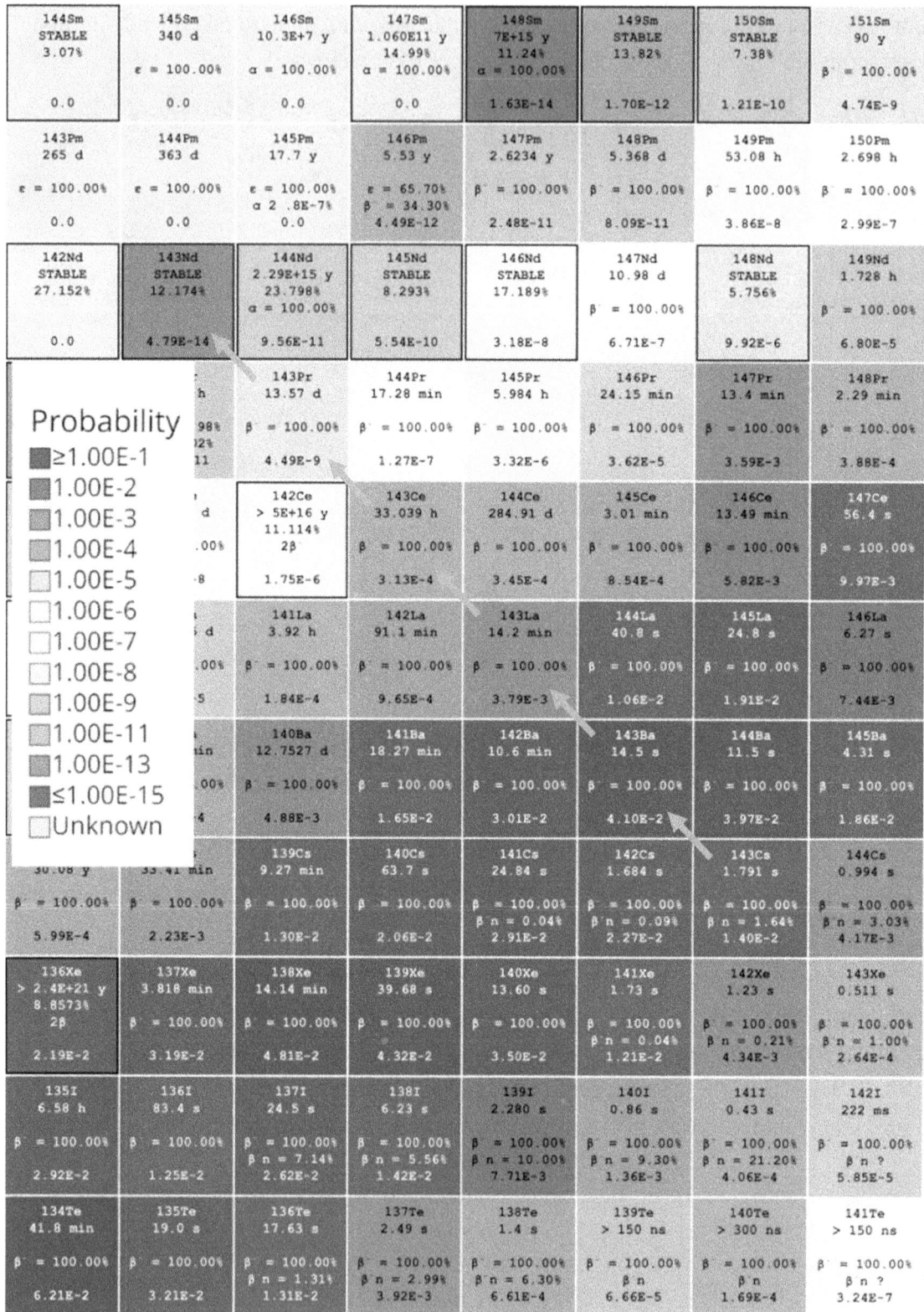

Figure 6-11. Distribution of fission products from U-235 on the chart of the nuclides.

6.7.6.2   Recoverable Energy

The energy from capture γ-rays somewhat compensates for the energy lost to neutrino emissions. The typical recoverable energy per fission is estimated to be around 200MeV, which is a standard value used for preliminary calculations in reactor design and operation.

6.7.6.3   Practical Application in Reactor Operations

Consider a reactor where the fission of $^{235}_{92}U$ releases energy at a rate of P megawatts, equating to a thermal output of P megawatts. Given a recoverable energy of 200MeV per fission, the rate of fission per second throughout the reactor can be calculated as:

$$\text{Fission rate} = \frac{P \times 10^6 \text{ joules}}{MW - \text{sec}} \times \frac{\text{fission}}{200 \text{MeV}} \times \frac{\text{MeV}}{1.60 \times 10^{-13} \text{ joule}} \times \frac{86,400 \text{sec}}{\text{day}}$$

$$\text{Fission rate} = 2.70 \times 10^{21} P \text{ fissions/day}$$

To determine the daily burnup rate in grams, divide the fission rate by Avogadro's number and multiply by the atomic mass of $^{235}_{92}U$ (235.0 g):

$$\text{Burnup rate} = 1.05P \text{ g/ day}$$

To determine the daily burnup rate in grams, divide the fission rate by Avogadro's number and multiply by the atomic mass of $^{235}_{92}U$ (235.0 g) :

$$\text{Burnup rate} = 1.05P \text{ g/ day}$$

For a reactor operating at 1MW, approximately 1 g of $^{235}_{92}U$ is fissioned per day, equivalent to releasing 1 megawatt-day of energy. It is also important to consider that $^{235}_{92}U$ is consumed both through fission and radiative capture, with the total absorption rate being $(1 + \alpha)$ times the fission rate, where α is 0.169 for $^{235}_{92}U$ when fissions are primarily induced by thermal neutrons. This results in a consumption rate of about 1.23 g/ day per megawatt of power.

Table 6-2 shows the distribution of energy from various components during the fission of U-235. The table includes both the actual energy values in MeV and their respective percentages of the total energy released in a typical fission event. The total energy released per fission event for $^{235}_{92}U$ is generally about 200 MeV.

Table 6-2. Distribution of energies from fission of U-235.

Component	Energy (MeV)	Percentage of Total
Fission Fragments	168 MeV	84%
Prompt Neutrons	5 MeV	2.5%
Prompt Betas	8 MeV	4%
Prompt Gammas	7 MeV	3.5%
Neutrinos	12 MeV	6%
Delayed Neutrons	0.1 MeV	0.05%
Delayed Betas	6 MeV	3%
Delayed Gammas	6 MeV	3%
Total Energy Released	212.1 MeV	100%

Notes on the Table:

Fission Fragments: This is the kinetic energy transferred to the heavy fission fragments. It is the largest portion of the energy distribution, primarily converted into heat as the fragments slow down in the reactor fuel.

Prompt Neutrons: These are neutrons released immediately during the fission event. Their kinetic energy contributes directly to sustaining the chain reaction in the reactor.

Prompt Betas and Gammas: These are beta particles and gamma rays emitted immediately during the fission process. They contribute to the radiation field within the reactor and are important for reactor control and heat generation.

Neutrinos: Although neutrinos carry away a significant amount of energy, this energy is not usable as neutrinos interact very weakly with matter and escape the reactor almost completely.

Delayed Betas and Gammas: These emissions come from the decay of fission products and are crucial for the decay heat that persists even after the reactor is shut down.

Delayed Neutrons: These neutrons come from the decay of some fission products and not directly from the fission event. They comprise a very small percentage of the total number of neutrons released during fission but are critical for the reactor's ability to be controlled over time, particularly during changes in reactor power.

The actual percentages might vary slightly based on the precise energies attributed to each category. The energy values provided in the table are average approximations used for general calculations in reactor physics.

Example 6-31. Calculate the energy content of a cylindrical UO2 (Uranium Dioxide) fuel pellet with dimensions of 9.5 mm in diameter and 15.9 mm in height, containing 5% enriched Uranium-235. Determine the equivalent amount of energy in terms of liters of gasoline (with a heat content of 34.2×10^6 J/kg and a density of 750 kg/m^3) and kilograms of coal (with a heat content of $24/10^6$ J/kg).

Given Data and Assumptions:
- Dimensions of UO2 pellet: 9.5 mm (diameter) × 15.9 mm (height)
- Enrichment: 5% U-235
- Energy from one U-235 fission: 200MeV
- Density of UO2: 10.97 g/cm^3
- Energy content of gasoline: 34.2MJ/L
- Energy content of coal: 24MJ/kg
- Fraction of U-235 that will fission: 0.85

Detailed Steps and Calculations:

Step 1: Calculate the volume and mass of the UO2 fuel pellet

- Volume and mass calculation:

$$\text{Volume } = \pi \times \left(\frac{0.95 \text{ cm}}{2}\right)^2 \times 1.59 \text{ cm} = 1.13 \text{ cm}^3$$

$$\text{Mass of UO2 } = 10.97 \text{ g/cm}^3 \times 1.13 \text{ cm}^3 = 12.4 \text{ g}$$

Step 2: Calculate the mass of uranium in the fuel pellet

- Mass of U calculation:

- Molar mass of UO2 $= 238$ g/mol $+ 2 \times 16$ g/mol $= 270$ g/mol

$$\text{Mass of U} = \left(\frac{12.4 \text{ g} \times 238 \text{ g/mol}}{270 \text{ g/mol}}\right) = 10.93 \text{ g}$$

Step 3: Calculate the mass of U-235 in the fuel pellet

- Mass of U-235 calculation:

$$\text{Mass of U-235} = 10.93 \text{ g} \times 0.05 = 0.547 \text{ g}$$

Step 4: Calculate the number of U-235 atoms that will fission in the fuel pellet

- Number of U-235 atoms that will fission:

Number of U-235 atoms that will fission
$$= \left(\frac{0.547 \text{ g}}{235 \text{ g/mol}} \right) \times 6.022 \times 10^{23} \text{ atoms /mol} \times 0.85 = 1.19 \times 10^{21} \text{ atoms}$$

Step 5: Calculate the energy released by the fissioning of the U-235 atoms in the fuel pellet

- Energy released calculation:

- Energy released
$$= 1.19 \times 10^{21} \text{ atoms} \times \frac{200 \text{MeV}}{\square} \text{ fission} \times 1.602 \times \frac{10^{-13} \text{ J}}{\text{MeV}}$$
$$= 3.81 \times 10^{10} \text{ J}$$

Step 6: Calculate the equivalent amount of energy in liters of gasoline

- Gasoline equivalent calculation:

$$\text{Gasoline equivalent} = \frac{3.81 \times 10^{10} \text{ J}}{34.2 \times 10^{6} \text{ J/L}} = 1{,}115 \text{ L}$$

Step 7: Calculate the equivalent amount of energy in kilograms of coal

- Coal equivalent calculation:

$$\text{Coal equivalent} = \frac{3.81 \times 10^{10} \text{ J}}{24 \times 10^{6} \text{ J/kg}} = 1{,}590 \text{ kg}$$

Considering that only 85% of the U-235 will fission, the energy content of a single UO2 fuel pellet with 5% U-235 enrichment is equivalent to approximately 1,115 liters of gasoline or 1,590 kilograms of coal. This analysis provides a demonstration of the high energy density of nuclear fuel compared to conventional energy sources such as gasoline and coal.

6.8 Interaction of Gamma Radiation with Matter

Gamma rays, along with X-rays, are forms of electromagnetic radiation. Although traditionally, gamma rays are defined as radiation emitted from nuclear reactions and X-rays from electronic transitions, for our discussion, we will refer to both as gamma rays. The primary interactions of gamma rays with matter that are relevant in nuclear engineering are the photoelectric effect, Compton scattering, and pair production.

6.8.1 The Photoelectric Effect

The photoelectric effect describes emission of electrons when a material absorbs electromagnetic radiation, such as light or gamma rays. Heinrich Hertz first observed it in 1887, but it was Albert Einstein who later provided a complete theoretical explanation of the phenomenon in 1905.

Einstein's theory of the photoelectric effect was based on the concept of light quanta, or photons. He proposed that light consists of individual packets of energy called photons, and each photon carries energy determined by its frequency. When a photon of sufficiently high energy interacts with an atom, it can transfer its energy to an electron in the atom's inner shell. If the photon's energy is greater than the electron's binding energy, the electron is ejected from the atom, creating a photoelectron. The photoelectron's kinetic energy is equal to the difference between the photon's energy and the electron's binding energy.

Einstein's theory of the photoelectric effect successfully explained the experimental observations that could not be accounted for by classical physics, such as the existence of a threshold frequency for electron emission and the instantaneous nature of the process. His theory on the photoelectric effect earned him the Nobel Prize in Physics in 1921. It was a tremendous contribution in the development of quantum mechanics.

6.8.1.1 Gamma Energy Attenuation in Lead

Gamma radiation is emitted by radioactive decay or in nuclear reactions. When gamma rays pass through matter, they can interact with the atoms in the material through various processes, including the photoelectric effect, Compton scattering, and pair production. These interactions result in the attenuation of the gamma radiation, reducing its intensity as it travels through the material.

Lead is a common material used for shielding against gamma radiation because of its high density and atomic number. The high atomic number of lead (82) makes it an effective material for the photoelectric effect, which is the dominant interaction mechanism for low-energy gamma rays. As a result, lead is particularly effective at attenuating low-energy gamma radiation.

When gamma rays interact with lead atoms through the photoelectric effect, the photons are absorbed, and photoelectrons are emitted. These photoelectrons have a very short range and are quickly absorbed by the surrounding lead atoms, further contributing to the attenuation of the gamma radiation. The probability of the photoelectric effect occurring in lead increases with decreasing gamma ray energy, making lead an excellent choice for shielding against low-energy gamma radiation.

The attenuation of gamma radiation in lead follows an exponential relationship, described by the equation

$$I = I_0 \times e^{-\mu x} \tag{6-79}$$

where I is the intensity of the gamma radiation after passing through a thickness x of lead, I_0 is the initial intensity, and μ is the linear attenuation coefficient. μ depends on the energy of the gamma radiation.

6.8.1.2   Linear Attenuation Coefficient (μ)

The linear attenuation coefficient, denoted as μ, is a probability of attenuation of an incident photon by a material per unit thickness of that material. It is expressed in units of inverse length (e.g., cm^{-1}). The higher the value of μ, the opaquer the material is to the radiation, indicating a greater likelihood that the photons will interact with the material over a given distance.

$$I = I_0 e^{-\mu x}$$

Where:

I is the intensity of the radiation after passing through a material of thickness x,
I_0 is the initial intensity of the radiation,
μ is the linear attenuation coefficient,
x is the thickness of the material.

6.8.1.3   Mass Attenuation Coefficient (μ/ρ)

The mass attenuation coefficient, denoted as μ/ρ, relates the linear attenuation coefficient to the density of the material (ρ), providing a measure of the attenuation per unit mass. This coefficient is particularly useful because it is independent of the physical state of the material, allowing comparisons between different materials or the same material in different states (solid, liquid, gas). The mass attenuation coefficient is expressed in units of cm^2/g.

The relationship between the mass attenuation coefficient and the linear attenuation coefficient is given by:

$$\mu = (\mu/\rho)\rho \tag{6-80}$$

Where:

μ is the linear attenuation coefficient,

μ/ρ is the mass attenuation coefficient,

ρ is the density of the material in g/cm^3.

This relationship allows for the conversion of the mass attenuation coefficient, which is often tabulated and cited in literature and databases, into the linear attenuation coefficient, which is directly applicable in practical scenarios involving material thickness. It provides a standardized method to compare and evaluate the effectiveness of various materials in radiation shielding applications regardless of their density.

Example 6-32. A gamma radiation source emits a beam with an intensity of 10^8 gamma rays per second, each with an energy of 1.7MeV. This beam is directed through a lead shield that has a thickness of 2 inches. Determine the intensity of the gamma beam after it has passed through this lead shield. Use the linear attenuation coefficient (μ) of lead for 1.7MeV gamma rays, which is 0.661 cm^{-1}.

Given Data:
- Initial intensity, $I_0 = 10^8$ gammas/second
- Gamma ray energy = 1.7MeV
- Thickness of lead shield = 2 inches
- Linear attenuation coefficient, $\mu = 0.661$ cm^{-1}

Solution:

Step 1: Convert the thickness of the lead shield from inches to centimeters.

Since 1 inch equals 2.54 cm, the thickness in centimeters is:

$$t = 2 \text{ inches} \times 2.54 \text{ cm/ inch} = 5.08 \text{ cm}$$

Step 2: Apply the attenuation law to calculate the final intensity of the gamma rays. The attenuation of gamma rays as they pass through a material can be described by the exponential attenuation law:

$$I = I_0 e^{-\mu t}$$

where

I_0 is the initial intensity of the gamma rays,

μ is the linear attenuation coefficient,

t is the thickness of the material in cm.

Step 3: Substitute the given values into the attenuation formula to find the final intensity.

$$I = 10^8 \text{ gammas/second } \times e^{-0.661 \text{ cm}^{-1} \times 5.08 \text{ cm}}$$

$$I = 10^8 \times e^{-3.35668}$$

$$I = 10^8 \times 0.03498 \text{ (using } e^{-3.35668} \approx 0.03498 \text{)}$$

Step 4: Calculate the final intensity.

$$I = 10^8 \times 0.03498 \approx 3.498 \times 10^6 \text{ gammas/second}$$

6.8.2   Compton Scattering

Compton Scattering is a quantum mechanical phenomenon and a key type of interaction between electromagnetic radiation (such as X-rays or gamma rays) and matter, specifically with free or loosely bound electrons. It was first observed by Arthur Holly Compton in 1923, for which he won the Nobel Prize in Physics in 1927. Compton scattering is important in many fields including astrophysics, radiation therapy, and radiography, as it explains the scattering of high-energy photons when they interact with matter.

Compton scattering occurs when a high-energy photon (like an X-ray or gamma ray) collides with a target electron. During the scattering, the photon transfers some of its energy to the electron and, as a result, the photon is scattered with reduced energy and a change in its direction. This scattering results in an increase in the wavelength of the photon, which is inversely proportional to its energy due to the conservation of energy and momentum.

Physical Description
- Initial Interaction: A photon with initial energy E_γ and wavelength λ strikes a loosely bound or free electron.
- Energy Transfer: The photon gives part of its energy to the electron, causing the electron to recoil while the photon itself is scattered with less energy and consequently a longer wavelength.
- Scattering Angle: The change in the photon's energy-and thus its wavelength-depends on the angle at which it is scattered, denoted as θ. The greater the scattering angle, the more significant the increase in wavelength.

The change in the wavelength of the photon can be described by the Compton equation:

$$\Delta\lambda = \lambda' - \lambda = \frac{h}{m_e c}(1 - \cos\theta) \qquad (6\text{-}81)$$

Where:

λ' is the wavelength of the scattered photon,
λ is the initial wavelength of the photon,
h is Planck's constant (6.626×10^{-34}Js),
m_e is the rest mass of the electron (9.109×10^{-31} kg),
c is the speed of light in vacuum (3×10^8 m/s),
θ is the scattering angle.

The term, $\frac{h}{m_e c}$ is known as the Compton wavelength of the electron, approximately 0.00243 nm.

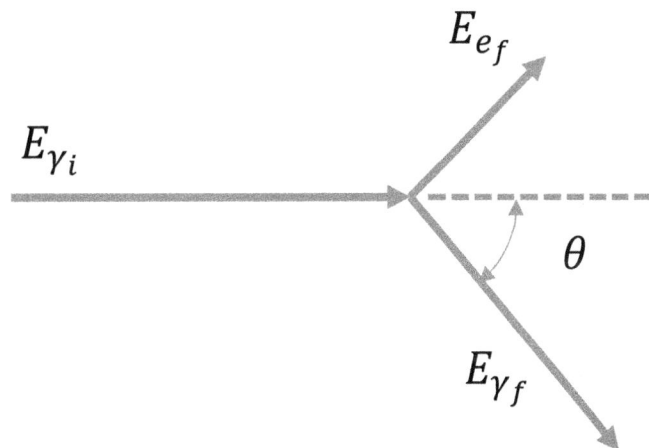

Figure 6-12. Compton scattering vector diagram.

During Compton scattering, a photon collides with an electron, causing the photon to lose energy and scatter at an angle, while the electron gains kinetic energy and recoils. Figure 6-12. shows the vector diagram for analysis of Compton scattering.

Initial Conditions:
- Photon initial energy: E_{γ_i}
- Photon final energy: E_{γ_f}
- Electron initial (rest) energy: $E_{e_i} = m_e c^2$, where m_e is the electron rest mass, and c is the speed of light.
- Electron final energy: E_{e_f}

Conservation Laws:
1. Conservation of Energy:

$$E_{\gamma_i} + m_e c^2 = E_{\gamma_f} + E_{e_f} \tag{6-82}$$

2. Conservation of Momentum:

Considering the photon's momentum and the electron's momentum in relativistic terms, we utilize the vector components:

$$\vec{p}_{\gamma_i} = \frac{E_{\gamma_i}}{c}\hat{\imath} \tag{6-83}$$

$$\vec{p}_{\gamma_f} = \frac{E_{\gamma_f}}{c}(\cos\theta\hat{\imath} + \sin\theta\hat{\jmath}) \tag{6-84}$$

$$\vec{p}_{e_f} = \vec{p}_{e_i} + \vec{p}_{\gamma_i} - \vec{p}_{\gamma_f} \tag{6-85}$$

Here, θ is the scattering angle of the photon, and $\hat{\imath}, \hat{\jmath}$ are the unit vectors in the horizontal and vertical directions, respectively. Since the initial momentum of the electron is zero (it is at rest), $\vec{p}_{e_i} = 0$.

Squaring the Momentum Equation:

To find an expression for E_{e_f}, square both sides of the conservation of momentum equation (vector equation) and solve for E_{e_f} :

$$\left(\frac{E_{\gamma_i}}{c} - \frac{E_{\gamma_f}}{c}\cos\theta\right)^2 + \left(-\frac{E_{\gamma_f}}{c}\sin\theta\right)^2 = \left(\frac{p_{e_f}}{c}\right)^2$$

$$\left(\frac{E_{\gamma_i} - E_{\gamma_f}\cos\theta}{c}\right)^2 + \left(\frac{E_{\gamma_f}\sin\theta}{c}\right)^2 = \left(\frac{p_{e_f}}{c}\right)^2$$

$$\frac{E_{\gamma_i}^2 - 2E_{\gamma_i}E_{\gamma_f}\cos\theta + E_{\gamma_f}^2\cos^2\theta + E_{\gamma_f}^2\sin^2\theta}{c^2} = \left(\frac{p_{e_f}}{c}\right)^2$$

$$\frac{E_{\gamma_i}^2 + E_{\gamma_f}^2 - 2E_{\gamma_i}E_{\gamma_f}\cos\theta}{c^2} = \left(\frac{p_{e_f}}{c}\right)^2$$

Since $p_{e_f} = \sqrt{E_{e_f}^2 - m_e^2 c^4}/c$, substitute and solve for E_{e_f}:

$$E_{e_f}^2 = m_e^2 c^4 + c^2\left(E_{\gamma_i}^2 + E_{\gamma_f}^2 - 2E_{\gamma_i}E_{\gamma_f}\cos\theta\right) \qquad (6\text{-}86)$$

$$E_{e_f} = \sqrt{m_e^2 c^4 + E_{\gamma_i}^2 + E_{\gamma_f}^2 - 2E_{\gamma_i}E_{\gamma_f}\cos\theta} \qquad (6\text{-}87)$$

This formula gives the final energy of the electron after a Compton scattering event, based on the initial and final energies of the gamma ray and the scattering angle.

The photon energy is inversely proportional to its wavelength, given by the equation:

$$E = \frac{hc}{\lambda} \qquad (6\text{-}88)$$

For the initial and final energies of the photon, the corresponding equations are:

$$E_{\gamma_i} = \frac{hc}{\lambda}$$

$$E_{\gamma_f} = \frac{hc}{\lambda'}$$

From the Compton equation, the change in wavelength can be related back to the change in energy:

$$\lambda' = \lambda + \frac{h}{m_e c}(1 - \cos\theta)$$

Using the relation $\lambda = \frac{hc}{E_{\gamma_i}}$ and $\lambda' = \frac{hc}{E_{\gamma_f}}$, we can substitute:

$$\frac{hc}{E_{\gamma_f}} = \frac{hc}{E_{\gamma_i}} + \frac{h}{m_e c}(1 - \cos \theta)$$

Solving for E_{γ_f} :

$$\frac{1}{E_{\gamma_f}} = \frac{1}{E_{\gamma_i}} + \frac{1}{m_e c^2}(1 - \cos \theta)$$

$$\frac{1}{E_{\gamma_f}} = \frac{1}{E_{\gamma_i}} + \frac{1}{m_e c^2}(1 - \cos \theta)$$

$$\frac{1}{E_{\gamma_f}} = \frac{1}{E_{\gamma_i}} + \frac{1}{511 \text{ keV}}(1 - \cos \theta) \tag{6-89}$$

where $m_e c^2 = 511 \text{ keV}$ is the rest mass energy of the electron.

Isolate E_{γ_f}:

$$E_{\gamma_f} = \left(\frac{1}{E_{\gamma_i}} + \frac{1}{511 \text{ keV}}(1 - \cos \theta) \right)^{-1} \tag{6-90}$$

This is the final formula to calculate the energy of the gamma ray after it has been scattered by an electron in Compton scattering, given the initial energy E_{γ_i}, the electron rest mass energy, and the scattering angle θ.

Example 6-33. If a gamma ray with an initial energy of 1.7MeV collides with an electron and is scattered at an angle of 45°, calculate the final gamma and recoil electron energies.

Given:

Initial energy of the gamma ray $E_{\gamma_i} = 1.7\text{MeV}$
Electron rest mass energy equivalent $m_e c^2 = 511 \text{ keV} = 0.511 \text{ MeV}$
Scattering angle $\theta = 45°$
Electron rest mass energy $m_e c^2 = 511 \text{ keV} = 0.511 \text{ MeV}$

$$\cos 45° = \frac{\sqrt{2}}{2}$$

Use:

$$E_{\gamma_f} = \left(\frac{1}{E_{\gamma_i}} + \frac{1}{511\ \text{keV}} (1 - \cos\theta) \right)^{-1}$$

Plug into the formula

$$E_{\gamma_f} = \left(\frac{1}{1.7\text{MeV}} + \frac{1}{0.511\ \text{MeV}} \left(1 - \frac{\sqrt{2}}{2}\right) \right)^{-1}$$

$$E_{\gamma_f} \approx 0.8606\ \text{MeV}$$

For electron final energy use:

:

$$E_{e_f} = \sqrt{m_e^2 c^4 + E_{\gamma_i}^2 + E_{\gamma_f}^2 - 2E_{\gamma_i}E_{\gamma_f}\cos\theta}$$

Step 1: Calculate each component

$$m_e^2 c^4 = (0.511\ \text{MeV})^2 = 0.261121\ \text{MeV}^2$$

$$E_{\gamma_i}^2 = (1.7\text{MeV})^2 = 2.89\ \text{MeV}^2$$

$$E_{\gamma_f}^2 = (0.8606\text{MeV})^2 \approx 0.74063\ \text{MeV}^2$$

$$2E_{\gamma_i}E_{\gamma_f}\cos\theta = 2 \times 1.7 \times 0.8606 \times \frac{\sqrt{2}}{2} \approx 2.06902\ \text{MeV}^2$$

Substitute and Solve

$$E_{e_f} = \sqrt{0.261121 + 2.89 + 0.74063 - 2.06902}$$

$$E_{e_f} \approx 1.35009\ \text{MeV}$$

Thus, after the collision, the gamma ray's energy decreases due to the energy transferred to the electron, and this can be calculated using the above approach for any given scattering angle and initial photon energy.

6.8.2.1   Coherent and Incoherent Scattering

Coherent and incoherent scattering are two types of interactions that occur when photons, such as X-rays or gamma rays, interact with matter. They differ in their physical processes and the outcomes of the scattering events. Here's a detailed comparison of the two:

6.8.2.1.1   Coherent Scattering (Rayleigh Scattering)

Coherent scattering involves the elastic scattering of photons by atoms or molecules where the incident photon is scattered without any loss of energy (i.e., no change in wavelength). In this process, the photon interacts with the entire atom or even multiple atoms simultaneously, causing the electrons of the atom to move in phase with each other.

Since the scattering is elastic, the energy (and hence the wavelength) of the photon remains unchanged. The only change that occurs is in the direction of the photon. The scattered radiation remains coherent with the incident beam, meaning the phase relation between the scattered and incident photons is maintained.

This type of scattering is important in diffraction experiments and is fundamental in studies involving crystallography and various forms of spectroscopy. Coherent scattering is more prominent in materials with higher atomic numbers because such atoms have more electrons that can scatter the radiation coherently.

6.8.2.1.2   Incoherent Scattering (Compton Scattering)

In incoherent scattering, the photon collides with an individual electron and transfers part of its energy to the electron resulting in the electron to be ejected from the atom. This process is inelastic, meaning the scattered photon has less energy and thus a longer wavelength than the incident photon.

The amount of energy lost by the photon depends on the scattering angle. A larger angle results in more significant energy transfer and a more substantial increase in wavelength. The scattered radiation is incoherent with the incident beam, leading to a loss of phase coherence.

This scattering is critical in understanding the behavior of X-rays and gamma rays passing through materials, and it has practical implications in radiology and radiation protection.

Incoherent scattering does not strongly depend on the atomic number of the atoms in the material. It mainly depends on the electron density of the material.

6.8.3 Pair Production

Pair production occurs when the gamma ray energy is above a threshold of 1.022 MeV, sufficient to create an electron-positron pair. This interaction can only happen near a nucleus, which conserves momentum during the creation of the particle pair. The excess energy of the photon, beyond the combined rest mass energy of the positron and the electron, is shared as the kinetic energy between the two particles.

The cross-section for pair production increases with the photon energy and the atomic number of the nearby nucleus. Thus, pair production becomes more probable in materials with high atomic numbers and for high-energy gamma rays.

6.8.4 The NIST XCOM database

The NIST XCOM database (https://www.nist.gov/pml/xcom-photon-cross-sections-database), developed and maintained by the National Institute of Standards and Technology (NIST), is a comprehensive resource for photon cross-sections and attenuation coefficients. It provides data on the interaction of photons with various materials over a wide range of energies, from 1 keV to 100 GeV. The database includes information on photoelectric absorption, Compton scattering, pair production, and total attenuation coefficients for elements, compounds, and mixtures. Figure 6-13. and Figure 6-14. show the attenuation coefficients in lead and aluminum, respectively.

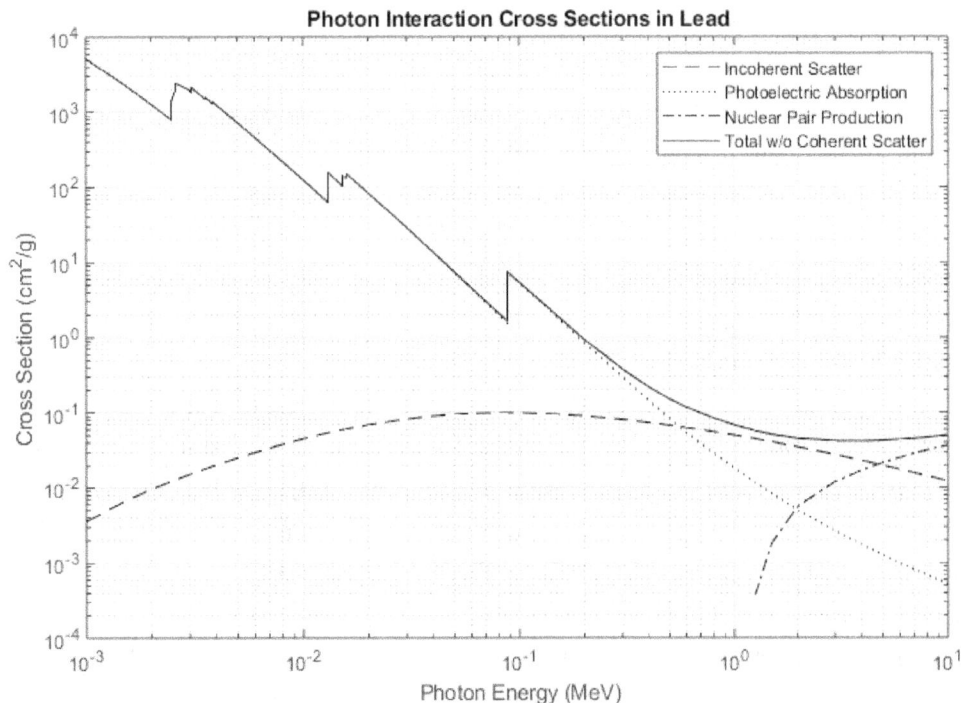

Figure 6-13. Attenuation coefficients in Lead.

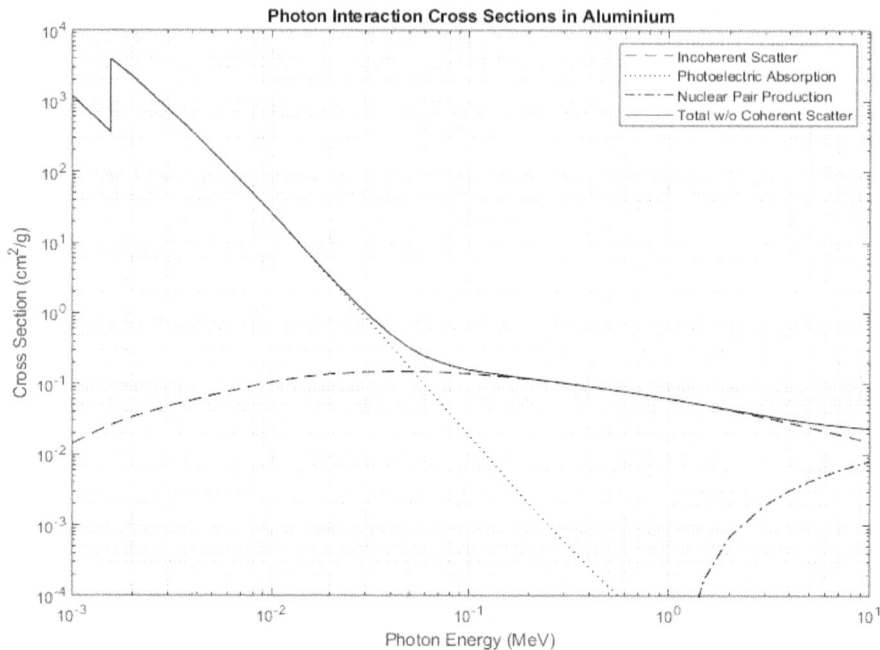

Figure 6-14. Attenuation coefficients in aluminum.

Key features and applications of the NIST XCOM database include:

1. Photon cross-section data: XCOM provides detailed information on photon cross-sections, which describe the probability of a photon interacting with a material through various processes such as photoelectric absorption, Compton scattering, and pair production. This data is essential for understanding and predicting the behavior of photons as they pass through matter.

2. Attenuation coefficients: The database includes mass attenuation coefficients and mass energy-absorption coefficients for a wide range of materials. These coefficients describe the reduction in photon intensity as a function of material thickness and are crucial for calculating radiation shielding, dose estimates, and material penetration depths.

3. Extensive material coverage: XCOM covers a vast array of materials, including elements, compounds, and mixtures. It provides data for all stable elements ($Z = 1$ to 92) and for a wide range of compounds and mixtures, such as air, water, and various biological tissues. This comprehensive material coverage makes XCOM a valuable resource for researchers working in diverse fields.

4. Energy range: The database spans a broad energy range from 1 keV to 100 GeV, covering the electromagnetic spectrum from low-energy X-rays to high-energy gamma rays. This wide energy range makes XCOM applicable to a variety of applications, including, nuclear physics, and astrophysics, radiation therapy, medical imaging.

5. Interpolation and calculation tools: XCOM includes tools for interpolating cross-section and attenuation coefficient data between tabulated energy values. It also provides functions for calculating the attenuation coefficients of mixtures and compounds based on their elemental composition.

Applications of the NIST XCOM database include:

1. Radiation shielding design: XCOM data is essential for designing effective radiation shielding for medical, industrial, and research facilities. By understanding the attenuation properties of various materials, engineers and physicists can optimize shielding designs to ensure the safety of personnel and equipment.
2. Medical imaging and radiation therapy: In medical applications, such as X-ray imaging, CT scans, and radiation therapy, XCOM data helps determine the appropriate photon energies and doses for diagnostic imaging and treatment planning. It also aids in the development of new imaging techniques and the improvement of existing ones.
3. Nuclear physics and astrophysics: Researchers in nuclear physics and astrophysics use XCOM data to model the interaction of photons with matter in various scenarios, such as stellar atmospheres, supernovae, and high-energy particle collisions.
4. Materials science: XCOM data is valuable for characterizing the properties of materials and studying their interaction with photons. This information is crucial for developing new materials with desired optical and radiation-related properties.

6.9   Decay Heat in Space Nuclear Reactors

The heat generated by the decay of radioactive isotopes must be effectively managed to ensure the safety and longevity of the reactor. In the vacuum of space, where convection and conduction are not available for heat dissipation, the primary method of rejecting decay heat is through thermal radiators.

6.9.1   Origin and Nature of Decay Heat in Reactors

In nuclear reactors, decay heat originates from the radioactive decay of fission products and actinides that remain after the reactor has been shut down. These isotopes continue to emit energy in the form of heat due to their unstable nature, even though the nuclear fission process has ceased. Managing this residual heat is crucial, especially in space, where the failure to reject decay heat effectively could lead to overheating, potentially damaging the reactor core or other critical systems.

6.9.2   Decay Heat Power and Its Time Dependence

The power generated by decay heat diminishes over time following reactor shutdown, but it remains significant in the immediate period post-shutdown. Accurate calculation and prediction of this decay heat are essential for the proper sizing and operation of thermal radiators in space applications.

6.9.2.1   Decay Heat as a Fraction of Reactor Power

For nuclear reactors, decay heat is often expressed as a fraction of the reactor's operational thermal power before shutdown. This fraction decreases with time and is typically highest immediately after the reactor is turned off. A commonly used empirical expression for decay heat power as a function of time is:

$$P_{\text{decay}}(t) = P_0 \cdot f(t) \tag{6-91}$$

where:

- $P_{\text{decay}}(t)$ is the decay heat power at time t after shutdown.
- P_0 is the reactor's thermal power before shutdown.
- $f(t)$ is the decay heat fraction as a function of time.

A typical empirical formula for $f(t)$ is:

$$f(t) = 0.06 \cdot t^{-0.2} \tag{6-92}$$

This relationship indicates that immediately after shutdown, decay heat might represent around 6% of the reactor's full power, decreasing over time as the radioactive isotopes decay.

6.9.2.2   Analytical Formulation Based on Isotopic Decay

A more detailed calculation of decay heat can be achieved by summing the contributions from individual fission products and actinides. The decay heat power is then expressed as:

$$P_{\text{decay}}(t) = \sum_i N_i \cdot E_i \cdot \lambda_i \cdot e^{-\lambda_i t} \tag{6-93}$$

where:

- N_i is the number of nuclei of isotope i produced during fission.
- E_i is the energy released per decay event of isotope i.
- λ_i is the decay constant of isotope i.

- t is the time since reactor shutdown.

This formula allows for precise calculations of decay heat over time, which is critical for designing the thermal management systems in space reactors.

6.9.3   Thermal Radiator Design for Decay Heat Rejection

In the vacuum of space, thermal radiators are the primary means of rejecting decay heat. These radiators work by emitting infrared radiation, thereby removing excess thermal energy from the reactor system. The design of thermal radiators is crucial in ensuring that the decay heat is efficiently managed, preventing overheating of the reactor core and associated systems.

6.9.3.1   Radiative Heat Transfer

The effectiveness of a thermal radiator is governed by the Stefan-Boltzmann law, which relates the power radiated by an object to its temperature:

$$P_{\text{radiator}} = \epsilon \sigma A T^4 \tag{6-94}$$

where:

- P_{radiator} is the power radiated by the thermal radiator.
- ϵ is the emissivity of the radiator surface (a value between 0 and 1).
- σ is the Stefan-Boltzmann constant (5.670×10^{-8} W/m^2 K^4).
- A is the surface area of the radiator.
- T is the absolute temperature of the radiator surface in Kelvin.

To reject decay heat efficiently, the radiator surface must have a high emissivity, and the surface area must be sufficiently large to radiate the required amount of heat. The temperature T is typically maintained within a range that balances effective heat rejection and material constraints.

6.9.3.2   Sizing the Radiator

The size of the thermal radiator must be determined based on the peak decay heat that needs to be rejected after reactor shutdown. Assuming that $P_{\text{decay}}(t)$ represents the decay heat at a particular time t, the radiator area A required can be calculated as:

$$A = \frac{P_{\text{decay}}(t)}{\epsilon \sigma T^4} \tag{6-95}$$

This equation underscores the relationship between the decay heat power, the desired radiator temperature, and the necessary surface area. For example, immediately after shutdown, when decay heat is highest, the radiator must have sufficient surface area to handle the peak load.

6.9.4   Time-Dependent Behavior and Radiator Performance

As decay heat decreases over time, the demand on the thermal radiators also diminishes. However, the radiators must be designed to operate effectively over a wide range of heat loads, from the initial high power immediately after shutdown to the much lower levels of decay heat that persist for extended periods.

6.9.4.1   Multi-Stage Radiator Systems

To manage the varying levels of decay heat, space reactors may employ multi-stage radiator systems that can adjust their active surface area or radiator orientation. For instance, additional radiator panels might be deployed during the initial high-decay-heat period and retracted as the heat load decreases. Alternatively, the coolant flow through the radiators can be adjusted to match the decay heat output.

6.9.4.2   Radiator Efficiency Over Time

The efficiency of the thermal radiators depends on their ability to maintain a stable operating temperature while rejecting varying levels of decay heat. Radiators designed for space applications must balance the need for sufficient surface area with the constraints of mass, volume, and deployment complexity.

6.9.5   Example Calculation: Radiator Sizing for Decay Heat Rejection

Consider a space nuclear reactor that operated at a thermal power of 1 MW before shutdown. Using the empirical decay heat formula, the decay heat power immediately after shutdown is:

$$P_{\text{decay}}(1) = 1\text{MW} \times 0.06 = 60 \text{ kW}$$

Assume the radiator surface has an emissivity $\epsilon = 0.9$ and is designed to operate at a temperature $T = 600$ K

Using the Stefan-Boltzmann law, the required radiator area A can be calculated as:

$$A = \frac{60 \text{ kW}}{0.9 \times 5.670 \times 10^{-8} \text{ W/m}^2 \text{ K}^4 \times (600 \text{ K})^4}$$

$$A \approx 9.072 \text{ m}^2$$

Thus, the radiator would require an area of approximately 9.072 m² square meters to effectively reject the decay heat immediately after reactor shutdown.

6.10   Problems

Example 6-34:

Polonium-210 has a half-life of 138 days. If you start with 100 grams of polonium-210, how much will remain after one year (365 days)?

Hint: Use the half-life formula $N = N_0 \times (1/2)^{t/T}$, where N_0 is the initial quantity, t is the time elapsed, and T is the half-life.

Solution:

Applying the half-life formula:

$$N = 100 \times (1/2)^{365/138} \approx 100 \times (1/2)^{2.64493} \approx 15.988 \text{ grams}$$

Example 6-35:

Calculate the kinetic energy released when uranium-238 undergoes alpha decay to form thorium-234, if the mass of uranium- 238 is $238.050788u$, the mass of thorium-234 is $234.043593u$, and the mass of the alpha particle is $4.002602u$.

Hint: Determine the mass defect from the decay and convert this mass into energy.

Solution:
- The mass defect Δm from alpha decay is:

$$\Delta m = 238.050788 - 234.043593 - 4.002602 = 0.004593u$$

$$E = 0.004593 \times 931.5 = 4.278 \text{ MeV}$$

Example 6-36:

If a neutron decays into a proton, calculate the change in mass. Assume the mass of a neutron is 1.008665u and the mass of a proton is 1.007276u.

Solution:

$$\Delta m = 1.008665u - 1.007276u = 0.001389u$$

$$E = \Delta m \times 931.5 \text{MeV/u} = 1.294 \text{MeV}$$

The change in mass results in an energy equivalent of approximately 1.294MeV being released.

Example 6-37:

Calculate the energy released when uranium-235 undergoes fission to produce krypton-90 and barium-144, with two neutrons emitted. Use the atomic masses: U-235 (235.043930 u), Kr-90 (89.9195238 u), Ba-144 (143.9229405 u), and neutron (1.008665 u).

Solution Hint: Compute the mass defect from the reaction and convert this mass into energy using $E = \Delta mc^2$, where c is the speed of light (931.5MeV/u).

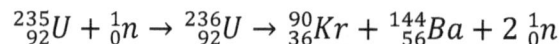

$$^{235}_{92}U + ^{1}_{0}n \rightarrow ^{236}_{92}U \rightarrow ^{90}_{36}Kr + ^{144}_{56}Ba + 2\,^{1}_{0}n$$

Solution:
- Masses:
- $U = 235.043930u$
- $Kr - 90 = 89.9195238\,u$
- $Ba - 144 = 143.9229405\,u$
- $2 \times n = 2 \times 1.008665 = 2.01733\,u$
- Mass Defect:

$$\Delta m = 235.043930 + 1.008665 - (89.9195238 + 143.9229405 + 2.01733)$$

$$\Delta m = 236.052595 - 235.8597943 = 0.1928007\,u$$

- Energy Released:

$$E = \Delta m \times \frac{931.5\text{MeV}}{u} = 0.1928007 \times \frac{931.5\text{MeV}}{u} = 179.59\text{ MeV}$$

The energy released is about 180 MeV.

Example 6-38:

Carbon-14 dating is used to determine the age of ancient artifacts. If a sample shows that the carbon-14 (half-life = 5730 years) activity is a quarter of what it is in a modern sample, how old is the artifact?

Solution Hint: Use the formula for radioactive decay $N = N_0 \times (1/2)^{t/T}$ to find the elapsed time t.

Solution:
- Half-life formula: $N = N_0 \times (1/2)^{t/T}$
- Given: $N/N_0 = 1/4$ (which is $(1/2)^2$), $T = 5730$ years

$$(1/2)^{t/5730} = 1/4$$

$$t/5730 = 2$$

$$t = 11460 \text{ years}$$

- The artifact is approximately 11,460 years old.

Example 6-39:
Polonium-210 undergoes alpha decay to lead-206. Calculate the kinetic energy of the emitted alpha particle, assuming the Q-value of the reaction is 5.407MeV and using conservation of momentum to estimate the energies of the daughter nucleus and the alpha particle. Given masses: Po-210 (209.9828737 u), Pb-206 (205.9744653 u), alpha particle (4.002603 u).

Solution Hint: Use the mass-energy equivalence to find the total kinetic energy released, then apply conservation of momentum to distribute the kinetic energy between the alpha particle and the lead nucleus.

Solution:
- Mass Defect:

$$\Delta m = 209.9828737 - (205.9744653 + 4.002603)$$

$$\Delta m = 209.9828737 - 209.9770683 = 0.0058054u$$

- Total Energy Released:

$$E = 0.0058054 \times 931.5 = 5.4077\text{MeV}$$

- Applying Conservation of Momentum (assuming the recoil of the lead nucleus and the alpha particle):

- Let v_α and v_{Pb} be the velocities of the alpha particle and lead nucleus.

$$M_\alpha v_\alpha = M_{\text{Pb}} v_{\text{Pb}}$$

$$E_\alpha = \frac{1}{2} m_\alpha v_\alpha^2$$

$$E_{\text{Pb}} = \frac{1}{2} m_{\text{Pb}} v_{\text{Pb}}^2$$

Assuming kinetic energy splits based on inverse mass ratio:

$$E_\alpha = \frac{M_{\text{Pb}}}{M_\alpha + M_{\text{Pb}}} \times E$$

$$E_\alpha = \frac{205.9744653}{205.9744653 + 4.002603} \times 5.4077$$

$$E_\alpha = \frac{205.9744653}{209.9770683} \times 5.4077 \approx 5.3046 \text{ MeV}$$

7 Multiplication, Criticality, and Reactivity

Nuclear reactors are fascinating technological marvels that transform the enormous energy released during nuclear fission reactions to generate electricity. At the heart of a nuclear reactor, a carefully controlled chain reaction takes place, where neutrons collide with fissile atoms, causing them to split and release energy in the form of heat. The nuclear heat is then used to produce steam, which drives turbines connected to generators, ultimately converting the thermal energy into electrical energy.

The development of nuclear reactors has been a remarkable achievement in human history. It has enabled us to tap into the vast potential of nuclear energy and provided a reliable, low-carbon alternative to fossil fuels. Nuclear reactors can generate large amounts of electricity with minimal greenhouse gas emissions, making nuclear power a vital technology in the fight against climate change.

However, operating nuclear reactors also comes with significant responsibilities and challenges. The safety and security of nuclear facilities are of paramount importance, as the consequences of accidents or incidents can be severe. Stringent regulations, advanced safety systems, and rigorous operational protocols are in place to minimize risks and protect both the public and the environment.

The design and functioning of nuclear reactors involve complex physics, engineering, and materials science principles. From the selection of suitable nuclear fuels and moderators to the management of reactor cooling systems and the handling of radioactive waste, every aspect of a nuclear reactor must be carefully considered and optimized to ensure efficient and safe operation.

As we embark on this journey of understanding nuclear reactors, it is essential to recognize their potential as a powerful tool in meeting our energy needs while also acknowledging the challenges and responsibilities that come with their use. By gaining a deeper knowledge of nuclear reactors, we can understand and appreciate their application in a sustainable energy future for generations to come.

7.1   Neutron Chain Reactions

Neutron chain reactions are the fundamental process that enables nuclear reactors to generate energy. Understanding how these reactions occur and how they are controlled is crucial for the safe and efficient operation of nuclear power plants. At the core of neutron chain reactions lies the concept of the effective multiplication factor. It plays a critical role in describing the behavior and stability of the reactor.

Neutron collision with a fissile nucleus, such as uranium-235 (^{235}U), can cause the nucleus to undergo nuclear fission, splitting into two fission fragments, releasing energy, and emitting more neutrons. The fission neutrons can then go on to interact with other fissile nuclei, causing further fissions and releasing more neutrons in the process. This series of events, where each fission leads to the production of more neutrons that can trigger subsequent fissions, is known as a neutron chain reaction.

For a neutron chain reaction to be self-sustaining, it is necessary to have a critical mass of fissile material. The critical mass is the minimum amount of fissile material required to maintain a stable chain reaction, where the number of neutrons produced by fission is equal to the number of neutrons lost through absorption and leakage. If the mass of fissile material is below the critical mass, the chain reaction will gradually die out, as there will be insufficient neutrons to sustain the reaction. Conversely, if the mass exceeds the critical mass, the chain reaction can grow exponentially, leading to an uncontrolled release of energy.

7.2   The Effective Multiplication Factor

The effective multiplication factor, denoted as k_{eff}, is a crucial parameter that quantifies the behavior of a neutron chain reaction in a nuclear reactor. It is defined as the ratio of the number of neutrons in one generation to the number of neutrons in the previous generation. In other words, k_{eff} represents the average number of neutrons produced by each fission that go on to cause another fission.

Mathematically, the effective multiplication factor can be expressed as:

$$k_{eff} = \frac{\text{Number of neutrons in current generation}}{\text{Number of neutrons in previous generation}} \qquad (7\text{-}1)$$

The value of k_{eff} determines the state of the reactor:

1.   If $k_{eff} = 1$, the reactor is said to be critical. In this state, the chain reaction is self-sustaining, and the number of neutrons produced by fission is equal to the number of neutrons lost. The reactor operates at a steady power level, and the neutron population remains constant over time.

2. If $k_{eff} < 1$, the reactor is subcritical. In this state, the number of neutrons produced by fission is less than the number of neutrons lost. The chain reaction will gradually diminish, and the reactor power will decrease over time.

3. If $k_{eff} > 1$, the reactor is supercritical. In this state, the number of neutrons produced by fission exceeds the number of neutrons lost. The chain reaction will grow exponentially, and the reactor power will increase rapidly, potentially leading to an uncontrolled situation if not promptly addressed.

To maintain a stable and controllable chain reaction, nuclear reactors are designed to operate near the critical state, with k_{eff} close to 1. This is achieved through the use of control rods, which are made of neutron-absorbing materials such as boron or cadmium. By inserting or withdrawing the control rods, reactor operators can fine-tune the neutron population and adjust the value of k_{eff} to maintain the desired power level.

The effective multiplication factor is influenced by several factors, including the geometry of the reactor core, the composition and enrichment of the fuel, the presence of neutron moderators and reflectors, and the efficiency of neutron absorption by the control rods. Reactor designers must carefully consider these factors to ensure that the reactor can operate safely and efficiently, with the ability to control the chain reaction under various conditions.

One important aspect of neutron chain reactions is the concept of delayed neutrons. While most neutrons are emitted promptly during fission, a small fraction (typically less than 1%) are released with a delay of several seconds to minutes. These delayed neutrons play a crucial role in the controllability of the reactor. Without delayed neutrons, the chain reaction would be highly sensitive to changes in reactivity, making it difficult to control and potentially leading to rapid power excursions.

The presence of delayed neutrons allows for a more gradual response to changes in reactivity, providing a larger margin of safety and enabling reactor operators to make necessary adjustments to maintain stable operation. The fraction of delayed neutrons, known as the delayed neutron fraction (β), is a key parameter in reactor physics calculations and is factored into the determination of the effective multiplication factor.

In addition to the effective multiplication factor, other important parameters in neutron chain reactions include the neutron flux, which represents the number of neutrons passing through a unit area per unit time, and the neutron energy spectrum, which describes the distribution of neutron energies within the reactor. The flux and reactivity parameters are closely monitored and controlled to ensure optimal reactor operations and safety.

The study of neutron chain reactions and the effective multiplication factor is a complex and multifaceted field, involving advanced concepts in nuclear physics, reactor physics, and engineering. Researchers and scientists continue to refine our understanding of these phenomena, developing new models, simulation tools, and experimental techniques to improve the efficiency, safety, and reliability of nuclear reactors.

7.2.1 Neutron Life Cycle and the Six-Factor Formula

Enrico Fermi, a renowned Italian-American physicist, developed a mathematical model to describe the neutron life cycle in a nuclear reactor. This model, known as the Fermi age theory, provides a foundation for understanding neutron behavior and reactor criticality. The six-factor formula, derived from the neutron life cycle, is a widely used expression for determining a reactor's effective multiplication factor (k_{eff}) of a reactor.

The neutron life cycle in a nuclear reactor can be described in several stages, each characterized by specific processes and probabilities. Let's define the following parameters:

- v: The average number of neutrons released per fission
- Σ_f: The macroscopic fission cross-section of the fuel
- Σ_a: The macroscopic absorption cross-section of the fuel
- Σ_c: The macroscopic capture cross-section of the fuel $\Sigma_c = \Sigma_a - \Sigma_f$
- Σ_s: The macroscopic scattering cross-section of the fuel
- L: The diffusion length of neutrons in the fuel
- B^2: The geometric buckling of the reactor (a measure of neutron leakage)

7.2.1.1 Fast Fission Factor (ε)

The fast fission factor represents the contribution of fast neutrons to the fission process. It accounts for the fissions caused by high-energy neutrons before they are slowed down to thermal energies. Because of the high resonance cross sections in U-238 in the epithermal neutron energies, fast fissions mostly come from U-238.

ε = (Number of fissions caused by fast neutrons) / (Number of thermal neutrons absorbed in fuel)

7.2.1.2 Resonance Escape Probability (p)

The resonance escape probability is the likelihood that a neutron will avoid capture by resonance absorption while being slowed down to thermal energies.

$$p = e^{-\int \frac{\Sigma_c(E)}{\xi \Sigma_s(E)} dE} \tag{7-2}$$

where ξ is the average logarithmic energy decrement per collision, and the integral is taken over the resonance energy range.

7.2.1.3   Thermal Utilization Factor (f)

The thermal utilization factor represents the probability that a thermal neutron will be absorbed in the fuel rather than in other reactor materials.

$$f = \frac{\Sigma_a(\text{fuel})}{\Sigma_a(\text{fuel}) + \Sigma_a(\text{non} - \text{fuel})} \tag{7-3}$$

7.2.1.4   Thermal Fission Factor (η)

The thermal fission factor is the average number of fast neutrons produced per thermal neutron absorbed in the fuel.

$$\eta = \frac{\nu\Sigma_f(\text{fuel})}{\Sigma_a(\text{fuel})} \tag{7-4}$$

Typical values for ν and η for various nuclear fuels is given in Table 7-1.

Table 7-1. Average Number of Neutrons Liberated in Fission.

Fissile Nucleus	Thermal Neutrons		Fast Neutrons	
	ν	η	ν	η
U-233	2.49	2.29	2.58	2.40
U-235	2.42	2.07	2.51	2.35
Pu-239	2.93	2.15	3.04	2.90

7.2.1.5   Fast Non-Leakage Probability (P_{NL_f})

The probability of a fast neutron not leaking out of the reactor before being slowed down to thermal energies is called the fast non-leakage probability.

$$P_{NL_f} = e^{-B^2\tau} \tag{7-5}$$

where τ is the Fermi age, a measure of the average distance a neutron travels from its birth as a fast neutron until it reaches thermal energies.

7.2.1.6   Thermal Non-Leakage Probability ($P_{NL_{th}}$)

The probability that a thermal neutron will not leak out of the reactor before being absorbed is called the thermal non-leakage probability.

$$P_{NL_{th}} = \frac{1}{1 + L^2 B^2} \qquad (7\text{-}6)$$

7.2.1.7 Derivation of the Six-Factor Formula:

The six-factor formula combines the six factors described above to determine the effective multiplication factor (k_{eff}) of a reactor.

$$k_{eff} = \epsilon \times p \times f \times \eta \times P_{NL_{th}} \times P_{NL_f}$$

Step 1: Start with N neutrons, multiply by the fast fission factor to account for total number of neutrons scattering and loosing energy towards the epi-thermal energy range. N_1 represents the number of neutrons after this event.

$$N_1 = N \times \epsilon$$

Step 2: Multiply N_1 by the fast non-leakage probability, P_{NL_f}, to account for neutrons that remain in the core and do not leak out at high energies:

$$N_2 = N_1 \times P_{NL_f} = N \times \epsilon \times P_{NL_f}$$

Number of neutrons that leak out at this step is $N \times \epsilon \times \left(1 - P_{NL_f}\right)$.

Step 3: Multiply N_2 by the resonance scape probability, p, to account for neutrons that scape the resonance peaks of U-238.

$$N_3 = N_2 \times p = N \times \epsilon \times P_{NL_f} \times p$$

Number of neutrons that are absorbed in the resonance peaks is given by:

$$N \times \epsilon \times P_{NL_f} \times (1 - p)$$

Step 4: Multiply N_3 by the thermal non-leakage probability, P_{NL_f}, to account for neutrons that remain in the core and do not leak out at thermal energies:

$$N_4 = N_3 \times P_{NL_{th}} = N \times \epsilon \times P_{NL_f} \times p \times P_{NL_{th}}$$

Number of neutrons that leak out of the reactor at thermal energies is given by:

$$N \times \epsilon \times P_{NL_f} \times p \times \left(1 - P_{NL_{th}}\right)$$

Step 5: Multiply N_4 by the fuel utilization factor to account for the thermal neutrons that are absorbed in the fuel and all other contents of the core.

$$N_5 = N_4 \times f = N \times \epsilon \times P_{NL_f} \times p \times P_{NL_{th}} \times f$$

Number of neutrons that are not absorbed in the fuel and therefore absorbed in all other non-fuel nuclei in the core is given by:

$$N \times \epsilon \times P_{NL_f} \times p \times P_{NL_{th}} \times (1 - f)$$

Step 6: Multiply N_5 by the thermal fission factor to account for the number fission neutrons produced from absorption and fission caused by thermal neutrons.

$$N_6 = N_5 \times \eta = N \times \epsilon \times P_{NL_f} \times p \times P_{NL_{th}} \times f \times \eta$$

Step 7: Finally, utilize the definition of k_{eff}.to achieve the formula for neutron multiplication in a nuclear chain-reacting system.

$$k_{eff} = \frac{\text{Number of neutrons in current generation}}{\text{Number of neutrons in previous generation}}$$

$$k_{eff} = \frac{N \times \epsilon \times P_{NL_f} \times p \times P_{NL_{th}} \times f \times \eta}{N}$$

$$k_{eff} = \epsilon \times p \times f \times \eta \times P_{NL_f} \times P_{NL_{th}} \tag{7-7}$$

The resulting six-factor formula represents the effective multiplication factor, which determines the criticality of the reactor. When $k_{eff} = 1$, the reactor is critical, and the neutron population remains stable. If $k_{eff} < 1$, the reactor is subcritical, and the neutron population declines. If $k_{eff} > 1$, the reactor is supercritical, and the neutron population increases.

The derivation of the neutron life cycle and the six-factor formula provides a comprehensive understanding of the various processes and probabilities involved in the behavior of neutrons within a nuclear reactor. This knowledge is essential for reactor design, safety analysis, and optimization of reactor performance.

It is important to note that the six-factor formula is a simplified model and does not account for all the complexities of neutron behavior in a reactor. More advanced models, such as multigroup diffusion theory, neutron transport theory, and Monte Carlo simulations, are used for more accurate and detailed analyses.

Nevertheless, the neutron life cycle and the six-factor formula remain foundational concepts in nuclear reactor engineering and continue to serve as valuable tools for understanding and predicting the behavior of neutrons in nuclear reactors.

Example 7-1. Calculate the k_{eff} for a typical reactor with the given six-factor parameter values.

Given Parameters:
- Fast Fission Factor (ϵ): 1.04
- Fast Non-Leakage Probability (P_{NL_f}): 0.865
- Resonance Escape Probability (p): 0.8
- Thermal Non-Leakage Probability ($P_{NL_{th}}$): 0.861
- Thermal Utilization Factor (f): 0.799
- Thermal Fission Factor (η): 2.02

Solution:

$$k_{eff} = \epsilon \times p \times f \times \eta \times P_{NL_f} \times P_{NL_{th}}$$

$$k_{eff} = 1.04 \times 0.8 \times 0.799 \times 2.02 \times 0.865 \times 0.861$$

$$k_{eff} = 1.275$$

Therefore, this reactor is supercritical which is a desirable situation when a reactor is loaded with fresh fuel. Neutron absorbers (burnable poison and control elements) are used to lower the value of f to adjust the k_{eff} value exactly to 1 for normal reactor operation.

Example 7-2. Calculate the prompt neutron population in a reactor with k_{eff} of 1.001 starting with 1000 neutrons after 100 generations. Assume that all neutrons are produced from fission.

Solution:

Assuming that this reactor starts with $N_0 = 1000$ neutrons, we can calculate the number of neutrons after one generation as

$$N_1 = N_0 \times k_{eff}$$

After another generation,

$$N_2 = N_1 \times k_{eff} = N_0 \times k_{eff} \times k_{eff} = N_0 \times \left(k_{eff}\right)^2$$

Extending this to n generations:

$$N_n = N_0 \times \left(k_{eff}\right)^n$$

So, for this problem

$$N_{100} = N_0 \times (1.001)^{100} = 1.105 \times N_0 = 1.105 \times 1000 = 1105$$

Considering that the fission neutron life cycle is of the order of 10^{-4} seconds, 100 generations amounts to only 0.01 seconds.

7.3   Conversion and Breeding in Nuclear Reactors

In nuclear fission reactors, criticality is achieved when on average, each neutron absorbed by a fissile or fissionable nucleus leads to at least one new fission neutron being released that goes on to cause another fission reaction. The number of new fission neutrons released per neutron absorbed is denoted by the Greek letter η. For a self-sustaining chain reaction to occur, η must be greater than 1 to compensate for the inevitable losses of neutrons that escape from the reactor core or are absorbed in non-fission reactions.

Checking fission data reveals that η substantially exceeds 1 for the key fissile isotopes U-235, U-233 and Pu-239 at all incident neutron energies. It is also greater than 1 for the fissionable but non-fissile isotopes Th-232 and U-238, but only for neutrons with energies above their fission threshold. However, in a reactor fueled solely with Th-232 or U-238, only a fraction of the neutrons absorbed have sufficient energy to actually induce fission. As a result, the "effective" value of η, equal to the actual η multiplied by this fission-inducing fraction, is always less than 1 for non-fissile fuels. This renders it impossible for reactors to achieve criticality using fertile isotopes alone - fissile isotopes are an essential ingredient.

Uranium-235 is the only naturally occurring fissile isotope, present at an abundance of just 0.72% in natural uranium. The other 99.27% is almost entirely U-238, apart from a trace of U-234. So, despite U-235's low natural concentration, it is possible to fuel certain reactor designs using natural uranium as the early reactors did. However, most modern power reactors require enriched uranium fuel in which the U-235 fraction has been boosted above its natural level.

While U-235 is essential for fueling reactors, its supply in nature is ultimately finite and limited. If nuclear power relied on U-235 fission alone, the nuclear age would likely span no more than a century before viable uranium resources were depleted. Fortunately, it is possible to artificially produce fissile isotopes from more abundant fertile materials through a process called conversion (also known as breeding). The two primary fissile products are U-233 bred from thorium-232, and plutonium-239 bred from uranium-238.

Uranium-233 is produced from thorium via a two-step neutron absorption and beta decay process:

$$\text{Th-232} + \text{n} \rightarrow \text{Th-233} \downarrow \beta\text{-} \rightarrow \text{Pa-233} \downarrow \beta\text{-} \rightarrow \text{U-233} \qquad (7\text{-}8)$$

Thorium-232, the only naturally occurring thorium isotope, is transformed into the fissile U-233 isotope after absorbing a neutron and undergoing two beta decays through the intermediate protactinium-233. Isotopes like Th-232 that are not directly fissile but can be converted to a fissile form via neutron absorption are classified as fertile.

The thorium conversion process is relatively straightforward in practice. Thorium, in metallic form or as an oxide, is simply placed into an operating reactor core where it is exposed to neutrons. After sufficient irradiation time for U-233 to build up, the thorium is removed and the bred U-233 is chemically separated from it. As of yet, U-233 has been produced only in limited quantities since there is almost no demand for it as a reactor fuel, with one notable exception discussed in Section 4.5. Nevertheless, thorium remains an important potential nuclear fuel resource for the future.

Plutonium-239 is produced from U-238 through a very similar two-step process:

$$\text{U-238} + \text{n} \rightarrow \text{U-239} \downarrow \beta\text{-} \rightarrow \text{Np-239} \downarrow \beta\text{-} \rightarrow \text{Pu-239} \qquad (7\text{-}9)$$

In this case, the abundant isotope U-238 serves as the fertile material that is transmuted to fissile Pu-239, with neptunium-239 as the intermediate step.

In practice, this plutonium breeding occurs naturally in most modern power reactors. These reactors are typically fueled with low-enriched uranium (LEU) in which the U-235 content is boosted to 3-5% above the natural 0.72% level. This leaves the bulk of the fuel as U-238 which steadily converts to Pu-239 during normal reactor operation. This plutonium is later extracted from the spent fuel via a chemical process called reprocessing. Uranium and plutonium are different elements which allows their chemical separation.

The extracted plutonium actually contains a mix of isotopes. Some of the bred Pu-239 absorbs additional neutrons and either fissions or is transmuted to Pu-240, which though not fissile, can become fissile Pu-241 upon absorbing yet another neutron. The Pu-241 then either fissions or becomes Pu-242. So reprocessed plutonium contains Pu-239, Pu-240, Pu-241 and Pu-242 in proportions that depend on the burnup of the fuel (the thermal energy extracted per unit mass).

To quantify the conversion concept, we define a parameter C called the conversion ratio or breeding ratio. It represents the average number of new fissile atoms produced per fissile atom consumed in the reactor. So, when N atoms of fissile fuel are consumed, NC atoms of fertile material are transmuted into new fissile atoms.

However, if the newly bred fissile species is the same as the driver fuel, those new fissile atoms may themselves later be consumed to breed another generation of NC × C = NC^2 new fissile atoms. Subsequently consuming that generation produces NC^3 new fissile atoms, and so on. Following this progression, we see that starting with N fissile atoms, fertile conversion will yield a total of:

$$NC + NC^2 + NC^3 + ... = NC / (1 - C) \qquad (7\text{-}10)$$

new fissile atoms over the life of the reactor, assuming the conversion ratio C is less than 1. In the special case where C reaches 1, an infinite amount of fertile material can theoretically be converted into new fuel starting from a given inventory of fissile material.

Summarizing the key points:

- Fissile isotopes U-235, U-233 and Pu-239 are essential for fueling reactors as they can maintain a fission chain reaction. Fertile isotopes Th-232 and U-238 alone cannot achieve criticality.
- U-235 is the only natural fissile isotope. Most reactors today require enriching natural uranium to increase its U-235 fraction.
- U-235 resources are limited. Breeding fissile U-233 from thorium and Pu-239 from U-238 can greatly extend fission energy resources.
- U-233 is bred by neutron irradiation of Th-232 which undergoes two beta decays. Pu-239 is similarly produced from U-238.
- Pu-239 breeding occurs inherently in modern reactors fueled by U-238 containing LEU. The plutonium is later extracted by reprocessing spent fuel.
- The conversion ratio C quantifies new fissile atoms produced per fissile atom consumed. A total of NC/(1-C) fissile atoms can be bred from N starter fissile atoms if C<1. If C reaches 1, fertile material can be converted indefinitely.

Example 7-3. In a critical reactor fueled with slightly enriched uranium (3% U-235), it is observed that for every neutron absorbed in U-235, 0.8 neutrons are absorbed in resonances of U-238, and 0.7 neutrons are absorbed by U-238 at thermal energies. The reactor has a neutron leakage fraction of 0.05. (a) Calculate the conversion ratio for the reactor. (b) Determine the mass of Pu-239 produced (in grams) when 1 kg of U-235 is consumed.

Solution:

(a) To find the conversion ratio, we need to calculate the number of fertile U-238 atoms converted to fissile Pu-239 atoms per U-235 atom consumed.

Given:
- For each neutron absorbed in U-235, 0.8 neutrons are absorbed in U-238 resonances.
- For each neutron absorbed in U-235, 0.7 neutrons are absorbed by U-238 at thermal energies.
- The neutron leakage fraction is 0.05.

Step 1: Calculate the total number of neutrons absorbed in U-238 per U-235 absorption. Total U-238 absorptions = 0.8 (resonance) + 0.7 (thermal) = 1.5

Step 2: Account for the neutron leakage. U-238 absorptions with leakage = 1.5 × (1 - 0.05) = 1.425
Therefore, the conversion ratio C = 1.425 Pu-239 atoms produced per U-235 atom consumed.

(b) To determine the mass of Pu-239 produced, we need to find the number of U-235 atoms in 1 kg and multiply it by the conversion ratio.

Given:

- Atomic mass of U-235 = 235 g/mol
- Atomic mass of Pu-239 = 239 g/mol
- Avogadro's number = 6.022 × 10^23 atoms/mol

Step 1: Calculate the number of U-235 atoms in 1 kg. Number of U-235 atoms

$$= 1000\,\text{g} \times \frac{1\,\text{mol}}{235\,\text{g}} \times 6.022 \times 10^{23}\,\frac{\text{atoms}}{\text{mol}} = 2.563 \times 10^{24}\,\text{atoms}$$

Step 2: Calculate the number of Pu-239 atoms produced. Number of Pu-239 atoms

$$= 2.563 \times 10^{24}\,\text{atoms} \times 1.425\,(\text{conversion ratio}) = 3.652 \times 10^{24}\,\text{atoms}$$

Step 3: Convert the number of Pu-239 atoms to mass in grams. Mass of Pu-239 = (3.652 × 10^24 atoms) × (1 mol / 6.022 × 10^23 atoms) × (239 g / 1 mol) = 1451 g

$$= 3.652 \times 10^{24}\,\text{atoms} \times \frac{1\,\text{mol}}{6.022 \times 10^{23}\,\text{atoms}} \times \frac{239\,\text{g}}{\text{mol}}$$

Therefore, when 1 kg of U-235 is consumed in the reactor, approximately 1451 grams (1.451 kg) of Pu-239 is produced.

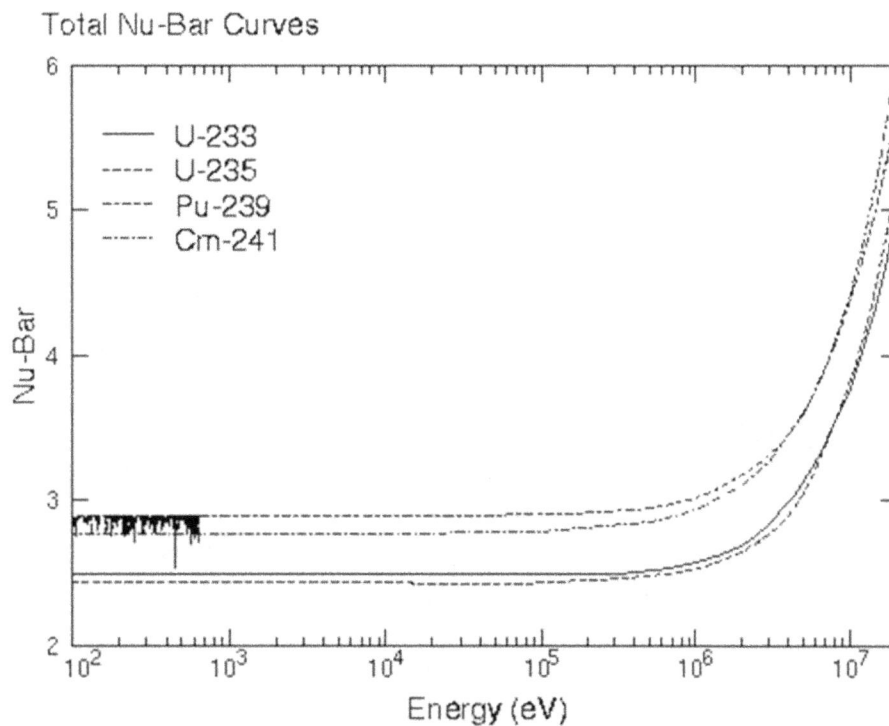

Figure 7-1. Average number of fission neutrons produced as a function of incoming neutron energies.

7.3.1   Breeding: Producing More Fuel than Consumed

An especially important scenario arises when the conversion ratio C of a reactor exceeds 1. In this case, more than one new fissile atom is produced for each fissile atom destroyed - a situation referred to as breeding. Reactors intentionally designed to achieve breeding are called breeder reactors. In contrast, reactors that convert fertile to fissile material but have a conversion ratio less than 1 are called converters. Those that do not convert fertile material at all and simply consume fissile fuel are known as burner reactors.

Breeder reactors are remarkable devices. In addition to generating useful power from the fission energy release, they actually create more fissile fuel than they consume during operation. However, designing a practical breeder reactor is significantly more challenging than creating a converter or burner reactor.

Neutron Economy and Breeding

One key requirement for breeding is that the parameter η, the average number of fission neutrons produced per neutron absorbed in the fuel, must comfortably exceed 2. This is because in any critical reactor, regardless of type, one of those fission neutrons must be absorbed by the fuel to sustain the chain reaction. Only "excess" neutrons beyond that one are available for fertile-to-fissile conversion. And since some neutrons are inevitably lost to non-fuel absorption or leakage from the reactor, η really needs to be substantially greater than 2 to make breeding feasible.

It's crucial to recognize that η is not a fixed value for a given fuel isotope, but rather depends on the energy of the neutron inducing fission. Figure 7-1 plots the variation of η with neutron energy for the three primary fissile fuels: U-233, U-235 and Pu-239.

For U-233, η is about 2.29 for fission induced by thermal neutrons (those in thermal equilibrium with the reactor structural materials, typically around 0.025 eV). This value is sufficiently above 2 that a well-designed thermal reactor fueled with U-233 could achieve breeding. In contrast, the thermal neutron η values for U-235 and Pu-239 are 2.07 and 2.14 respectively - not enough margin over 2 to overcome neutron losses and make a thermal breeder reactor practical with those fuels.

In the intermediate neutron energy range from about 1 eV to 100 keV for U-235, and 10 eV to 20 keV for Pu-239, η actually dips below 2 as seen in Figure 7-1. Breeding is therefore impossible when these isotopes fuel a reactor with predominantly intermediate energy neutrons inducing fission. Recent work has shown that U-233 fueled reactors can achieve at least a modest degree of breeding in the intermediate spectrum thanks to its η staying above 2 in that energy range.

Lastly, at fast neutron energies above about 100 keV, η climbs to values substantially exceeding 2 for all three fuels according to Figure 7-1.Figure 7-2. So, from an η standpoint, it should be possible to breed with fast fission neutrons using any of these fissile isotopes, as long as the reactor is designed to ensure most fissions are induced by high energy neutrons. Such reactors are termed fast reactors, or more commonly fast breeder reactors.

One point to note is that while U-235 has a sufficiently high fast η to permit breeding, there is no fertile isotope that can actually be converted into new U-235 fuel. A fast reactor starting on U-235 may be able to produce more Pu-239 from U-238 than the U-235 it consumes, but the U-235 is depleted without replacement. Some fast breeders have been started up using U-235 fuel in lieu of an initial plutonium inventory, but they are meant to transition to Pu-239 fuel for the long term. Plutonium is also generally preferred over U-233 as fast breeder fuel owing to its higher η value in the fast spectrum. As a result, fast breeders are nearly always Pu-239/U-238 fueled devices.

Quantifying Breeding Gain and Doubling Time

The degree of breeding that a reactor achieves is quantified by the breeding gain, denoted as G. It is simply defined as the net increase in fissile atoms per fissile atom consumed in the reactor. The breeding gain is directly related to the conversion ratio C as:

$$G = C - 1 \tag{7-11}$$

So, for a reactor with a conversion ratio of 1.2, a net 0.2 fissile atoms are produced per fissile atom burned, giving a breeding gain of 0.2.

Another common metric for breeding performance is the doubling time - defined as the time it would take for the fissile inventory of the reactor to double. To derive an expression for doubling time, consider a reactor operating at a steady thermal power of P_0 MW. It consumes fissile fuel at a constant rate of wP_0 grams per day, where w is the fuel consumption rate per unit thermal power. For U-235 fuel, w is 1.23 g/MWday. This consumption rate is equivalent to burning $wP\_0 * N\_A / M\_f$ fissile atoms per day, where N_A is Avogadro's number and M_f the atomic weight of the fuel.

7.3.2   Linear Growth of Fissile Material

In a nuclear reactor, especially one designed for breeding fissile material, the fissile inventory grows over time as a result of neutron absorption by fertile material (like U-238 or Th-232) converting it into fissile isotopes (like Pu-239 or U-233). The growth of fissile material can be modeled as a linear function of time if the rate of production remains constant and the reactor operates under steady-state conditions. This scenario can be described by:

$$m(t) = m_0 + GwP_0t \tag{7-12}$$

Where:

- $m(t)$ is the total fissile mass at time t,
- m_0 is the initial fissile inventory,
- Gw is the conversion factor representing the efficiency of converting fertile material to fissile material,
- P_0 is the initial power output of the reactor.

Linear Doubling Time (t_{Dl}) is then defined as the time required for the fissile inventory to double from its initial mass m_0 :

$$t_{Dl} = \frac{m_0}{GwP_0}$$
(7-13)

This doubling time gives a basic metric for how quickly the reactor can produce useful fissile material. However, this simplistic approach does not typically reflect realistic operational strategies because it assumes that all produced fissile material remains within the reactor, potentially leading to suboptimal and unsafe conditions.

7.3.3   Exponential Growth of Fissile Material:

Realistically, allowing fissile material to accumulate indefinitely within a reactor is neither safe nor practical. Instead, breeder reactors are designed to periodically extract the bred fissile material and replace it with fresh fertile material. This cycle optimizes the reactor's operation and safety while also supporting the start-up of additional reactors.

When the produced fissile material is removed and used efficiently, the growth rate of the remaining fissile inventory can be modeled more accurately with an exponential function:

$$\frac{dm}{dt} = Gw\beta m$$

Where β is a proportionality constant linking the reactor's power output to its fissile mass:

$$P = \beta m$$
(7-14)

This leads to the differential equation:

$$\frac{dm}{dt} = Gw\beta m$$
(7-15)

The solution to this differential equation is:

$$m = m_0 \cdot \exp\left(Gw\beta t\right) \tag{7-16}$$

Where m_0 is the initial fissile mass.

Exponential Doubling Time (t_{De}), or compound doubling time, is then the time required for the fissile inventory to double under this exponential growth model:

$$t_{De} = \frac{\ln(2)}{Gw\beta}$$

Given the initial condition $P_0 = \beta m_0$, we can relate β to the initial conditions and express the exponential doubling time as:

$$t_{De} = \frac{m_0 \cdot \ln(2)}{GwP_0}$$

This shows that:

$$t_{De} = t_{Dl} \cdot \ln(2) = 0.693 \cdot t_{Dl} \tag{7-17}$$

Comparing t_{De} with t_{Dl}, we observe that the exponential model predicts a faster doubling time due to the efficient use and management of the fissile material. This comparison underlines the importance of effective fuel cycle management in nuclear reactor operations, emphasizing that continuous buildup of fissile material within the reactor is less practical than a cyclic refresh strategy which aligns with actual reactor operations and safety protocols.

The distinction between linear and exponential growth of fissile inventory in nuclear reactors is depicted in theoretical models, which outline the differences in doubling times, t_{Dl} and t_{De}, respectively. Historically, t_{Dl} has been used to estimate the time required to double the fissile inventory through linear growth. However, in contemporary discussions and applications, t_{De} is more frequently referenced as it better captures the dynamics of modern fuel cycle practices where fissile material is not simply accumulated but cyclically processed and reused.

It is crucial to understand that the actual time required for a reactor operator to double their fissile inventory typically surpasses both t_{Dl} and t_{De}. These theoretical doubling times do not consider the practical aspects such as the time needed to extract bred fuel, separate it chemically from other substances, and reform it into new fuel elements. Despite this, the concept of doubling time remains a valuable benchmark for comparing the breeding efficiency across different reactor designs. Reactors with shorter doubling times are generally regarded as more efficient breeders, all other factors being constant.

Key Concepts in Breeding:

- Breeding Ratio: Breeding occurs effectively when the conversion ratio exceeds 1, indicating that more fissile atoms are being produced from fertile material than are being consumed. This process allows a reactor to generate more fissile material than it uses, thereby sustaining or expanding its fuel supply.
- Neutron Economy: Successful breeding relies on having an average neutron economy factor, η (neutrons produced per neutron absorbed), that is significantly higher than 2. This compensates for any neutron losses during the process. The value of η achievable is dependent on both the type of fissile isotope used and the neutron energy spectrum initiating the fission reactions. For instance, only U-233, among commonly used fissile isotopes, has an η value greater than 2, making it suitable for thermal breeding. In contrast, all fissile isotopes can achieve the necessary η for effective fast breeding.
- Breeding Gain: Quantified by the breeding gain G, which is defined as the excess fissile atoms produced per fissile atom consumed (i.e., $G = C - 1$, where C is the conversion ratio).
- Doubling Time: This is a measure of how quickly a breeder reactor can increase its fissile inventory. The linear model assumes that the bred fuel accumulates within the reactor itself, while the exponential model, which is more commonly utilized today, assumes that bred fuel is continuously recycled and used in new or existing breeders. Real-world doubling times will always extend beyond these theoretical estimates due to the additional time required for fuel processing.
- Fissile Fuel Choice: Fast breeders typically use plutonium (Pu-239) due to its high fast η value. Although these reactors may initially use U-235, they are intended to transition to predominantly using Pu-239. It is noteworthy that breeding U-235 directly from U-238 is not feasible.

This more detailed explanation clarifies the theoretical frameworks used to understand fissile material growth in reactors and emphasizes the practical challenges and strategic considerations involved in managing nuclear fuel cycles efficiently.

Example 7-4. A conceptual fast breeder reactor is initially fueled with 1500 kg of U-233 and operates at full power, consuming U-233 at a rate of 1200 grams per day. It achieves a breeding gain of 0.20.

(a) Calculation of U-233 Breeding Rate:
   Breeding gain (G) is defined as the ratio of the rate of production of new fissile

material (U-233 in this case) to the rate of its consumption.

Given:

- U-233 consumption rate = 1200 g/ day
- Breeding gain $(G) = 0.20$

Breeding rate calculation:

The breeding rate can be calculated by rearranging the formula for breeding gain:

Rate of U-233 production = $G \times$ Rate of U-233 consumption

Rate of U-233 production = 0.20×1200 g/ day = 240 g/ day

(b) Calculation of Doubling Times:

Doubling times are measures of how quickly the fissile inventory can double in mass.

Linear Doubling Time (t_{Dl}) :

The linear doubling time is calculated using the formula:

$$t_{Dl} = \frac{m_0}{G \times w \times P_0}$$

Where:

- m_0 = initial U-233 inventory = 1500 kg
- w = fissile consumption rate per unit power (assumed as 1.2 g/MW-day for U-233)
- P_0 = reactor power (assumed as 1000MW th)

Converting m_0 from kg to grams for consistency:

$$t_{Dl} = \frac{1500 \times 10^3 \text{ g}}{0.20 \times 1.2 \text{ g/MW} - \text{ day } \times 1000\text{MW}}$$

$$t_{Dl} = 6{,}250{,}000 \text{ days} \approx 17{,}123 \text{ years}$$

Exponential Doubling Time (t_{De}):

The exponential doubling time, accounting for continuous compound growth, is calculated as:

$$t_{De} = \frac{m_0 \times \ln(2)}{G \times w \times P_0}$$

$$t_{De} = \frac{1500 \times 10^3 \text{ g} \times 0.693}{0.20 \times 1.2 \frac{\text{g}}{\text{MW}} - \text{day} \times 1000\text{MW}}$$

$$t_{De} = 4{,}331{,}250 \text{ days} \approx 11{,}866 \text{ years}$$

In this theoretical fast breeder reactor setup:

- (a) U-233 is being bred at a rate of 240 grams per day.
- (b) The linear doubling time is approximately 17,123 years, while the exponential doubling time is approximately 11,866 years.

Note: These extremely long doubling times are a result of the low breeding gain and large initial inventory of fissile material used in this hypothetical example. Practical breeder reactors typically aim for much shorter doubling times, often on the order of years or decades.

7.3.4   Nuclear Fuel Performance: Burnup and Specific Burnup

In the context of nuclear fuel performance, two key metrics are used to quantify the energy extracted from the fuel: burnup and specific burnup. These parameters provide essential information about the efficiency and longevity of the fuel in a nuclear reactor.

Burnup

Burnup is defined as the total energy released through fission by a given quantity of nuclear fuel. It is typically measured in megawatt-days (MWd), which represents the thermal energy output of the fuel over time. Burnup is a cumulative measure, meaning it increases as the fuel continues to undergo fission in the reactor.

The concept of burnup is crucial for assessing the overall performance and economic viability of nuclear fuel. Higher burnup values indicate that more energy has been extracted from the fuel, making it more cost-effective. However, as the fuel undergoes fission and accumulates fission products, its physical and chemical properties change, which can limit the maximum achievable burnup.

Specific Burnup

Specific burnup, on the other hand, is a measure of the fission energy released per unit mass of the nuclear fuel. It is typically expressed in megawatt-days per metric ton (MWd/t) or megawatt-days per kilogram (MWd/kg) of the initial heavy metal content in the fuel. Specific burnup allows for a standardized comparison of fuel performance across different reactor designs and fuel types.

To understand the theoretical maximum specific burnup, let's consider the fissioning of U-235, the most common fissile isotope used in nuclear fuel. The complete fission of 1 gram of U-235 releases approximately 1 MW-day of energy. Therefore, the maximum theoretical specific burnup for U-235 fuel is:

$$(1 \text{ MW-day} / 1 \text{ g U-235}) \times (1{,}000{,}000 \text{ g} / 1 \text{ metric ton}) = 1{,}000{,}000 \text{ MWd/t}$$

or

$$(1 \text{ MW-day} / 1 \text{ g U-235}) \times (1000 \text{ g} / 1 \text{ kg}) = 1{,}000 \text{ MWd/kg}$$

However, it's important to note that achieving this theoretical maximum is not practical due to various limiting factors, such as the accumulation of fission products, structural changes in the fuel, and reactor safety considerations.

Fractional Burnup

Another way to express fuel burnup is through the concept of fractional burnup, denoted as β. Fractional burnup is defined as the ratio of the number of fissions that have occurred in a specific mass of fuel to the total number of heavy atoms initially present in that fuel. It is often expressed as a percentage.

Mathematically, fractional burnup can be represented as:

$$\beta = (\text{Number of fissions}) / (\text{Total number of initial heavy atoms})$$

By considering the energy released per fission, the relationship between fractional burnup and specific burnup can be derived. For U-235 fuel, if we assume that all the heavy atoms are U-235 and undergo complete fission, the specific burnup at any given time is proportional to the fractional burnup:

$$\text{Specific Burnup} = (\text{Fractional Burnup}) \times (\text{Maximum Theoretical Specific Burnup})$$

In practice, the maximum achievable specific burnup for a particular reactor fuel depends on several factors, including:

1. Fuel composition: The initial enrichment of fissile isotopes in the fuel and the presence of burnable poisons affect the maximum burnup.
2. Fuel design: The physical form of the fuel (e.g., pellets, rods, plates) and its cladding material influence its performance and burnup limits.
3. Reactor operating conditions: The power density, neutron flux, and coolant temperature in the reactor core impact the fuel's behavior and burnup.
4. Fuel integrity: As the fuel undergoes fission, it experiences changes in its mechanical, thermal, and chemical properties. These changes can lead to fuel swelling, cracking, or other forms of damage that limit the maximum burnup.
5. Reactor safety margins: Regulatory requirements and safety considerations, such as maintaining the integrity of the fuel cladding and preventing the release of radioactive materials, also impose limits on the maximum allowable burnup.

The interplay of these factors determines the practical maximum specific burnup that can be achieved in a given reactor system. Research and development aim to enhance fuel performance, increase burnup limits, and optimize reactor operations while ensuring safety and reliability.

Burnup and specific burnup are essential metrics for evaluating nuclear fuel performance. They quantify the energy extracted from the fuel and provide insights into its efficiency and limitations. Fractional burnup relates the number of fissions to the initial heavy metal content and is proportional to the specific burnup. Understanding these concepts is crucial for the design, operation, and economics of nuclear power plants.

Example 7-5. Example: A pressurized water reactor (PWR) is loaded with 110 metric tons of uranium dioxide (UO_2) fuel, where the uranium is enriched to 4.5% U-235. The reactor operates at a steady power level of 3,800 MW for 900 days before it is shut down for refueling. (a) Calculate the specific burnup of the fuel at shutdown. (b) Determine the fractional burnup at shutdown.

Solution: (a) To calculate the specific burnup, we use the formula: Specific Burnup = (Reactor Power × Operating Time) / (Mass of Heavy Metal)

Given:
- Reactor Power: 3,800MW
- Operating Time: 900 days
- Total Mass of UO_2 Fuel: 110 metric tons

Conversion of UO_2 Mass to Uranium Mass:
- Molecular Weight of UO_2: $U(238 \text{ g/mol}) + O_2(2 \times 16 \text{ g/mol}) = 270 \text{ g/mol}$
- Molecular Weight of Uranium (U): 238 g/mol

The mass of uranium (heavy metal) in the UO_2 can be calculated by:

$$\text{Mass of U} = 110 \text{ metric tons} \times \left(\frac{238}{270}\right) \approx 97.04 \text{ metric tons}$$

Calculation of Specific Burnup:

$$\text{Specific Burnup} = \frac{\text{Reactor Power} \times \text{Operating Time}}{\text{Mass of Uranium}}$$

$$\text{Specific Burnup} = \frac{3,800\text{MW} \times 900 \text{ days}}{97.04 \text{ metric tons}} \approx 35,285\text{MWd/t}$$

(b) Determine the Fractional Burnup at Shutdown:

Fractional burnup is the ratio of the specific burnup to the maximum theoretical specific burnup of U235, providing a measure of how much of the potential energy in the fuel was utilized.

Given Maximum Theoretical Specific Burnup for U-235: 1,000,000 MWd/t

$$\text{Fractional Burnup} = \frac{\text{Specific Burnup}}{\text{Maximum Theoretical Specific Burnup}}$$

$$\text{Fractional Burnup} = \frac{35,285\text{MWd/t}}{1,000,000\text{MWd/t}} = 0.03528 \text{ or } 3.528\%$$

For this PWR operation cycle:
- (a) The specific burnup of the fuel at shutdown is 35,285MWd/t.
- (b) The fractional burnup at shutdown is **3.528%**.

Note: This calculation simplifies by focusing on the burnup of U-235, which is part of an enriched UO_2 fuel matrix. While it's an approximation (since not all heavy metal is U-235), it's a common practice in nuclear engineering to provide a good estimate of fuel utilization and efficiency.

In a nuclear reactor fueled with low-enriched uranium (LEU), the majority of the fuel consists of uranium-238 (U-238 or "28"). LEU typically contains only 3-5% of the fissile isotope uranium-235 (U-235 or "25"), with the remainder being U-238. Although U-238 is not directly fissionable by thermal neutrons, it plays a crucial role in the reactor's operation and contributes significantly to the overall energy production.

As the reactor operates, some of the U-238 undergoes neutron capture reactions, leading to the formation of various plutonium isotopes. The most significant of these isotopes is plutonium-239 (Pu-239 or "49"), which is a fissile nuclide. The process of converting fertile U-238 into fissile Pu-239 is known as breeding, and it occurs through the following reaction chain:

$$\text{U-238 (n,}\gamma) \rightarrow \text{U-239 (}\beta^-) \rightarrow \text{Np-239 (}\beta^-) \rightarrow \text{Pu-239} \qquad (7\text{-}18)$$

In this process, U-238 captures a neutron, becoming U-239, which then undergoes two beta decays, first to neptunium-239 (Np-239 or "39") and then to Pu-239. The bred Pu-239 can then participate in the fission process, contributing to the reactor's power output.

In addition to the breeding of Pu-239, some of the U-238 in the reactor can also undergo fission directly when interacting with fast neutrons. Although the fission cross-section of U-238 for thermal neutrons is nearly zero, it has a small but non-negligible fission cross-section for high-energy neutrons. These fast fissions of U-238 contribute to the overall energy production and neutron economy of the reactor.

As a result of the breeding and fast fission processes, a significant portion of the fissions in an LEU-fueled reactor actually occur in non-U-235 atoms. In the example given, over 35% of the fissions may take place in plutonium and U-238. This fact explains why the fractional burnup (β) of 3.528% is larger than the initial fuel enrichment of 4.5% U-235.

The composition of the fuel evolves throughout the reactor's operation, with the U-235 content gradually decreasing and the plutonium content increasing. By the time the fuel is removed from the reactor, the spent fuel may still have a U-235 enrichment of around 1%, even though a significant amount of the initial U-235 has been consumed. This remaining U-235, along with the bred plutonium, contributes to the residual radioactivity and decay heat of the spent fuel.

To simplify the notation and discussion of various fissile and fertile nuclides, a numerical code system was introduced during the Manhattan Project of World War II. In this code, each nuclide is represented by a two-digit number derived from its atomic number (Z) and atomic mass number (A).

The code is constructed as follows:

- The atomic number (Z) is represented by the second digit of Z. For example, if Z = 92 (uranium), the corresponding digit is 2.
- The atomic mass number (A) is represented by the last digit of A. For example, if A = 235, the corresponding digit is 5.

Using this code, some common fissile and fertile nuclides are represented as:

- U-235: "25"
- U-238: "28"
- Pu-239: "49"
- Pu-241: "41"

This numerical code provides a concise way to refer to the various nuclides involved in nuclear reactions and is widely used in the field of nuclear engineering.

The presence of U-238 in LEU-fueled reactors leads to the breeding of fissile Pu-239 and the occurrence of fast fissions, which contribute significantly to the reactor's energy production and result in a higher fractional burnup than the initial U-235 enrichment. The composition of the fuel changes throughout the reactor's operation, with spent fuel containing a reduced but still significant amount of U-235 and bred plutonium. The Manhattan Project numerical code system provides a convenient way to refer to the various fissile and fertile nuclides involved in these processes.

7.4   Reactivity

Reactivity is a fundamental concept in nuclear reactor physics, encapsulating the deviation of a reactor's operational state from criticality. Represented by the Greek letter ρ (rho), reactivity measures the relative change in neutron population per neutron generation, serving as an indicator of the reactor's response behavior. This dimensionless quantity closely correlates with the reactor's effective multiplication factor (k_{eff}).

7.4.1   Derivation of Reactivity

Consider a nuclear reactor operating without any external neutron sources. If we denote the number of neutrons in one generation as N_0, the number in the subsequent generation can be expressed as

The difference in neutron numbers between these two generations, representing the change in neutron population, is calculated as:

$$\Delta N = N_0 \times k_{\text{eff}} - N_0$$

To express this as a fractional change per generation, we normalize ΔN by the number of neutrons in the generation that follows:

$$\rho = \frac{N_0 \times k_{\text{eff}} - N_0}{N_0 \times k_{\text{eff}}} \tag{7-19}$$

By simplifying the expression, the N_0 terms cancel out, leading to the formula for reactivity:

$$\rho = \frac{k_{eff} - 1}{k_{eff}} \qquad (7\text{-}20)$$

This equation reveals that reactivity is fundamentally a function of $k_{eff.}$.

7.4.2   Interpretation of Reactivity Values

The value of k_{eff} directly influences the sign and magnitude of ρ :

- If $k_{eff} > 1$: Reactivity is positive ($\rho > 0$), indicating a supercritical state. In this condition, the neutron population increases with each generation, suggesting a reactor that is generating more neutrons than it consumes.
- If $k_{eff} = 1$: Reactivity is zero ($\rho = 0$), indicating a critical state where the neutron population remains stable across generations. This state is ideal for steady power generation in a controlled environment.
- If $k_{eff} < 1$: Reactivity is negative ($\rho < 0$), indicating a subcritical state where the neutron population diminishes with each generation, implying that the reactor is consuming more neutrons than it produces.

7.4.3   Practical Implications of Reactivity

The absolute value of reactivity provides a quantitative measure of how far a reactor is from its critical state. Higher absolute values of reactivity indicate a greater deviation from criticality, which can have significant operational implications:

- Positive high reactivity requires swift corrective actions to prevent potential safety hazards due to an uncontrolled increase in power.
- Negative high reactivity may necessitate intervention to boost reactor performance and return to criticality, ensuring effective power generation.

Understanding reactivity and its relationship with k_{eff} is crucial for effective reactor design, operation, and safety management. This knowledge allows reactor operators and engineers to predict the behavior of the reactor core under different operational scenarios and implement necessary control strategies to maintain desired performance and safety standards.

Example 7-6. Example: Calculate the reactivity in a reactor core when k_{eff} is equal to 1.002 and 0.998.

In this example, we calculate the reactivity ρ for two specific values of the effective multiplication factor $k_{eff} = 1.002$ and $k_{eff} = 0.998$.

Calculation Process

Reactivity ρ is defined as the relative change in neutron population per neutron generation and is expressed mathematically as:

$$\rho = \frac{k_{\text{eff}} - 1}{k_{\text{eff}}}$$

1. For $k_{\text{eff}} = 1.002$:

$$\rho = \frac{1.002 - 1}{1.002} \approx 0.002 = 2 \times 10^{-3} \Delta k/k$$

2. For $k_{\text{eff}} = 0.998$:

$$\rho = \frac{0.998 - 1}{0.998} \approx -0.002 = -2 \times 10^{-3} \Delta k/k$$

Units of Reactivity

Reactivity is inherently a dimensionless quantity since it arises from a ratio of changes in the multiplication factor. Various units are commonly used to make reactivity values more interpretable:

- **Δk/k** : Base unit derived directly from the formula.
- **%Δk/k** : Percentage of $\Delta k/k$. Convert by multiplying by 100.
- pcm (percent milli-rho): Defined as $10^{-5} \Delta k/k$. Convert by multiplying by 10^5.
- Cents: Related to the delayed neutron fraction (β), typically discussed in more advanced reactor physics contexts.

Converting Reactivity Values to Different Units

Let's apply these conversions to specific examples:

1. Converting $0.0025 \Delta k/k$ to %Δk/k :

$$0.0025 \Delta k/k \times 100 = 0.25\% \Delta k/k$$

2. Converting -150pcm to Δk/k :

$$-150 \text{pcm} \times 10^{-5} = -0.0015 \Delta k/k$$

Determining k_{eff} from Reactivity

If reactivity ρ is known, the effective multiplication factor k_{eff} can be calculated by rearranging the formula for reactivity:

$$k_{\mathrm{eff}} = \frac{1}{1-\rho}$$

Example: Calculate k_{eff} given a reactivity of $-20.0 \times 10^{-4}\Delta k/k$:

$$\rho = -\frac{0.0020\Delta k}{k}$$

$$k_{\mathrm{eff}} = \frac{1}{1-(-0.0020)} \approx 0.998$$

This example highlights how changes in k_{eff} impact reactivity, and it underscores the importance of understanding these relationships for effective reactor management and safety assessments.

Reactivity is a fundamental concept in nuclear reactor physics that quantifies the deviation of a reactor from the critical state. It is a dimensionless quantity that depends on the effective multiplication factor (k_{eff}) and can be expressed in various units such as $\Delta k/k$, $\%\Delta k/k$, and pcm. The sign and magnitude of reactivity indicate the state of the reactor (supercritical, critical, or subcritical) and the extent of its departure from criticality. Understanding reactivity is essential for the safe operation and control of nuclear reactors.

7.5   Reactivity Coefficients and Reactivity Defects

Reactivity (ρ) is a pivotal factor in nuclear reactor physics, fundamentally determining the behavior of neutron populations and thus directly influencing reactor power dynamics. Reactivity can be affected by a range of operational conditions including fuel consumption, changes in temperature, variations in pressure, and the introduction of neutron-absorbing materials (neutron poisons).

Reactivity Coefficients (α_x) quantify how reactivity ($\Delta\rho$) changes in response to variations in specific reactor parameters (Δx). These coefficients provide critical insight into how sensitive the reactor's reactivity is to different operational changes. The formula for calculating a reactivity coefficient is given by:

$$\alpha_x = \frac{\Delta\rho}{\Delta x} \qquad (7\text{-}21)$$

Here, the subscript x indicates the particular reactor parameter influencing reactivity. This could include moderator temperature, fuel temperature, reactor pressure, or control rod positions. The units of measurement for reactivity coefficients will vary based on the parameter in question; for instance, the moderator temperature coefficient might be measured in pcm/ °F (percent millirho per degree Fahrenheit).

7.5.1 Significance of the Sign in Reactivity Coefficients

The sign of a reactivity coefficient is a critical indicator of how reactivity will respond to changes in the associated parameter:

- Positive Reactivity Coefficient ($\alpha_x > 0$) : An increase in the parameter results in an increase in reactivity. This indicates a direct relationship where changes that raise the parameter's value enhance the reactor's reactivity, potentially leading to higher power output. However, in safetycritical systems, a positive reactivity coefficient can sometimes signal increased risk, as it may lead to runaway reactions if not properly managed.
- Negative Reactivity Coefficient ($\alpha_x < 0$) : An increase in the parameter leads to a decrease in reactivity. This inverse relationship is generally favorable for reactor safety, as it implies that any rise in the parameter naturally dampens the reactor's reactivity, contributing to a self-stabilizing system behavior. Negative coefficients are particularly important in scenarios involving temperature increases; they help ensure that the reactor remains under control even under adverse conditions.

Example 7-7. Consider a reactor with a moderator temperature coefficient of -7.5pcm/ °F. If the temperature of the moderator increases by $10°$F, we can calculate the resulting change in reactivity using the given coefficient.

Calculation Steps:

1. Identify the reactivity coefficient (α_x): -7.5pcm/ °F
2. Measure the change in moderator temperature (Δx): $10°$F

Using the formula for reactivity defects:

$$\Delta\rho = \alpha_x \times \Delta x$$

Substitute the values into the formula:

$$\Delta\rho = -7.5\text{pcm}/\,°\text{F} \times 10°\text{F} = -75\text{pcm}$$

The calculated reactivity change of -75pcm indicates a decrease in reactivity as a result of the temperature increase. The negative sign confirms that an increase in moderator temperature leads to a reduction in reactivity, enhancing the reactor's safety by moderating its response to heat increases.

7.5.2   Reactivity Defects

A reactivity defect ($\Delta\rho$) represents the overall change in reactivity that results from a specific alteration in a reactor parameter. To determine a reactivity defect, you multiply the change in the parameter (Δx) by the mean value of the associated reactivity coefficient (α_x) across the extent of that parameter's alteration. The formula used to calculate a reactivity defect is straightforward:

$$\Delta\rho = \alpha_x \times \Delta x \qquad\qquad (7\text{-}22)$$

This equation helps quantify how variations in operational conditions, such as temperature, pressure, or control rod adjustments, impact the reactor's overall reactivity, facilitating precise adjustments to maintain reactor stability and safety.

Example 7-8. Consider a scenario where a reactor's fuel temperature coefficient is -1.2pcm$/\,°$C. If the fuel temperature in this reactor increases by $50°$C, we can calculate the resulting change in reactivity, referred to as the reactivity defect.

Calculation Steps:

1. Reactivity Coefficient (α_x): -1.2pcm$/\,°$C
2. Change in Fuel Temperature (Δx): $50°$C

Using the formula for reactivity defects:

$$\Delta\rho = \alpha_x \times \Delta x$$

Substitute the known values into the formula:

$$\Delta\rho = -1.2\text{pcm}/\,°\text{C} \times 50\,°\text{C} = -60\text{pcm}$$

The result, -60pcm, represents a decrease in reactivity as a consequence of the increase in fuel temperature. The negative sign in the reactivity defect confirms that the reactor's reactivity decreases with a rise in fuel temperature, which contributes to stabilizing the reactor's operational state under increased thermal conditions.

The importance of reactivity coefficients and defects in nuclear reactor physics cannot be overstated. These concepts are fundamental to understanding, predicting, and controlling the behavior of a nuclear reactor, ensuring its safe and efficient operation. Reactivity coefficients quantify the change in reactivity due to variations in parameters such as temperature, pressure, or void fraction, while reactivity defects represent the overall reactivity change caused by specific conditions like fuel burnup or xenon poisoning.

One of the primary applications of reactivity coefficients is in the design and optimization of reactor control systems. By understanding how the reactor responds to changes in various parameters, engineers can develop effective control mechanisms, such as control rods and chemical shim, to maintain the desired reactivity and power level. This ensures that the reactor operates within its designed limits and prevents unintended fluctuations in power output.

Reactivity coefficients also play a vital role in assessing the inherent safety features of a reactor. Negative reactivity coefficients, particularly the negative moderator temperature coefficient, provide a self-regulating effect that helps stabilize the reactor during transients or accidents. If the reactor temperature rises, the negative coefficient causes a decrease in reactivity, slowing down the fission process and preventing further temperature escalation. This inherent safety feature is crucial for preventing reactor runaway and maintaining control in emergency situations.

Another important application of reactivity coefficients and defects is in predicting reactor behavior during various operational stages, such as startup, shutdown, and power maneuvers. By accurately determining these values, reactor operators can anticipate the required control rod movements and other adjustments needed to compensate for reactivity changes. This knowledge is essential for ensuring smooth and safe transitions between different power levels and operating conditions.

Furthermore, reactivity coefficients and defects are integral to fuel management strategies. They influence decisions regarding fuel loading patterns, burnup optimization, and the overall efficiency of the reactor. By considering these factors, nuclear engineers can design fuel cycles that maximize energy output while minimizing waste and ensuring safe operation throughout the entire fuel lifecycle.

7.6   Reactivity Coefficients and Moderator Effects in Nuclear Reactors

Reactivity coefficients are crucial parameters in nuclear reactor physics that quantify the change in reactivity because of variations in the physical properties of the reactor materials. These coefficients are essential for understanding the behavior and safety of nuclear reactors, as they determine how the reactor responds to changes in temperature, pressure, and other variables. One of the most important factors affecting reactivity is the moderator, which plays a vital role in slowing down neutrons to sustain the fission chain reaction.

7.6.1 Moderator Characteristics and Effectiveness

An effective moderator should possess the following desirable characteristics: a. Large neutron scattering cross section: This ensures efficient slowing down of neutrons through elastic collisions. b. Low neutron absorption cross section: Minimizes the loss of neutrons due to absorption in the moderator. c. Large neutron energy loss per collision: Allows neutrons to reach thermal energies rapidly, reducing the likelihood of loss through resonance absorption.

Light moderators, such as hydrogen and deuterium, are generally more effective than heavy moderators because they remove more energy per collision, enabling neutrons to reach thermal energies more quickly. The effectiveness of a moderator can be quantified using two key parameters: the macroscopic slowing down power (MSDP) and the moderating ratio (MR).

The macroscopic slowing down power is defined as the product of the logarithmic energy decrement per collision (ξ) and the macroscopic scattering cross section (Σs):

$$MSDP = \xi \, \Sigma s \qquad (7\text{-}23)$$

MSDP indicates the rate at which neutrons are slowed down in the moderator material. However, it does not fully account for the moderator's effectiveness, as it does not consider neutron absorption. To incorporate this aspect, the moderating ratio (MR) is introduced:

$$MR = \xi \, \Sigma s \, / \, \Sigma a \qquad (7\text{-}24)$$

The moderating ratio is the ratio of the slowing down power to the macroscopic absorption cross section (Σa). A higher moderating ratio indicates a more effective moderator material.

7.6.2 Moderator-to-Fuel Ratio and Its Impact on Reactor Performance

The moderator-to-fuel ratio (N_m/N_u) is a critical parameter that significantly influences the behavior and performance of a nuclear reactor. As the amount of moderator in the core increases (i.e., N_m/N_u increases), neutron leakage decreases, while neutron absorption in the moderator increases, leading to a decrease in the thermal utilization factor (f). On the other hand, having insufficient moderator in the core (i.e., N_m/N_u decreases) results in a longer slowing down time, which leads to a greater loss of neutrons through resonance absorption and an increase in neutron leakage.

The effects of varying the moderator-to-fuel ratio on the thermal utilization factor and the resonance escape probability (p) are illustrated in Figure 7-2.. There exists an optimum point above which increasing the moderator-to-fuel ratio decreases the effective multiplication factor (k-eff) because decreasing thermal utilization factor dominates. Below the optimum point, decreasing the moderator-to-fuel ratio decreases k-eff because the increased resonance absorption in the fuel dominates. A reactor core with a moderator-to-fuel ratio above the optimum point is said to be over-moderated, while a core with a ratio below the optimum point is considered under-moderated.

Under-moderation and Its Advantages In practice, many water-moderated reactors are designed to operate in an under-moderated condition. This design choice is based on the fact that if the reactor were over-moderated, an increase in temperature would decrease the moderator-to-fuel ratio due to the expansion of water and the reduction in its density. This decrease in Nm/Nu would result in a positive reactivity addition, increasing k-eff and further raising power and temperature in a potentially dangerous feedback cycle.

By designing the reactor to be under-moderated, the same increase in temperature leads to the addition of negative reactivity, making the reactor more self-regulating. This negative feedback helps to stabilize the reactor power and prevents uncontrolled power excursions.

7.6.3   Temperature Coefficient of Reactivity

The temperature coefficient of reactivity measures how the reactivity of the reactor changes with respect to variations in temperature. It is defined as the change in reactivity ($\Delta\rho$) per unit change in temperature (ΔT):

$$\text{Temperature Coefficient} = \Delta\rho \ / \ \Delta T \qquad (7\text{-}25)$$

A negative temperature coefficient of reactivity is highly desirable for reactor safety and stability. When the temperature of the reactor increases, a negative temperature coefficient causes a decrease in reactivity, which in turn reduces the reactor power and helps to mitigate the temperature rise. This self-regulating behavior is crucial for preventing runaway power excursions and maintaining the reactor within safe operating limits.

The temperature coefficient of reactivity consists of two components: the fuel temperature coefficient and the moderator temperature coefficient. The fuel temperature coefficient is generally more effective than the moderator temperature coefficient in terminating a rapid power rise. This is because the fuel temperature responds more quickly to changes in power due to its lower heat capacity and higher power density compared to the moderator.

7.6.4 Doppler Broadening and Resonance

Absorption One of the key mechanisms contributing to the negative fuel temperature coefficient is the Doppler broadening of resonance absorption peaks. Resonance absorption occurs when neutrons have energies that closely match the energy levels of certain nuclides in the fuel, such as U-238 and Th-232. At these specific energies, the absorption cross section of these nuclides increases significantly, leading to enhanced neutron absorption.

As the fuel temperature increases, the thermal motion of the fuel atoms also increases, causing the resonance absorption peaks to broaden. This broadening results in a larger energy range over which neutrons can be absorbed, effectively increasing the overall absorption cross section. The increased absorption of neutrons in the fuel leads to a decrease in reactivity, contributing to the negative fuel temperature coefficient.

7.6.5 Pressure Coefficient of Reactivity

The pressure coefficient of reactivity describes the change in reactivity due to variations in the system pressure. It is defined as the change in reactivity ($\Delta\rho$) per unit change in pressure (ΔP):

$$\text{Pressure Coefficient} = \Delta\rho \ / \ \Delta P \tag{7-26}$$

In most cases, the pressure coefficient of reactivity is negligible in reactors cooled and moderated by a subcooled liquid. This is because the density of the liquid moderator does not change significantly with pressure variations under subcooled conditions. As a result, the moderator-to-fuel ratio remains relatively constant, and the impact on reactivity is minimal.

7.6.6 Void Coefficient of Reactivity

The void coefficient of reactivity is a measure of how the reactivity changes with the formation of vapor voids in the moderator. It is defined as the change in reactivity ($\Delta\rho$) per unit change in void fraction ($\Delta\alpha$):

$$\text{Void Coefficient} = \Delta\rho/\Delta\alpha \tag{7-27}$$

The void coefficient of reactivity becomes significant when the moderator conditions approach saturation, such as in boiling water reactors (BWRs). As the moderator temperature increases and reaches the saturation point, vapor bubbles (voids) begin to form. The presence of these voids reduces the moderator density and the moderator-to-fuel ratio, which can have a substantial impact on reactivity.

In under-moderated reactors, the formation of voids typically leads to a negative void coefficient of reactivity. As the void fraction increases, the moderator density decreases, resulting in a reduction of reactivity. This negative feedback helps to stabilize the reactor power and prevents excessive void formation.

However, in over-moderated reactors, the void coefficient of reactivity can become positive. In this case, an increase in void fraction leads to a positive reactivity insertion, which can further increase power and void formation, potentially leading to instability and safety concerns.

Reactivity coefficients, particularly those related to the moderator, are essential parameters in understanding the behavior and safety of nuclear reactors. The moderator plays a crucial role in slowing down neutrons and sustaining the fission chain reaction. The effectiveness of a moderator is determined by its macroscopic slowing down power and moderating ratio.

The moderator-to-fuel ratio significantly influences the reactor's performance, with under-moderated reactors exhibiting desirable self-regulating properties. The temperature coefficient of reactivity, especially the negative fuel temperature coefficient enhanced by Doppler broadening, is crucial for reactor safety and stability.

While the pressure coefficient of reactivity is usually negligible in subcooled liquid-moderated reactors, the void coefficient of reactivity becomes significant in reactors operating near saturation conditions. Understanding and carefully designing reactors considering these reactivity coefficients is essential for ensuring safe and reliable operation.

7.6.7 Moderator Temperature Coefficient (MTC)

The moderator temperature coefficient (MTC) is a measure of how the reactivity of a nuclear reactor changes with respect to variations in the moderator temperature. It is defined as the change in reactivity ($\Delta\rho$) per unit change in moderator temperature (ΔT_m):

$$MTC = \Delta\rho / \Delta T_m \tag{7-28}$$

The magnitude and sign (positive or negative) of the MTC are primarily determined by the moderator-to-fuel ratio in the reactor core. In an under-moderated reactor, where the moderator-to-fuel ratio is below the optimum value, the MTC is typically negative. This means that an increase in moderator temperature leads to a decrease in reactivity. Conversely, in an over-moderated reactor, where the moderator-to-fuel ratio is above the optimum value, the MTC tends to be positive, indicating that an increase in moderator temperature results in an increase in reactivity.

A negative MTC is highly desirable from a reactor safety and stability perspective due to its self-regulating effect. When the reactor power increases, the moderator temperature rises, and the negative MTC introduces negative reactivity feedback, which helps to slow down or turn the power rise. This self-regulating behavior is crucial for preventing uncontrolled power excursions and maintaining the reactor within safe operating limits.

The MTC is influenced by several factors, including the moderator-to-fuel ratio, the type of moderator, and the presence of neutron poisons in the reactor core. In light water reactors (LWRs), which use water as both a moderator and coolant, the MTC is affected by the changes in water density with temperature. As the moderator temperature increases, the water density decreases, leading to a reduction in the moderator-to-fuel ratio and a negative reactivity contribution.

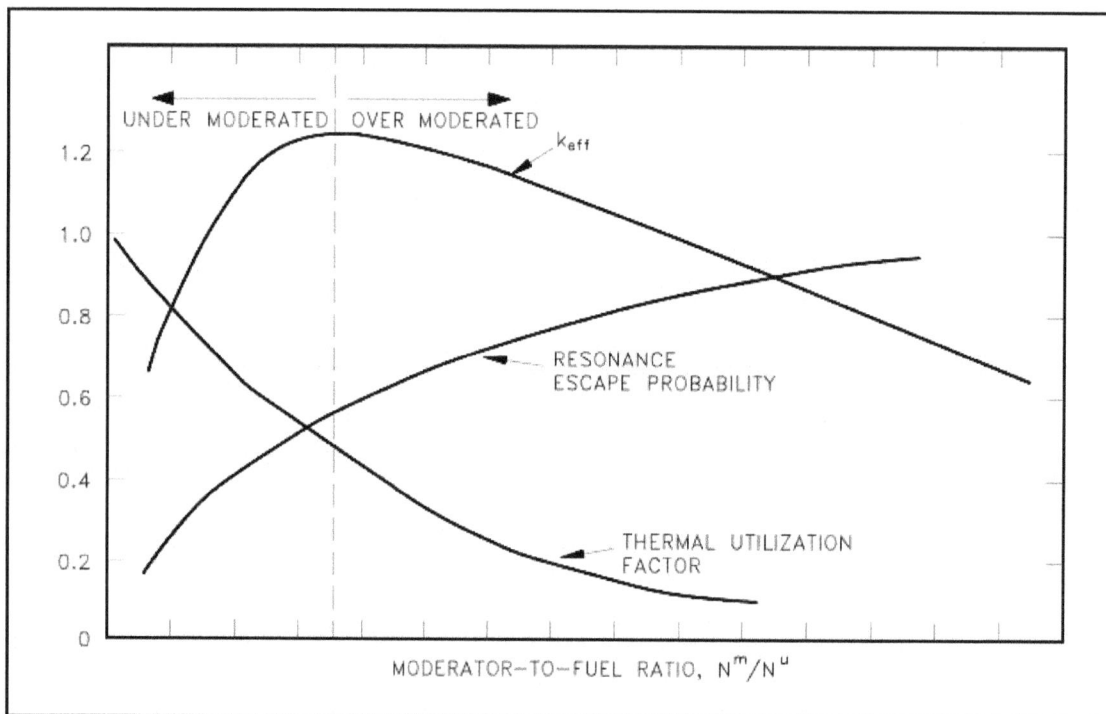

Figure 7-2. Changes in k-eff as a function of moderator-to-fuel ratio.**Fuel Temperature Coefficient (FTC) and Doppler Broadening**

The fuel temperature coefficient (FTC), also known as the prompt temperature coefficient, is a measure of how the reactivity changes with variations in the fuel temperature. It is defined as the change in reactivity ($\Delta\rho$) per unit change in fuel temperature (ΔT_f):

$$FTC = \Delta\rho \: / \: \Delta T_f \qquad (7\text{-}29)$$

The FTC is often considered more important than the MTC in terms of reactor safety because the fuel temperature responds more quickly to changes in reactor power compared to the moderator temperature. When the reactor power increases, the fuel temperature immediately rises, and a negative FTC provides an instant negative reactivity feedback, helping to mitigate the power rise.

In low-enriched uranium (LEU) fueled thermal reactors, the negative FTC is primarily attributed to the Doppler broadening effect. Doppler broadening refers to the apparent broadening of the resonance absorption peaks in the fuel material because of the thermal motion of the nuclei. As the fuel temperature increases, the thermal vibrations of the nuclei become more energetic, effectively broadening the energy range over which neutrons can be resonantly absorbed.

Two key nuclides present in significant quantities in the fuel of some reactors, namely uranium-238 (U-238) and plutonium-240 (Pu-240), exhibit large resonance absorption peaks that dominate the Doppler effect. The broadening of these resonance peaks with increasing fuel temperature leads to enhanced neutron absorption, resulting in a negative reactivity contribution.

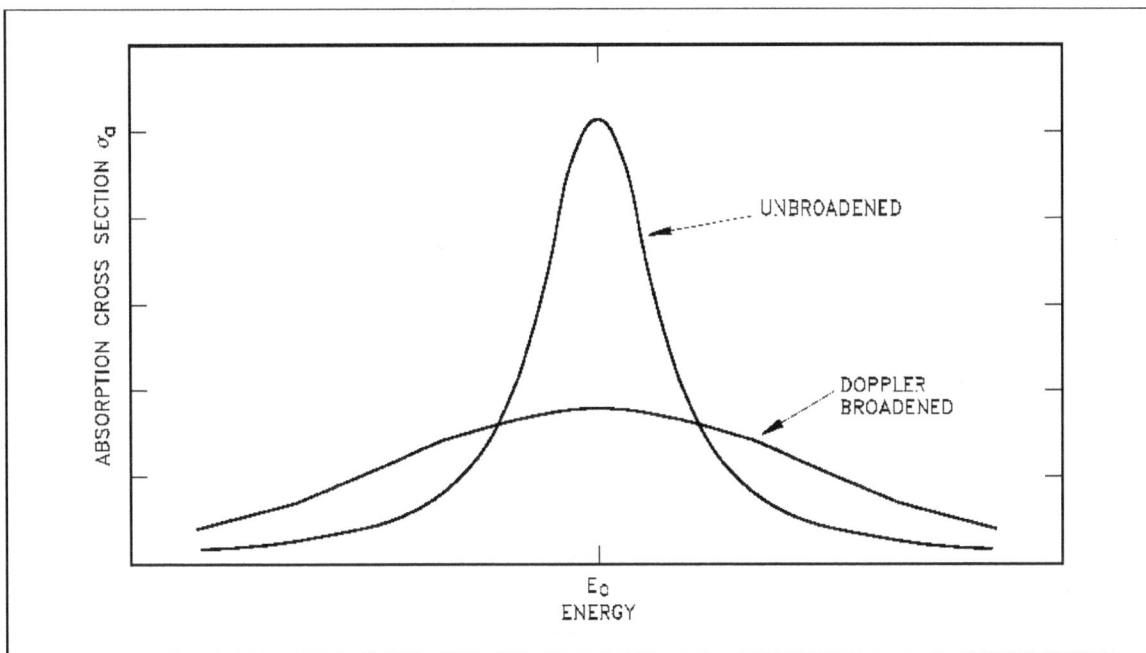

Figure 7-3. Doppler broadening. The Doppler effect is particularly important in ensuring the inherent safety of the reactor. During a rapid power increase, the fuel temperature rises quickly, and the Doppler broadening of the resonance peaks provides an immediate negative reactivity feedback, helping to terminate the power excursion and maintain the reactor within safe limits.

7.6.9   Pressure Coefficient and Moderator Density Effects

The pressure coefficient of reactivity describes how the reactivity changes with variations in the system pressure. It is defined as the change in reactivity ($\Delta\rho$) per unit change in pressure (ΔP): Pressure Coefficient = $\Delta\rho$ / ΔP

The pressure coefficient is primarily influenced by the effect of pressure on the moderator density. As the pressure increases, the moderator density typically increases, leading to a higher moderator-to-fuel ratio in the reactor core. In an under-moderated core, this increase in the moderator-to-fuel ratio results in a positive reactivity contribution.

However, in most light water reactors (LWRs), the absolute value of the pressure coefficient is relatively small compared to the moderator temperature coefficient. This is because the changes in moderator density due to pressure variations are generally less significant than the density changes caused by temperature variations, especially when the moderator is in a subcooled state.

7.6.10 Void Coefficient

In reactor systems that operate under boiling conditions, such as boiling water reactors (BWRs), the void coefficient becomes a crucial parameter. The void coefficient is a measure of how the reactivity changes with the formation of steam voids in the moderator. It is defined as the change in reactivity ($\Delta\rho$) per unit change in void fraction ($\Delta\alpha$):

$$\text{Void Coefficient} = \Delta\rho \, / \, \Delta\alpha \qquad (7\text{-}30)$$

As the reactor power increases and reaches the point where steam voids start to form, the voids displace moderator from the coolant channels in the core. This displacement reduces the moderator-to-fuel ratio, which, in an under-moderated core, leads to a negative reactivity contribution. The negative void coefficient acts as a power-limiting mechanism, preventing excessive void formation and helping to stabilize the reactor power.

In reactors that operate at or near saturated conditions and use water as a moderator, the void coefficient becomes particularly significant. In BWRs, the void coefficient plays a crucial role in the reactor's inherent safety characteristics. As the reactor power rises, the increased void formation introduces negative reactivity feedback, which helps to mitigate further power increases and maintains the reactor within safe operating limits.

It is important to note that the void coefficient can be affected by various factors, such as the core geometry, fuel design, and the presence of neutron poisons. In some cases, particularly in over-moderated cores, the void coefficient can become positive, leading to potential instability and safety concerns. Therefore, careful design and analysis are necessary to ensure that the void coefficient remains negative. A negative void coefficient contributes to the overall stability and safety of the reactor.

Reactivity coefficients, including the moderator temperature coefficient, fuel temperature coefficient, pressure coefficient, and void coefficient, are essential parameters in understanding the behavior and safety characteristics of nuclear reactors. These coefficients quantify how the reactivity of the reactor changes in response to variations in the corresponding physical properties.

The moderator temperature coefficient and fuel temperature coefficient, particularly the negative fuel temperature coefficient enhanced by Doppler broadening, play crucial roles in ensuring the inherent safety and stability of the reactor. They provide negative reactivity feedback mechanisms that help to mitigate power excursions and maintain the reactor within safe operating limits.

While the pressure coefficient is generally less significant in light water reactors, the void coefficient becomes a critical parameter in reactors operating under boiling conditions, such as BWRs. The negative void coefficient acts as a power-limiting mechanism, preventing excessive void formation and contributing to the reactor's inherent safety.

Understanding and carefully designing reactors considering these reactivity coefficients is essential for ensuring safe, stable, and reliable operation. The interplay of these coefficients and their impact on reactor behavior must be thoroughly analyzed and optimized to maintain the reactor within acceptable safety margins and achieve efficient power production.

7.7 Control Rods in Nuclear Reactors

Control rods are crucial components of most nuclear reactors, playing a pivotal role in managing the core's reactivity. Composed of materials that have a high capacity to absorb neutrons, control rods effectively adjust the neutron population within the reactor core, thereby controlling the rate of the nuclear chain reaction.

7.7.1 Composition and Functionality of Control Rods

Control rods are typically made from materials known for their strong neutron-absorbing properties, such as boron, cadmium, hafnium, or silver-indium-cadmium alloys. These materials are chosen because of their ability to capture neutrons without becoming radioactive or degrading under the intense conditions inside the reactor core.

The primary function of control rods is to manage the reactor's reactivity, which is the capability of a reactor to maintain a controlled, sustained nuclear chain reaction. By absorbing neutrons, control rods reduce the number of free neutrons available to continue the chain reaction, thus lowering the reactor's reactivity.

7.7.2 Roles of Control Rods

Control rods serve several distinct functions within the reactor control strategy:
- **Coarse Control:**
    - Control rods are configured for coarse control by adjusting the overall reactivity of the reactor to help in maintaining the desired power output

over extended periods. This adjustment is typically gradual and involves either inserting or withdrawing the rods over a range of positions to achieve the required balance of reactivity.

- **Fine Control:**
  - For more precise reactivity adjustments, control rods are used for fine control. This involves minor, precise movements of the rods to fine-tune the reactor's power output, ensuring optimal performance and response to varying operational demands. Fine control is essential during changes in reactor load and to compensate for fuel burn-up over time.
- **Fast Shutdowns (SCRAM):**
  - In emergency situations or for planned shutdowns, control rods can be fully and rapidly inserted into the reactor core to halt the nuclear chain reaction almost instantaneously. This process, often referred to as a SCRAM, is a critical safety mechanism designed to quickly reduce the reactor's reactivity to subcritical levels, effectively stopping the power generation process and ensuring the safety of the reactor.

7.7.3   Strategic Placement and Design

The design and placement of control rods within the reactor core are strategically planned to ensure even absorption of neutrons and effective management of the core's reactivity profile. The configuration of the control rods can vary depending on the type of reactor and specific operational requirements, but generally, they are inserted vertically into the reactor core, where they can be moved up and down to modulate the neutron flux.

Overall, control rods are indispensable for the safe and efficient operation of nuclear reactors. Their ability to precisely control the rate of the nuclear reaction makes them fundamental to achieving a balance between power generation and nuclear safety, highlighting their critical role in the field of nuclear energy.

7.7.4   Selection of Control Rod Materials in Nuclear Reactors

Control rods are critical components in nuclear reactors, tasked with managing the reactor core's reactivity through the absorption of neutrons. These rods, which can be inserted into or removed from the reactor core, are composed of materials with high neutron-absorption capabilities such as silver, indium, cadmium, boron, and hafnium. The choice of material for control rods is pivotal and varies based on reactor design, operational needs, and desired longevity and efficiency of the rods.

7.7.4.1   Material Properties and Selection Criteria

The primary criteria for selecting materials for control rods include:

1. **High Neutron Absorption Cross Section**: The material should have a robust capability to absorb neutrons. This property is critical for effectively controlling the rate of the nuclear chain reaction within the reactor core.

2. **Durability and Longevity**: Materials chosen should not only be effective in absorbing neutrons but also resistant to "burning out" or losing their absorption capacity rapidly. This ensures that the control rods remain effective over longer operational periods without frequent replacements.

3. **Adjustability in Manufacturing**: Control rod materials can be engineered to vary in their neutron absorption properties. During the manufacturing process, control rods can be designed to be "black" absorbers, which absorb nearly all incident neutrons, or "grey" absorbers, which absorb only a portion of incident neutrons.

7.7.4.2 Black versus Grey Absorbers

- **Black Absorbers:** These materials are used when maximum neutron absorption is required. They are highly efficient at capturing neutrons, but their use can result in significant local reductions in neutron flux, potentially causing uneven power distribution within the reactor core.

- **Grey Absorbers:** Commonly favored in modern reactor designs, grey absorbers provide a more balanced approach to neutron absorption. By absorbing only a portion of the incident neutrons, grey rods help maintain a more uniform neutron flux and power distribution across the reactor core. Although more grey rods may be needed to achieve the same level of reactivity control as fewer black rods, their use generally leads to improved reactor stability and efficiency.

7.7.4.3 Consideration of Resonance Absorption

Another crucial consideration in selecting control rod materials is the preference for resonance absorbers rather than materials that mainly absorb thermal neutrons. Resonance absorbers are particularly effective at capturing neutrons in the epithermal energy range, which includes neutron energies that are higher than thermal neutrons but lower than fast neutrons..

- **Advantages of Resonance Absorbers**: The path lengths of epithermal neutrons are generally longer than those of thermal neutrons within the reactor. This characteristic allows resonance absorbers to influence a larger area around the control rod, helping to maintain a flatter and more uniform flux profile throughout the reactor. This wider area of influence is beneficial for achieving more consistent reactor performance and minimizing the effects of localized power peaks.

The selection of materials for control rods is a nuanced decision that impacts the reactor's control capabilities, operational lifespan, and safety. By choosing appropriate absorber types and materials, nuclear engineers can optimize reactor control and ensure that the reactor operates safely and efficiently. The balance between robust neutron absorption, material longevity, and the maintenance of a stable and uniform power distribution within the reactor core is central to the effectiveness of control rods in nuclear reactor operations.

7.7.5   Types of Control Rods in Nuclear Reactors

Control rods are essential components in nuclear reactors, playing a crucial role in managing the reactor core's reactivity and ensuring safe and efficient operation. These rods can be categorized based on their specific functions within the reactor. Each type of control rod is engineered to fulfill distinct tasks that are vital for maintaining operational control and adhering to safety protocols. In this discussion, we will delve into the prevalent types of control rods and examine their respective purposes in detail.

7.7.5.1   1. Shim Rods

- **Purpose**: Shim rods are primarily used for coarse control of the reactor's reactivity. They are capable of absorbing or adding substantial amounts of reactivity, which is crucial during significant changes in reactor power levels or operational states.
- **Functionality**: These rods adjust the overall neutron flux within the core to manage large-scale reactivity changes, such as those required during startup, changes in power output, or compensating for fuel burnup over time.

7.7.5.2   2. Regulating Rods

- **Purpose**: Regulating rods are essential for making fine adjustments to the reactor's power output or temperature to maintain stability and achieve desired operational conditions.
- **Functionality**: They provide precise control over the reactor's reactivity on a smaller scale than shim rods. This fine tuning is critical for maintaining the desired power level and ensuring the reactor operates within safe and efficient parameters.

7.7.5.3   3. Safety Rods

- **Purpose**: Safety rods are designed for emergency use to achieve a rapid shutdown—known as a "scram" or "trip"—of the reactor in case of unsafe conditions.
- **Functionality**: These rods can introduce a large amount of negative reactivity very quickly by fully inserting into the reactor core. This rapid insertion

effectively stops the nuclear chain reaction, ensuring the reactor's immediate and safe shutdown to prevent potential accidents.

7.7.5.4 Dual-Purpose and Multi-Function Control Rods

In many modern reactors, the design and utility of control rods are optimized to serve multiple functions. This flexibility is particularly advantageous in reactors with limited space or those requiring highly integrated control systems. Examples include:

- **Dual-Purpose Rods**: Some rods are designed to function both as shim and regulating rods, allowing them to control reactivity over a broad range while also making finer adjustments as needed.
- **Triple-Function Rods**: In more sophisticated designs, control rods can serve as shim, regulating, and safety rods. They can be inserted slowly for routine adjustments, partially for moderate reactivity control, or fully for emergency scrams.

This versatility enhances the reactor's operational flexibility and efficiency, providing operators with robust tools for managing reactivity under various conditions. Additionally, these multi-functional rods can reduce the complexity and cost of the control rod drive mechanisms and associated reactor control systems.

The classification and design of control rods are critical factors in the overall safety, efficiency, and operational capability of nuclear reactors. By understanding and appropriately implementing various types of control rods, nuclear engineers can ensure that reactors not only meet power generation needs but also adhere to stringent safety standards. The ability to perform multiple functions with fewer rods also reflects advancements in nuclear technology, aiming for simpler, more reliable reactor designs.

7.8   Control Rod Effectiveness in Nuclear Reactors

The efficacy of control rods in nuclear reactors is paramount for ensuring safe and efficient operation. The main factor that determines the effectiveness of these rods is their strategic positioning in relation to the neutron flux distribution throughout the reactor core.

7.8.1 Influence of Neutron Flux on Control Rod Effectiveness

The effectiveness of a control rod is significantly influenced by the ratio of the neutron flux at the rod's location to the average neutron flux throughout the reactor core. Ideally, a control rod will have the maximum impact—inserting the most negative reactivity—when it is positioned where the neutron flux is at its peak.

In reactors with a single control rod, optimal placement is typically at the center of the core, where the flux is generally highest. This positioning ensures that the rod impacts the reactor's reactivity most significantly, leading to more efficient control of the nuclear chain reaction. The effect of this strategic placement can dramatically alter the radial flux distribution within the core.

7.8.2 Optimal Placement of Multiple Control Rods

When multiple control rods are employed, each should ideally be placed in regions of maximum flux to maximize their collective effectiveness. This strategic placement is particularly important in reactors designed with excess reactivity, which is the additional reactivity beyond what is necessary for the reactor to reach criticality. Managing this excess reactivity effectively requires precise control, which is facilitated by optimally positioned control rods.

The concept of "control rod worth" refers to the specific amount of reactivity that each rod contributes to the reactor. This contribution varies depending on several factors including the rod's material, size, and exact placement within the flux profile of the reactor.

7.8.3 Reactor Design and Control Rod Configuration

The design of the reactor significantly influences how control rods are configured. In complex reactor designs, especially those with large cores and significant excess reactivity, numerous control rods may be required to achieve desired control levels. These rods are arranged to ensure that there are no areas within the reactor core that are under-regulated, maintaining a balanced and uniform control over the nuclear reaction process.

The configuration and adjustment of control rods are often done dynamically during reactor operation to respond to changes in operating conditions or to compensate for fuel burnup and other factors that affect neutron economy. This dynamic capability allows for continuous fine-tuning of reactivity and ensures that the reactor operates within safe and optimal parameters at all times.

The placement and configuration of control rods within a nuclear reactor are critical for effective reactor management. These rods must be strategically positioned to maximize their impact on the reactor's neutron flux and overall reactivity, ensuring that the reactor remains safe, stable, and efficient throughout its operational lifecycle. Understanding and optimizing control rod worth and placement are key aspects of nuclear engineering, highlighting the complex interplay between reactor physics and design in the pursuit of energy security and safety.

7.9   Integral and Differential Control Rod Worth

The effectiveness of control rods in adjusting reactor reactivity is a fundamental aspect of nuclear reactor control and safety. To quantitatively assess this effectiveness, two key metrics are used: integral control rod worth and differential control rod worth. These metrics are often determined experimentally and provide essential data for optimizing reactor operations.

7.9.1 Integral Control Rod Worth

Integral control rod worth refers to the cumulative change in reactivity as a control rod is withdrawn from the reactor core. To measure this, a control rod is typically withdrawn in small increments (e.g., 0.5 inch), and the change in reactivity ($\Delta\rho$) is measured after each increment. Plotting these reactivity changes against rod positions yields a graph that shows how the rod's reactivity contribution varies with its position within the reactor. See Figure 7-4. for an illustration of integral control rod worth.

Integral Rod Worth=$\sum(\Delta\rho)$

This value is highest when the rod is fully withdrawn, indicating the maximum reactivity worth the rod can offer when it is least engaged with the reactor core.

7.9.2 Differential Control Rod Worth

Differential control rod worth measures the rate of change of reactivity per unit of rod movement. It is essentially the derivative of the integral control rod worth with respect to rod position. This metric is crucial for understanding how small movements of the rod impact reactor reactivity, especially in regions of high neutron flux.

Differential Rod Worth $= \dfrac{d\rho}{dz}$

Where $d\rho$ is the change in reactivity and dz is the change in control rod position.

Graphical Representation: The slope of the curve on a plot of integral rod worth versus rod position (as seen in hypothetical Figure 7-4.) represents the differential rod worth. The steepest slope—and thus the highest differential rod worth—is typically found when the rod is around the mid-point of the core, where neutron flux is highest.

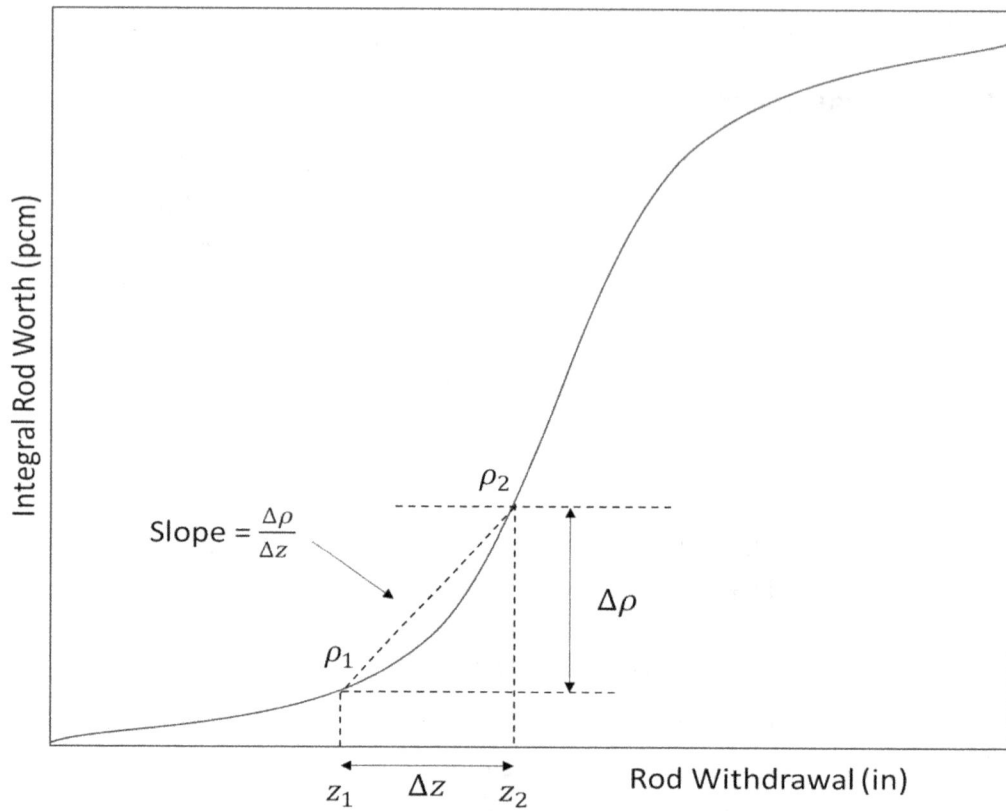

Figure 7-4. Illustration of integral control rod worth.

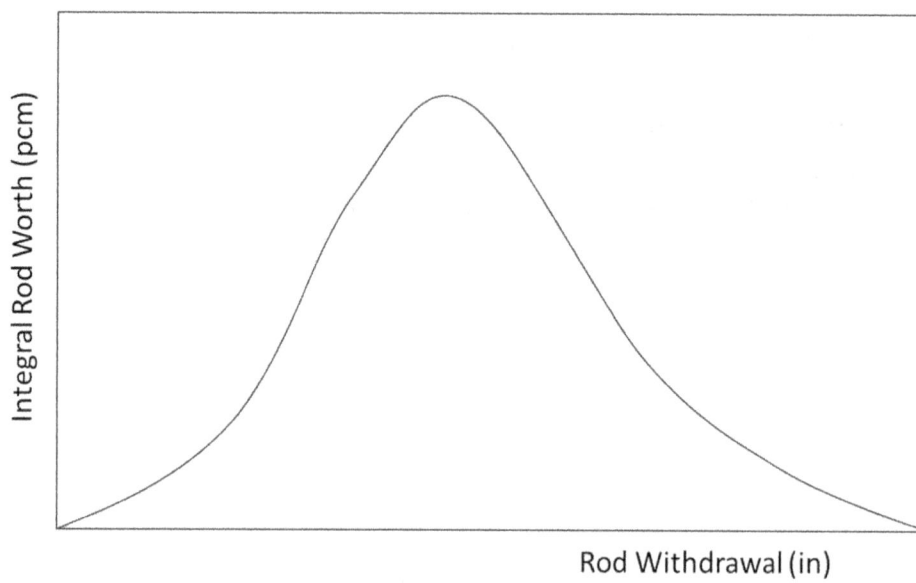

Figure 7-5. Illustration of a typical differential rod worth curve.

7.9.3   Implications of Rod Worth Measurements

- **At the Core's Center:** The differential rod worth tends to peak where the neutron flux is greatest, often near the center of the core. Here, even small movements of the rod have a large impact on reactivity.
- **Near the Core's Top or Bottom:** Towards the top or bottom of the core where neutron flux is lower, movements of the control rod result in smaller changes in reactivity. Thus, the differential rod worth is lower.
- **Directional Symmetry:** The pattern of rod worth from the center to the top of the core often mirrors that from the center to the bottom. This symmetry is useful for predicting rod effectiveness and planning rod movements during reactor operations.

7.9.4   Application in Reactor Operation

Understanding both the integral and differential rod worth is crucial for reactor operators, especially when configuring control rods for both routine operations and emergency scenarios. These measurements help ensure that control rods are placed and moved optimally within the core to effectively manage the reactor's reactivity under various conditions.

7.9.5   Control Rod Mechanisms in Nuclear Reactors

Control rods are essential safety and regulatory components in nuclear reactors, designed to manage the reactor's reactivity and ensure safe operations under various conditions. The design and operational mechanisms of control rods must account for a range of normal and extraordinary scenarios that might occur during the reactor's lifespan.

7.9.5.1   Scram and Emergency Insertion Rates

A "scram" or emergency shutdown is a critical safety feature in nuclear reactors, where control rods are inserted fully and rapidly into the reactor core to halt the nuclear chain reaction immediately. The insertion rates during a scram are carefully calculated to be fast enough to protect the reactor against any potential damage from expected transients throughout its operational life. This rapid insertion counteracts any excessive positive reactivity, thereby preventing uncontrolled increases in power that could lead to overheating or damage.

- **Design Considerations for Scram**: The mechanisms controlling scram actions are robustly designed to ensure fail-safe operation under emergency conditions. These systems often involve gravity-driven mechanisms enhanced by springs or hydraulic systems to ensure that control rods can be deployed instantly and reliably, even if other reactor systems fail.

7.9.5.2   Normal Operation and Control Rod Speed

During regular operations, the movement of control rods must be finely tuned to respond adequately to changes in the reactor's condition. Control rod speed during normal operation is set based on the fastest expected rates of positive reactivity increase, such as those caused by the burnout of xenon-135 at full power.

- **Xenon-135 Burnout**: Xenon burnout can lead to rapid shifts in reactivity. Xenon-135, a fission product, has a significant neutron absorption capability, and its transient behavior can critically impact reactor operations. As xenon decays, particularly after a period of low power operation or shutdown, it can cause a sharp increase in reactivity as its neutron-absorbing effect diminishes. Control rods must be able to move swiftly to compensate for these rapid changes and stabilize the reactor.

- **Setting Minimum Rod Speed**: The minimum speed at which control rods must operate is often determined by scenarios like xenon burnout during full power operation, which is considered one of the most rapid and critical non-accident transients. Adequate control rod speed ensures that reactivity can be effectively managed to maintain stable and safe reactor conditions.

7.9.5.3   Limits on Maximum Rod Speed

While rapid movement of control rods is essential in certain scenarios, the maximum speed of rod movement is typically restricted to prevent potential risks associated with their continuous withdrawal. If control rods are withdrawn too quickly or too far, particularly in a non-uniform manner, it could lead to localized areas of high reactivity, which might escalate into dangerous conditions or accidents.

- **Safety Considerations**: Limiting the maximum speed and extent of control rod withdrawal helps mitigate the potential severity of accidents. By ensuring that control rods cannot be removed too rapidly, reactor safety systems help maintain a controlled and gradual response to changes in operational conditions, thus enhancing overall safety.

Control rod mechanisms are designed to provide a balance between rapid response capabilities for safety and finely-tuned adjustments for operational control. These mechanisms are critical for ensuring that nuclear reactors can respond effectively to both expected transients and emergency situations, thereby maintaining safe and stable operations throughout the reactor's lifetime. The engineering behind these systems involves a deep understanding of reactor kinetics, materials science, and safety engineering, underscoring the complex and critical nature of control rod operations in nuclear technology.

8 Nuclear Reactor Theory

Nuclear reactor theory is a fundamental aspect of reactor physics that deals with understanding and predicting the behavior of neutrons within a reactor. This understanding is crucial for the design, operation, and safety of nuclear reactors. The primary objective in reactor theory is to ensure that the reactor operates in a critical state, where the number of neutrons balances the number of neutrons produced in fission lost either through absorption within the reactor materials or by leakage from the reactor boundary. Achieving and maintaining this delicate balance is vital for the stable and sustained operation of a nuclear reactor.

In a critical reactor, there is an equilibrium between neutron production and neutron loss. This balance is achieved when each fission event on average produces one neutron that goes on to induce another fission event. The design challenge is to calculate the appropriate size, shape, and material composition of the reactor core to sustain this balance under various operating conditions. This problem is addressed through a series of theoretical and computational methods that form the core of reactor theory.

This chapter delves into the methodologies used to calculate the conditions necessary for reactor criticality. One of the primary tools for these calculations is the group-diffusion method, which was introduced at the end of the previous chapter. The group-diffusion method involves simplifying the complex interactions of neutrons within the reactor into manageable calculations by grouping neutrons into energy categories and treating their movements and interactions statistically within those groups.

To lay the groundwork for more complex analyses, we begin our discussion with the simplest case: the one-group diffusion calculation. This method, while most directly applicable to fast reactors where neutron energy spectra are relatively simple, provides a foundational understanding that can be extended and modified for more complex systems, including thermal reactors. One-group calculations assume that all neutrons have the same average behavior, allowing us to use a single set of parameters to describe their interactions within the reactor.

Through these discussions, we aim to equip you with the theoretical foundation and practical tools necessary to analyze and design nuclear reactors. By understanding the principles of neutron behavior and the methods used to control and utilize these behaviors, we can ensure that reactors operate safely and efficiently, providing a reliable source of nuclear energy.

8.1   One-Group Reactor Equation

Consider a critical fast reactor containing a homogeneous mixture of fuel and coolant. It is assumed that the reactor consists of only one region and has neither a blanket nor a reflector. Such a system is said to be a bare reactor.

This reactor is described in a one-group calculation by the one-group time-dependent diffusion equation:

$$D\nabla^2\phi - \Sigma_a\phi + s = \frac{1}{v}\frac{\partial\phi}{\partial t} \tag{8-1}$$

Here, ϕ is the one-group flux, D and Σ_a are the one-group diffusion coefficient and macroscopic absorption cross section for the fuel-coolant mixture, s is the source density (i.e., the number of neutrons emitted per cm^3/sec), and v is the neutron speed. Note that we are making no assumptions as to the state of the reactor, so there may be time dependence involved. If this were not the case, there would be no time derivative term, and a balance between the source and the absorption and leakage would exist.

In a reactor at a measurable power, the source neutrons are emitted in fission. To determine s, let Σ_f be the fission cross section for the fuel. If there are v neutrons produced per fission, then the source is given by:

$$s = v\Sigma_f\phi$$

Here, we are assuming that there is no non-fission source of neutrons present in the reactor, which is a good approximation for a reactor except at very low power levels.

If the fission source does not balance the leakage and absorption terms, then the right-hand side of the diffusion equation is nonzero. To balance the equation, we multiply the source term by a constant $\frac{1}{k}$, where k is an unknown constant. If the source is too small, then k is less than 1. If it is too large, then k is greater than 1. The diffusion equation may now be written as:

$$D\nabla^2\phi - \Sigma_a\phi + \frac{1}{k}v\Sigma_f\phi = 0 \tag{8-2}$$

The equation may be rewritten as an eigenvalue equation by letting:

$$B^2 = \frac{1}{D}\left(\frac{1}{k}v\Sigma_f - \Sigma_a\right) \tag{8-3}$$

where B^2 is defined as the geometric buckling. Then, the diffusion equation may be written as:

$$\nabla^2 \phi = -B^2 \phi \qquad (8\text{-}4)$$

Finally, this expression for the leakage term may be substituted into the diffusion equation, and the equation can be rewritten as:

$$-DB^2\phi - \Sigma_a\phi + \frac{1}{k}\nu\Sigma_f\phi = 0$$

or

$$\nabla^2\phi + B^2\phi = 0 \qquad (8\text{-}5)$$

This important result is known as the one-group reactor equation. The one-group equation may be solved for the constant k :

$$k = \frac{\nu\Sigma_f\phi}{DB^2\phi + \Sigma_a\phi} = \frac{\nu\Sigma_f}{DB^2 + \Sigma_a} \qquad (8\text{-}6)$$

Note that this equation does not give a value for k since B^2 is still unknown.

Physically, this equation may be interpreted as follows. The numerator is the number of neutrons that are born in fission in the current generation, whereas the denominator represents those that were lost from the previous generation. Since all neutrons in a generation either are absorbed or leak from the reactor, the denominator must also equal the number born in the previous generation. This is the definition of the multiplication factor for a finite reactor. We may now define the multiplication factor for a reactor as the birth rate, $\nu\Sigma_f\phi$, divided by the leakage rate plus the absorption rate, $DB^2\phi + \Sigma_a\phi$, of the neutrons.

The source term in the one-group equation may be written in terms of the fuel absorption cross section. Let $\Sigma_{a,F}$ be the one-group absorption cross section of the fuel, and let η be the average number of fission neutrons emitted per neutron absorbed in the fuel. The source term is then given by:

$$s = \eta\Sigma_{a,F}\phi$$

This may also be written as:

$$s = \eta \left(\frac{\Sigma_{a,F}}{\Sigma_a} \right) \Sigma_a \phi = \eta f \Sigma_a \phi \qquad (8\text{-}7)$$

where $f = \frac{\Sigma_{a,F}}{\Sigma_a}$ is called the fuel utilization. Since Σ_a is the cross section for the mixture of fuel and coolant, whereas $\Sigma_{a,F}$ is the cross section of the fuel only, it follows that f is equal to the fraction of the neutrons absorbed in the reactor that are absorbed by the fuel.

The source term can also be written in terms of the multiplication factor for an infinite reactor. For this purpose, consider an infinite reactor having the same composition as the bare reactor under discussion. With such a reactor, there can be no escape of neutrons as there is from the surface of a bare reactor. All neutrons eventually are absorbed, either in the fuel or the coolant. Furthermore, the neutron flux must be a constant, independent of position.

The number of fissions in one generation is simply related to the number of neutrons born in one generation divided by ν, the number of neutrons produced per fission. Since all neutrons born must eventually be absorbed in the system, this must be equal to $\Sigma_a \phi$, the number of neutrons absorbed per cm^3/sec everywhere in the system. Of these neutrons, the number $f \Sigma_a \phi$ are absorbed in fuel, and these release $\eta f \Sigma_a \phi$ fission neutrons in the next generation.

The number of neutrons born in this new generation can again be related to the fission rate multiplied by ν. Sooner or later, all of these neutrons must, in turn, be absorbed in the reactor. Thus, the absorption of $\Sigma_a \phi$ neutrons in one generation leads to the absorption of $\eta f \Sigma_a \phi$ in the next.

Table 8-1. One-group cross sections for materials of interest in average fast neutron spectrum.

Material	σ_γ (barns)	σ_f (barns)	σ_a (barns)	σ_tr (barns)
Na	0.0008	0.0008	3.3	3.3
Al	0.0020	0.002	3.1	3.1
Fe	0.0060	0.006	2.7	2.7
U-235	0.25	1.4	1.65	6.8
U-238	0.16	0.095	0.255	6.9
Pu-239	0.26	1.85	2.11	6.8
C	0.00032	0.00032	2.3	2.3
K	0.032	0.032	2.2	2.2
He	0.0000007	0.0000007	0.8	0.8

According to the definition of the multiplication factor, it is the number of fissions in one generation divided by the number in the preceding generation. Since in an infinite medium, these are related to the absorption rate, it follows that:

$$k_\infty = \frac{\eta f \Sigma_a \phi}{\Sigma_a \phi} = \eta f \qquad (8\text{-}8)$$

where the subscript on k_∞ signifies that this result is only valid for the infinite reactor.

Table 8-1 shows some nominal values of one-group averaged cross sections for a one-group fast reactor. Table 7-2 shows nominal values for one-group η (reproduction factor) and v (average number of neutrons produced per fission) in fast reactor for some fuel materials of interest.

Table 8-2. One-group η (reproduction factor) and v (average number of neutrons produced per fission) in average fast neutron spectrum.

Material	η	v
U-235	2.20	2.51
U-238	0.96	2.50
Pu-239	2.90	3.04
Pu-241	2.70	3.10
U-233	2.45	2.50

Example 8-1. Calculate f and k_∞ for a mixture of ^{235}U and carbon in which the uranium is present to 3w/o.

Solution:

$$f = \frac{\Sigma_{aF}}{\Sigma_a} = \frac{\Sigma_{aF}}{\Sigma_{aF} + \Sigma_{aS}}$$

where Σ_{aF} and Σ_{aS} are the macroscopic absorption cross sections of the uranium and carbon, respectively. Dividing numerator and denominator by Σ_{aF} gives:

$$f = \frac{1}{1 + \dfrac{\Sigma_{aS}}{\Sigma_{aF}}} = \frac{1}{1 + \dfrac{N_S \sigma_{aS}}{N_F \sigma_{aF}}}$$

where N_F and N_S are the atom densities of uranium and carbon, respectively. Next, let ρ_F and ρ_S be the number of grams of uranium and carbon per cm^3 in the mixture. Then:

$$\frac{N_S}{N_F} = \frac{\rho_S}{\rho_F} \cdot \frac{M_F}{M_S}$$

where M_F and M_S are the gram atomic weights of uranium and carbon. Since 3w/o of the mixture is fuel, this means that:

$$\frac{\rho_F}{\rho_F + \rho_S} = 0.03$$

or

$$\rho_S = \frac{97}{3}\rho_F \approx 32.33\rho_F$$

Using the values of σ_a given in the table below, the value of f is given by:

$$f^{-1} = 1 + \frac{N_S \sigma_{aS}}{N_F \sigma_{aF}} = 1 + \frac{\rho_S}{\rho_F} \times \frac{M_F}{M_S} \times \frac{\sigma_{aS}}{\sigma_{aF}}$$

$$f^{-1} = 1 + \frac{32.33 \times 235}{12} \times \frac{0.00032}{0.68}$$

$$f^{-1} = 1.368$$

Values for Calculation

Material	σ_a (barns)	M(g/mol)
U^{235}	0.68	235
Carbon	0.00032	12

The value of k_∞ is:

$$k_\infty = \eta f = 2.35 \times 0.731 \approx 1.72$$

Therefore, an infinite reactor with this composition is supercritical.

8.2   The Slab Reactor

As the first example of a bare reactor, consider a critical system consisting of an infinite bare slab of thickness a. This model helps illustrate the principles of neutron flux distribution and reactor buckling in a simple geometry. Figure 8-1. shows the geometry of the infinite slab reactor.

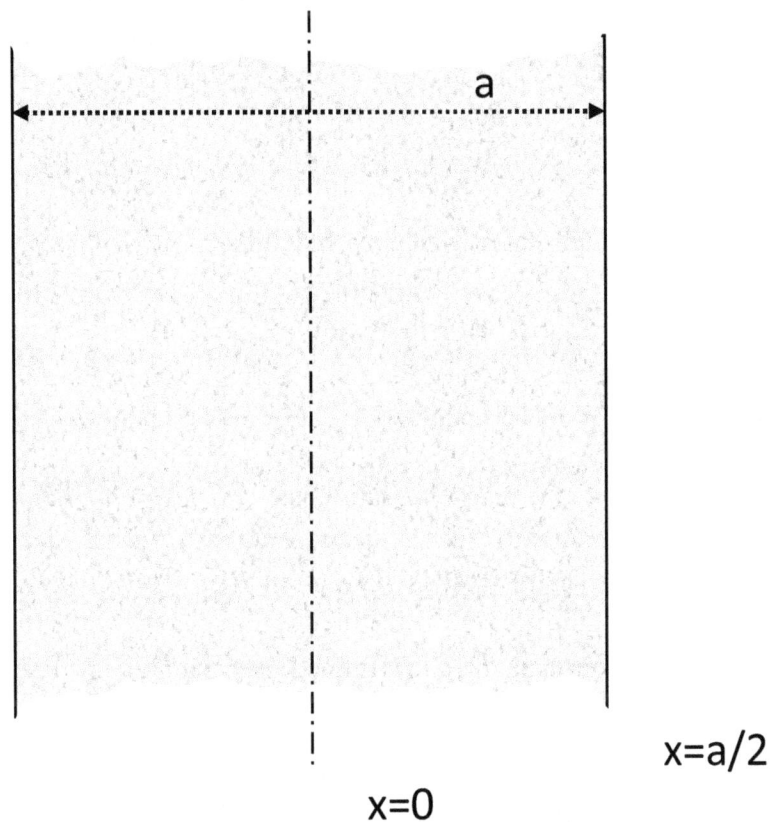

Figure 8-1. The infinite slab reactor. **The Reactor Equation**

For a slab reactor, the one-group diffusion equation in one dimension is given by:

$$\frac{d^2\phi}{dx^2} + B^2\phi = 0 \tag{8-9}$$

where x is measured from the center of the slab, ϕ is the neutron flux, and B is the buckling.

8.2.2 Boundary Conditions

To determine the flux within the reactor, this equation must be solved with the boundary condition that ϕ vanishes at the extrapolated faces of the slab, i.e., at $x = \pm\tilde{a}/2$, where $\tilde{a} = a + 2d$. Therefore, the boundary conditions are:

$$\phi\left(\frac{\tilde{a}}{2}\right) = \phi\left(-\frac{\tilde{a}}{2}\right) = 0$$

Additionally, due to the symmetry of the problem, there can be no net flow of neutrons at the center of the slab. This implies that the derivative of ϕ is zero at $x = 0$:

$$\frac{d\phi}{dx} = 0 \text{ at } x = 0$$

This condition indicates that ϕ is an even function:

$$\phi(-x) = \phi(x)$$

General Solution

The general solution to the reactor equation is:

$$\phi(x) = A\cos Bx + C\sin Bx$$

where A and C are constants to be determined. Applying the symmetry condition $\frac{d\phi}{dx} = 0$ at $x = 0$:

$$\frac{d\phi}{dx} = -AB\sin Bx + CB\cos Bx$$

At $x = 0$:

$$\frac{d\phi}{dx} = CB = 0$$

Thus, $C = 0$, and the solution reduces to:

$$\phi(x) = A\cos Bx$$

Next, applying the boundary condition at $x = \pm \tilde{a}/2$:

$$\phi \left(\frac{\tilde{a}}{2}\right) = A\cos \left(\frac{B\tilde{a}}{2}\right) = 0$$

This equation can be satisfied either by $A = 0$, leading to the trivial solution $\phi(x) = 0$, or by requiring that:

$$\cos \left(\frac{B\tilde{a}}{2}\right) = 0$$

This is satisfied if B assumes any of the values B_n, where:

$$B_n = \frac{n\pi}{\tilde{a}}$$

and n is an odd integer, as can be readily seen by direct substitution into the boundary condition equation.

8.2.3 Eigenvalues and Eigenfunctions

The various constants B_n are known as eigenvalues, and the corresponding functions $\cos B_n x$ are called eigenfunctions. For a critical reactor, all eigenfunctions except the fundamental mode die out in time, leaving:

$$\phi(x) = A\cos \left(\frac{\pi x}{\tilde{a}}\right) \tag{8-10}$$

8.2.4 Buckling

The square of the lowest eigenvalue B_1^2 is called the buckling of the reactor. Solving the reactor equation for B_1^2:

$$B_1^2 = \left(\frac{\pi}{\tilde{a}}\right)^2 \tag{8-11}$$

As a increases, B_1^2 decreases. In the limit as a becomes infinite, $B_1^2 = 0$, and ϕ is a constant, indicating no curvature.

8.2.5 Determining the Magnitude of ϕ

The constant A in the fundamental mode equation, which determines the magnitude of ϕ, is not established by the prior analysis. This is because the reactor equation is homogeneous. The magnitude of the flux in a reactor is determined by the power at which the system operates, not by its material properties.

8.2.6 Calculating Reactor Power

To find an expression for A, a separate calculation of reactor power is necessary. The number of fissions per cm^3/sec at the point x is $\Sigma_f \phi(x)$, where Σ_f is the macroscopic fission cross section. If the recoverable energy per fission is E_R joules (with a recoverable energy of $200 MeV$, $E_R = 3.2 \times 10^{-11}$ joules), then the total power per unit area of the slab, in watts $/cm^2$, is:

$$P = E_R \Sigma_f \int_{-\frac{a}{2}}^{\frac{a}{2}} \phi(x) dx \tag{8-12}$$

Inserting the expression for $\phi(x)$ and performing the integration gives:

$$P = \frac{2\tilde{a} E_R \Sigma_f A \sin\left(\frac{\pi a}{2\tilde{\pi}}\right)}{\pi}$$

Thus, the final formula for the thermal flux in a slab reactor is:

$$\phi(x) = \frac{\pi P}{2\bar{a} E_R \Sigma_f \sin\left(\frac{\pi a}{2\bar{a}}\right)} \cos\left(\frac{\pi x}{\bar{a}}\right)$$

If d is small compared to a, this reduces to:

$$\phi(x) = \frac{\pi P}{2a E_R \Sigma_f} \cos\left(\frac{\pi x}{a}\right) \tag{8-13}$$

The one-group reactor equation for a slab reactor provides a detailed understanding of neutron flux distribution and reactor buckling. The criticality condition, $B_m^2 = B_g^2$, must be satisfied for the reactor to maintain a steady state. The geometric buckling B_g^2 depends on the reactor's dimensions and shape, and the relationship between material and geometric buckling offers insights into the reactor's criticality and stability.

8.3   The Bare Spherical Reactor

Consider a critical spherical reactor of radius R. The neutron flux in this reactor is a function of the radial distance r from the center of the sphere. The one-group diffusion equation for this spherical geometry is:

$$\frac{1}{r^2}\frac{d}{dr}\left(r^2\frac{d\phi}{dr}\right) + B^2\phi = 0 \tag{8-14}$$

where the Laplacian in spherical coordinates has been used, and B represents the reactor buckling. Figure 8-2. shows the geometry of the bare sphere reactor.

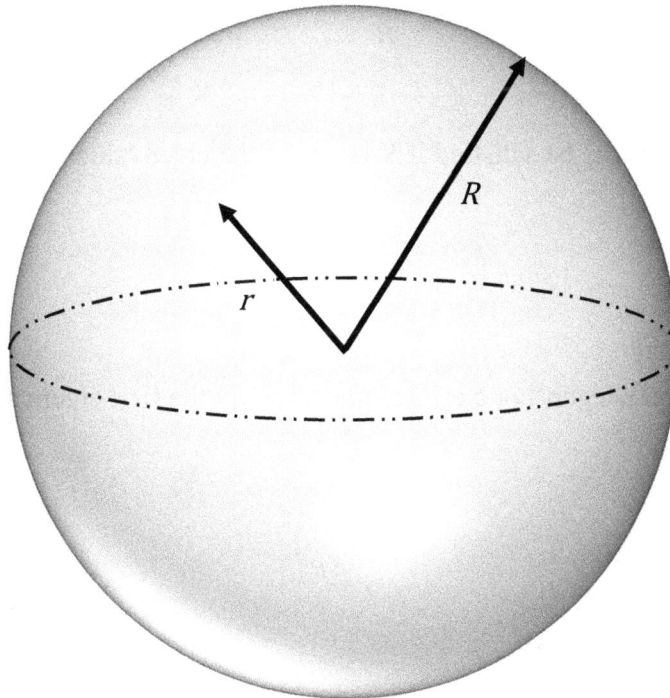

Figure 8-2. The bare sphere reactor.

8.3.1 Boundary Conditions

The flux must satisfy the boundary condition $\phi(\tilde{R}) = 0$, where $\tilde{R} = R + d$ is the extrapolated radius, and must remain finite throughout the interior of the reactor.

Solving the Differential Equation

By substituting $\phi = \frac{w}{r}$ into the diffusion equation and solving the resulting equation for w, we find the general solution:

$$\phi(r) = \frac{A\sin(Br)}{r} + \frac{C\cos(Br)}{r}$$

where A and C are constants. To ensure that the flux is finite at $r = 0$, we must set

$C = 0$. Thus, the solution simplifies to:

$$\phi(r) = \frac{A \sin(Br)}{r}$$

8.3.2 Determining Eigenvalues

The boundary condition $\phi(\tilde{R}) = 0$ leads to:

$$\frac{A \sin(B\tilde{R})}{\tilde{R}} = 0 \tag{8-15}$$

This condition can be satisfied if B is one of the eigenvalues:

$$B_n = \frac{n\pi}{\tilde{R}} \tag{8-16}$$

where n is an integer. For a critical reactor, only the first eigenvalue is relevant, so with $n = 1$:

$$B_1^2 = \left(\frac{\pi}{\tilde{R}}\right)^2$$

8.3.3 Flux Distribution

The neutron flux in the spherical reactor is then given by:

$$\phi(r) = \frac{A \sin\left(\frac{\pi r}{\tilde{B}}\right)}{r} \tag{8-17}$$

Determining the Constant A

The constant A is determined by the reactor's operating power. The total power P generated by the reactor is:

$$P = E_R \Sigma_f \int \phi(r) dV$$

where E_R is the recoverable energy per fission, Σ_f is the macroscopic fission cross section, and dV is the differential volume element. For a sphere, the volume element is:

$$dV = 4\pi r^2 dr$$

So, the power integral becomes:

$$P = 4\pi E_R \Sigma_f \int_0^R r^2 \phi(r) dr$$

Integrating the Flux

Substituting the expression for $\phi(r)$:

$$P = 4\pi E_R \Sigma_f \int_0^R r^2 \frac{A\sin\left(\frac{\pi r}{\tilde{R}}\right)}{r} dr$$

$$P = 4\pi E_R \Sigma_f A \int_0^R r \sin\left(\frac{\pi r}{\tilde{R}}\right) dr$$

Performing the integration:

$$P = 4\pi E_R \Sigma_f A \left[-\frac{\tilde{R}}{\pi} r\cos\left(\frac{\pi r}{\tilde{R}}\right) + \frac{\tilde{R}^2}{\pi^2} \sin\left(\frac{\pi r}{\tilde{R}}\right) \right]_0^R$$

Evaluating at the limits:

$$P = 4\pi E_R \Sigma_f A \left[-\frac{\tilde{R}}{\pi} R\cos\left(\frac{\pi R}{\tilde{R}}\right) + \frac{\tilde{R}^2}{\pi^2} \sin\left(\frac{\pi R}{\tilde{R}}\right) \right]$$

For d small compared to a :

$$P = 4\pi E_R \Sigma_f A \left[-\frac{\tilde{R}}{\pi} R\cos\left(\frac{\pi R}{\tilde{R}}\right) + \frac{\tilde{R}^2}{\pi^2} \sin\left(\frac{\pi R}{\tilde{R}}\right) \right]$$

Simplifying for $\tilde{R} \approx R$:

$$P = 4\pi E_R \Sigma_f A \left[\frac{R^2}{\pi^2} \sin\left(\frac{\pi R}{R}\right) - \frac{R^2}{\pi^2} \sin\left(\frac{\pi R}{R}\right) \right]$$

Since $\sin(\pi) = 0$, the dominant term simplifies to:

$$P = 4\pi E_R \Sigma_f A \left[\frac{R^2}{\pi^2} \sin\left(\frac{\pi R}{R}\right) \right] \tag{8-18}$$

Finally, the thermal flux in the spherical reactor becomes:

$$\phi(r) = \frac{P}{4 E_R \Sigma_f R^2} \frac{\sin\left(\frac{\pi r}{R}\right)}{r} \tag{8-19}$$

8.4   The Infinite Cylindrical Reactor

Consider an infinite critical cylindrical reactor of radius R. In this reactor, the neutron flux ϕ depends only on the radial distance r from the axis of the cylinder. The one-group diffusion equation for this cylindrical geometry, using the Laplacian in cylindrical coordinates, is:

$$\frac{1}{r} \frac{d}{dr}\left(r \frac{d\phi}{dr} \right) + B^2 \phi = 0 \tag{8-20}$$

When the differentiation in the first term is carried out, it becomes:

$$\frac{d^2\phi}{dr^2} + \frac{1}{r}\frac{d\phi}{dr} + B^2 \phi = 0$$

Figure 8-3 shows the geometry of an infinite cylinder reactor.

8.4.1 Boundary Conditions

The flux must satisfy the boundary condition $\phi(\tilde{R}) = 0$, where $\tilde{R} = R + d$ is the extrapolated radius. Additionally, ϕ must remain finite throughout the interior of the reactor.

Solving the Differential Equation

This equation is a special case of Bessel's equation:

$$\frac{d^2\phi}{dr^2} + \frac{1}{r}\frac{d\phi}{dr} + \left(B^2 - \frac{m^2}{r^2} \right)\phi = 0$$

where m is a constant. For the cylindrical reactor, m is equal to zero.

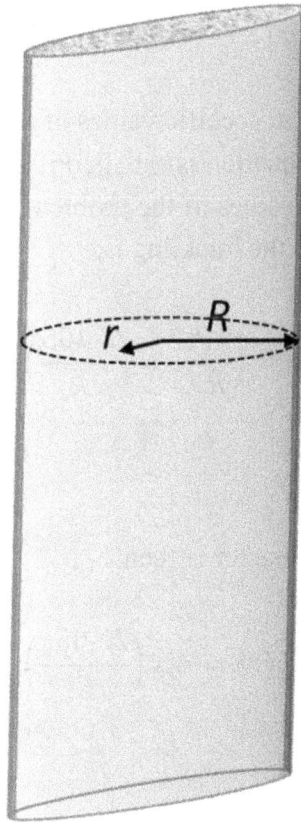

Figure 8-3. The infinite cylinder reactor geometry.

Therefore, the general solution to the reactor equation can be written as:

$$\phi(r) = AJ_0(Br) + CY_0(Br)$$

where A and C are constants, and J_0 and Y_0 are Bessel functions of the first and second kind, respectively.

Ensuring Finite Flux

The function $Y_0(x)$ is infinite at $x = 0$, while $J_0(0) = 1$. Since the neutron flux ϕ must remain finite within the reactor, C must be zero. Thus, the flux reduces to:

$$\phi(r) = AJ_0(Br)$$

8.4.2   Determining Eigenvalues

The boundary condition $\phi(\tilde{R}) = 0$ gives:

$$\phi(\tilde{R}) = AJ_0(B\tilde{R}) = 0$$

The function $J_0(x)$ is zero at specific values of x, labeled $x_1, x_2, ...,$ such that $J_0(x_n) = 0$. This means that the equation is satisfied if B is one of the values:

These values are the eigenvalues of the problem. For a critical reactor, only the lowest eigenvalue is important, so the buckling is:

$$B_1^2 = \left(\frac{x_1}{\tilde{R}}\right)^2 = \left(\frac{2.405}{\tilde{R}}\right)^2 \tag{8-21}$$

8.4.3   Flux Distribution

The one-group flux in the reactor is then:

$$\phi(r) = AJ_0\left(\frac{2.405r}{\tilde{R}}\right)$$

Determining the Constant A

The constant A is determined by the reactor's operating power. The total power P generated by the reactor per unit length is:

$$P = 2\pi E_R \Sigma_f \int_0^R \phi(r)rdr$$

where E_R is the recoverable energy per fission, Σ_f is the macroscopic fission cross section, and $dV = 2\pi r dr$ is the differential volume element.

Integrating the Flux

Substituting the expression for $\phi(r)$:

$$P = 2\pi E_R \Sigma_f A \int_0^R J_0\left(\frac{2.405r}{R}\right)rdr$$

Using the integral formula for Bessel functions:

$$\int_0^x J_0(x')x'dx' = xJ_1(x)$$

Integrating the Flux

Substituting the expression for $\phi(r)$:

$$P = 2\pi E_R \Sigma_f A \int_0^R J_0\left(\frac{2.405r}{R}\right)rdr$$

Using the integral formula for Bessel functions:

$$\int_0^x J_0(x')x'dx' = xJ_1(x)$$

Evaluating the integral:

$$P = \frac{2\pi E_R \Sigma_f R^2 A J_1(2.405)}{2.405} \tag{8-22}$$

Given that $J_1(2.405) \approx 0.519$, we have:

$$P \approx 1.36 E_R \Sigma_f R^2 A$$

Finally, the flux in the infinite cylindrical reactor is:

$$\phi(r) = \frac{0.737P}{E_R \Sigma_f R^2} J_0\left(\frac{2.405r}{R}\right) \tag{8-23}$$

8.5   The Finite Cylindrical Reactor

An interesting flux distribution is obtained for a finite cylindrical reactor of height H and radius R. In this reactor, the neutron flux ϕ depends on both the radial distance r from the axis and the axial distance z from the midpoint of the cylinder. The reactor equation, using the Laplacian appropriate to cylindrical coordinates, is:

$$\frac{1}{r}\frac{\partial}{\partial r}\left(r\frac{\partial \phi}{\partial r}\right) + \frac{\partial^2 \phi}{\partial z^2} + B^2 \phi = 0 \tag{8-24}$$

When the differentiation in the first term is carried out, it becomes:

$$\frac{\partial^2 \phi}{\partial r^2} + \frac{1}{r}\frac{\partial \phi}{\partial r} + \frac{\partial^2 \phi}{\partial z^2} + B^2 \phi = 0$$

Figure 8-4. shows the geometry of the finite cylinder reactor.

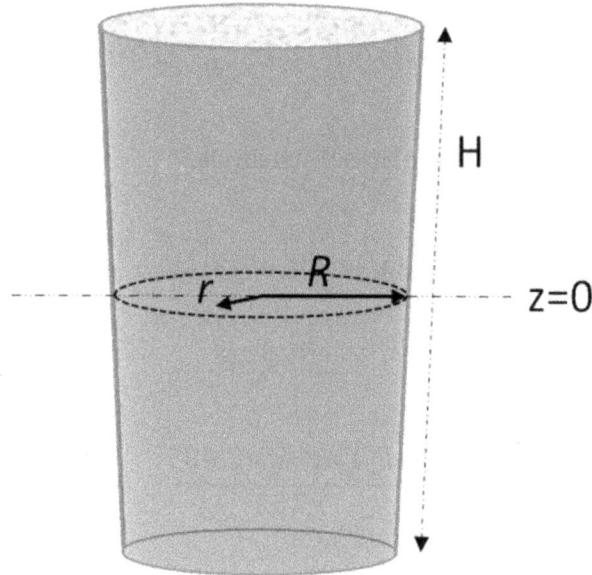

Figure 8-4. The finite cylinder reactor.**Boundary Conditions**

The flux ϕ must satisfy the usual boundary conditions, including $\phi(\tilde{R}, z) = 0$ and $\phi(r, \tilde{H}/2) = 0$, where $\tilde{R} = R + d$ and $\tilde{H} = H + 2d$ are the extrapolated radius and height, respectively.

8.5.2   Separation of Variables

The solution is obtained by assuming separation of variables:

$$\phi(r, z) = R(r)Z(z)$$

Substituting this into the reactor equation gives:

$$\frac{1}{R}\left(\frac{1}{r}\frac{\partial}{\partial r}\left(r\frac{\partial R}{\partial r}\right)\right) + \frac{1}{Z}\frac{\partial^2 Z}{\partial z^2} = -B^2$$

Since this equation must be satisfied for any combination of r and z, each term must be a constant. Letting the first term equal B_r^2 and the second B_z^2, we can write:

$$\frac{d^2R}{dr^2} + \frac{1}{r}\frac{dR}{dr} + B_r^2 R = 0$$

This is Bessel's equation of the first kind, order zero. Additionally, we have:

$$\frac{d^2Z}{dz^2} + B_z^2 Z = 0$$

The total buckling B^2 is given by:

$$B^2 = B_r^2 + B_z^2$$

8.5.3 Solving the Axial Component

The axial component $Z(z)$ is solved similarly to the infinite slab reactor. The differential equation:

$$\frac{d^2Z}{dz^2} + B_z^2 Z = 0$$

has the general solution:

$$Z(z) = A\cos\left(B_z z\right) + C\sin\left(B_z z\right)$$

To satisfy the boundary condition $\phi(r, \tilde{H}/2) = 0$, we impose $Z(\tilde{H}/2) = 0$:

$$A\cos\left(B_z\frac{\tilde{H}}{2}\right) + C\sin\left(B_z\frac{\tilde{H}}{2}\right) = 0$$

For a non-trivial solution, $\cos\left(B_z\tilde{H}/2\right)$ must be zero, implying:

$$B_z = \frac{n\pi}{\tilde{H}}$$

where n is an integer. For the fundamental mode $n = 1$, we have:

$$B_z = \frac{\pi}{\bar{H}}$$

Thus, the axial flux distribution is:

$$Z(z) = A\cos\left(\frac{\pi z}{\bar{H}}\right)$$

8.5.4   Solving the Radial Component

The radial component $R(r)$ is governed by Bessel's equation:

$$\frac{d^2R}{dr^2} + \frac{1}{r}\frac{dR}{dr} + B_r^2 R = 0$$

The general solution to this equation is:

$$R(r) = AJ_0(B_r r) + CY_0(B_r r)$$

Given that $Y_0(x)$ becomes infinite as x approaches zero and the flux must remain finite, $C = 0$. Therefore, the solution simplifies to:

$$R(r) = AJ_0(B_r r)$$

To satisfy the boundary condition $\phi(\tilde{R}, z) = 0$:

$$J_0\left(B_r \tilde{R}\right) = 0$$

The function $J_0(x)$ is zero at specific values x_n. For the fundamental mode, the smallest root x_1 is 2.405. Thus, the radial buckling is:

$$B_r = \frac{2.405}{\bar{R}}$$

8.5.5   Total Flux Distribution

Combining the radial and axial solutions, the total neutron flux distribution in the reactor is:

$$\phi(r,z) = AJ_0\left(\frac{2.405r}{\tilde{R}}\right)\cos\left(\frac{\pi z}{\tilde{H}}\right) \tag{8-25}$$

Determining the Constant A

The constant A is determined by the reactor's operating power. The total power P generated by the reactor is:

$$P = E_R\Sigma_f \int \phi(r,z)dV$$

where E_R is the recoverable energy per fission, Σ_f is the macroscopic fission cross section, and dV is the differential volume element. For a finite cylinder:

$$dV = 2\pi r\,dr\,dz$$

Thus, the power integral becomes:

$$P = 2\pi E_R\Sigma_f A \int_0^R \int_{-H/2}^{H/2} J_0\left(\frac{2.405r}{\tilde{R}}\right)\cos\left(\frac{\pi z}{\tilde{H}}\right)r\,dr\,dz$$

Integrating the Flux

Separate the integrals over r and z :

$$P = 2\pi E_R\Sigma_f A \left(\int_0^R rJ_0\left(\frac{2.405r}{\tilde{R}}\right)dr\right)\left(\int_{-H/2}^{H/2}\cos\left(\frac{\pi z}{\tilde{H}}\right)dz\right)$$

Using the integral formula for Bessel functions:

$$\int_0^x J_0(x')x'dx' = xJ_1(x)$$

and evaluating:

$$\int_0^R rJ_0\left(\frac{2.405r}{\tilde{R}}\right)dr = \frac{\tilde{R}^2}{2.405}J_1(2.405)$$

Given $J_1(2.405) \approx 0.519$:

$$\int_0^R r J_0\left(\frac{2.405r}{\tilde{R}}\right) dr = \frac{\tilde{R}^2 \cdot 0.519}{2.405}$$

For the axial integral:

$$\int_{-H/2}^{H/2} \cos\left(\frac{\pi z}{\tilde{H}}\right) dz = \frac{\tilde{H}}{\pi} \sin\left(\frac{\pi H/2}{\tilde{H}}\right)$$

Assuming $H \approx \tilde{H}$:

$$\int_{-H/2}^{H/2} \cos\left(\frac{\pi z}{\tilde{H}}\right) dz = \frac{\tilde{H}}{\pi}$$

Combining these results, the power becomes:

$$P = 2\pi E_R \Sigma_f A \left(\frac{\tilde{R}^2 \cdot 0.519}{2.405}\right)\left(\frac{\tilde{H}}{\pi}\right) \tag{8-26}$$

Solving for A :

$$A = \frac{P}{1.36 E_R \Sigma_f \tilde{R}^2 \tilde{H}}$$

Finally, the neutron flux distribution in the finite cylindrical reactor is:

$$\phi(r,z) = \frac{P}{1.36 E_R \Sigma_f \tilde{R}^2 \tilde{H}} J_0\left(\frac{2.405r}{\tilde{R}}\right) \cos\left(\frac{\pi z}{\tilde{H}}\right) \tag{8-27}$$

8.6   The Rectangular Parallelepiped Reactor

Consider a rectangular parallelepiped reactor with dimensions a, b, and c along the $x-, y-$, and z axes, respectively. The neutron flux $\phi(x, y, z)$ within this reactor depends on the spatial coordinates in all three dimensions. The one-group diffusion equation for this geometry is given by:

$$\nabla^2 \phi + B^2 \phi = 0 \tag{8-28}$$

where ∇^2 is the Laplacian operator, and B^2 is the buckling.
Reactor Equation in Rectangular Coordinates
In rectangular coordinates, the Laplacian operator ∇^2 is:

$$\nabla^2 = \frac{\partial^2}{\partial x^2} + \frac{\partial^2}{\partial y^2} + \frac{\partial^2}{\partial z^2}$$

Thus, the reactor equation becomes:

$$\frac{\partial^2 \phi}{\partial x^2} + \frac{\partial^2 \phi}{\partial y^2} + \frac{\partial^2 \phi}{\partial z^2} + B^2 \phi = 0 \tag{8-29}$$

Figure 8-5. shows the geometry of a rectangular parallelepiped reactor.

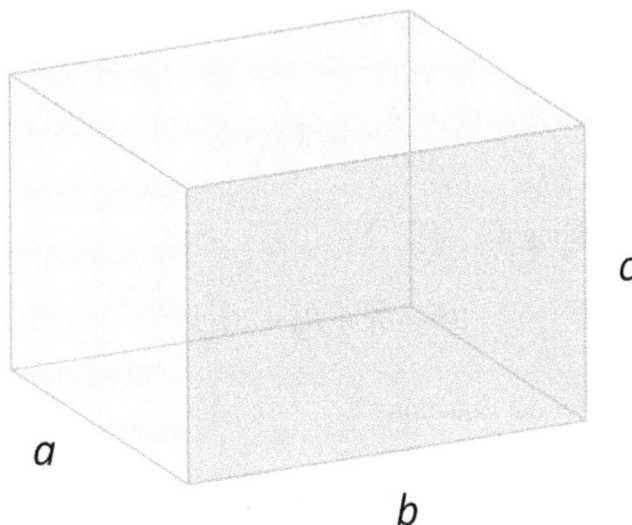

Figure 8-5. The rectangular parallelepiped reactor.

8.6.1 Separation of Variables

To solve this equation, we assume a solution of the form:

$$\phi(x, y, z) = X(x)Y(y)Z(z)$$

Substituting this into the reactor equation gives:

$$Y(y)Z(z)\frac{d^2 X}{dx^2} + X(x)Z(z)\frac{d^2 Y}{dy^2} + X(x)Y(y)\frac{d^2 Z}{dz^2} + B^2 X(x)Y(y)Z(z) = 0$$

Dividing through by $X(x)Y(y)Z(z)$, we obtain:

$$\frac{1}{X}\frac{d^2 X}{dx^2} + \frac{1}{Y}\frac{d^2 Y}{dy^2} + \frac{1}{Z}\frac{d^2 Z}{dz^2} + B^2 = 0$$

Since each term depends on a different variable, each must be equal to a constant.

Let these constants be $-B_x^2$, $-B_y^2$, and $-B_z^2$. Thus, we have three separate equations:

$$\frac{d^2X}{dx^2} + B_x^2 X = 0$$

$$\frac{d^2Y}{dy^2} + B_y^2 Y = 0$$

$$\frac{d^2Z}{dz^2} + B_z^2 Z = 0$$

And the total buckling is:

$$B^2 = B_x^2 + B_y^2 + B_z^2 \tag{8-30}$$

Solving the Differential Equations

Each of these differential equations has the general solution of a sinusoidal form. Let's solve them individually.

8.6.2   Solution for $X(x)$

$$\frac{d^2X}{dx^2} + B_x^2 X = 0$$

The general solution is:

$$X(x) = A\cos(B_x x) + B\sin(B_x x)$$

Applying the boundary conditions $\phi(\pm a/2, y, z) = 0$:

$$X\left(\frac{a}{2}\right) = A\cos\left(B_x \frac{a}{2}\right) + B\sin\left(B_x \frac{a}{2}\right) = 0$$

$$X\left(-\frac{a}{2}\right) = A\cos\left(-B_x \frac{a}{2}\right) + B\sin\left(-B_x \frac{a}{2}\right) = 0$$

Since cos is an even function and sin is an odd function, these simplify to:

$$A\cos\left(B_x\frac{a}{2}\right) = 0$$

To avoid the trivial solution $A = 0$, we must have:

$$B_x\frac{a}{2} = n\pi$$

where n is an integer. Thus, the eigenvalues are:

$$B_x = \frac{2n\pi}{a}$$

And the solution is:

$$X(x) = A\cos\left(\frac{2n\pi x}{a}\right)$$

8.6.3 Solution for $Y(y)$

$$\frac{d^2Y}{dy^2} + B_y^2Y = 0$$

The general solution is:

$$Y(y) = C\cos\left(B_y y\right) + D\sin\left(B_y y\right)$$

Applying the boundary conditions $\phi(x, \pm b/2, z) = 0$:

$$Y\left(\frac{b}{2}\right) = C\cos\left(B_y\frac{b}{2}\right) + D\sin\left(B_y\frac{b}{2}\right) = 0$$

$$Y\left(-\frac{b}{2}\right) = C\cos\left(-B_y\frac{b}{2}\right) + D\sin\left(-B_y\frac{b}{2}\right) = 0$$

These simplify to:

$$C\cos\left(B_y\frac{b}{2}\right) = 0$$

To avoid the trivial solution $C = 0$, we must have:

$$B_y \frac{b}{2} = m\pi$$

where m is an integer. Thus, the eigenvalues are:

$$B_y = \frac{2m\pi}{b}$$

And the solution is:

$$Y(y) = C\cos\left(\frac{2m\pi y}{b}\right)$$

8.6.4   Solution for $Z(z)$

$$\frac{d^2 Z}{dz^2} + B_z^2 Z = 0$$

The general solution is:

$$Z(z) = E\cos(B_z z) + F\sin(B_z z)$$

Applying the boundary conditions $\phi(x, y, \pm c/2) = 0$:

$$Z\left(\frac{c}{2}\right) = E\cos\left(B_z \frac{c}{2}\right) + F\sin\left(B_z \frac{c}{2}\right) = 0$$

$$Z\left(-\frac{c}{2}\right) = E\cos\left(-B_z \frac{c}{2}\right) + F\sin\left(-B_z \frac{c}{2}\right) = 0$$

These simplify to:

$$E\cos\left(B_z \frac{c}{2}\right) = 0$$

To avoid the trivial solution $E = 0$, we must have:

$$B_z \frac{c}{2} = l\pi$$

where l is an integer. Thus, the eigenvalues are:

$$B_z = \frac{2l\pi}{c}$$

And the solution is:

$$Z(z) = E\cos\left(\frac{2l\pi z}{c}\right)$$

8.6.5 Total Flux Solution

Combining the solutions for $X(x), Y(y)$, and $Z(z)$, the total flux distribution is:

$$\phi(x,y,z) = A\cos\left(\frac{2n\pi x}{a}\right)C\cos\left(\frac{2m\pi y}{b}\right)E\cos\left(\frac{2l\pi z}{c}\right) \qquad (8\text{-}31)$$

Rewriting this, we have:

$$\phi(x,y,z) = A'\cos\left(\frac{2n\pi x}{a}\right)\cos\left(\frac{2m\pi y}{b}\right)\cos\left(\frac{2l\pi z}{c}\right) \qquad (8\text{-}32)$$

where $A' = A \times C \times E$.

Determining the Constant A'

The constant A' is determined by the reactor's operating power. The total power P generated by the reactor is:

$$P = E_R\Sigma_f \int \phi(x,y,z)dV$$

where E_R is the recoverable energy per fission, Σ_f is the macroscopic fission cross section, and dV is the differential volume element. For a rectangular parallelepiped:

$$dV = dxdydz$$

Thus, the power integral becomes:

$$P = E_R \Sigma_f A' \int_{-a/2}^{a/2} \cos\left(\frac{2n\pi x}{a}\right) dx \int_{-b/2}^{b/2} \cos\left(\frac{2m\pi y}{b}\right) dy \int_{-c/2}^{c/2} \cos\left(\frac{2l\pi z}{c}\right) dz$$

Each integral is of the form:

$$\int_{-L/2}^{L/2} \cos\left(\frac{2k\pi x}{L}\right) dx$$

This integral is zero for $k \neq 0$, and L for $k = 0$. Thus, only the fundamental mode $n = m = l = 1$ contributes to the power:

$$P = E_R \Sigma_f A' \left(\int_{-a/2}^{a/2} \cos\left(\frac{2\pi x}{a}\right) dx\right)\left(\int_{-b/2}^{b/2} \cos\left(\frac{2\pi y}{b}\right) dy\right)\left(\int_{-c/2}^{c/2} \cos\left(\frac{2\pi z}{c}\right) dz\right)$$

Evaluating these integrals:

$$\int_{-L/2}^{L/2} \cos\left(\frac{2\pi x}{L}\right) dx = \frac{L}{2\pi}\left[\sin\left(\frac{2\pi x}{L}\right)\right]_{-L/2}^{L/2} = 0$$

For the fundamental mode $n = m = l = 1$:

$$\phi(x, y, z) = A' \cos\left(\frac{\pi x}{a}\right) \cos\left(\frac{\pi y}{b}\right) \cos\left(\frac{\pi z}{c}\right) \qquad (8\text{-}33)$$

8.6.6   Final Power Profile

The power per unit volume is:

$$p(x, y, z) = E_R \Sigma_f \phi(x, y, z)$$

Using the expression for $\phi(x, y, z)$, we have:

$$p(x, y, z) = E_R \Sigma_f A' \cos\left(\frac{\pi x}{a}\right) \cos\left(\frac{\pi y}{b}\right) \cos\left(\frac{\pi z}{c}\right)$$

To determine A', integrate over the reactor volume:

$$P = \int_{-a/2}^{a/2} \int_{-b/2}^{b/2} \int_{-c/2}^{c/2} p(x, y, z) dx dy dz$$

$$P = E_R \Sigma_f A' \int_{-a/2}^{a/2} \cos\left(\frac{\pi x}{a}\right) dx \int_{-b/2}^{b/2} \cos\left(\frac{\pi y}{b}\right) dy \int_{-c/2}^{c/2} \cos\left(\frac{\pi z}{c}\right) dz$$

Each integral is:

$$\int_{-L/2}^{L/2} \cos\left(\frac{\pi x}{L}\right) dx = L$$

Therefore:

$$P = E_R \Sigma_f A' abc$$

Thus:

$$A' = \frac{P}{E_R \Sigma_f abc}$$

The final neutron flux distribution is:

$$\phi(x, y, z) = \frac{P}{E_R \Sigma_f abc} \cos\left(\frac{\pi x}{a}\right) \cos\left(\frac{\pi y}{b}\right) \cos\left(\frac{\pi z}{c}\right)$$

And the power profile is:

$$p(x, y, z) = \frac{P}{abc} \cos\left(\frac{\pi x}{a}\right) \cos\left(\frac{\pi y}{b}\right) \cos\left(\frac{\pi z}{c}\right) \tag{8-34}$$

8.7 Maximum-to-Average Flux Ratio

The maximum-to-average flux ratio (Ω) is a measure of the extent to which the neutron flux (and hence power density) at the center of the reactor exceeds the average flux throughout the reactor. This ratio is crucial for reactor safety and efficiency. Below, we derive Ω for different reactor geometries: slab, infinite cylinder, finite cylinder, sphere, and rectangular parallelepiped.

8.7.1 Slab Reactor

For a slab reactor of thickness a, the neutron flux $\phi(x)$ is given by:

$$\phi(x) = \phi_{max}\cos\left(\frac{\pi x}{a}\right)$$

where x ranges from $-a/2$ to $a/2$.

Maximum Flux:

$$\phi_{max} = \phi(0) = \phi_{max}\cos(0) = \phi_{max}$$

Average Flux:

The average flux ϕ_{av} is obtained by integrating $\phi(x)$ over the volume and dividing by the volume:

$$\phi_{av} = \frac{1}{a}\int_{-a/2}^{a/2}\phi(x)dx$$

$$\phi_{av} = \frac{\phi_{max}}{a}\int_{-a/2}^{a/2}\cos\left(\frac{\pi x}{a}\right)dx$$

Using the integral of cosine:

$$\phi_{av} = \frac{\phi_{max}}{a}\left[\frac{\sin(\pi x/a)}{\pi/a}\right]_{-a/2}^{a/2}$$

$$\phi_{av} = \frac{\phi_{max}}{a}\cdot\frac{a}{\pi}[\sin(\pi/2) - \sin(-\pi/2)]$$

$$\phi_{av} = \frac{\phi_{max}}{\pi}\cdot 2$$

Maximum-to-Average Flux Ratio

$$\Omega = \frac{\phi_{max}}{\phi_{av}} = \frac{\phi_{max}}{\frac{\phi_{max}}{\pi}\cdot 2} = \frac{\pi}{2} \approx 1.57 \tag{8-35}$$

8.7.2   Infinite Cylinder

For an infinite cylinder of radius R, the neutron flux $\phi(r)$ is given by:

$$\phi(r) = \phi_{max} J_0\left(\frac{2.405r}{R}\right)$$

where J_0 is the Bessel function of the first kind.

Maximum Flux:

$$\phi_{max} = \phi(0) = \phi_{max} J_0(0) = \phi_{max}$$

Average Flux:

The average flux ϕ_{av} is obtained by integrating $\phi(r)$ over the volume and dividing by the volume:

$$\phi_{av} = \frac{1}{\pi R^2} \int_0^R 2\pi r \phi(r)dr$$

$$\phi_{av} = \frac{2\phi_{max}}{R^2} \int_0^R r J_0\left(\frac{2.405\, r}{R}\right) dr$$

Using the integral for Bessel functions:

$$\int_0^x x J_0(x)dx = \frac{x^2}{2} J_1(x)$$

$$\phi_{av} = \frac{2\phi_{max}}{R^2} \left[\frac{R^2}{2.405^2} J_1(2.405)\right]$$

$$\phi_{av} = \frac{2\phi_{max}}{2.405} J_1(2.405)$$

$$\phi_{av} = 0.43175\, \phi_{max}$$

Maximum-to-Average Flux Ratio:

$$\Omega = \frac{\phi_{max}}{\phi_{av}} = \frac{\phi_{max}}{0.43175\, \phi_{max}} = 2.316 \tag{8-36}$$

8.7.3 Finite Cylinder

For a finite cylinder of height H and radius R, the neutron flux $\phi(r,z)$ is given by:

$$\phi(r,z) = \phi_{max}J_0\left(\frac{2.405r}{R}\right)\cos\left(\frac{\pi z}{H}\right)$$

Maximum Flux:

$$\phi_{max} = \phi(0,0) = \phi_{max}J_0(0)\cos(0) = \phi_{max}$$

Average Flux:

The average flux ϕ_{av} is obtained by integrating $\phi(r,z)$ over the volume and dividing by the volume:

$$\phi_{av} = \frac{1}{\pi R^2 H}\int_0^R\int_{-H/2}^{H/2} 2\pi r\phi(r,z)dzdr$$

$$\phi_{av} = \frac{2\phi_{max}}{R^2 H}\int_0^R rJ_0\left(\frac{2.405r}{R}\right)dr\int_{-H/2}^{H/2}\cos\left(\frac{\pi z}{H}\right)dz$$

Using the integral for Bessel functions:

$$\int_0^x xJ_0(x)dx = \frac{x^2}{2}J_1(x)$$

And the integral of cosine:

$$\int\cos(ax)\,dx = \frac{\sin(ax)}{a}$$

$$\phi_{av} = \frac{2\phi_{max}}{R^2 H}\left[\frac{R^2}{2.405}J_1(2.405)\right]\left[\frac{H}{\pi}\sin\left(\frac{\pi H}{H}\right)\right]$$

$$\phi_{av} = 0.43175\,\phi_{max}\times\frac{2}{\pi}$$

$$\phi_{av} = 0.2749\,\phi_{max}$$

Maximum-to-Average Flux Ratio:

$$\Omega = \frac{\phi_{max}}{\phi_{av}} = \frac{\phi_{max}}{0.2749\ \phi_{max}} = 3.638 \tag{8-37}$$

8.7.4 Sphere

For a spherical reactor of radius R, the neutron flux $\phi(r)$ is given by:

$$\phi(r) = A\frac{\sin(\pi r/R)}{r}$$

Maximum Flux:

$$\phi_{max} = \lim_{r \to 0} \phi(r) = \frac{\pi A}{R}$$

Average Flux:

The average flux ϕ_{av} is obtained by integrating $\phi(r)$ over the volume and dividing by the volume:

$$\phi_{av} = \frac{1}{V}\int_0^R \phi(r)4\pi r^2 dr$$

$$\phi_{av} = \frac{A}{4\pi R^3/3}\int_0^R \sin\left(\frac{\pi r}{R}\right)4\pi r^2 dr$$

$$\phi_{av} = \frac{\phi_{max}}{4\pi R^3/3}\left[4\pi\left[-\frac{r R \cos\left(\frac{\pi r}{R}\right)}{\pi} + \frac{R^2 \sin\left(\frac{\pi r}{R}\right)}{\pi^2}\right]\right]_0^R$$

$$\phi_{av} = \frac{3 A}{\pi R}$$

Maximum-to-Average Flux Ratio:

$$\Omega = \frac{\phi_{max}}{\phi_{av}} = \frac{\dfrac{\pi A}{R}}{\dfrac{3 A}{\pi R}} = \frac{\pi^2}{3} = 3.2899 \tag{8-38}$$

8.7.5   Rectangular Parallelepiped

For a rectangular parallelepiped reactor with dimensions a, b, and c, the neutron flux $\phi(x, y, z)$ is given by:

$$\phi(x, y, z) = A \cos\left(\frac{\pi x}{a}\right) \cos\left(\frac{\pi y}{b}\right) \cos\left(\frac{\pi z}{c}\right)$$

Maximum Flux:

$$\phi_{\max} = \phi(0,0,0) = A$$

Average Flux:

The average flux ϕ_{av} is obtained by integrating $\phi(x, y, z)$ over the volume and dividing by the volume:

$$\phi_{av} = \frac{1}{V} \int_{-a/2}^{a/2} \int_{-b/2}^{b/2} \int_{-c/2}^{c/2} \phi(x, y, z) \, dx \, dy \, dz$$

$$\phi_{av} = \frac{A}{abc} \int_{-a/2}^{a/2} \cos\left(\frac{\pi x}{a}\right) dx \int_{-b/2}^{b/2} \cos\left(\frac{\pi y}{b}\right) dy \int_{-c/2}^{c/2} \cos\left(\frac{\pi z}{c}\right) dz$$

$$\int_{-L/2}^{L/2} \cos\left(\frac{\pi x}{L}\right) dx = \frac{2L}{\pi}$$

$$\phi_{av} = \frac{A}{abc} \cdot \frac{2a}{\pi} \cdot \frac{2b}{\pi} \cdot \frac{2c}{\pi} = \frac{8A}{\pi^3}$$

Maximum-to-Average Flux Ratio:

$$\Omega = \frac{\phi_{\max}}{\phi_{av}} = \frac{A}{\frac{8A}{\pi^3}} = \frac{\pi^3}{8} = 3.876 \qquad (8\text{-}39)$$

8.8   Problems

7-1. Consider an infinite, homogeneous slab reactor with thickness $2a = 60$ cm. The reactor is composed of a mixture of ^{235}U and graphite. The reactor operates under critical conditions.

1. Derive the neutron flux distribution $\phi(x)$ within the reactor.
2. Determine the reactor buckling B_1^2.
3. Calculate the neutron flux at $x = 15$ cm if the reactor power per unit area is $P = 500$ W/cm^2, the macroscopic fission cross section $\Sigma_f = 0.1$ cm^{-1}, and the recoverable energy per fission $E_R = 3.2 \times 10^{-11}$ joules.

Solution:

1. Derive the Neutron Flux Distribution:

The one-group diffusion equation for the slab reactor is:

$$\frac{d^2\phi}{dx^2} + B^2\phi = 0$$

The general solution to this differential equation is:

$$\phi(x) = A\cos(Bx) + C\sin(Bx)$$

Applying the symmetry condition $\frac{d\phi}{dx} = 0$ at $x = 0$:

$$\frac{d\phi}{dx} = -AB\sin(Bx) + CB\cos(Bx)$$

At $x = 0$:

$$\frac{d\phi}{dx} = 0 \Longrightarrow C = 0$$

Thus, the solution reduces to:

$$\phi(x) = A\cos(Bx)$$

Applying the boundary conditions $\phi(\pm\tilde{a}/2) = 0$, where $\tilde{a} = a + 2d$:

$$\phi(\tilde{a}/2) = A\cos\left(\frac{B\tilde{a}}{2}\right) = 0$$

This implies:

$$\cos\left(\frac{B\tilde{a}}{2}\right) = 0$$

The non-trivial solution requires:

$$\frac{B\tilde{a}}{2} = \frac{(2n+1)\pi}{2}$$

$$B_n = \frac{(2n+1)\pi}{\tilde{a}}$$

For the fundamental mode $n = 0$:

$$B_1 = \frac{\pi}{\tilde{a}}$$

Thus, the neutron flux distribution is:

$$\phi(x) = A\cos\left(\frac{\pi x}{\breve{a}}\right)$$

2. Determine the Reactor Buckling:

The square of the lowest eigenvalue B_1^2 is the reactor buckling:

$$B_1^2 = \left(\frac{\pi}{\tilde{a}}\right)^2$$

Given $\tilde{a} = a + 2d$ and assuming d is small compared to a :

$$\tilde{a} \approx a$$

Thus:

$$B_1^2 = \left(\frac{\pi}{a}\right)^2$$

For $a = 30$ cm :

$$B_1^2 = \left(\frac{\pi}{30}\right)^2 \approx 0.011 \text{ cm}^{-2}$$

Calculate the Neutron Flux at $x = 15$ cm :

Given:

$$P = 500 \text{ W/cm}^2$$

$$\Sigma_f = 0.1 \text{ cm}^{-1}$$

$$E_R = 3.2 \times 10^{-11} \text{ J}$$

The neutron flux is given by:

$$\phi(x) = \frac{\pi P}{2aE_R\Sigma_f} \cos\left(\frac{\pi x}{a}\right)$$

For $a = 30$ cm :

$$\phi(15) = \frac{\pi \times 500}{2 \times 30 \times 3.2 \times 10^{-11} \times 0.1} \cos\left(\frac{\pi \times 15}{30}\right)$$

$$\phi(15) = \frac{3.14 \times 500}{6 \times 3.2 \times 10^{-11} \times 0.1} \cos\left(\frac{\pi}{2}\right)$$

$$\phi(15) = \frac{1570}{1.92 \times 10^{-11}} \times 0$$

$$\phi(15) = 0$$

The neutron flux at $x = 15$ cm is zero due to the cosine term.

7-2. Consider an infinite, homogeneous slab reactor with thickness $2a = 40$ cm. The reactor is composed of a mixture of ^{239}Pu and graphite. The reactor operates under critical conditions.
  1. Derive the neutron flux distribution $\phi(x)$ within the reactor.
  2. Determine the reactor buckling B_1^2.
  3. Calculate the total reactor power per unit area if the macroscopic fission cross

section $\Sigma_f = 0.15$ cm^{-1}, the recoverable energy per fission $E_R = 3.2 \times 10^{-11}$ joules, and the neutron flux at the center $\phi(0) = 2 \times 10^{12}$ neutrons /cm^2/s.

Solution:

1. Derive the Neutron Flux Distribution:

The one-group diffusion equation for the slab reactor is:

$$\frac{d^2\phi}{dx^2} + B^2\phi = 0$$

The general solution to this differential equation is:

$$\phi(x) = A\cos(Bx) + C\sin(Bx)$$

Applying the symmetry condition $\frac{d\phi}{dx} = 0$ at $x = 0$:

$$\frac{d\phi}{dx} = -AB\sin(Bx) + CB\cos(Bx)$$

At $x = 0$:

$$\frac{d\phi}{dx} = 0 \Longrightarrow C = 0$$

Thus, the solution reduces to:

$$\phi(x) = A\cos(Bx)$$

Applying the boundary conditions $\phi(\pm\tilde{a}/2) = 0$, where $\tilde{a} = a + 2d$:

$$\phi(\tilde{a}/2) = A\cos\left(\frac{B\tilde{a}}{2}\right) = 0$$

This implies:

$$\cos\left(\frac{B\tilde{a}}{2}\right) = 0$$

The non-trivial solution requires:

$$\frac{B\tilde{a}}{2} = \frac{(2n+1)\pi}{2}$$

$$B_n = \frac{(2n+1)\pi}{\tilde{a}}$$

For the fundamental mode $n = 0$:

$$B_1 = \frac{\pi}{\tilde{a}}$$

Thus, the neutron flux distribution is:

$$\phi(x) = A\cos\left(\frac{\pi x}{\breve{a}}\right)$$

2. Determine the Reactor Buckling:

The square of the lowest eigenvalue B_1^2 is the reactor buckling:

$$B_1^2 = \left(\frac{\pi}{\tilde{a}}\right)^2$$

Given $\tilde{a} = a + 2d$ and assuming d is small compared to $a : \tilde{a} \approx a$
Thus:

$$B_1^2 = \left(\frac{\pi}{a}\right)^2$$

For $a = 20$ cm :

$$B_1^2 = \left(\frac{\pi}{20}\right)^2 \approx 0.025 \text{ cm}^{-2}$$

Calculate the Total Reactor Power per Unit Area:
Given:

$$\phi(x) = A\cos\left(\frac{\pi x}{a}\right)$$

At $x = 0$:

$$\phi(0) = A$$

$$A = 2 \times 10^{12} \text{ neutrons /cm}^2/\text{s}$$

At $x = 0$:

The total power per unit area P is given by:

$$P = E_R \Sigma_f \int_{-a/2}^{a/2} \phi(x) dx$$

Substitute $\phi(x)$:

$$P = E_R \Sigma_f \int_{-a/2}^{a/2} 2 \times 10^{12} \cos\left(\frac{\pi x}{a}\right) dx$$

Substitute $\phi(x)$:

$$P = E_R \Sigma_f \int_{-a/2}^{a/2} 2 \times 10^{12} \cos\left(\frac{\pi x}{a}\right) dx$$

Evaluating the integral:

$$P = 2 \times 10^{12} E_R \Sigma_f \left[\frac{a}{\pi} \sin\left(\frac{\pi x}{a}\right)\right]_{-a/2}^{a/2}$$

$$P = 2 \times 10^{12} E_R \Sigma_f \left(\frac{a}{\pi}\right) \left[\sin\left(\frac{\pi}{2}\right) - \sin\left(-\frac{\pi}{2}\right)\right]$$

$$P = 2 \times 10^{12} E_R \Sigma_f \left(\frac{a}{\pi}\right) (1 - (-1))$$

$$P = 2 \times 10^{12} E_R \Sigma_f \left(\frac{a}{\pi}\right) \times 2$$

$$P = 4 \times 10^{12} E_R \Sigma_f \left(\frac{a}{\pi}\right)$$

For $a = 20$ cm :

$$P = 4 \times 10^{12} \times 3.2 \times 10^{-11} \times 0.15 \times \left(\frac{20}{3.14}\right)$$

$$P = 4 \times 10^{12} \times 3.2 \times 10^{-11} \times 0.15 \times 6.37$$

$$P = 4 \times 3.2 \times 0.15 \times 6.37$$

$$P \approx 12.23$$

The total reactor power per unit area is approximately 12.23 W/cm^2.

7-3. Consider a bare spherical reactor with a radius of 50 cm. The reactor is critical, and the neutron flux within the reactor is given by the expression:

$$\phi(r) = A\frac{\sin\left(\frac{\pi r}{R}\right)}{r}$$

where $R = 50$ cm. If the reactor's total power output is 2×10^7 watts, the recoverable energy per fission E_R is 200×10^6 eV (where $1\text{eV} = 1.602 \times 10^{-19}$ joules), and the macroscopic fission cross-section $\Sigma_f = 0.025$ cm^{-1}, calculate the neutron flux $\phi(r)$ at a distance of 20 cm from the center of the reactor.

Solution Steps:

1. Convert E_R from eV to joules:

$$E_R = 200 \times 10^6 \text{eV} \times 1.602 \times 10^{-19} \text{ joules /eV} = 3.204 \times 10^{-11} \text{ joules}$$

2. Determine the constant A using the given power P :

$$A = \frac{P}{8E_R\Sigma_f R^2}$$

3. Substitute the given values into the equation for A :

$$A = \frac{2 \times 10^7 \text{ watts}}{8 \times 3.204 \times 10^{-11} \text{ joules } \times 0.025 \text{ cm}^{-1} \times (50 \text{ cm})^2}$$

$$A = \frac{2 \times 10^7}{8 \times 3.204 \times 10^{-11} \times 0.025 \times 2500}$$

$$A \approx 1.56 \times 10^{17} \text{ neutrons /cm}^3 \text{ per second}$$

4. Calculate the neutron flux $\phi(r)$ at $r = 20$ cm :

$$\phi(20) = A \frac{\sin\left(\frac{\pi \times 20}{50}\right)}{20}$$

$$\phi(20) = 1.56 \times 10^{17} \times \frac{\sin\left(\frac{2\pi}{5}\right)}{20}$$

$$\phi(20) \approx 1.56 \times 10^{17} \times \frac{0.951}{20}$$

$$\phi(20) \approx 1.56 \times 10^{17} \times \frac{0.951}{20}$$

$$\phi(20) \approx 7.41 \times 10^{15} \text{ neutrons /cm}^2 \text{ per second}$$

Final Answer:

The neutron flux at a distance of 20 cm from the center of the reactor is approximately 7.41×10^{15} neutrons /cm^2 per second.

7-4. A bare spherical reactor of radius 40 cm has a neutron flux distribution given by:

$$\phi(r) = A \frac{\sin\left(\frac{\pi r}{R}\right)}{r}$$

The recoverable energy per fission E_R is 200×10^6 eV (where $1\text{eV} = 1.602 \times 10^{-19}$ joules), and the macroscopic fission cross-section $\Sigma_f = 0.03$ cm^{-1}. If the constant A is 2×10^{17} neutrons /cm^3 per second, determine the total power output P of the reactor.

Solution Steps:

1. Convert E_R from eV to joules:

$$E_R = 200 \times 10^6 \text{eV} \times 1.602 \times 10^{-19} \text{ joules /eV} = 3.204 \times 10^{-11} \text{ joules}$$

2. The total power P generated by the reactor is given by:

$$P = 4\pi E_R \Sigma_f A \int_0^R r\sin\left(\frac{\pi r}{R}\right) dr$$

3. Evaluate the integral:

$$\int_0^R r\sin\left(\frac{\pi r}{R}\right) dr = \left[-\frac{R^2}{\pi}\cos\left(\frac{\pi r}{R}\right)\right]_0^R = -\frac{R^2}{\pi}(\cos(\pi) - \cos(0)) = \frac{2R^2}{\pi}$$

4. Substitute the integral and given values into the power equation:

$$P = 4\pi E_R \Sigma_f A \frac{2R^2}{\pi}$$

$$P = 8 E_R \Sigma_f A R^2$$

5. Substitute the given values:

$$P = 8 \times 3.204 \times 10^{-11} \text{ joules } \times 0.03 \text{ cm}^{-1} \times 2 \times 10^{17} \times (40 \text{ cm})^2$$

$$P = 8 \times 3.204 \times 10^{-11} \times 0.03 \times 2 \times 10^{17} \times 1600$$

$$P \approx 2.46 \times 10^8 \text{ watts}$$

Final Answer:
The total power output of the reactor is approximately 2.46×10^8 watts.

7-5. Consider a finite cylindrical reactor with a height $H = 100$ cm and a radius $R = 50$ cm. The neutron flux within the reactor is given by:

$$\phi(r,z) = A J_0\left(\frac{2.405r}{R'}\right)\cos\left(\frac{\pi z}{H'}\right)$$

where $R' = R + d$ and $H' = H + 2d$ are the extrapolated radius and height, respectively, and $d = 2$ cm is the extrapolation distance. If the reactor's total power output is 1×10^7 watts, the recoverable energy per fission $E_R = 200 \times 10^6$ eV (where $1\text{eV} = 1.602 \times 10^{-19}$ joules), and the macroscopic fission cross-section $\Sigma_f = 0.025$ cm^{-1}, calculate the constant A and the neutron flux $\phi(r,z)$ at $r = 20$ cm and

$z = 0.$

Solution Steps:

1. Convert E_R from eV to joules:

$$E_R = 200 \times 10^6 \text{eV} \times 1.602 \times 10^{-19} \text{ joules /eV} = 3.204 \times 10^{-11} \text{ joules}$$

2. Calculate the extrapolated dimensions R' and H' :

$$R' = 50 \text{ cm} + 2 \text{ cm} = 52 \text{ cm}$$

$$H' = 100 \text{ cm} + 2 \times 2 \text{ cm} = 104 \text{ cm}$$

3. Use the given power P to determine the constant A :

$$A = \frac{P}{1.36 E_R \Sigma_f R'^2 H'}$$

$$A = \frac{1 \times 10^7 \text{ watts}}{1.36 \times 3.204 \times 10^{-11} \text{ joules} \times 0.025 \text{ cm}^{-1} \times (52 \text{ cm})^2 \times 104 \text{ cm}}$$

$$A \approx 1.51 \times 10^{16} \text{ neutrons /cm}^3 \text{ per second}$$

4. Calculate the neutron flux $\phi(r, z)$ at $r = 20$ cm and $z = 0$:

$$\phi(20,0) = A J_0 \left(\frac{2.405 \times 20}{52} \right) \cos \left(\frac{\pi \times 0}{104} \right)$$

$$\phi(20,0) = 1.51 \times 10^{16} \times J_0(0.924) \times 1$$

Given $J_0(0.924) \approx 0.779$:

$$\phi(20,0) \approx 1.51 \times 10^{16} \times 0.779$$

$$\phi(20,0) \approx 1.18 \times 10^{16} \text{ neutrons /cm}^3 \text{ per second}$$

Final Answer:
The constant A is approximately 1.51×10^{16} neutrons /cm^3 per second, and the neutron flux at $r = 20$ cm and $z = 0$ is approximately 1.18×10^{16} neutrons /cm^3 per second.

7-6. A finite cylindrical reactor has a height $H = 80$ cm and a radius $R = 40$ cm. The neutron flux distribution is given by:

$$\phi(r, z) = A J_0 \left(\frac{2.405r}{R'} \right) \cos \left(\frac{\pi z}{H'} \right)$$

where $R' = R + d$ and $H' = H + 2d$ with $d = 3$ cm. If the constant A is 1.2×10^{16} neutrons $/\text{cm}^3$ per second, the recoverable energy per fission $E_R = 200 \times 10^6 \text{eV}$ (where $1\text{eV} = 1.602 \times 10^{-19}$ joules), and the macroscopic fission cross-section $\Sigma_f = 0.02$ cm^{-1}, determine the total power output P of the reactor.

Solution Steps:

1. Convert E_R from eV to joules:

    $$E_R = 200 \times 10^6 \text{eV} \times 1.602 \times 10^{-19} \text{ joules /eV} = 3.204 \times 10^{-11} \text{ joules}$$

2. Calculate the extrapolated dimensions R' and H' :

    $$R' = 40 \text{ cm} + 3 \text{ cm} = 43 \text{ cm}$$

    $$H' = 80 \text{ cm} + 2 \times 3 \text{ cm} = 86 \text{ cm}$$

3. Determine the total power P :

    $$P = 1.36 E_R \Sigma_f A R'^2 H'$$

3. Determine the total power P :

    $$P = 1.36 E_R \Sigma_f A R'^2 H'$$

    $$P = 1.36 \times 3.204 \times 10^{-11} \text{ joules } \times 0.02 \text{ cm}^{-1} \times 1.2 \\ \times 10^{16} \text{ neutrons /cm}^3 \text{ per secon}'$$

    $$P = 1.36 \times 3.204 \times 10^{-11} \times 0.02 \times 1.2 \times 10^{16} \times 1849 \times 86$$

    $$P \approx 1.59 \times 10^8 \text{ watts}$$

Final Answer:

The total power output of the reactor is approximately 1.59×10^8 watts.

7-7. Given the neutron flux distribution $\phi(x, y, z) = A' \cos\left(\frac{\pi x}{a}\right) \cos\left(\frac{\pi y}{a}\right) \cos\left(\frac{\pi z}{a}\right)$ in a cubical reactor with side length a, determine the constant A' if the total power P generated by the reactor is known. Assume the total power is given by:

$$P = E_R \Sigma_f \int_{-a/2}^{a/2} \int_{-a/2}^{a/2} \int_{-a/2}^{a/2} \phi(x, y, z) \, dx \, dy \, dz$$

where E_R is the recoverable energy per fission and Σ_f is the macroscopic fission cross section.

Solution:

The total power can be expressed as:

$$P = E_R \Sigma_f A' \int_{-a/2}^{a/2} \cos\left(\frac{\pi x}{a}\right) dx \int_{-a/2}^{a/2} \cos\left(\frac{\pi y}{a}\right) dy \int_{-a/2}^{a/2} \cos\left(\frac{\pi z}{a}\right) dz$$

Evaluating each integral:

$$\int_{-a/2}^{a/2} \cos\left(\frac{\pi x}{a}\right) dx = \frac{a}{\pi}\left[\sin\left(\frac{\pi x}{a}\right)\right]_{-a/2}^{a/2} = \frac{a}{\pi}\left[\sin\left(\pi/2\right) - \sin\left(-\pi/2\right)\right] = \frac{2a}{\pi}$$

$$\int_{-a/2}^{a/2} \cos\left(\frac{\pi y}{a}\right) dy = \frac{2a}{\pi}$$

$$\int_{-a/2}^{a/2} \cos\left(\frac{\pi z}{a}\right) dz = \frac{2a}{\pi}$$

Therefore:

$$P = E_R \Sigma_f A' \left(\frac{2a}{\pi}\right)^3 = E_R \Sigma_f A' \frac{8a^3}{\pi^3}$$

Solving for A' :

$$A' = \frac{P\pi^3}{8E_R\Sigma_f a^3}$$

The final neutron flux distribution is:

$$\phi(x,y,z) = \frac{P\pi^3}{8E_R\Sigma_f a^3}\cos\left(\frac{\pi x}{a}\right)\cos\left(\frac{\pi y}{a}\right)\cos\left(\frac{\pi z}{a}\right)$$

7-8. Consider a rectangular parallelepiped reactor with dimensions 20 cm × 40 cm × 80 cm made of graphite and 3 weight percent enriched uranium metal dispersed in the graphite. The neutron flux $\phi(x,y,z)$ within this reactor depends on the spatial coordinates in all three dimensions. The onegroup diffusion equation for this geometry is given by:

$$\nabla^2\phi + B^2\phi = 0$$

where ∇^2 is the Laplacian operator and B^2 is the buckling. The Laplacian operator in rectangular coordinates is:

$$\nabla^2 = \frac{\partial^2}{\partial x^2} + \frac{\partial^2}{\partial y^2} + \frac{\partial^2}{\partial z^2}$$

Assuming a solution of the form $\phi(x,y,z) = X(x)Y(y)Z(z)$, and using the method of separation of variables, find the neutron flux profile $\phi(x,y,z)$. Additionally, calculate the total power produced in watts if the reactor operates at a power density corresponding to a fission rate of 2.5×10^{13} fissions/ cm³-s. Assume the recoverable energy per fission is 200MeV and the macroscopic fission cross-section Σ_f is 0.1 cm^{-1}.

Solution:

1. Flux Profile:
    The reactor dimensions are $a = 20$ cm, $b = 40$ cm, and $c = 80$ cm.
    The separated solutions for $X(x), Y(y)$, and $Z(z)$ are:

$$X(x) = A\cos\left(\frac{2n\pi x}{a}\right)$$

$$Y(y) = C\cos\left(\frac{2m\pi y}{b}\right)$$

$$Z(z) = E\cos\left(\frac{2l\pi z}{c}\right)$$

Applying the boundary conditions $\phi(\pm a/2, y, z) = 0$, $\phi(x, \pm b/2, z) = 0$, and $\phi(x, y, \pm c/2) = 0$, we get:

$$B_x = \frac{2n\pi}{a}, \; B_y = \frac{2m\pi}{b}, \; B_z = \frac{2l\pi}{c}$$

Thus, the total flux distribution is:

$$\phi(x, y, z) = A'\cos\left(\frac{\pi x}{10}\right)\cos\left(\frac{\pi y}{20}\right)\cos\left(\frac{\pi z}{40}\right)$$

where A' is a constant determined by the reactor's operating power.

2. Total Power Calculation:

Given the fission rate $\dot{f} = 2.5 \times 10^{13}$ fissions $/cm^3 - s$

Recoverable energy per fission $E_R = 200 MeV$

Total volume $V = a \times b \times c = 20\text{ cm} \times 40\text{ cm} \times 80\text{ cm} = 64000\text{ cm}^3$

Converting E_R to joules:

$$E_R = 200 MeV \times 1.60218 \times 10^{-13}\text{ J/MeV} = 3.20436 \times 10^{-11}\text{ J}$$

The power density P_d in W/cm^3 is:

$$P_d = \dot{f} \times E_R$$

$$= 2.5 \times 10^{13}\text{ fissions }/cm^3 \cdot s \times 3.20436 \times 10^{-11}\text{ J/ fission } = 0.80109\text{ W,}$$

The total power P is:

$$P = P_d \times V = 0.80109\text{ W/cm}^3 \times 64000\text{ cm}^3 = 51269.76\text{ W} = 51.27\text{ kW}$$

Thus, the neutron flux profile is:

$$\phi(x, y, z) = A' \cos\left(\frac{\pi x}{10}\right) \cos\left(\frac{\pi y}{20}\right) \cos\left(\frac{\pi z}{40}\right)$$

And the total power produced by the reactor is approximately 51.27 kW.

7-9. For a fixed volume V, determine which shape of a rectangular parallelepiped reactor will have the minimum neutron leakage. Prove your answer using the concept of geometric buckling.

Solution:

1. Geometric Buckling:

For a rectangular parallelepiped reactor with dimensions $a \times b \times c$, the geometric buckling B_g^2 is given by:

$$B_g^2 = \left(\frac{\pi}{a}\right)^2 + \left(\frac{\pi}{b}\right)^2 + \left(\frac{\pi}{c}\right)^2$$

2. Volume Constraint:

The volume V of the reactor is given by $V = a \times b \times c$.

3. Objective:

We need to minimize B_g^2 under the constraint $V = abc$.

4. Setting Up the Problem:

Using the method of Lagrange multipliers, we define the Lagrangian function:

$$\mathcal{L}(a, b, c, \lambda) = \left(\frac{\pi}{a}\right)^2 + \left(\frac{\pi}{b}\right)^2 + \left(\frac{\pi}{c}\right)^2 + \lambda(abc - V)$$

5. Taking Partial Derivatives:
Taking partial derivatives with respect to a, b, and c, and setting them to zero:

$$\frac{\partial \mathcal{L}}{\partial a} = -2\frac{\pi^2}{a^3} + \lambda bc = 0$$

$$\frac{\partial \mathcal{L}}{\partial b} = -2\frac{\pi^2}{b^3} + \lambda ac = 0$$

$$\frac{\partial \mathcal{L}}{\partial c} = -2\frac{\pi^2}{c^3} + \lambda ab = 0$$

6. Solving for λ :

Solving these equations simultaneously, we get:

$$\frac{-2\pi^2}{a^3} = \frac{-2\pi^2}{b^3} = \frac{-2\pi^2}{c^3} = \lambda abc$$

This implies:

$$\frac{1}{a^3} = \frac{1}{b^3} = \frac{1}{c^3}$$

Therefore, $a = b = c$.

7. Cubical Shape:

This solution indicates that for the minimal geometric buckling, the dimensions of the reactor must be equal, meaning the reactor is a cube.

8. Volume Constraint for Cube:

For a cube with side length s, the volume constraint $V = s^3$ implies:

$$s = \sqrt[3]{V}$$

9. Geometric Buckling for Cube:
The geometric buckling for the cube is:

$$B_{g,\,\text{cube}}^2 = 3\left(\frac{\pi}{s}\right)^2 = \frac{3\pi^2}{s^2} = \frac{3\pi^2}{(\sqrt[3]{V})^2} = \frac{3\pi^2}{V^{2/3}}$$

Comparison:
For any other shape of a rectangular parallelepiped with the same volume, the terms in B_g^2 will be larger because $a, b,$ and c will not be equal, increasing the sum of

squares in the buckling formula.

Therefore, we have shown that a cubical reactor, with all sides equal, will have the smallest geometric buckling and, consequently, the minimum neutron leakage for a fixed volume V.

7-10. Consider a cylindrical reactor with a fixed volume V. Determine the shape of the cylinder (i.e., the ratio of height h to radius r) that will have the lowest leakage of neutrons through the surface. Prove your answer by minimizing the surface area for the given volume.

Solution:

1. Volume of the Cylinder:
- The volume V of a cylindrical reactor is given by:

$$V = \pi r^2 h$$

2. Surface Area of the Cylinder:
- The surface area SA of a cylindrical reactor is given by:

$$SA = 2\pi rh + 2\pi r^2$$

3. Expressing Height h in Terms of Volume and Radius:
- From the volume equation:

$$h = \frac{V}{\pi r^2}$$

4. Substitute h into the Surface Area Equation:
- Substituting h in the surface area equation:

$$SA = 2\pi r \left(\frac{V}{\pi r^2}\right) + 2\pi r^2$$

Simplifying:

$$SA = \frac{2V}{r} + 2\pi r^2$$

5. Minimize Surface Area:
- To find the minimum surface area, take the derivative of SA with respect to r and

set it to zero:

$$\frac{d(SA)}{dr} = -\frac{2V}{r^2} + 4\pi r$$

Setting the derivative to zero:

$$-\frac{2V}{r^2} + 4\pi r = 0$$

$$4\pi r^3 = 2V$$

$$r^3 = \frac{V}{2\pi}$$

$$r = \left(\frac{V}{2\pi}\right)^{1/3}$$

6. Finding the Corresponding Height h :
*   Using the volume equation:

$$h = \frac{V}{\pi r^2}$$

Substituting $r = \left(\frac{V}{2\pi}\right)^{1/3}$:

$$h = \frac{V}{\pi \left(\frac{V}{2\pi}\right)^{2/3}}$$

$$h = \frac{V}{\pi \left(\frac{V^{2/3}}{(2\pi)^{2/3}}\right)}$$

$$h = \frac{V}{\frac{V^{2/3}}{(2^{2/3}\pi^{2/3})}}$$

$$h = \frac{V \cdot 2^{2/3}\pi^{2/3}}{V^{2/3}}$$

$$h = 2^{2/3} \pi^{2/3} V^{1/3}$$

Simplifying, we get:

$$h = 2^{2/3} \left(\frac{V}{\pi}\right)^{1/3}$$

7. Verify the Ratio of Height to Radius:

The ratio $\frac{h}{r}$ is:

$$\frac{h}{r} = \frac{2^{2/3} \left(\frac{V}{\pi}\right)^{1/3}}{\left(\frac{V}{2\pi}\right)^{1/3}}$$

$$\frac{h}{r} = 2^{2/3} \left(\frac{V}{\pi}\right)^{1/3} \cdot \left(\frac{2\pi}{V}\right)^{1/3}$$

$$\frac{h}{r} = 2^{2/3} \cdot 2^{1/3} = 2$$

Therefore, the cylindrical reactor with the lowest surface leakage of neutrons will have a radius $r = \left(\frac{V}{2\pi}\right)^{1/3}$ and a height $h = 2\left(\frac{V}{2\pi}\right)^{1/3}$, where the height is twice the radius. This configuration minimizes the surface area for a given volume, thus minimizing the neutron leakage.

7-11. Consider three different reactor shapes: cylindrical, rectangular parallelepiped, and spherical. For each shape, derive the surface area to volume (S/V) ratio formula. Assume that the volume is fixed at V for all shapes. Rank the geometries from the lowest S/V ratio to the highest S/V ratio. Calculate the percentage increase in S/V for each shape compared to the shape with the lowest S/V ratio.

Solution:

1. Cylindrical Reactor:

Let r be the radius and h be the height of the cylinder.

Volume $V = \pi r^2 h$

Surface Area $SA = 2\pi rh + 2\pi r^2$

S/V ratio:

$$\frac{SA}{V} = \frac{2\pi rh + 2\pi r^2}{\pi r^2 h} = \frac{2}{r} + \frac{2}{h}$$

For a fixed volume V, the S/V ratio is minimized when $h = 2r$.

$$V = \pi r^2 (2r) = 2\pi r^3 \Rightarrow r = \left(\frac{V}{2\pi}\right)^{1/3}$$

Optimized S/V ratio for a cylinder:

$$\frac{SA}{V} = \frac{3}{r} = \frac{3}{\left(\frac{V}{2\pi}\right)^{1/3}} = 3\left(\frac{2\pi}{V}\right)^{1/3}$$

2. Rectangular Parallelepiped Reactor:

Let a, b, and c be the lengths of the three sides of the parallelepiped.

Volume $V = abc$

Surface Area $SA = 2(ab + bc + ac)$

S/V ratio:

$$\frac{SA}{V} = \frac{2(ab + bc + ac)}{abc} = \frac{2}{a} + \frac{2}{b} + \frac{2}{c}$$

For a fixed volume V, the S/V ratio is minimized when $a = b = c = V^{1/3}$.

$$V = \left(V^{1/3}\right)^3 = V \Rightarrow a = b = c = V^{1/3}$$

Optimized S/V ratio for a parallelepiped:

$$\frac{SA}{V} = \frac{6}{a} = \frac{6}{V^{1/3}}$$

3. Spherical Reactor:

Let r be the radius of the sphere.

Volume $V = \frac{4}{3}\pi r^3$

Surface Area $SA = 4\pi r^2$

S/V ratio:

$$\frac{SA}{V} = \frac{4\pi r^2}{\frac{4}{3}\pi r^3} = \frac{3}{r}$$

For a fixed volume V :

$$V = \frac{4}{3}\pi r^3 \Rightarrow r = \left(\frac{3V}{4\pi}\right)^{1/3}$$

Optimized S/V ratio for a sphere:

$$\frac{SA}{V} = \frac{3}{r} = \frac{3}{\left(\frac{3V}{4\pi}\right)^{1/3}} = 3\left(\frac{4\pi}{3V}\right)^{1/3}$$

4. Ranking from Lowest S/V Ratio to Highest S/V Ratio:
   Spherical Reactor:
$$\frac{SA}{V} = 3\left(\frac{4\pi}{3V}\right)^{1/3}$$
   Cylindrical Reactor:
$$\frac{SA}{V} = 3\left(\frac{2\pi}{V}\right)^{1/3}$$

   Rectangular Parallelepiped Reactor:

$$\frac{SA}{V} = \frac{6}{V^{1/3}}$$

5. Calculating Percentage Increase in S/V:
   Spherical Reactor S/V:

$$\text{Spherical S/V} = 3\left(\frac{4\pi}{3V}\right)^{1/3}$$

- Cylindrical Reactor S/V:

$$\text{Cylindrical S/V} = 3\left(\frac{2\pi}{V}\right)^{1/3}$$

- Rectangular Parallelepiped Reactor S/V:

$$\text{Parallelepiped S/V} = \frac{6}{V^{1/3}}$$

- Percentage Increase for Cylinder:

$$\text{Percentage Increase} = \frac{3\left(\frac{2\pi}{V}\right)^{1/3} - 3\left(\frac{4\pi}{3V}\right)^{1/3}}{3\left(\frac{4\pi}{3V}\right)^{1/3}} \times 100\%$$

Simplifying:

$$\text{Percentage Increase} = \frac{\left(\frac{2\pi}{V}\right)^{1/3} - \left(\frac{4\pi}{3V}\right)^{1/3}}{\left(\frac{4\pi}{3V}\right)^{1/3}} \times 100\%$$

$$\text{Percentage Increase} = \frac{(\frac{3}{\pi})^{1/3}\left(-2^{2/3}(\frac{\pi}{3})^{1/3}(\frac{1}{V})^{1/3} + (2\pi)^{1/3}(\frac{1}{V})^{1/3}\right)}{2^{2/3}(\frac{1}{V})^{1/3}} \times 100\%$$

$$\text{Percentage Increase} = \left[-1 + \left(\frac{3}{2}\right)^{1/3}\right] \times 100\%$$

$$\text{Percentage Increase} = 14.5\%$$

Simplifying:

- Percentage Increase for Parallelepiped:

$$\text{Percentage Increase} = \frac{\dfrac{6}{V^{\frac{1}{3}}} - 3\left(\dfrac{4\pi}{3V}\right)^{\frac{1}{3}}}{3\left(\dfrac{4\pi}{3V}\right)^{\frac{1}{3}}} \times 100\%$$

$$\text{Percentage Increase} = \frac{-6^{2/3}\pi^{1/3}\left(\dfrac{1}{V}\right)^{1/3} + \dfrac{6}{V^{1/3}}}{6^{2/3}\pi^{1/3}\left(\dfrac{1}{V}\right)^{1/3}} \times 100\%$$

$$\text{Percentage Increase} = -1 + \left(\dfrac{6}{\pi}\right)^{1/3} \times 100\%$$

$$\text{Percentage Increase} \approx 24.1\%$$

In conclusion, for a fixed volume, the spherical reactor has the lowest surface area to volume ratio. The cylindrical reactor has a 14.5% higher S/V ratio, and the rectangular parallelepiped reactor has a 24.1% higher S/V ratio compared to the spherical reactor.

7-12. For a reactor with a fixed volume V, find the reactor shape that has the lowest maximum-to-average flux ratio. Compare the shapes of a cube, a sphere, and a square cylinder (where the height equals the diameter) and rank them from the lowest maximum-to-average flux ratio to the highest.

Solution:

To solve this problem, we will use the known formulas for the maximum-to-average flux ratio for each geometry.

1. Cube (Rectangular Parallelepiped with $a = b = c$):

Let the side length of the cube be a.

The volume of the cube $V = a^3$.

The maximum-to-average flux ratio for a cube is approximately 2.47.

2. Sphere:

Let the radius of the sphere be r.

The volume of the sphere $V = \frac{4}{3}\pi r^3$.

Solving for r :

$$r = \left(\frac{3V}{4\pi}\right)^{1/3}$$

The maximum-to-average flux ratio for a sphere is approximately 1.64.

3. Square Cylinder (Finite Cylinder with $h = 2r$):

Let the radius of the cylinder be r and the height be $h = 2r$.

The volume of the cylinder $V = \pi r^2 h = \pi r^2 (2r) = 2\pi r^3$.

Solving for r :

$$r = \left(\frac{V}{2\pi}\right)^{1/3}$$

The maximum-to-average flux ratio for a square cylinder is approximately 1.85.

Comparing Maximum-to-Average Flux Ratios:

Cube:

Maximum-to-Average Flux Ratio $\approx$ 2.47

Sphere:

Maximum-to-Average Flux Ratio $\approx$ 1.64

Square Cylinder:

Maximum-to-Average Flux Ratio $\approx$ 1.85

Ranking from Lowest Maximum-to-Average Flux Ratio to Highest:

1. Sphere: 1.64
2. Square Cylinder: 1.85
3. Cube: 2.47

Percentage Increase in Maximum-to-Average Flux Ratio:

- Cube compared to Sphere:

$$\text{Percentage Increase} = \frac{2.47 - 1.64}{1.64} \times 100\% \approx 50.6\%$$

- Square Cylinder compared to Sphere:

$$\text{Percentage Increase} = \frac{1.85 - 1.64}{1.64} \times 100\% \approx 12.8\%$$

Conclusion:

For a fixed volume reactor, the spherical shape has the lowest maximum-to-average flux ratio, followed by the square cylinder, and lastly, the cube. The spherical reactor shape is therefore the most efficient in terms of minimizing the maximum-to-average flux ratio.

9 Radio-Isotope Generators

9.1 Introduction

Space nuclear power systems provide reliable and long-lasting energy sources essential for deep space exploration where solar power is insufficient. This chapter delves into the technical intricacies of the two primary types of space nuclear power systems: Radioisotope Thermoelectric Generators (RTGs) and Space Nuclear Reactors (SNRs). We will explore the principles, design, historical development, mathematical models, and recent innovations driving these technologies.

9.1.1 Radioisotope Thermoelectric Generators (RTGs)

9.1.1.1 Principles and Design

RTGs exploit the heat generated from the radioactive decay of isotopes to produce electricity via thermoelectric conversion. Plutonium-238 (Pu-238) is the preferred isotope due to its half-life of 87.7 years and high thermal power density. The decay process for $\text{Pu} - 238$ is described by:

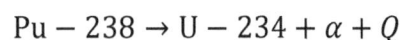

$$\text{Pu} - 238 \rightarrow \text{U} - 234 + \alpha + Q$$

where Q represents the decay energy, primarily carried by the alpha particle. The power output from the decay can be modeled using:

$$P(t) = P_0 e^{-\lambda t} \tag{9-1}$$

where P_0 is the initial power, λ is the decay constant $\left(\lambda = \ln (2)/t_{1/2}\right)$, and t is time. The thermoelectric conversion in RTGs relies on the Seebeck effect. The voltage generated V is given by:

$$V = \alpha(T_h - T_c) \tag{9-2}$$

where α is the Seebeck coefficient, T_h is the temperature of the hot junction (radioisotope heat source), and T_c is the temperature of the cold junction (heat sink).

The efficiency η of a thermoelectric generator is:

$$\eta = \frac{P_e}{P_t} = \frac{\eta_c \cdot ZT}{1 + \eta_c \cdot ZT} \tag{9-3}$$

where P_e is the electrical power output, P_t is the thermal power input, η_c is the Carnot efficiency ($\eta_c = 1 - \frac{T_c}{T_h}$, and ZT is the dimensionless figure of merit of the thermoelectric material.

9.1.1.2   Historical Development and Applications

RTGs have been integral to space missions since the 1960s. The Transit IV-A satellite first utilized an RTG in 1961. Notable RTG-powered missions include:

- Voyager 1 and 2: Each spacecraft carried three RTGs, producing approximately 470 watts at launch.
- Curiosity and Perseverance Rovers: These rovers utilize multi-mission RTGs (MMRTGs), each generating about 110 watts at launch with a nominal degradation rate of 0.8% per year.

9.1.1.3   Advantages and Challenges

Advantages:

- Longevity: With half-lives of decades, isotopes like Pu-238 provide long-term power.
- Reliability: The solid-state nature of thermoelectric converters ensures high reliability.
- Autonomy from Solar Power: RTGs provide power in environments where solar flux is inadequate.

Challenges:

- Radioactive Material: Pu3-238 production and handling require stringent safety protocols.
- Power Density: RTGs have relatively low power density compared to reactors, limiting their applications to low-power missions.
- Conversion Efficiency: Thermoelectric conversion efficiency remains low, around 6-7%.

9.2 Space Missions Powered by Radioisotope Thermoelectric Generators (RTGs)

Radioisotope Thermoelectric Generators (RTGs) have been critical in providing reliable, long-term power for various space missions, especially those exploring distant regions where solar power is insufficient. This chapter provides an in-depth examination of the RTG-powered space missions, detailing the specific RTG type, size, and power levels for each mission.

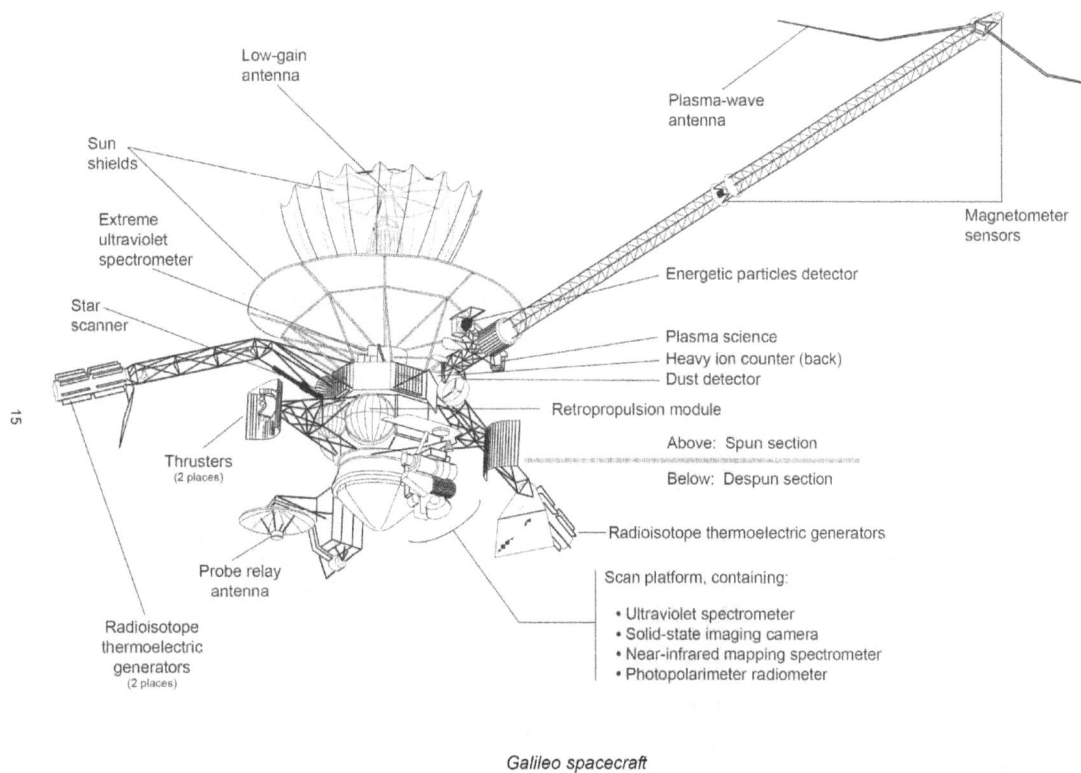

Figure 9-1. Galileo spacecraft powered by 2 RTGs
(source: https://science.nasa.gov/mission/galileo)

9.2.1 Historical Missions

9.2.1.1 Transit IV-A

- **Launch Date**: June 29, 1961
- **RTG Type**: SNAP-3B (Systems for Nuclear Auxiliary Power)
- **Size**: Compact cylindrical unit, ~20 cm in length, ~10 cm in diameter
- **Power Level**: ~2.7 watts at launch
- **Significance**: The first satellite to carry an RTG demonstrates the feasibility of using nuclear power in space.

Transit IV-A, with its companion payloads, was put in orbit on June 29th. The sphere at the top is Greb III, designed to measure solar X-rays. In the middle is Injun, an experiment to record the flux of charged particles responsible for the aurora and airglow. Solar cells (the rectangular elements covering much of the payloads' exteriors) extract energy from sunlight for supplementary power. The Transit series of satellites is sponsored by the U. S. Navy, its purpose being to provide exact navigational positions regardless of surface weather. Johns Hopkins University photo.

Figure 9-2. Transit IV-A RTG.

9.2.1.2   Apollo Lunar Surface Experiments Package (ALSEP)

- o **Launch Dates**: 1969-1972 (Apollo 12, 14, 15, 16, and 17)
- o **RTG Type**: SNAP-27
- o **Size**: Approximately 46 cm in diameter, 40 cm in height
- o **Power Level**: ~70 watts at launch
- o **Significance**: Provided power for scientific instruments deployed on the lunar surface, enabling long-term data collection.

Figure 9-3. SNAP-27.

9.2.1.3   Pioneer 10 and 11

- o **Launch Dates**: March 2, 1972 (Pioneer 10), April 5, 1973 (Pioneer 11)
- o **RTG Type**: SNAP-19
- o **Size**: Approximately 56 cm in diameter, 58 cm in height
- o **Power Level**: ~160 watts at launch (4 RTGs, ~40 watts each)
- o **Significance**: Enabled extended missions through the asteroid belt and provided data on Jupiter and Saturn.

Figure 9-4. SNAP-19 (Pioneer 11).

9.2.1.4   Viking 1 and 2

- o **Launch Dates**: August 20, 1975 (Viking 1), September 9, 1975 (Viking 2)
- o **RTG Type**: SNAP-19 (two units per lander)
- o **Size**: Approximately 56 cm in diameter, 58 cm in height
- o **Power Level**: ~100 watts per lander (50 watts each RTG)
- o **Significance**: Powered instruments and transmitters on the Martian surface, facilitating the first successful U.S. Mars landings.

Figure 9-5. SNAP-19 (Viking 1 and 2).

9.2.1.5 Voyager 1 and 2

- o **Launch Dates**: September 5, 1977 (Voyager 1), August 20, 1977 (Voyager 2)
- o **RTG Type**: MHW-RTG (Multi-Hundred Watt RTG)
- o **Size**: Approximately 114 cm in length, 43 cm in diameter
- o **Power Level**: ~470 watts at launch (3 RTGs per spacecraft, ~157 watts each)
- o **Significance**: Provided continuous power for instruments and communication systems, enabling long-term exploration of the outer planets and beyond.

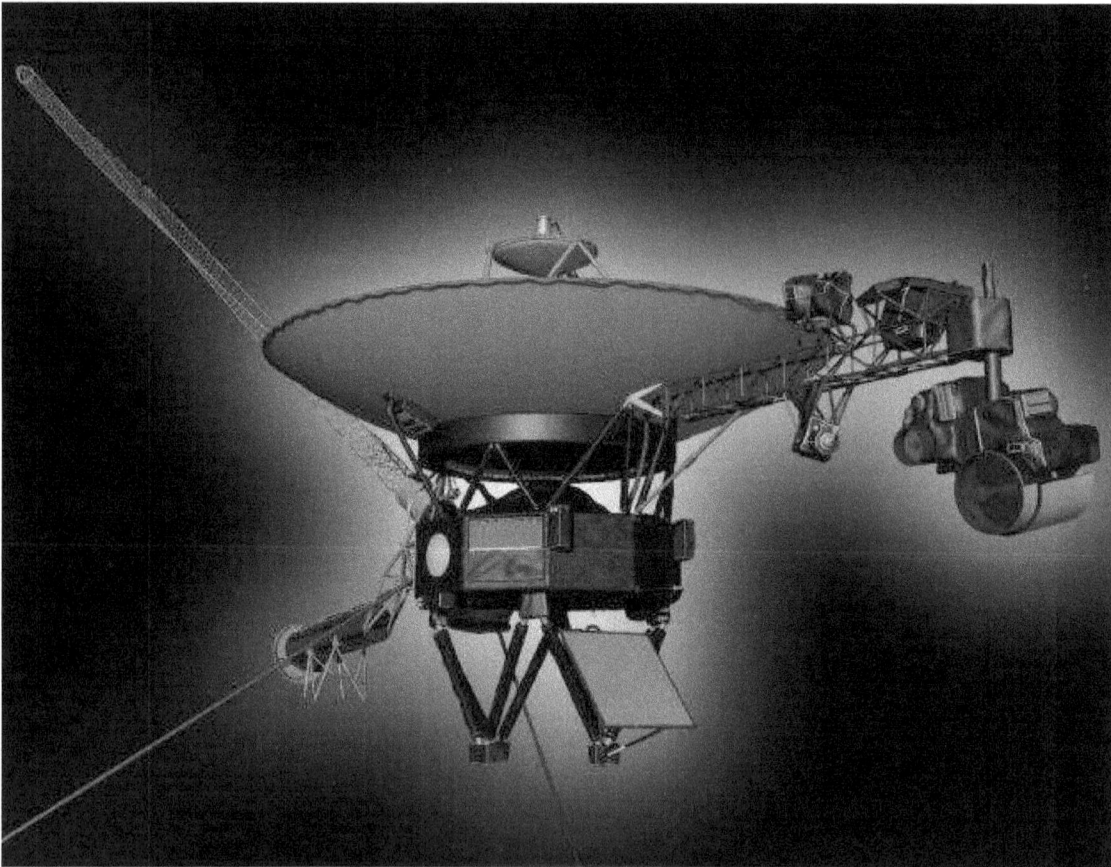

Figure 9-6. MHW-RTG: Voyager 2 & Voyager 1.

9.2.1.6   Galileo

- o **Launch Date**: October 18, 1989
- o **RTG Type**: GPHS-RTG (General Purpose Heat Source RTG)
- o **Size**: Approximately 114 cm in length, 43 cm in diameter
- o **Power Level**: ~570 watts at launch (2 RTGs, ~285 watts each)
- o **Significance**: Powered the spacecraft's extensive study of the Jupiter system, including detailed observations of its moons and magnetosphere.

Figure 9-7. 1989 -- Galileo: 2 RTG, 288.4 We -120 RHU, W (Interplanetary Missions).

9.2.1.7 Ulysses

- **Launch Date**: October 6, 1990
- **RTG Type**: GPHS-RTG
- **Size**: Approximately 114 cm in length, 43 cm in diameter
- **Power Level**: ~285 watts at launch
- **Significance**: Enabled the mission to study the Sun's polar regions and solar wind over a prolonged period.

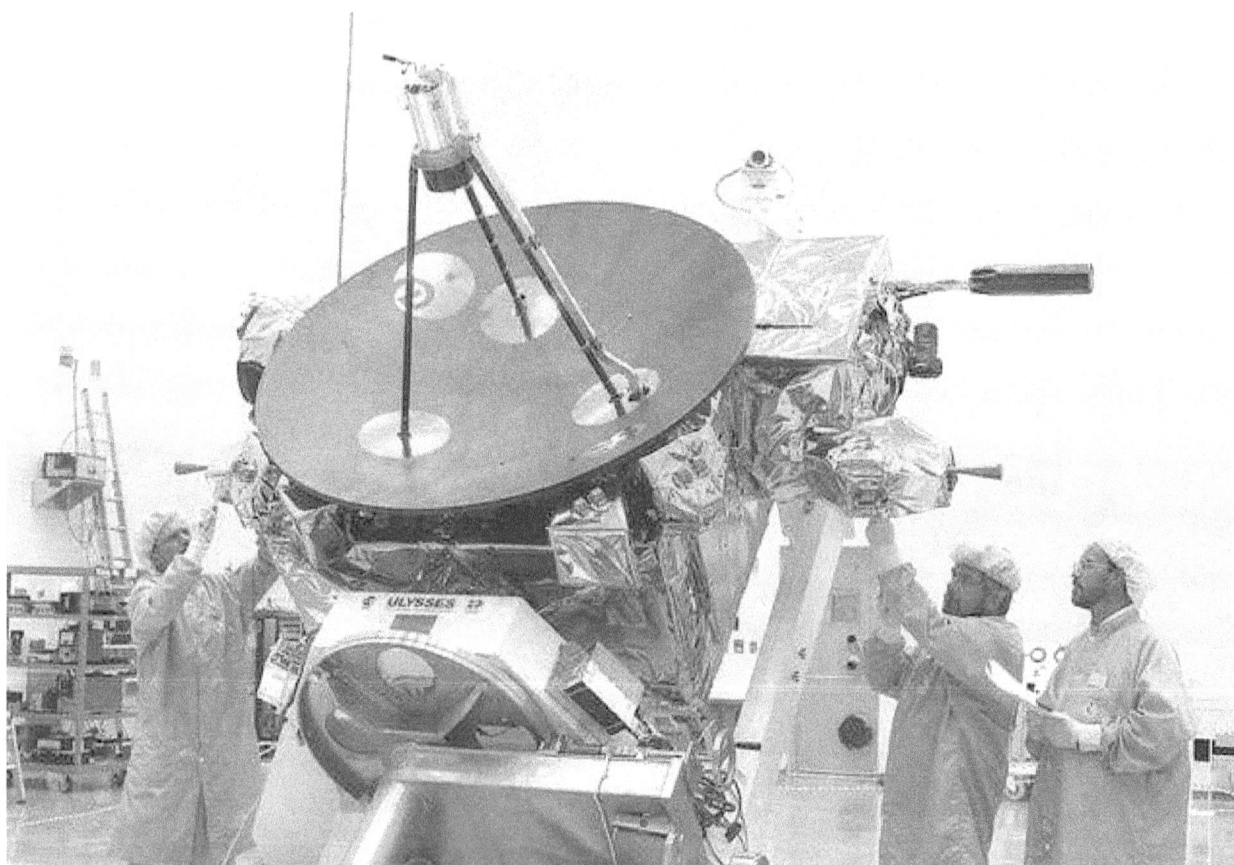

Figure 9-8. GPHS-RTG: Ulysses – 1990 -- Ulysses: 1 RTG, 283 We (Interplanetary Missions)).

9.2.1.8 Cassini-Huygens

- o **Launch Date**: October 15, 1997
- o **RTG Type**: GPHS-RTG
- o **Size**: Approximately 114 cm in length, 43 cm in diameter
- o **Power Level**: ~885 watts at launch (3 RTGs, ~295 watts each)
- o **Significance**: Provided power for the spacecraft's 13-year mission studying Saturn and its moons, including the landing of the Huygens probe on Titan.

Figure 9-9. GPHS-RTG: Cassini – 1997 -- Cassini: 3 RTG, 295.7 We - 117 RHU.

9.2.1.9   New Horizons

- o **Launch Date**: January 19, 2006
- o **RTG Type**: GPHS-RTG
- o **Size**: Approximately 114 cm in length, 43 cm in diameter
- o **Power Level**: ~250 watts at launch
- o **Significance**: Enabled the first flyby of Pluto and continues to power the spacecraft as it explores the Kuiper Belt.

Figure 9-10. RHU: Spirit (rover) & Opportunity (rover) – 2003 -- Spirit (rover): 8 RHU, 140 W (Planetary Missions) Opportunity (rover): 8 RHU, 140 W (Planetary Missions).

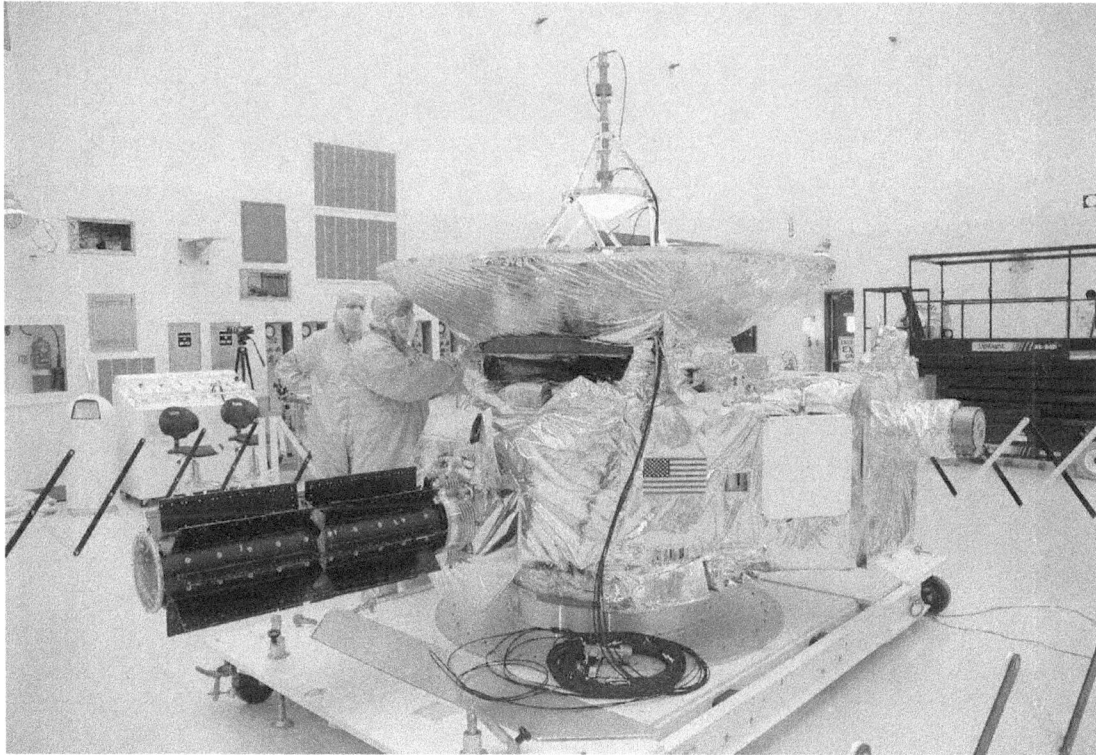

Figure 9-11. GPHS-RTG: New Horizons – 2006 -- New Horizons: 1 RTG, 249.6 We (Interplanetary Missions).

9.2.1.10 Mars Science Laboratory (Curiosity)

- o **Launch Date**: November 26, 2011
- o **RTG Type**: MMRTG (Multi-Mission RTG)
- o **Size**: Approximately 64 cm in length, 66 cm in diameter
- o **Power Level**: ~110 watts at launch
- o **Significance**: Powers the rover's extensive suite of scientific instruments, allowing continuous operation regardless of sunlight availability.

Figure 9-12. MMRTG: Curiosity (rover) – 2011 -- 1 RTG, 113 We (Planetary Missions).

9.2.1.11 Mars 2020 (Perseverance)

o   **Launch Date**: July 30, 2020
o   **RTG Type**: MMRTG
o   **Size**: Approximately 64 cm in length, 66 cm in diameter
o   **Power Level**: ~110 watts at launch
o   **Significance**: Supports the rover's mission to search for signs of ancient life and collect samples for potential return to Earth.

9.2.2 Recent Developments

9.2.2.1 Dragonfly

o   **Planned Launch Date**: 2027
o   **RTG Type**: MMRTG
o   **Size**: Approximately 64 cm in length, 66 cm in diameter
o   **Power Level**: ~110 watts at launch
o   **Significance**: Will be the first rotorcraft to explore Titan, Saturn's largest moon, utilizing RTG power to conduct scientific investigations in Titan's complex environment.

Figure 9-13. MMRTG: Dragonfly rotorcraft lander mission – 2027 -- Dragonfly rotorcraft lander mission: 1 MMRTG, EXP. 70 We (Planetary Missions).

9.2.2.2 Upcoming NASA Missions

o **Artemis Program**: The Artemis missions aim to establish a sustainable human presence on the Moon. RTGs and other nuclear power systems are considered to support lunar habitats, scientific stations, and surface operations.
o **Europa Clipper**: Scheduled for launch in the 2020s, this mission will investigate Jupiter's moon Europa. An RTG will power the spacecraft, enabling it to operate in the extreme cold and radiation-heavy environment of the outer solar system.

9.2.3 Technical Performance and Innovations

RTGs have seen significant advancements in efficiency, power output, and safety since their inception:

9.2.3.1 Power Output and Efficiency

o **Early RTGs**: Pioneer 10 and 11's SNAP-19 RTGs provided around 160 watts of electrical power.
o **Modern RTGs**: MMRTGs, like those on Curiosity and Perseverance,

generate approximately 110 watts at the start of the mission, with a nominal decay rate of about 0.8% per year.

9.2.3.2 Thermoelectric Material Advances

- o   Initial RTGs utilized silicon-germanium (SiGe) thermocouples.
- o   Recent research into materials like skutterudites and other high-efficiency thermoelectric materials has the potential to increase conversion efficiency from 6-7% to 10-15%.

9.2.3.3 Safety Enhancements

- o   RTG designs incorporate multiple layers of shielding to contain radioactive material in case of a launch accident.
- o   Innovations in encapsulation and containment have further improved the safety profile of RTGs, making them more robust against potential mishaps.

RTGs have been a cornerstone of space exploration, providing reliable and long-lasting power for missions to the Moon, Mars, and beyond. Their ability to operate in harsh environments and provide continuous power makes them invaluable for long-duration missions. As technology advances, RTGs will continue to evolve, offering higher efficiencies and safer designs, thereby enabling even more ambitious explorations of our solar system and beyond. The integration of RTGs in upcoming missions like Dragonfly and the Artemis program underscores their continued importance in the future of space exploration.

9.3   The Next Generation Radioisotope Generators

A The paper, authored by Jean-Pierre Fleurial, et. al discusses the Next Generation Radioisotope Thermoelectric Generator (Next Gen RTG) Project. This project, part of NASA's Radioisotope Power Systems (RPS) Program, aims to develop a new RTG system to support future deep space missions.

The Next Gen RTG Project seeks to re-establish the capability to manufacture Silicon Germanium (SiGe) unicouple-based thermoelectric converters and associated hardware, leveraging the heritage design of the General-Purpose Heat Source-RTG (GPHS-RTG). The paper details a multi-phase effort to refurbish existing hardware, including the GPHS-RTG Flight Unit #5 (F-5), and to produce new RTGs with minimal deviations from the heritage design.

The project faces several technical challenges, including the re-establishment of SiGe unicouple production, the refurbishment of legacy hardware, and the integration of modern standards with the heritage design. The authors discuss the project's approach to overcoming these challenges, including the use of legacy hardware and components stored at Idaho National Laboratory, and the assessment of potential enhancements to improve power output and efficiency.

The Next Gen RTG Project represents a critical effort to ensure the availability of reliable power sources for future NASA missions, with the aim of delivering a fully flight-qualified RTG by 2030. The collaboration between NASA, JPL, and INL highlights the importance of leveraging existing expertise and infrastructure to achieve the project's goals.

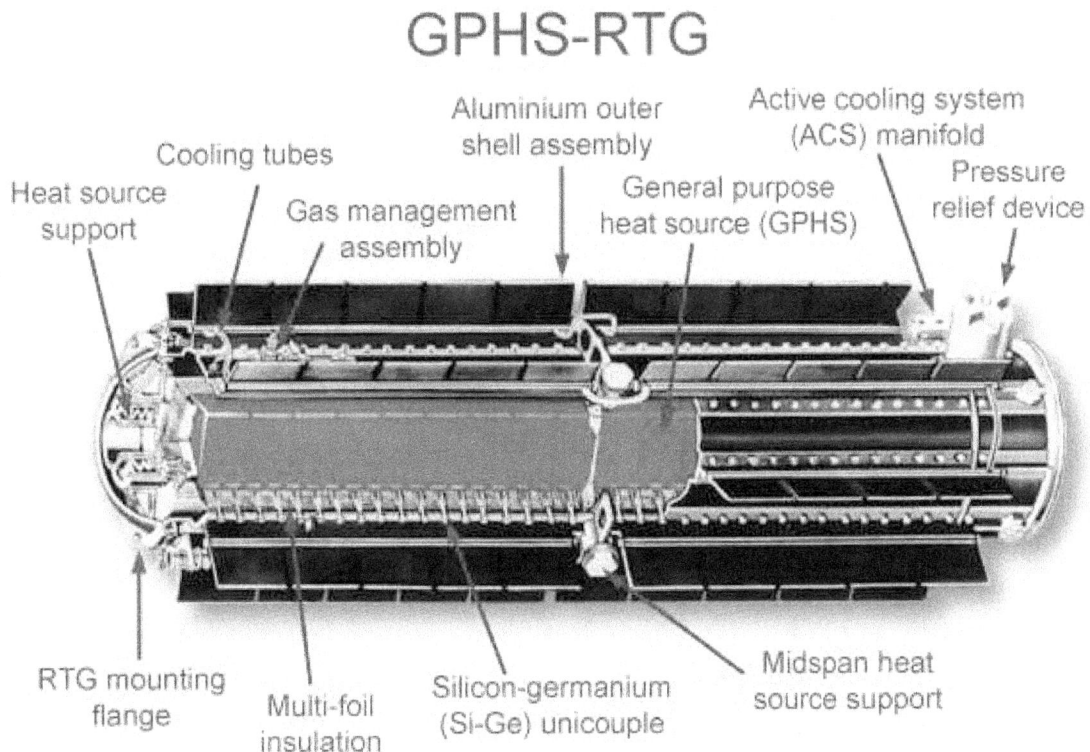

Figure 9-14. General-Purpose Heat Source Radioisotope Thermoelectric Generator.

9.4   Fuel Considerations in Radioisotope Power Systems

During the production, fabrication, testing, and pre-launch handling of radioisotope power systems (RPS), various radiation hazards must be carefully managed. For example, during the Apollo missions, astronauts were protected while handling radioisotope generators on the Moon. It is crucial to ensure that, in the event of an accident, the general public is shielded from radiation exposure above safe levels. Additionally, scientific instruments aboard some missions may include highly sensitive particle and photon detectors, which must not be adversely affected by the chosen radioisotope heat source.

9.4.1 Mission-Derived Design Requirements and Plutonium-238 Production

The choice of radioisotope fuel is influenced by the power conversion system's design and the need to dissipate waste heat through radiation. The efficiency of this process is proportional to the fourth power of the temperature, thus necessitating a high temperature to minimize the size and mass of the radiator. The selected fuel must be stable in its form—whether as a compound, alloy, or matrix—and compatible with the power conversion system's temperatures. Moreover, it must be chemically compatible with its containment material, typically metallic cladding, throughout the heat source's operational life. The fuel should also withstand high temperatures that might occur during accidents, such as fires or atmospheric reentry. Importantly, the fuel form must exhibit low solubility in the human body and the environment to minimize contamination risks.

9.4.2 Daughter Products and Fuel Integrity

The decay of radioisotopes generates daughter products, which must not compromise the integrity of the fuel form or degrade its properties. Alpha particle decay, in particular, can cause helium gas buildup, potentially leading to the creation of inhalable fine particles within the fuel or fuel cavity, which could be released during an accident. It is crucial that the decay process does not break the chemical bonds within the fuel form, thereby maintaining the fuel's stability and safety.

9.4.3 Metrics for Suitability of Radioisotope Fuels

Two key metrics for determining the suitability of a radioisotope as a heat source are power density (watts per cubic centimeter) and specific power (watts per gram). These metrics are directly proportional to the energy released per disintegration and inversely proportional to the isotope's half-life. Higher power density allows for smaller heat sources, while higher specific power results in lighter heat sources. Alpha-emitting fuels typically have smaller and lighter heat sources compared to beta-emitting fuels of similar half-lives.

To manage helium gas release from alpha-emitting fuels, the design may include thin-walled, vented capsules with minimal void volumes or sufficient void space to accommodate the maximum pressure buildup during the heat source's life. As the fuel ages, daughter products accumulate. While chemical separation can remove these products, it is preferable to process the fuel as late as possible before use to maintain high specific power and density. However, the presence of other isotopes in the fuel, which cannot be chemically separated, can dilute these properties, posing a challenge to maintaining optimal performance.

Ensuring the safety and effectiveness of radioisotope power systems involves a comprehensive approach to managing radiation hazards and optimizing fuel properties.

These considerations are critical for protecting both mission personnel and the general public, as well as ensuring the reliable operation of sensitive scientific instruments on space missions.

9.5 Power Conversion Systems in Radioisotope Power Systems

Radioisotope power systems (RPS) are essential for providing continuous electrical power to spacecraft, especially for missions that venture far from the Sun where solar energy is insufficient. The energy produced by radioisotope heat sources comes in the form of heat. This thermal energy can be utilized directly for environmental control within spacecraft or on planetary surfaces. However, most often, it is necessary to convert this thermal energy into electricity to power various onboard systems and instruments. The selection and design of power conversion systems are influenced by several critical factors, each of which plays a significant role in ensuring the mission's success.

9.5.1 Key Factors Influencing Power Conversion Systems

9.5.1.1 1. Conversion Efficiency

Conversion efficiency is a crucial consideration because it determines how much of the thermal energy can be converted into electrical energy. Higher efficiency means more usable power is available for spacecraft operations. Thermoelectric generators (TEGs) are commonly used in RPS due to their simplicity and reliability, despite having relatively lower efficiencies compared to dynamic systems like Stirling engines. The efficiency of a TEG is largely dependent on the temperature gradient between the hot and cold junctions of the thermoelectric material. Therefore, achieving high operating temperatures is desirable, as it enhances the thermal gradient and, consequently, the efficiency.

9.5.1.2 Weight and Size

In space missions, every kilogram counts, as launch costs are directly related to the payload's mass. Power conversion systems need to be compact and lightweight to maximize the payload capacity for scientific instruments and other mission-critical equipment. The choice of materials and the design of the conversion system are optimized to reduce weight without compromising performance and reliability. This is particularly important for missions requiring long-duration power supplies, as the power system must remain functional for the entire mission lifetime, which can span decades.

9.5.1.3 Operating Temperature

The operating temperature of the power conversion system is another vital factor.

Higher temperatures generally lead to improved efficiency and reduced system mass. However, the materials used must withstand these high temperatures without degrading. The selection of materials that can operate reliably over the mission's lifetime is crucial. Advanced materials with high thermal stability and low degradation rates are preferred. Moreover, the system must manage waste heat effectively, as excess heat can affect other spacecraft systems and instruments.

9.5.1.4   Reliability and Durability

Reliability is arguably the most critical factor in the design of RPS. The success of a mission depends on the continuous and predictable supply of electrical power. Any failure in the power system could jeopardize the mission, especially for deep space missions where repair or replacement is not possible. Therefore, the power conversion system must be designed to operate flawlessly under extreme conditions, including exposure to high levels of radiation, vacuum, and temperature extremes. Thermoelectric systems are favored because they have no moving parts, reducing the risk of mechanical failure and increasing overall reliability.

9.5.1.5   Resilience to Environmental Conditions

Spacecraft are subjected to various environmental stresses, including shock and vibration during launch, radiation exposure, and extreme temperature variations. The power conversion system must be rugged enough to withstand these conditions without compromising functionality. This includes being able to survive the harsh particle and radiation environments found in space or on certain planetary surfaces. The system's design must also account for the potential impact of radiation on electronic components and materials.

9.5.1.6   Scalability and Integration Flexibility

The ability to scale the power output and integrate the system with different spacecraft and launch vehicles is another important consideration. Power conversion systems need to be adaptable to varying mission requirements, whether for small scientific probes or large exploratory spacecraft. This flexibility extends to integration with various spacecraft designs and launch configurations, ensuring compatibility with the mission's specific needs.

9.5.2   Material Considerations and System Design

The choice of materials for power conversion systems is governed by their ability to operate at high temperatures, their mechanical properties, and their resistance to radiation damage. For example, thermoelectric materials must maintain their structural integrity and electrical properties at elevated temperatures. The development of new thermoelectric materials with higher efficiency and durability is an ongoing area of research, aimed at improving the overall performance of RPS.

The design of the power conversion system also includes considerations for thermal management. Effective thermal management ensures that waste heat is dissipated efficiently, preventing overheating and maintaining optimal operating conditions for the power system and other spacecraft components. This is particularly challenging in the vacuum of space, where traditional cooling methods like convection are not available.

9.5.3 Long-Term Performance and Redundancy

Given the lengthy duration of many space missions, it is acceptable for power conversion systems to experience gradual power degradation, provided it remains within predictable limits. This degradation is typically due to the natural decay of the radioisotope fuel and minor wear in the system's materials and components. Redundancy is a key strategy to mitigate the impact of potential failures. For instance, thermoelectric systems can continue to function even if individual thermoelectric couples fail, thanks to their series-parallel circuit design. This inherent redundancy ensures that power generation continues even in the event of partial system degradation.

The design and selection of power conversion systems for radioisotope power systems involve a complex interplay of factors, including efficiency, weight, temperature management, reliability, environmental resilience, and scalability. The success of space missions, particularly those venturing into deep space, relies heavily on the robustness and reliability of these systems. As technology advances, continued improvements in materials and system design will further enhance the capabilities and performance of RPS, supporting the exploration and study of our solar system and beyond.

9.6   Nuclear Safety Considerations in Space Missions

The approach to nuclear safety for space missions significantly differs from that of terrestrial nuclear power plants, particularly in defining postulated accidents. Terrestrial nuclear safety is based on a set of design basis accidents and environmental conditions that a system must survive. These postulated events have evolved through consensus among designers and reviewers, focusing resources and making the safety process manageable. However, extreme events like core meltdowns are typically not part of the required safety analyses.

In contrast, space nuclear safety has evolved to include a wide range of postulated accidents, regardless of their likelihood. This comprehensive approach is due to the unique and varied nature of space environments and the high stakes involved. Probabilistic risk assessment, pioneered by the space nuclear community, spans a broad spectrum of potential accident scenarios. Safety is a prime consideration in designing and operating radioisotope systems, with thorough failure analysis supported by testing. These analyses cover ground operations before launch, launch operations, in-orbit failures, and other mission-related failures that might lead to radiological exposure.

9.6.1   System Design and Safety Features

Key safety features incorporated into the design of space nuclear systems include using surrounding structures to protect the heat source from explosion fragments and over-pressure scenarios during launch pad aborts. Additionally, the design allows the system to disassemble upon atmospheric reentry, freeing the heat source aeroshells to reenter independently, reducing the risk of contamination.

The power system must also manage the heat generated by the radioisotope heat source under all operating and failure conditions. This includes ensuring that the heat rejection system can function without exceeding the temperature limits within the heat source, thereby preventing potential overheating and subsequent failure.

9.6.2   Safety Review Process

All U.S. nuclear-fueled power supplies considered for space use must undergo a rigorous safety review process. This process ensures that the potential risks associated with using nuclear energy sources are justified by the anticipated mission benefits. The Interagency Nuclear Safety Review Panel (INSRP) oversees this process, comprising representatives from the Department of Energy (DOE), NASA, and the Department of Defense (DOD). These agencies evaluate mission safety for each launch, leveraging their expertise in nuclear safety and space mission operations.

The safety evaluation process involves several steps:

1. The sponsoring agency directs the manufacturer to prepare a Preliminary Safety Analysis Report (PSAR) or Updated Safety Analysis Report (USAR), detailing all aspects of mission safety.
2. The INSRP reviews these reports, with each agency conducting independent evaluations.
3. A meeting is held where the INSRP members, along with contractors responsible for mission hardware, discuss the review findings. Action items are generated to resolve any outstanding issues.
4. The power system contractor, with input from other agencies, revises the report into a Final Safety Analysis Report (FSAR), incorporating feedback and new information.
5. The FSAR undergoes further review, leading to the generation of a Safety Evaluation Report (SER), which accompanies the request for Presidential approval for the launch.

9.6.3   Ground Operations and Radiation Exposure

Safety during ground operations at the launch facility is critical. RTGs emit radiation, posing potential risks to both personnel and equipment. The goal is to keep radiation exposure to ground workers as low as reasonably achievable (ALARA), complying with Nuclear Regulatory Commission standards. The occupational dose limit for radiation exposure is 1.25 rem per calendar quarter. Dose assessments are conducted for all ground and initial launch operations phases, including detailed time and motion studies to estimate exposure rates.

The installation of RTGs on the spacecraft is delayed until the final stages of launch preparation to minimize radiation exposure. This timing necessitates a spacecraft design that facilitates the late installation of the heat source. Additionally, contingency plans must be in place to address unplanned situations requiring proximity to the RTGs, such as replacing components after the RTG has been installed.

9.6.4 Equipment and Environmental Considerations

Equipment at the launch site, including electronics and organic materials like cable insulation and Teflon, must be evaluated for potential radiation effects. While the exposure levels are generally below thresholds that could damage these materials, the risk of single event upsets (bit-flips) in computer microchips caused by neutrons from the RTG is also assessed. Experimental programs have shown a low probability of such events, indicating minimal impact.

The comprehensive nuclear safety review process for space missions ensures that all potential risks are thoroughly assessed and managed. This includes evaluating possible accidents, their probabilities, and the impact on the nuclear system. By incorporating robust safety features and detailed contingency planning, the risk to both personnel and the public is minimized, allowing for the safe use of nuclear power in space exploration.

9.7　Plutonium-238 Production and Processing

Plutonium-238 (Pu-238) is a critical material for radioisotope power systems (RPS) used in space exploration, providing a reliable heat source for electricity generation. Unlike naturally occurring elements, Pu-238 must be artificially produced through the irradiation of Neptunium-237 (Np-237) in a nuclear reactor. The production process involves several stages, from the fabrication of Np-237 targets to the extraction and purification of Pu-238, requiring specialized facilities to manage the material's radioactivity and heat.

9.7.1 Production Process Overview

1. **Target Fabrication**: Np-237, derived from highly enriched uranium, is processed into oxide form and mixed with aluminum powder. The mixture is then pressed into pellets and encased in aluminum cladding to form targets.
2. **Irradiation**: These targets are loaded into a nuclear reactor, such as the High Flux Isotope Reactor (HFIR) at Oak Ridge National Laboratory (ORNL) or the

Advanced Test Reactor (ATR) at the Idaho National Laboratory (INL). Neutron capture by Np-237 converts it into Np-238, which decays into Pu-238.

3. **Chemical Processing**: After irradiation, the targets are cooled and chemically processed to extract Pu-238. This involves dissolving the targets in acid, separating the aluminum cladding, and purifying the solution to isolate Pu-238. The purification process typically includes multiple cycles to ensure high purity, resulting in Pu-238 dioxide (PuO_2) powder with an isotopic composition of over 80% Pu-238.

4. **Pellet and Capsule Fabrication**: The purified PuO_2 is processed into pellets, which are then encapsulated in iridium cladding to form heat sources. The encapsulation process involves precise welding and handling in shielded glove boxes to protect workers from alpha radiation. The final encapsulated pellets are then assembled into the RTG units.

9.7.2   Historical and Current Production Challenges

The United States ceased domestic production of Pu-238 after the shutdown of the K-Reactor at the Savannah River Site in the late 1980s. Subsequently, the U.S. procured Pu-238 from Russia until 2010, securing approximately 30-40 kilograms for NASA missions. The need for Pu-238 has continued to grow, with an estimated annual requirement of around 5 kilograms to support future missions.

The restart of Pu-238 production in the U.S. involves significant infrastructure development, estimated to cost several hundred million dollars. The process encompasses the fabrication of Np-237 targets, irradiation, and chemical processing to extract Pu-238, followed by fuel encapsulation and assembly into RTGs. The DOE has designated ORNL's Radiochemical Engineering Development Center and INL as the primary sites for these operations. Np-237, previously stored at the Savannah River Site, is now stored at INL.

9.7.3   Safety and Handling Considerations

Facilities handling Pu-238 must provide extensive shielding to protect personnel from radiation exposure and manage the significant heat generated by the isotope. Despite Pu-238's strong alpha emission, criticality control is not a concern due to the large mass required to reach criticality. Instead, managing the thermal load and preventing contamination are primary safety considerations. All processing steps are conducted in shielded environments, such as glove boxes and hot cells, equipped with negative pressure systems to prevent the spread of radioactive materials.

The handling and processing of Pu-238 require stringent safety protocols to protect workers and the environment. The final encapsulated product is robust and designed to withstand harsh conditions, including potential reentry scenarios, ensuring the safe containment of radioactive material throughout the mission lifecycle.

230Pu 107 s	231Pu 8.6 min	232Pu 33.1 min	233Pu 20.9 min	234Pu 8.8 h	235Pu 25.3 min	236Pu 2.858 y	237Pu 45.64 d	238Pu 87.7 y	239Pu 2.41e14 y
α=100%	ε+β+=90% α=10%	ε=89% α=11%	ε+β+=99.88% α=0.12%	ε=94% α=6%	ε+β+=99.997% α=2.8e-3%	α=100% SF=1.9e-7% 28MG=2.7e-12%	ε=99.996% α=3.2e-3%	α=100% SF=1.9e-7% SI=1.4e-14%	α=100% SF=3.01e-10%

229Np 4 min	230Np 4.63 min	231Np 48.8 min	232Np 14.7 min	233Np 36.2 min	234Np 4.4 d	235Np 396.1 d	236Np 1.55e5 y	237Np 2.14e6 y	238Np 2.10 d
α=68% ε+β+=32%	ε+β+=97% α≥3%	ε+β+=98% α=2%	ε+β+=100%	ε=100% α≤7e-4%	ε+β+=100%	ε=99.997% α=2.6e-3%	ε=87.84% β-=12% α=0.16%	α=100%	β-=100%

228U 9.1 min	229U 57.8 min	230U 20.23 d	231U 4.2 d	232U 68.9 y	233U 1.59e5 y	234U 2.46e5 y 0.0054%	235U 7.04e8 y 0.7204%	236U 2.34e7 y	237U 6.752 d
α>95% ε<5%	ε=80% α=20%	α=100% 22NE=4.8e-12%	ε=99.996% α=0.004%	α=100% 24NE=8.9e-10% SF=2.7e-12%	α=100% 24NE=7.2e-11%	α=100% SF=1.64e-9% 28MG=1.4e-11%	α=100% SF=7e-9% 20NE=8e-10%	α=100% SF=9.4e-8%	β-=100%

227Pa 38.3 min	228Pa 19.8 h	229Pa 1.55 d	230Pa 17.4 d	231Pa 3.27e4 y 100%	232Pa 1.31 d	233Pa 26.975 d	234Pa 6.671 h	235Pa 24.4 min	236Pa 9.1 min
α=85% ε=15%	ε+β+=98.15% α=1.85%	ε=99.49% α=0.51%	ε+β+=92.2% β-=7.8% α=3.5e-3%	α=100% 24NE=1.34e-9%	β-=100%	β-=100%	β-=100%	β-=100%	β-=100%

226Th 30.69 min	227Th 18.69 d	228Th 1.91 y	229Th 7907.5 y	230Th 7.54e4 y 0.02%	231Th 25.52 h	232Th 1.41e10 y 99.98%	233Th 21.83 min	234Th 24.107 d	235Th 7.2 min
α=100%	α=100%	α=100% 20O=1.13e-11%	α=100%	α=100% 24NE=5.8e-11%	β-=100%	α=100% SF=1.1e-9%	β-=100%	β-=100%	β-=100%

225Ac 9.919 d	226Ac 29.37 h	227Ac 21.77 y	228Ac 6.15 h	229Ac 62.7 min	230Ac 122 s	231Ac 7.5 min	232Ac 119 s	233Ac 143 s	234Ac 44 s
α=100% 14C=5.3e-10%	β-=83% ε=17% α=0.006%	β-=98.62% α=1.38%	β-=100%	β-=100%	β-=100%	β-=100% B-SF=1.2e-6%	β-=100%	β-=100%	β-=100%

224Ra 3.63 d	225Ra 14.8 d	226Ra 1600 y	227Ra 42.2 min	228Ra 5.75 y	229Ra 4 min	230Ra 93 min	231Ra 103.9 s	232Ra 4.2 min	233Ra 30 s
α=100% 14C=4e-9%	β-=100%	α=100% 14C=2.6e-9%	β-=100%	β-=100%	β-=100%	β-=100%	β-=100%	β-=100%	β-=100%

223Fr 21.99 min	224Fr 3.33 min	225Fr 4 min	226Fr 49 s	227Fr 2.47 min	228Fr 36 s	229Fr 50.2 s	230Fr 19.1 s	231Fr 17.6 s	232Fr 5.5 s
β-=99.994% α=0.006%	β-=100%	β-=100%	β-=100%	β-=100%	β-=100%	β-=100%	β-=100%	β-=100%	β-=100%

222Rn 3.82 d	223Rn 24.3 min	224Rn 114 min	225Rn 4.66 min	226Rn 7.4 min	227Rn 20.8 s	228Rn 65 s	229Rn 12 s	230Rn	231Rn
α=100%	β-=100%	β-=100%	β-=100%	β-=100%	β-=100%	β-=100%	β-=100%	β-=100%	β-=100%

Figure 9-15. Decay chain of Np-238 to Rn-222

9.7.4 Future Directions and Alternatives

The reestablishment of Pu-238 production includes not only reviving old facilities but also enhancing security and incorporating modern technological advancements. While current plans focus on utilizing existing reactors like HFIR and ATR, other production methods and facilities are also being explored to increase efficiency and security.

The production and processing of Pu-238 are essential for powering long-duration space missions. The complex and specialized nature of these operations, coupled with stringent safety and security measures, underscores the importance of a reliable and sustainable supply of Pu-238. As new missions are planned, the need for Pu-238 will continue to drive advancements in production technology and infrastructure development.

Figure 9-16. Decay chain of Rn-222 to Pb-206.

9.8   Radioactive Decay and Power Production in RTGs

Radioactive decay is a fundamental nuclear process in which unstable atomic nuclei lose energy by emitting radiation, leading to the transformation of these nuclei into more stable configurations. This process is exponential and can be mathematically expressed by the law of radioactive decay:

$$\frac{dN}{dt} = -\lambda N \tag{9-4}$$

Where:

- N is the number of atoms of a particular radionuclide at a given time.
- λ is the decay constant, representing the probability of decay of an individual nucleus per unit time.
- t is time.

The decay constant λ is specific to each radionuclide and is related to its half-life $T_{1/2}$, the time required for half of the radioactive atoms in a sample to decay. The relationship between λ and $T_{1/2}$ is given by:

$$T_{1/2} = \frac{\ln 2}{\lambda} = \frac{0.693}{\lambda} \qquad (9\text{-}5)$$

Exponential Decay and Activity

The decay law can be integrated to show that the number of remaining radioactive atoms decreases exponentially over time:

$$N(t) = N_0 e^{-\lambda t} \qquad (9\text{-}6)$$

Where:

- N_0 is the initial number of radioactive atoms.

The activity A of a radioactive substance, which is the number of decays per unit time, is proportional to the number of atoms present:

$$A = \lambda N \qquad (9\text{-}7)$$

As the radioactive material decays, the activity decreases exponentially, mirroring the reduction in the number of atoms.

Energy Release and Thermal Power Production

In the context of RTGs, the energy released from radioactive decay, primarily in the form of alpha particles, is converted into heat. This heat is then converted into electricity using thermoelectric materials. The energy released in each decay event is a crucial factor in determining the amount of thermal power that can be generated. The total thermal power output $Q(t)$ of a radioisotope source at any given time is:

$$Q(t) = Q_0 e^{-\lambda t} \qquad (9\text{-}8)$$

Where Q_0 is the initial thermal power output.

For Plutonium-238 (^{238}Pu), the primary isotope used in space RTGs, the decay process can be represented as:

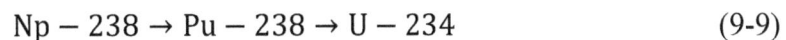

$$Np - 238 \rightarrow Pu - 238 \rightarrow U - 234 \qquad (9\text{-}9)$$

Here, Plutonium-238 decays into Uranium-234 with a half-life of 87.75 years. The decay of ^{238}Pu primarily emits alpha particles with an energy of approximately 5.593 MeV per decay. The high energy of these alpha particles is converted into heat within the fuel material.

9.8.1   Conversion to Electrical Power

The heat generated by radioactive decay is converted into electrical power using thermoelectric generators. The efficiency of this conversion depends on the thermoelectric materials used, typically silicon-germanium (SiGe) alloys in space applications. The efficiency η of converting thermal power to electrical power is generally low, around 6-7%. The electrical power output P_{elec} is therefore:

$$P_{\text{elec}} = \eta Q(t) \qquad\qquad (9\text{-}10)$$

This efficiency accounts for the inherent limitations in thermoelectric materials, which involve converting a fraction of the thermal gradient into electrical energy while the rest is dissipated as waste heat.

9.8.2   Practical Considerations for RTGs

1. Decay Chain and Daughter Products: In RTGs, it is often assumed that the decay chain beyond the primary decay can be ignored for power calculations, especially when the daughter products have extremely long half-lives. For example, the decay product Uranium-234 from Plutonium238 decay has a half-life of 244,500 years, effectively remaining stable for the duration of the RTG's operational life.
2. Mass Calculation: The amount of fuel required depends on the desired power output, the efficiency of the thermoelectric conversion, and the decay rate of the fuel. For instance, to produce 200 W of electrical power continuously, the initial mass of ^{238}Pu must account for the decay over the mission lifetime (e.g., 10 years) and the conversion efficiency.
3. Encapsulation and Safety: The radioactive material in an RTG is encapsulated in multiple layers, including a liner, strength member, oxidation-resistant cladding, and a reentry aeroshell. This design ensures containment of radioactive material under normal and potential accident conditions, such as reentry into Earth's atmosphere or a launch failure.
4. Thermal Management: RTGs are designed to dissipate the waste heat effectively while maintaining a stable temperature gradient across the thermoelectric elements. This thermal management is crucial for ensuring consistent power output and the longevity of the RTG.

The production of electrical power in an RTG involves the conversion of heat generated by the radioactive decay of isotopes like Plutonium-238. This process is governed by the exponential decay law, and the efficiency of power conversion depends on the properties of the thermoelectric materials and the design of the RTG system. The careful management of radioactive materials, along with robust safety measures, ensures the reliable and safe operation of these power systems in space applications.

Using the law of radioactive decay, an RTG system fueled by Plutonium-238 (^{238}Pu), encapsulated as plutonium dioxide (PuO_2), starts with an initial thermal power output of 400 watts-thermal per kilogram of fuel. After ten years of continuous operation in space, the total thermal power output decreases to approximately 370 watts-thermal, representing about 92.5% of the original thermal power. This decline is due to the radioactive decay of ^{238}Pu, which has a half-life of 87.75 years, ensuring sustained power generation over extended missions.

9.9   Encapsulation and Safety Measures

9.9.1   Liner

The plutonium fuel is housed in a multilayered encapsulation system to ensure safety and manage the radioactive material. The innermost layer, known as the liner, is a thin metallic capsule that serves as a primary containment barrier. Tantalum is frequently used for the liner due to its chemical compatibility with $PuO2PuO\_2PuO2$ across the range of temperatures expected during the fuel's operational life. This layer helps prevent contamination of subsequent encapsulation layers during assembly and operation.

9.9.2   Strength Member

The next critical component is the strength member, designed to contain the radioactive fuel under normal and extreme conditions. It must withstand high-velocity impacts, fragmentation, overpressure from explosions, and exposure to fire. Additionally, the strength member handles the internal pressure from helium gas buildup, a byproduct of alpha decay from $238Pu^\{238\}Pu238Pu$. For lower temperature systems intended for atmospheric burn-up, superalloys such as Haynes-25 are utilized. In contrast, high-temperature systems, designed to survive reentry and impact, use refractory metals like iridium, capable of enduring the intense thermal environment up to 1675 K.

9.9.3   Oxidation-Resistant Cladding and Vents

To protect the strength member from oxidation and environmental factors, a thin cladding layer of noble metals is applied. This oxidation-resistant layer is crucial for preventing degradation of the strength member and ensuring long-term containment integrity. Additionally, to manage the internal pressure from helium gas, selective or non-selective vents are incorporated. Selective vents allow only helium to escape, preventing the release of any radioactive particles or other gaseous byproducts, while non-selective vents may also allow other gases to escape, including uncondensed fuel and impurity vapors.

9.9.4   Aeroshell

An aeroshell, made from materials like graphite, surrounds the entire assembly for heat management and structural integrity during potential atmospheric reentry. The aeroshell must have low sublimation or ablation rates to withstand the aerodynamic forces and thermal stresses during reentry. It also serves as a thermal conduit, transferring heat from the fuel to the thermoelectric conversion system during normal operation. In cases where the aeroshell cannot sufficiently insulate the system, additional materials like pyrolytic graphite may be used to manage thermal gradients, ensuring that the RTG components operate within their safe temperature limits.

9.9.5   Safety and Design Evolution

Nuclear heat source designs have evolved significantly, guided by stringent aerospace nuclear safety philosophies. Early RTG designs, such as the SNAP-3B and SNAP-9A systems, were based on a "burn-up and disperse" strategy, where nuclear fuel would burn up and disperse at high altitudes in case of atmospheric reentry. This approach utilized superalloys suitable for lower operating temperatures due to the use of lead telluride (PbTe) thermoelectric converters.

The transition to higher operating temperature systems, such as those using silicon germanium (SiGe) thermoelectric converters, necessitated a shift to materials with higher melting points and superior thermal properties. Modern RTG designs, such as SNAP-19 and beyond, have incorporated containment strategies to ensure that the fuel remains intact during all mission phases, including potential reentry and impact scenarios. The development of iridium alloys and advanced graphite composites has been crucial in this regard, providing the necessary structural integrity and thermal management capabilities.

9.9.6   Considerations for Fuel Selection and Encapsulation

When designing a radioisotope generator, several key factors are considered:
1. **Power Density**: The specific power (W/cm^3) of the radioisotope fuel is crucial for determining the overall size and weight of the RTG.
2. **Half-Life**: A longer half-life ensures sustained power output over extended missions, reducing the need for frequent replacements or replenishments.
3. **Fuel Availability and Cost**: The production, availability, and cost of the radioisotope are significant considerations, especially given the specialized facilities required for handling and processing these materials.
4. **External Nuclear Radiations**: The type and intensity of radiations (neutrons, gamma rays) emitted must be manageable within the shielding capabilities of the RTG design.
5. **Advancements in Material Science**: Ongoing developments in materials science can influence the choice of encapsulation materials and structural components, enhancing the safety and efficiency of the RTG.

Alpha-emitting radioisotopes like ^{238}Pu are preferred for space applications due to their high-power density and relatively low shielding requirements. However, the generation of helium gas during alpha decay presents challenges in managing internal pressure, necessitating robust design solutions like venting systems or strong containment structures.

The design and construction of RTGs involve intricate considerations of radioactive decay behavior, thermal management, and robust safety measures. These systems are engineered to provide reliable power for space missions, ensuring the safe containment of radioactive materials throughout their operational life.

9.10 Thermoelectric Power Conversion

Thermoelectric power conversion is a well-established technology used in all U.S. radioisotope power systems (RPS). The primary advantage of thermoelectric (TE) devices lies in their simplicity and reliability, as they have no moving parts and are thus passive power conversion generators. However, a significant limitation of TE systems is their relatively low conversion efficiency, typically less than 10%. Despite this drawback, they are favored in space missions due to their durability and ability to operate without maintenance.

9.10.1 Fundamentals of Thermoelectric Conversion

The conversion of thermal energy into electricity in a TE device is based on the Seebeck effect, discovered by Thomas Seebeck in 1821. He observed that an electromotive force (emf) is generated when the junctions of two dissimilar metals are maintained at different temperatures. This phenomenon occurs because a temperature gradient in a conducting material causes charge carriers (electrons and holes) to diffuse from the hot region to the cold region, creating a voltage difference.

An ideal thermoelectric material possesses a high Seebeck coefficient (a measure of the voltage generated per unit temperature difference), high electrical conductivity (to minimize resistive losses), and low thermal conductivity (to maintain a temperature gradient). These properties are often found in semiconductors, which are used in modern TE devices.

9.10.2 Materials and Design

In a practical thermoelectric converter, semiconductor materials are connected electrically in series and thermally in parallel. A typical TE converter consists of two semiconductor legs, one p-type and one n-type, connected between two heat transfer surfaces known as the hot and cold junctions. The p-type material, where holes are the majority carriers, becomes positively charged at the cold junction due to the migration of holes. Conversely, the n-type material, where electrons are the majority carriers, accumulates electrons at the cold junction.

When a temperature difference is applied across these materials, a voltage is generated across the cold junctions. By connecting an external load across these junctions, a current can flow, thus converting thermal energy into electrical energy. The power output of a TE device depends on the temperature difference across the junctions, the Seebeck coefficient of the materials, and the internal and external resistances.

9.10.3 Performance Metrics and Optimization

The figure of merit Z is a critical parameter for evaluating TE materials, defined as:

$$Z = \frac{a^2}{k\,p} \tag{9-11}$$

Where:
- a is the Seebeck coefficient
- k is the thermal conductivity
- p is the electrical resistivity

Higher values of Z indicate better performance and higher efficiency of the TE converter. The efficiency of a TE device is also influenced by the choice of materials. The materials with higher Z values are more desirable for efficient power conversion.

9.10.4 Practical Applications and System Integration

Thermoelectric materials used in U.S. RTGs have evolved over time. Early systems used tellurides, such as lead telluride (PbTe), while later systems, such as the Multi-Hundred-Watt (MHW) generators and the General-Purpose Heat Source (GPHS)-RTG, employed silicon-germanium (SiGe) alloys. These materials were chosen for their ability to operate at higher temperatures and provide better performance.

In RTG systems, TE modules are typically assembled between the heat source and the heat sink. Springs are often used to maintain good thermal and electrical contact and to accommodate thermal expansion. However, some designs, such as the hermetically sealed, close-packed tubular modules, do not use springs. The SiGe-based converters, for instance, use cantilevered thermocouples radiatively heated from the heat source.

9.10.5 Dynamic Power Conversion Systems

In addition to TE systems, dynamic power conversion systems can be employed to achieve higher efficiencies. These systems, which involve moving parts, can convert thermal energy into mechanical energy and then into electrical energy. While they offer the potential for higher efficiency and power output, they come with additional complexity and challenges, such as the need for long-lived bearings and the management of rotational and vibrational forces.

Dynamic systems require careful integration with the spacecraft to handle these forces and to ensure safe operation, especially if the working fluid is lost or if there is a need for emergency cooling. Multiple radioisotope heat sources can be coupled to a common power conversion system or vice versa, providing flexibility and redundancy.

Thermoelectric power conversion remains a key technology for space missions due to its reliability and simplicity. While the efficiency of TE systems is relatively low, ongoing research into new materials and configurations holds promise for future improvements. Dynamic systems offer an alternative for higher power and efficiency but require more complex integration. As space exploration continues to advance, the development of efficient and reliable power conversion technologies will be crucial for the success of long-duration missions.

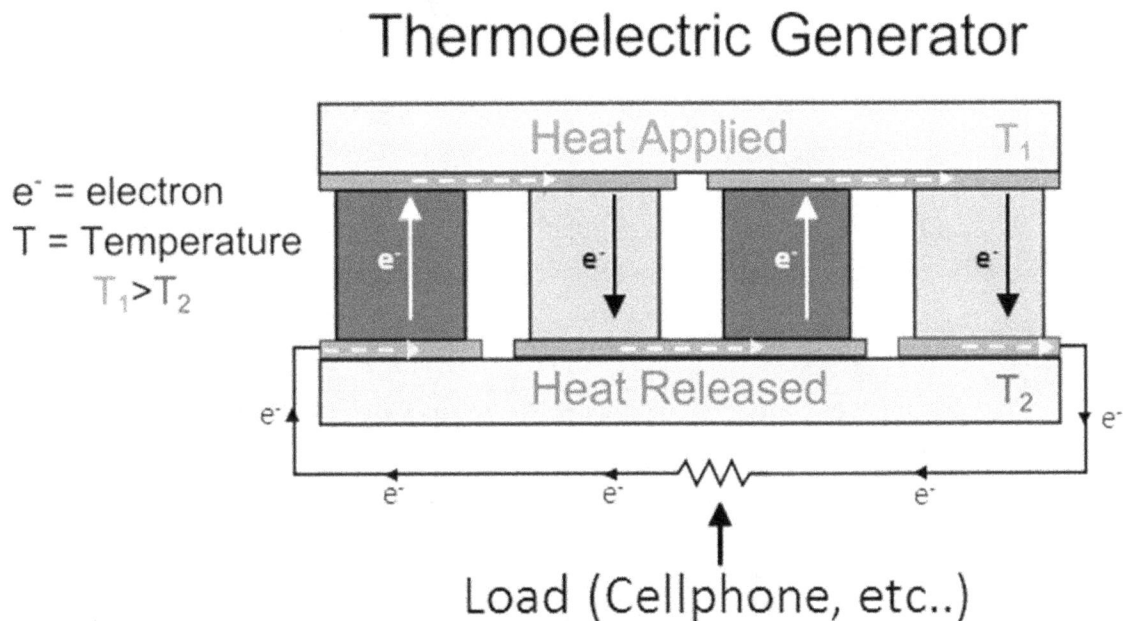

Figure 9-17. TEG physics.

9.11   Heat Rejection Subsystems

In space systems, the rejection of waste heat is crucial for maintaining optimal operating temperatures and ensuring system longevity. Unlike terrestrial systems, which can utilize convective and conductive heat transfer to the surrounding environment, space systems rely solely on thermal radiation. This limitation requires careful design and optimization of radiator systems to effectively manage the thermal energy produced by the system, particularly in the case of radioisotope thermoelectric generator (RTG) power systems.

9.11.1 Fundamentals of Radiator Design

The primary mechanism for heat rejection in space is thermal radiation, governed by the Stefan Boltzmann law. The law states that the power radiated per unit area of a black body is proportional to the fourth power of its absolute temperature (T). Mathematically, this can be expressed as:

$$P = \epsilon \sigma A T^4 \qquad\qquad (9\text{-}12)$$

where:

- P is the power radiated,
- ϵ is the emissivity of the surface,
- σ is the Stefan-Boltzmann constant (5.67×10^{-8} W/m^2 K^4),
- A is the surface area of the radiator, and
- T is the temperature of the radiating surface in Kelvin.

This fourth-power dependency on temperature incentivizes the design of power systems to operate at higher temperatures, thereby reducing the required radiator area and mass. For RTG systems, typical radiator temperatures are around 575 K.

9.11.2 Thermodynamic Efficiency and Heat Rejection

The efficiency of a thermodynamic cycle, such as those used in space power plants, is inherently tied to the temperatures of heat absorption and rejection. According to the Carnot efficiency, the maximum possible efficiency is achieved when the temperature difference between the heat source and the heat sink is maximized. In practice, however, space systems operate at lower efficiencies compared to terrestrial systems because the need to minimize radiator area and mass often necessitates higher heat rejection temperatures. This trade-off is critical, as lower heat rejection temperatures would require larger radiators, increasing the system's mass and complexity.

9.11.3 Thermal Environment and Energy Balance

A spacecraft's thermal environment in orbit includes several factors contributing to the heat load on the radiator. These factors include:

1. Direct Solar Radiation: The primary source of energy, with an intensity of approximately 1371 W/m^2 at one astronomical unit (AU) from the Sun.
2. Earth-Emitted Radiation: Infrared radiation emitted from the Earth, typically

around 240 W/m².

3. Earth-Reflected Solar Radiation (Earthshine): Solar energy reflected by the Earth's surface and atmosphere, which can vary but typically accounts for a significant portion of the thermal load.

The thermal energy balance for the radiator can be expressed as:

$$\alpha G_s F_s + \alpha_r G_r F_r + \alpha_e G_e F_e + P_i = \epsilon \sigma A T^4 \qquad (9\text{-}13)$$

where:
- α is the absorptivity of the surface,
- G_s, G_r, and G_e are the solar, reflected, and emitted Earth radiation, respectively,
- F_s, F_r, and F_e are the view factors for solar, reflected, and emitted Earth radiation, respectively,
- P_i is the internal waste heat load,
- ϵ is the emissivity of the radiator,
- σ is the Stefan-Boltzmann constant, and
- A and T are the area and temperature of the radiator, respectively.

9.11.4 Material Selection for Radiators

The choice of materials for radiator construction is critical, as it must withstand the high temperatures and the space environment's unique conditions. Potential radiator materials are selected based on their thermal conductivity, emissivity, and structural properties at high temperatures. Fig. 5 provides a range of materials suitable for different temperature regimes, including aluminum alloys, titanium alloys, and high-temperature materials like tantalum and molybdenum alloys.

The radiator's surface is typically coated with materials that have a high emissivity-to-absorptivity ratio, maximizing thermal radiation while minimizing solar energy absorption. This optimization helps maintain the efficiency of the heat rejection process.

9.11.5 Design Considerations and Optimization

The design of heat rejection systems in space involves optimizing the radiator's size, shape, and placement to ensure effective thermal management. Static conduction and radiation heat transfer fin designs are commonly used. The fins enhance the heat dissipation surface area and are integrated into the generator housing to facilitate thermal conduction from the power conversion system.

To further enhance performance, the system may use thermal insulation materials to minimize heat loss and ensure that the maximum amount of thermal energy is transferred to the power conversion thermocouples. These materials are selected based on their compatibility with the system's operating temperature range and their chemical stability in the space environment.

The heat rejection subsystem in space power systems, particularly those using RTGs, is a critical component that ensures the system's thermal stability and efficiency. The design and optimization of these systems are driven by the need to minimize mass and maximize power output, while also managing the unique thermal environment of space. Advances in materials science and thermal engineering continue to enhance the performance and reliability of these systems, enabling longer and more complex space missions.

9.12   SNAP-3B RTG

The SNAP-3B (Systems for Nuclear Auxiliary Power) radioisotope thermoelectric generator (RTG) represents a pioneering effort in the application of nuclear power for space exploration. Launched in 1961 aboard the Transit 4A and Transit 4B navigational satellites, the SNAP-3B marked the first use of nuclear power systems in space. These RTGs supplemented the spacecraft's solar cell arrays and provided a reliable power source, demonstrating the viability of nuclear systems for sustained space missions.

9.12.1 Design Philosophy and Safety Considerations

During the era of SNAP-3B's deployment, the aerospace nuclear safety philosophy emphasized minimizing potential radioactive contamination in the event of a mission abort resulting in atmospheric reentry. The design strategy was to ensure that the RTGs would burn up and disperse at high altitudes, reducing ground-level radioactivity to levels comparable to those from atmospheric nuclear weapons testing. This was a critical safety measure to protect both the environment and human health.

9.12.2 Heat Source and Fuel Composition

The SNAP-3B utilized plutonium-238 (Pu-238) as its primary heat source. Pu-238 was chosen for its favorable half-life and power density, making it ideal for long-duration missions. The specific isotopic composition of the plutonium metal used in the SNAP-3B was approximately 80 wt% Pu-238, 16 wt% Pu-239, 3 wt% Pu-240, and 1 wt% Pu-241. The fuel mass was 92.7 grams, contained within a total heat source mass of 359 grams. This configuration generated approximately 52.5 watts of thermal power at an operating temperature of 810 K.

9.12.3 Structural Components and Materials

To ensure the safe containment of the radioactive material, the SNAP-3B's design incorporated several layers of protective materials:

1. **Liner:** Tantalum was used as the liner material due to its chemical compatibility with the fuel and its ability to withstand high temperatures. The liner served as a barrier to prevent contamination of other components during assembly.
2. **Strength Member:** A cylindrical strength member made from Haynes Alloy No. 25 provided structural integrity. This material was chosen for its high-temperature resilience and mechanical strength, capable of withstanding high-velocity impacts and other stresses.
3. **Thermal Insulation:** The insulation consisted of materials like Min-K 1301, which helped maintain the required thermal gradient across the thermoelectric (TE) elements.

9.12.4 Thermoelectric Conversion and Efficiency

The SNAP-3B utilized lead telluride (PbTe) as the thermoelectric material. The thermoelectric elements were arranged to convert the thermal energy from the plutonium heat source into electrical energy through the Seebeck effect. The generator's overall efficiency was approximately 5%, a common characteristic of thermoelectric devices at the time. The efficiency was limited by the materials available and the inherent properties of thermoelectric conversion, which generally exhibits lower efficiency compared to dynamic systems.

Springs were integrated into the design to maintain thermal contact between the heat source and the TE converters, ensuring consistent heat transfer and electrical output. The system also employed gas filling to enhance thermal conduction within the RTG.

9.12.5 Encapsulation and Safety Features

The SNAP-3B's fuel capsule was designed with multiple layers of protection. The outer canister walls were made of Type 304 stainless steel, chosen for its durability and resistance to corrosion. The use of tapered plugs and blocks, often made from materials like molybdenum or Haynes Alloy No. 25, provided additional structural support and helped manage thermal expansion.

9.12.6 Performance and Operational Longevity

The performance data from the SNAP-3B RTGs, as telemetered from the Transit 4A and 4B spacecraft, indicated remarkable reliability and longevity. The Transit 4A generator remained operational for at least 15 years, until 1976, while the last signal from Transit 4B was reported in April 1971. This longevity highlighted the durability of the RTG design and the stability of Pu-238 as a power source.

The SNAP-3B RTG represented a significant technological achievement in the early days of space exploration. Its successful operation demonstrated the practicality and reliability of using radioisotope power systems in space, setting the stage for future advancements in nuclear power for spacecraft. The lessons learned from the SNAP-3B program have informed the design and development of subsequent RTGs, contributing to the evolution of space nuclear power systems.

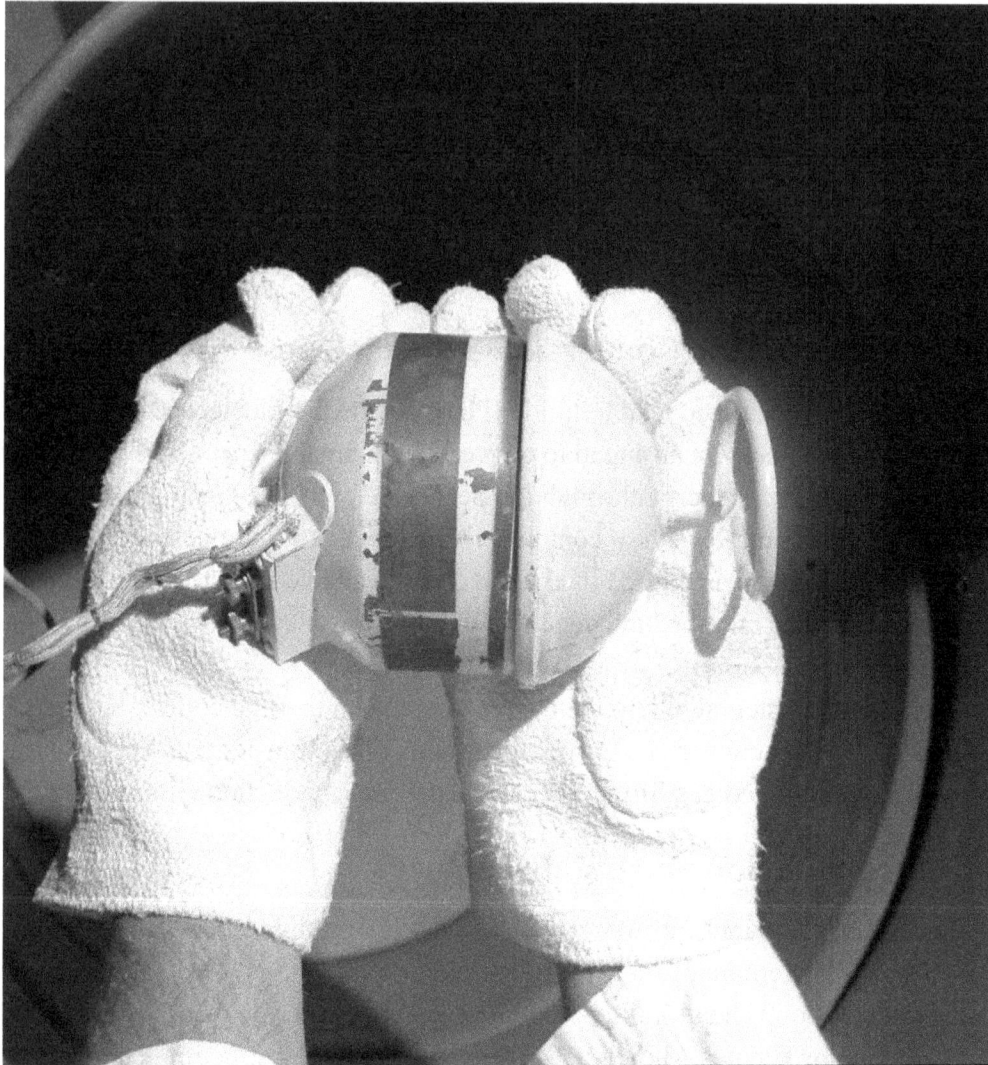

Figure 9-18. SNAP-3B (Mound Science and Energy Museum).

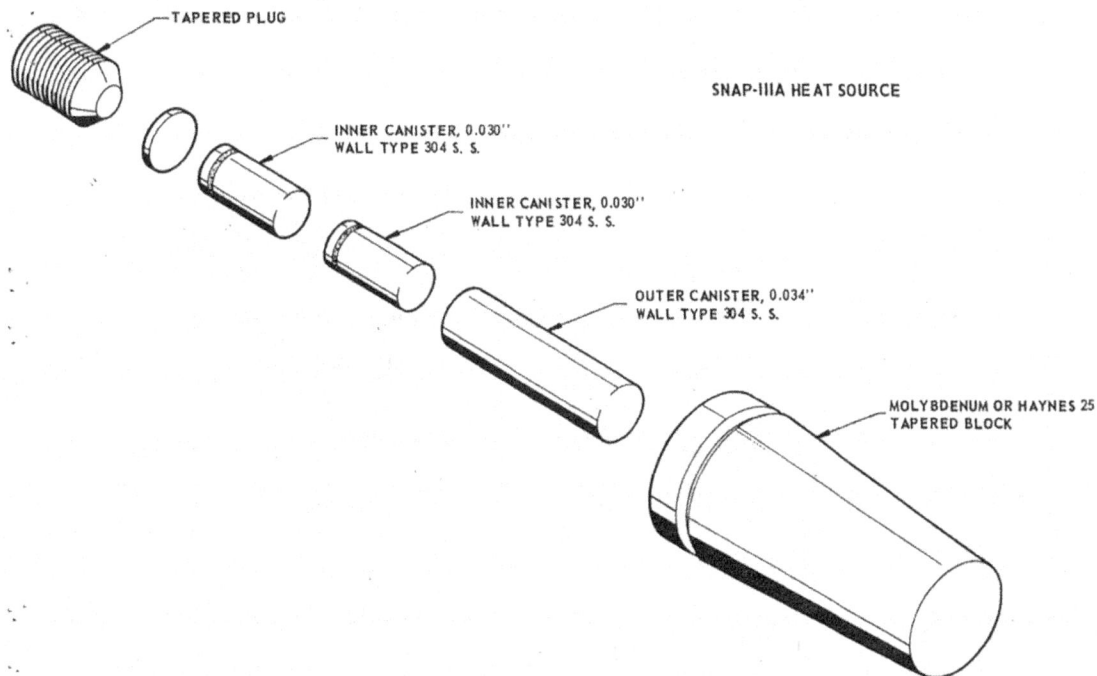

Figure 9-19. SNAP-3A/3B Fuel capsule (source:
https://beyondnerva.wordpress.com/radioisotope-power-sources/radioisotope-fuel-form-
and-containment/

9.13   SNAP-19 RTG

The Systems for Nuclear Auxiliary Power (SNAP)-19 series represents a
significant advancement in the use of radioisotope thermoelectric generators (RTGs) for
space missions. These systems were designed to provide reliable power for a variety of
missions, including the NASA Nimbus III meteorological satellite and the Pioneer and
Viking missions. SNAP-19 utilized an advanced heat source and thermoelectric materials
to convert thermal energy into electrical power, addressing both operational and safety
concerns.

9.13.1 Heat Source and Safety Philosophy

The SNAP-19 RTG employed a 645-watt thermal heat source known as the Intact
Impact Heat Source. This source was specifically designed to contain the radioisotope
fuel during normal operations and to limit the release of radioactive materials in the event
of an accident, such as launch failure or reentry. The fuel consisted of plutonium-238
dioxide, a potent alpha-emitter with a long half-life, providing a steady heat output.

The safety philosophy for SNAP-19 evolved from earlier designs, focusing on
containing or immobilizing the fuel to prevent contamination of the biosphere. This

approach was critical for ensuring that the RTG could withstand reentry and impact scenarios, preventing the dispersal of radioactive materials.

9.13.2 Thermoelectric Conversion System

The power conversion system in SNAP-19 consisted of an array of 90 thermoelectric (TE) elements made from lead telluride (PbTe) and silver antimony germanium telluride (PbTe-TAGS). These materials were chosen for their favorable thermoelectric properties, including high Seebeck coefficients and relatively low thermal conductivities. The conversion efficiency of the system was approximately 5%, which, while modest, was sufficient for the mission requirements at the time.

The thermoelectric modules in SNAP-19 were designed to be resilient under the harsh conditions of space. They were capable of withstanding significant temperature variations and mechanical stresses. The TE elements were arranged in series and parallel configurations to ensure redundancy and reliability. The degradation of performance over time, such as in the Nimbus III mission, was attributed to issues like sublimation of the TE material and loss of the hot junction bond due to internal cover gas depletion.

9.13.3 Mission Applications

SNAP-19 RTGs were first used on the NASA Nimbus III meteorological satellite, launched in April 1969. This mission demonstrated the durability and reliability of the system, operating well beyond its expected 20,000-hour lifespan.

For the Pioneer missions to Jupiter and Saturn, as well as the Viking missions to Mars, the SNAP-19 design was modified to meet specific mission requirements, such as extended operational life (up to six years) and the ability to function under various environmental conditions, including sterilization for planetary protection. The RTGs provided power for critical instruments and systems, supporting groundbreaking scientific discoveries about the outer planets and Mars.

9.13.4 Structural and Thermal Design

The SNAP-19 RTG structure consisted of several key components designed for durability and safety:

- **Fuel Capsules:** The fuel was encapsulated in a multi-layer structure, including tantalum and platinum-rhodium alloy layers, to contain the fuel and prevent oxidation. The capsules were also equipped with pressure relief valves to vent decay gases.
- **Graphite Heat Shield:** Surrounding the fuel capsules, the graphite heat shield provided protection during reentry and served as a thermal sink during normal operation. The shield was designed to ablate and dissipate heat, minimizing the temperature impact on the rest of the spacecraft.
- **Thermoelectric Modules:** The TE modules were arranged around the heat source

in a hexagonal configuration. This design maximized the use of available thermal energy while minimizing thermal losses. The modules included components such as springs, pistons, alignment buttons, and heat sink bars to maintain thermal and electrical contact.

- **Radiator System:** The radiator fins, made from magnesium-thorium alloy, were coated with a zirconium oxide/sodium silicate material. This coating provided high emissivity and low absorptivity, enhancing the heat rejection capability of the system. The radiator was critical for maintaining the operational temperature of the TE modules.

9.13.5 Performance and Legacy

The SNAP-19 RTGs provided reliable power throughout their missions. Notably, the Pioneer 10 and 11 spacecraft, powered by SNAP-19 units, continued to operate far beyond their expected lifespans, with Pioneer 10 becoming the first man-made object to leave the solar system. Similarly, the Viking landers, which also utilized SNAP-19 RTGs, conducted extensive scientific experiments on the Martian surface, including the search for life.

The design and operational success of SNAP-19 RTGs provided valuable insights and lessons for subsequent RTG designs, contributing to the evolution of space nuclear power systems. The continued use and development of RTGs underscore their importance as a reliable power source for long-duration space missions, where solar power may not be feasible.

9.14 Multi-Mission Radioisotope Thermoelectric Generator (MMRTG)

The Multi-Mission Radioisotope Thermoelectric Generator (MMRTG) is a pivotal technology in NASA's planetary exploration toolkit, offering a reliable and long-lasting power source for spacecraft and landers. It was notably employed in the Mars Science Laboratory (MSL) mission, which successfully launched on November 26, 2011, and delivered the Curiosity rover to the Martian surface on August 6, 2012. The MMRTG enabled Curiosity's extensive exploration of Mars, significantly contributing to our understanding of the Red Planet.

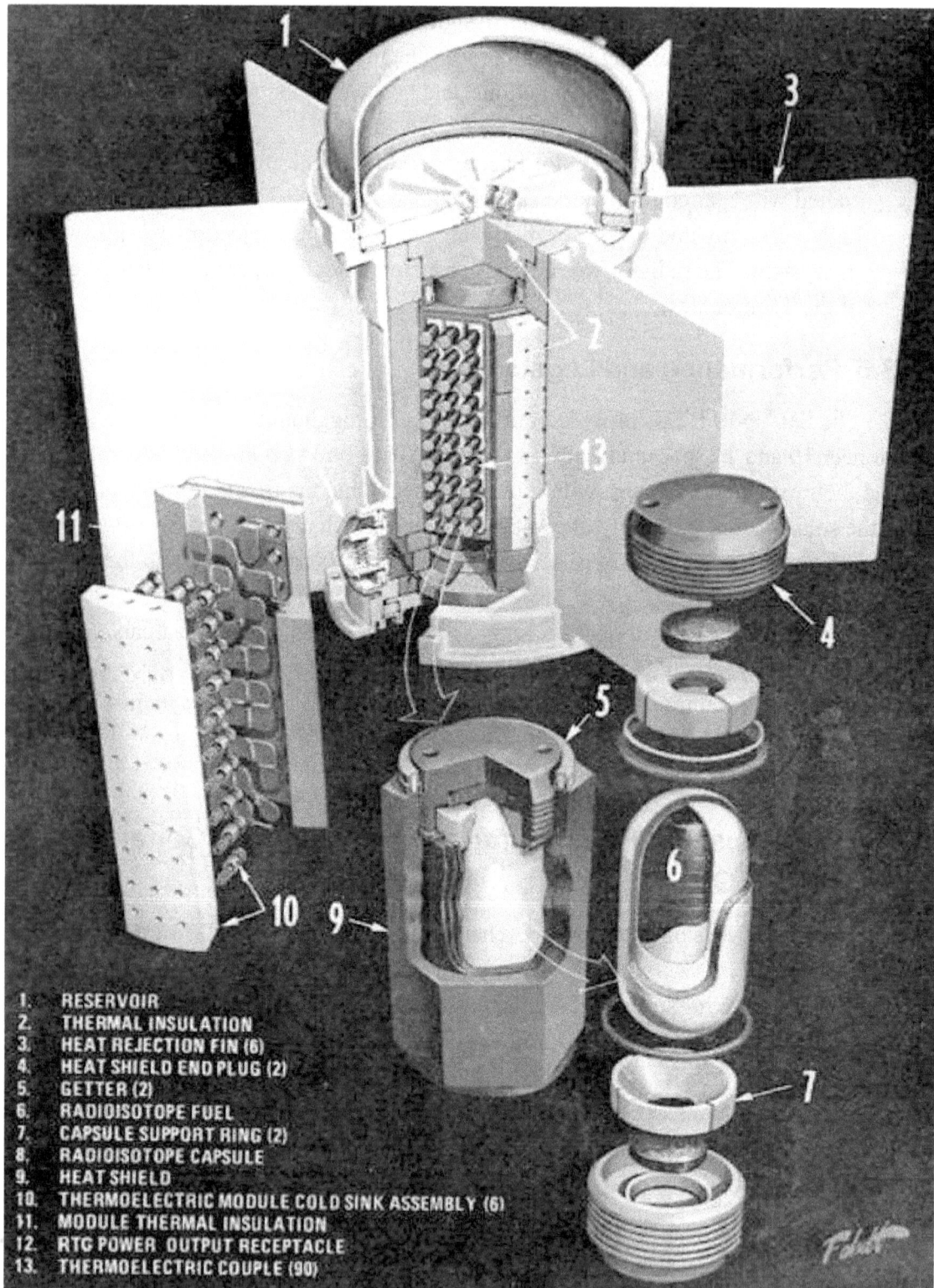

1.   RESERVOIR
2.   THERMAL INSULATION
3.   HEAT REJECTION FIN (6)
4.   HEAT SHIELD END PLUG (2)
5.   GETTER (2)
6.   RADIOISOTOPE FUEL
7.   CAPSULE SUPPORT RING (2)
8.   RADIOISOTOPE CAPSULE
9.   HEAT SHIELD
10.   THERMOELECTRIC MODULE COLD SINK ASSEMBLY (6)
11.   MODULE THERMAL INSULATION
12.   RTG POWER OUTPUT RECEPTACLE
13.   THERMOELECTRIC COUPLE (90)

Figure 9-20. Viking/SNAP-19 RTG.

9.14.1 Primary Objectives of the Mars Science Laboratory Mission

1. **Search for Life on Mars** The mission sought to explore Mars for signs of past or present life. Key goals included:
    o   Investigating the potential habitability of ancient Martian environments.
    o   Identifying organic compounds, the chemical building blocks of life, in Martian rocks and soil.
2. **Understanding Martian Climate and Geology** The MSL mission aimed to:
    o   Characterize the Martian climate, both past and present.
    o   Study the geological processes that shaped the Martian surface, including the planet's crust and atmospheric evolution.
    o   Investigate the planet's interior structure and history.
3. **Preparation for Human Exploration** The mission also focused on collecting data critical for future human missions to Mars, such as:
    o   Analyzing the radiation environment on the Martian surface.
    o   Demonstrating technologies that future missions might use for in situ resource utilization and other engineering challenges.

9.14.2 MMRTG: A Versatile Power Source

The MMRTG was selected for the Curiosity rover due to its ability to provide consistent power and heat in the challenging Martian environment. Unlike solar panels, which depend on sunlight and are affected by dust and seasonal changes, the MMRTG operates independently of external light conditions. It generates electricity through the natural decay of plutonium-238, which emits heat that is then converted into electricity via thermoelectric materials.

The MMRTG's design ensures operational efficiency across a broad range of environmental conditions, including the vacuum of space and the varied temperatures found on Mars. It provides approximately 110 watts of electrical power at the start of the mission, with a gradual decline over time. This power supports the rover's scientific instruments, mobility systems, and communication capabilities, making it a crucial component for long-duration missions.

9.14.3 Design and Integration on Curiosity Rover

The MMRTG is mounted on the rear of the Curiosity rover, as depicted in the mission's schematics. Its robust design ensures stable power output and helps maintain operational temperatures for the rover's systems during cold Martian nights and winters. The MMRTG's heat output not only generates electricity but also helps keep the rover's instruments and batteries warm, preventing damage from the extreme cold.

9.14.4 Impact and Legacy

Since its successful deployment, the MMRTG on the Curiosity rover has played an essential role in numerous scientific discoveries. It has enabled the collection of valuable data on Martian geology, climate, and potential habitability. For instance, Curiosity's findings in Gale Crater, such as evidence of ancient lakes and streams, have provided compelling clues about Mars' wetter and potentially habitable past.

The MMRTG's consistent performance has also set the stage for future missions. Its demonstrated reliability and efficiency make it a prime candidate for powering upcoming missions, including those that may involve human exploration. The technology offers the promise of sustained energy supply, which is crucial for long-term missions far from the Sun, where solar power becomes less viable.

The MMRTG has been instrumental in advancing our understanding of Mars and continues to support NASA's ambitious exploration goals, making it a cornerstone of current and future space missions.

Figure 9-21. Mars Curiosity Rover (2011-Present) - Mars Landing: Aug. 5, 2012 (source: https://science.nasa.gov/planetary-science/programs/radioisotope-power-systems/missions/)

9.16 Problems

Example 9-1:

Calculate the mass of $^{238}PuO_2$ required for a Radioisotope Thermoelectric Generator (RTG) to produce an initial electrical power output of 400 W with a thermoelectric conversion efficiency of 7%. Additionally, determine the electrical power output after 10 years.

Solution:

1. Initial Thermal Power Calculation

To generate 400 W of electrical power with 7% efficiency, the required thermal power can be calculated as:

$$P_{thermal,,initial} = \frac{P_{electrical,initial}}{\eta}$$

Where:

- $P_{electrical,initial}$ is the initial electrical power output (400 W)
- η is the thermoelectric conversion efficiency (7% or 0.07)

$$P_{thermal,initial} = \frac{400 \text{ W}}{0.07} = 5714.2857 \text{ W}$$

Thus, the required initial thermal power output is 5714.2857 W.

2. Number of Alpha Decays Required

The energy released per alpha decay of ^{238}Pu is 5.593 MeV .

To find the total number of decays per second needed to achieve the required thermal power, we use:

$$\text{Total energy per second} = P_{thermal,initial} (\text{ in MeV/s})$$

$$P_{thermal,initial} = 5714.2857 \text{ W} \times 6.242 \times 10^{12} \text{MeV/J}$$

$$P_{thermal,initial} = 3.567 \times 10^{16} \text{ MeV/s}$$

The number of decays per second (activity, A) required is:

$$A = \frac{3.567 \times 10^{16} \text{MeV/s}}{5.593 \text{MeV/ decay}}$$

$$A = 6.377 \times 10^{15} \text{ decays /s}$$

3. Number of ^{238}Pu Atoms Required

Using the decay law:

$$A = \lambda N$$

where:
- λ is the decay constant
- N is the number of atoms of ^{238}Pu

The decay constant λ is related to the half-life $T_{1/2}$ by:

$$\lambda = \frac{\ln(2)}{T_{1/2}}$$

$$T_{1/2} = 87.75 \text{ years } = 87.75 \times 3.154 \times 10^7 \text{ s}$$

$$\lambda = \frac{\ln(2)}{87.75 \times 3.154 \times 10^7 \text{ s}}$$

$$\lambda = 2.505 \times 10^{-10} \text{ s}^{-1}$$

The number of ^{238}Pu atoms N is:

$$N = \frac{A}{\lambda}$$

$$N = \frac{6.377 \times 10^{15} \text{ decays /s}}{2.505 \times 10^{-10} \text{ s}^{-1}}$$

$$N = 2.548 \times 10^{25} \text{ atoms}$$

4. Mass of ^{238}Pu Required

To calculate the mass of ^{238}Pu :

$$\text{Mass} = \frac{N \times m_{\text{Pu}}}{N_A}$$

Where:

- m_{Pu} is the atomic mass of $^{238}\text{Pu}(238 \text{ g/mol})$
- N_A is Avogadro's number (6.022×10^{23} atoms /mol)

$$\text{Mass} = \frac{2.548 \times 10^{25} \text{ atoms} \times 238 \text{ g/mol}}{6.022 \times 10^{23} \text{ atoms /mol}}$$

$$\text{Mass} = 10070.1 \text{ g} \approx 10.07 \text{ kg}$$

This mass represents the amount of pure ^{238}Pu. However, since the RTG uses $^{238}\text{PuO}_2$, we account for the molecular mass of PuO_2 :

$$\text{Molecular mass of PuO}_2 = 238 + 2 \times 16 = 270 \text{ g/mol}$$

$$\text{Mass of PuO}_2 = 10.07 \text{ kg} \times \frac{270}{238}$$

$$\text{Mass of PuO}_2 = 11.42 \text{ kg}$$

Thus, 11.42 kg of $^{238}\text{PuO}_2$ is needed to provide the required initial thermal power output.

5. Thermal Power Output After 10 Years

The thermal power after 10 years, accounting for radioactive decay, is given by:

$$Q(t) = Q_0 e^{-\lambda t}$$

where:

- Q_0 is the initial thermal power output
- λ is the decay constant
- t is the time elapsed (10 years)

$$Q(10 \text{ years }) = 5714.2857 \text{ W} \times e^{-2.505 \times 10^{-10} \text{ s}^{-1} \times 10 \times 3.154 \times 10^7 \text{ s}}$$

$$Q(10 \text{ years }) \approx 5280.2 \text{ W}$$

6. Electrical Power Output After 10 Years

With 7% efficiency:

$$P_{\text{electrical},10\text{ years}} = 5280.2\text{ W} \times 0.07$$

$$P_{\text{electrical},10\text{ years}} \approx 369.6\text{ W}$$

Conclusion:

After 10 years, the RTG will produce approximately 369.6 W of electrical power, starting with a 400 W output and gradually declining due to the decay of ^{238}Pu.

10 Space Nuclear Power Reactors

10.1 Introduction

The advancement of space exploration over the past several decades has pushed the boundaries of human ingenuity and technology, necessitating the development of reliable and long-duration power sources capable of operating in the harsh environment of space. Among the various technologies explored to meet these demands, space nuclear reactors have emerged as one of the most promising solutions. These reactors, harnessing the immense energy potential of nuclear reactions, offer the power and propulsion capabilities required for missions beyond Earth orbit, including deep-space exploration, long-term habitation on other celestial bodies, and the establishment of permanent outposts on the Moon and Mars.

Space nuclear reactors are based on the same fundamental principles as terrestrial nuclear power systems, where the energy released from nuclear fission is used to generate heat, which can then be converted into electricity or used directly for propulsion. However, the unique challenges of space environments—such as the vacuum of space, extreme temperature variations, radiation, and the need for long-term reliability without maintenance—require specialized designs and materials. The development of space nuclear reactors has thus been a complex and ongoing process, rooted in the broader history of nuclear energy and driven by the specific demands of space exploration.

10.1.1 Early Developments: The Genesis of Space Nuclear Power

The concept of using nuclear power for space applications dates back to the early days of the Space Age, when the United States and the Soviet Union were racing to establish their dominance in space. The idea of using nuclear energy to power spacecraft was appealing because of its ability to provide a steady, high-density power source over long durations—an essential capability for missions that could last for months or even years.

One of the earliest and most significant programs in this regard was the United States' SNAP (Systems for Nuclear Auxiliary Power) program, which began in the late 1950s. The SNAP program aimed to develop compact nuclear power systems that could be used in space. Among the various SNAP reactors, the SNAP-10A holds a special place in history as the first and only nuclear reactor to be launched and operated in space by the United States. Launched in 1965, SNAP-10A was designed to provide 500 watts of electrical power for satellite operations. The reactor operated successfully for 43 days before it was shut down due to a malfunction in the spacecraft's voltage regulator, unrelated to the reactor itself. Despite the early termination, the SNAP-10A mission demonstrated the feasibility of using nuclear reactors in space and provided valuable data for future developments.

Parallel to the U.S. efforts, the Soviet Union also invested heavily in the development of space nuclear reactors, driven by similar strategic and scientific goals. The Soviet space nuclear program developed several reactors under the BES-5 "Buk" and TOPAZ programs, which were used to power radar reconnaissance satellites in low Earth orbit. The BES-5 reactor, for example, was capable of generating 3 kW of electrical power and was used in over 30 satellites between 1970 and 1988. The TOPAZ reactors, which followed, were more advanced, with the TOPAZ-II capable of generating up to 6 kW of electrical power using a thermionic energy conversion system. These reactors were highly reliable and operated for extended periods in space, contributing to the Soviet Union's extensive space reconnaissance capabilities.

The development of these early space nuclear reactors was marked by a series of technical challenges, including the need for effective radiation shielding, the management of heat in the vacuum of space, and the miniaturization of reactor components to fit within the constraints of a spacecraft. Additionally, ensuring the safety of launching nuclear materials into space was a paramount concern, leading to the development of stringent safety protocols and design considerations to minimize the risk of radioactive contamination in the event of a launch failure.

10.1.2 The Rise of Nuclear Thermal Propulsion: NERVA and Beyond

While early space nuclear reactors were primarily focused on power generation, the potential of nuclear energy for spacecraft propulsion was also being explored. Nuclear Thermal Propulsion (NTP) emerged as a promising technology for enabling more efficient and faster travel within the solar system, particularly for missions to Mars and beyond. The concept of NTP involves using a nuclear reactor to heat a propellant, typically hydrogen, to extremely high temperatures. The heated propellant is then expelled through a rocket nozzle to produce thrust. The specific impulse of NTP systems—an indicator of propulsion efficiency—is roughly twice that of the best chemical rockets, offering the potential to significantly reduce travel times for interplanetary missions.

The most prominent NTP program was the United States' NERVA (Nuclear Engine for Rocket Vehicle Application), which began in the early 1960s as a joint effort between NASA and the Atomic Energy Commission. NERVA built upon earlier research conducted under the Rover program, which aimed to develop nuclear-powered rockets for space travel. The NERVA engines were designed to be robust and capable of multiple starts, making them ideal for deep-space missions. Throughout the 1960s and early 1970s, a series of ground tests were conducted, demonstrating the viability of NTP technology. The NERVA program successfully developed and tested several reactor designs, achieving significant milestones in engine performance and reliability.

One of the key advantages of NTP systems like NERVA is their ability to provide high thrust while maintaining high efficiency, making them well-suited for crewed missions to Mars. For example, an NTP system could potentially cut the travel time to Mars by nearly half compared to conventional chemical rockets, reducing the exposure of astronauts to harmful cosmic radiation and the physiological effects of prolonged microgravity. Despite its technical successes, the NERVA program was ultimately canceled in the early 1970s due to shifting priorities and budget constraints, particularly following the conclusion of the Apollo program.

10.1.3 Modern Developments and the Return to Space Nuclear Reactors

The cancellation of NERVA and the winding down of other early space nuclear programs in the 1970s and 1980s led to a period of relative inactivity in the field. However, the need for advanced power and propulsion systems never disappeared, and the concept of space nuclear reactors has seen a resurgence in recent years, driven by renewed interest in deep-space exploration and the goal of establishing a human presence on Mars and beyond.

In the early 21st century, NASA, in collaboration with other agencies and private industry, has revisited the idea of space nuclear reactors, leading to new projects and technological advancements. One notable project is Kilopower, a small fission reactor designed to generate between 1 and 10 kilowatts of electrical power. Kilopower was developed to support long-duration lunar and Martian missions, where solar power may not be viable due to long nights or dust storms. The Kilopower reactor uses a highly enriched uranium core and a Stirling engine to convert heat into electricity. The design emphasizes simplicity, reliability, and scalability, making it a potential candidate for powering future human habitats on the Moon or Mars.

In parallel, the concept of Nuclear Electric Propulsion (NEP) has gained traction as a potential method for long-duration, low-thrust missions to the outer planets and beyond. NEP systems generate electricity using a nuclear reactor, which then powers electric thrusters, such as ion engines. While NEP offers lower thrust compared to NTP, it is extremely efficient and capable of providing continuous propulsion over extended periods, making it ideal for deep-space exploration missions. Recent advancements in reactor design, power conversion technologies, and electric propulsion systems have brought NEP closer to practical implementation.

The ongoing development of space nuclear reactors is a testament to their enduring potential to transform space exploration. With the growing interest in returning to the Moon, sending humans to Mars, and exploring the outer reaches of the solar system, space nuclear reactors are poised to play a critical role in achieving these ambitious goals. As the technological, regulatory, and safety challenges are addressed, space nuclear reactors may soon become a cornerstone of humanity's efforts to explore and utilize the vast expanse of space.

10.1.4 Advantages and Challenges

Advantages:

- High Power Output: SNRs can produce kilowatts to megawatts of power, supporting large-scale and high-energy missions.
- Scalability: Reactor designs can be scaled to match mission requirements.
- Long-Duration Operation: SNRs can provide sustained power for years to decades without refueling.

Challenges:

- Complex Systems: Reactor operation involves intricate control systems and active cooling mechanisms.
- Safety: Ensuring safe operation and handling of fission materials is paramount.
- Thermal Management: Efficiently dissipating waste heat in space is crucial to maintaining reactor stability.

10.2   Thermal Power Generation in Space: Fission and Radioisotope Decay

Space exploration requires power systems that are both reliable and capable of functioning in the harsh environment of space, where traditional energy sources such as solar power may not be sufficient or practical. Among the various methods of generating thermal power in space, two nuclear-based technologies stand out: nuclear fission and radioisotope decay. Both of these methods convert nuclear energy into thermal energy, but they do so through fundamentally different processes, each with its own set of advantages and challenges.

10.2.1 Nuclear Fission

Nuclear fission is a process in which the nucleus of a heavy atom, typically uranium-235 or plutonium-239, is split into smaller nuclei when struck by a neutron.

Fission releases a significant amount of energy in the form of heat, as well as additional neutrons that can sustain a chain reaction. In a controlled environment, such as a nuclear reactor, this heat can be harnessed and converted into electrical power or used directly for propulsion.

In a space nuclear fission reactor, a controlled chain reaction is maintained within a core composed of nuclear fuel rods. Neutrons released during fission initiate further fission reactions in adjacent fuel atoms, maintaining a steady release of heat. This heat is transferred to a working fluid—often a liquid metal like sodium, potassium, or a gas like helium—that circulates through the reactor core. The heated fluid is then used to generate electricity via a power conversion system, such as a Stirling engine or a Brayton cycle turbine, or it can be used directly to heat a propellant in nuclear thermal propulsion systems.

Nuclear fission reactors are designed for missions requiring high power outputs over long durations. They are particularly suitable for powering spacecraft systems, scientific instruments, and habitats on the Moon or Mars, where solar power is either insufficient or unreliable. For example, NASA's Kilopower project aims to develop small fission reactors that can provide power for lunar and Martian bases, offering a scalable and long-term energy solution that operates independently of the sun.

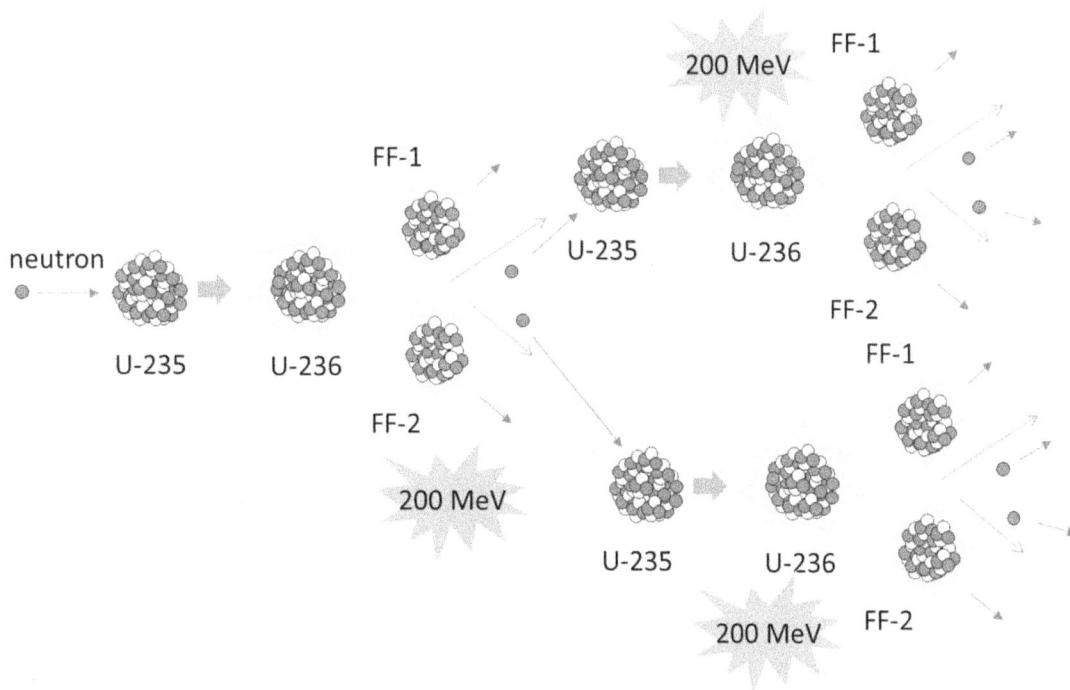

Figure 10-1. Fission chain reaction.

Advantages:

- **High Power Output:** Fission reactors can generate substantial amounts of power, from kilowatts to megawatts, making them ideal for high-demand applications.
- **Long Duration:** Fission reactors can operate for extended periods, potentially years or decades, without refueling, providing a reliable power source for long-term missions.
- **Scalability:** The design of fission reactors can be scaled to meet varying power requirements, from small reactors for individual spacecraft to larger systems for planetary bases.

Challenges:

- **Complexity:** Fission reactors are complex systems that require sophisticated control mechanisms to manage the chain reaction and ensure safety.
- **Radiation Shielding:** The operation of a fission reactor produces significant radiation, necessitating heavy shielding to protect spacecraft systems and crew members.
- **Thermal Management:** Managing the heat generated by the reactor, particularly in the vacuum of space, presents engineering challenges that must be addressed to prevent overheating.

10.2.2 Radioisotope Decay

Radioisotope decay, unlike fission, does not involve a chain reaction. Instead, it relies on the natural decay of radioactive isotopes to generate heat. The most commonly used isotope for space power applications is plutonium-238, which undergoes alpha decay to produce heat. This heat is then converted into electrical power using thermoelectric generators, a technology known as Radioisotope Thermoelectric Generators (RTGs).

In an RTG, the heat generated by the radioactive decay of plutonium-238 is converted directly into electricity using thermocouples—devices that generate voltage when there is a temperature difference between their hot and cold ends. The decay process is continuous and does not require any external control, making RTGs extremely reliable. The power output of an RTG decreases slowly over time as the radioactive material decays, but the half-life of plutonium-238 (about 87.7 years) ensures that RTGs can provide power for several decades.

RTGs are used in missions where long-term, low-maintenance power sources are required, particularly in environments where solar power is impractical. They have powered many of NASA's most notable missions, including the Voyager probes, the Curiosity rover on Mars, and the New Horizons mission to Pluto. RTGs are especially valuable for missions to the outer solar system, where sunlight is too weak to generate sufficient power with solar panels.

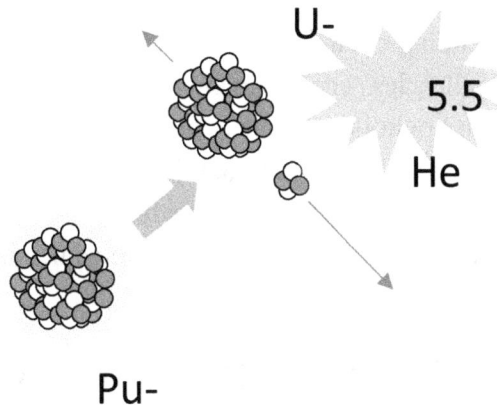

Figure 10-2. Radioactive alpha decay of Pu-238.

Advantages:
- **Simplicity and Reliability:** RTGs have no moving parts and require no active control, making them extremely reliable and capable of operating for decades without maintenance.
- **Independence from Environmental Conditions:** RTGs provide power regardless of the environmental conditions, such as the availability of sunlight, making them ideal for deep-space missions and shadowed regions like craters on the Moon.
- **Compactness:** RTGs are relatively small and lightweight, making them suitable for a variety of spacecraft and rover designs.

Challenges:
- **Low Power Output:** RTGs typically generate less power compared to fission reactors, limiting their use to low-power applications.
- **Radiation Safety:** While RTGs produce much less radiation than fission reactors, the handling and containment of radioactive material still pose safety concerns.
- **Power Decline Over Time:** The power output of an RTG decreases as the radioactive material decays, although this decline is gradual due to the long half-life of plutonium-238.

10.2.3 Comparison of Fission and Radioisotope Decay

While both fission and radioisotope decay processes provide critical thermal power for space missions, they serve different purposes and are suited to different mission profiles. Fission reactors are capable of delivering high power levels, making them suitable for missions with substantial energy needs, such as crewed bases on the Moon or Mars and high-power spacecraft. They offer scalability and long operational lifetimes but require more complex control systems and extensive radiation shielding.

In contrast, radioisotope decay systems like RTGs are simpler and more reliable, with no need for active control, making them ideal for long-duration, low-power missions, especially in remote or challenging environments. However, their lower power output limits their application to smaller, less energy-intensive tasks.

The choice between fission and radioisotope decay depends on the specific requirements of the mission, including power needs, duration, environmental conditions, and safety considerations. Both technologies are essential for enabling the continued exploration and utilization of space, each contributing uniquely to the diverse needs of modern space missions.

10.3   Classification of space nuclear power systems.

The development and deployment of nuclear power systems in space have been driven by the need for reliable, long-duration energy sources capable of functioning in the harsh environment beyond Earth. These systems are essential for missions where solar power is either insufficient, such as in deep-space exploration, or impractical, such as in high-power applications like crewed outposts on the Moon or Mars. Space nuclear power systems can be classified based on their design, power output, and power conversion methods. This section provides an integrated overview of the various design classes of space nuclear reactors, detailing their electric power ranges, types, and power conversion technologies, and exploring their specific mission applications.

10.3.1 1. Radioisotope Power Systems

Radioisotope power systems (RPS) are among the most mature and widely used nuclear power systems for space applications. These systems rely on the natural decay of radioactive isotopes to generate heat, which is then converted into electricity. RPS are known for their reliability and longevity, making them ideal for missions where long-term, maintenance-free power is essential.

Radioisotope Thermoelectric Generators (RTGs) are the simplest form of RPS. They provide electrical power in the range of up to 500 watts of electrical energy (We) using static thermoelectric conversion. RTGs operate based on the Seebeck effect, where the heat generated from the radioactive decay of isotopes like plutonium-238 is converted directly into electricity by thermoelectric materials. Given their lack of moving parts, RTGs are highly reliable and can function for decades without maintenance. This makes them particularly suited for deep-space missions where solar power is impractical. Examples of RTGs in action include the Voyager probes, the Curiosity rover on Mars, and the New Horizons mission to Pluto.

For missions requiring higher power output, **Radioisotope Dynamic Conversion Generators** are employed. These systems operate within an electric power range of 0.5 to 10 kilowatts of electrical energy (kWe) and utilize dynamic power conversion methods such as Brayton, Rankine, or Stirling cycles. These systems offer higher efficiency than RTGs by mechanically converting heat into electricity. For instance, the Brayton cycle uses a gas turbine to convert thermal energy into mechanical energy, which is then used to generate electricity. Similarly, the Rankine cycle operates by vaporizing a working fluid to drive a turbine, while the Stirling cycle uses temperature differences across a working gas to drive a piston. These dynamic systems are more complex but provide a significant boost in power output, making them suitable for applications that require more robust power generation, such as larger scientific instruments or systems supporting more complex missions.

10.3.2 Reactor-Based Power Systems

While radioisotope systems provide reliable low-power outputs, reactor-based power systems are necessary for missions demanding higher energy levels. These reactors, which harness the energy released from nuclear fission, can be classified based on their power output and the type of power conversion systems they employ.

Low-Power Reactor Systems, which generate between 10 kWe to 1,000 kWe, are designed for a variety of applications, from powering spacecraft to supporting crewed outposts on the Moon or Mars. These reactors can utilize both static and dynamic power conversion methods depending on the mission requirements.

Static systems include **Heat Pipe Thermoelectrics** and **Thermionics**. Heat pipe thermoelectric systems use heat pipes to efficiently transfer thermal energy from the reactor core to thermoelectric generators, which then convert this heat into electricity. Thermionics, on the other hand, convert heat into electricity through the thermionic emission of electrons from a hot surface to a cooler collector. These technologies are ideal for missions requiring compact, efficient power sources that can operate reliably over long periods.

Dynamic systems, such as those employing the Brayton, Rankine, or Stirling cycles, are also used in this power range. These systems are more efficient than static converters and are capable of handling higher power levels. The choice of cycle depends on the specific mission requirements, such as the desired power output, operational environment, and efficiency needs.

Medium-Power Reactor Systems are designed to produce between 1 to 10 megawatts of electrical energy (MWe). These systems are essential for larger-scale applications, such as sustaining human habitats on Mars or supporting high-power propulsion systems for interplanetary travel. Medium-power reactors exclusively use dynamic power conversion methods due to the higher power demands.

For instance, a Brayton cycle in a medium-power reactor might involve the use of a gas turbine to convert heat into electricity efficiently. The Rankine cycle, utilizing vaporized working fluids, and the Stirling cycle, which drives pistons with temperature-induced pressure differences, are also employed to maximize power output and efficiency. These systems are crucial for missions that require continuous, high-power energy sources over extended periods.

High-Power Reactor Systems, producing between 10 to 100 MWe, are at the cutting edge of space nuclear power technology. These reactors are designed for the most demanding space missions, where extremely high power levels are required, such as interplanetary spacecraft, asteroid mining operations, or large-scale human settlements on other planets.

These systems use advanced dynamic power conversion technologies to achieve the necessary power levels. For example, a **Brayton Cycle (Open Loop)** system might be used, where the working fluid is expelled after passing through the turbine, suitable for environments where the working fluid can be replenished or where disposal is manageable. **Stirling engines** remain an option for high-efficiency power generation even at these scales, though they are more often used in medium-power applications. Other advanced technologies include **Magnetohydrodynamic (MHD) Generators**, which convert thermal energy directly into electricity by moving an electrically conductive fluid through a magnetic field, and **Fluidized Bed Reactors**, which use a bed of particles to enhance heat transfer. **Gaseous Core Reactors** represent a more experimental approach, where the nuclear fuel is in a gaseous state, allowing for extremely high operating temperatures and power densities.

10.3.3 Applications Across Mission Scenarios

Space nuclear power systems are critical for a wide range of mission scenarios, each with specific power requirements. For instance, **Near-Earth missions** such as non-proliferation and treaty verification, underground measurements, or moveable surface sensors typically require up to 40 kWe and lifetimes of 7 to 10 years. These missions benefit from the reliability and compactness of low-power reactor systems or dynamic radioisotope systems, particularly in environments where rapid deployment and high reliability are essential.

In the realm of **environmental monitoring** and **aviation**, missions such as Earth observations, upper air turbulence monitoring, and anti-collision aircraft radar can demand over 10 kWe. These applications often operate in high elliptical or medium Earth orbits (HEEO/MEO) and require power systems that balance performance with long operational lifetimes.

Commercial applications like the electronic information highway and direct broadcast television necessitate power levels between 25 to 100 kWe. These applications often involve large, continuous power demands over extended periods, making them suitable candidates for reactor-based systems with dynamic power conversion.

For **solar system exploration** missions, such as Neptune or Pluto orbiter/probes, or the Jupiter grand tour, power levels can range from 40 to 100 kWe. These missions require reliable, high-power systems that can operate over long distances and durations, often relying on medium to high-power reactor systems with advanced dynamic power conversion technologies.

Finally, **lunar and Mars exploration** efforts, including the establishment of a lunar outpost, enhanced outpost with in-situ resource utilization (ISRU), or Mars transportation systems, require robust power systems that can provide between 15 to 200 kWe or more. These missions are likely to depend on the most advanced nuclear power systems available, including high-power reactors with efficient dynamic power conversion systems to ensure sustained human presence and resource utilization on other celestial bodies.

Table 10-1. Design classes of reactors for space applications.

Nuclear Power System	Electric Power Range	Type	Power Conversion (Module Size)
Radioisotope Thermoelectric Generator (RTG)	Up to 500 We	Static	Thermoelectric
Radioisotope Dynamic Conversion Generator	0.5 to 10 kWe	Dynamic	Brayton, Rankine, Stirling
Reactor Systems	10 kWe to 1,000 kWe		
- Heat Pipe Thermoelectrics		Static	Heat Pipe Thermoelectrics
- Solid Core Thermionics		Static	Thermionics
- Thermionics		Static	Thermionics
- Brayton		Dynamic	Brayton
- Rankine		Dynamic	Rankine
- Stirling		Dynamic	Stirling
Reactor Systems	1 to 10 MWe		
- Heat Pipe		Dynamic	Brayton, Rankine, Stirling
- Solid Core		Dynamic	Brayton, Rankine, Stirling
Reactor Systems	10 to 100 MWe		
- Brayton Cycle (Open Loop)		Dynamic	Brayton Cycle (Open Loop)
- Solid Core		Dynamic	Stirling
- Pellet Bed		Dynamic	MHD
- Fluidized Bed		Dynamic	Fluidized Bed
- Gaseous Core		Dynamic	Gaseous Core

10.4   Historical Developments

10.4.1 Early Development and SNAP Programs (1957-1967)

The development of space nuclear power reactors began in 1955 with the SNAP (Systems for Nuclear Auxiliary Power) program. The initial design aimed to produce a 3-5 kWe power system using uranium-zirconium-hydride fuel and a mercury-Rankine power conversion system, known as SNAP-2. Two reactors were tested under this program: the SNAP-2 Experimental Reactor (SER) and the SNAP-2 Developmental Reactor (S2DR). The SER achieved criticality in September 1959 and operated until

December 1961, accumulating 1,800 hours at 920 K and 5,300 hours at temperatures above 750 K. The S2DR, tested between April 1961 and December 1962, operated for 2,800 hours above 920 K and 7,700 hours above 750 K. Despite demonstrating significant technological advancements, the SNAP-2 program concluded in 1967.

Table 10-2. Space nuclear system application and power level.

Mission	Power Level	Lifetime	Deployment Mode	Payload Mass (tons)	Power System Mass (tons)	Key Requirements
Near-Earth						
Non-Proliferation and Treaty Verification	To 40 kWe	7 to 10 y				
Underground measurements	To 40 kWe	7 to 10 y				
Moveable surface sensors	To 40 kWe	7 to 10 y				
Chemical, biological, nuclear (CBN) effluent monitoring	To 40 kWe	7 to 10 y				
Environmental Monitoring						
Earth observations	>10 kWe	7 to 10 y				
Upper air turbulence	>10 kWe	7 to 10 y				
Aviation						
Anti-collision aircraft radar	20 to 40 kWe	7 to 10 y	HEEO/MEO			High elliptical/medium Earth orbit
Commercial						

Mission	Power Level	Lifetime	Deployment Mode	Payload Mass (tons)	Power System Mass (tons)	Key Requirements
Electronic information highway	25 to 100 kWe					
Direct broadcast television	25 to 100 kWe					
Near-Earth Defense						
Wide area surveillance	20 to 40 kWe	7 to 10 y	Rapid deployment, HEEO/MEO			
Battlefield communications	10 to 20 kWe	7 to 10 y	Rapid deployment, GEO			Geosynchronous Earth orbit
Electronic jammers	>10 kWe	7 to 10 y	Rapid deployment, HEEO/MEO			
Submarine communications	10 to 40 kWe	7 to 10 y	Rapid deployment, HEEO			
Near-Earth Resources						
In situ probes	30 kWe	3 y		1.5		
Solar System Exploration						
Neptune orbiter/probe	100 kWe			1.8	3.7	
Pluto orbiter/probe	56 kWe			1.4	2.8	
Uranus orbiter	100 kWe			1.4	3.7	
Jupiter grand tour	58 kWe			1.4	2.9	
Rendezvous	40 kWe			1.4	2.35	

Mission	Power Level	Lifetime	Deployment Mode	Payload Mass (tons)	Power System Mass (tons)	Key Requirements
Comet Sample/Return	100 kWe			1.8	3.7	
Lunar-Mars Exploration						
First lunar outpost	>12 kWe					
Enhanced outpost (ISRU)	>200 kWe					
Mars transportation				52		Flight time <180 d
Mars stationary (600 d)	15-150 kWe					
Mars in situ resources	>200 kWe					
Mars comsats	20 kWe					

The SNAP-10A program, initiated in 1958, built upon the SNAP-2 reactor design but incorporated a thermoelectric power conversion subsystem. This effort led to the SNAPSHOT mission, which launched the first reactor power system into space on April 3, 1965. The reactor achieved a nominal orbit and operated at a power level between 530-660 watts. Although a spacecraft malfunction terminated the reactor's operation after 43 days, the mission proved the feasibility of nuclear reactor power systems in space.

10.4.2 SNAP-8 and SNAP-50/SPUR Programs (1960-1973)

The SNAP-8 program, initiated in 1960, aimed to develop a 30-60 kWe system. The program utilized a modified SNAP-2 core with 211 fuel-moderated elements and a mercury Rankine power conversion system. Two reactors were ground tested: the SNAP-8 Experimental Reactor (S8ER) and the SNAP-8 Development Reactor (S8DR). The S8ER, tested from May 1963 to April 1965, demonstrated an outlet temperature exceeding 974 K but experienced significant fuel cladding cracking. The S8DR, a prototype flight system, operated from January to December 1969. Despite improvements, the program faced persistent issues with fuel cladding integrity, leading to its cancellation in 1973.

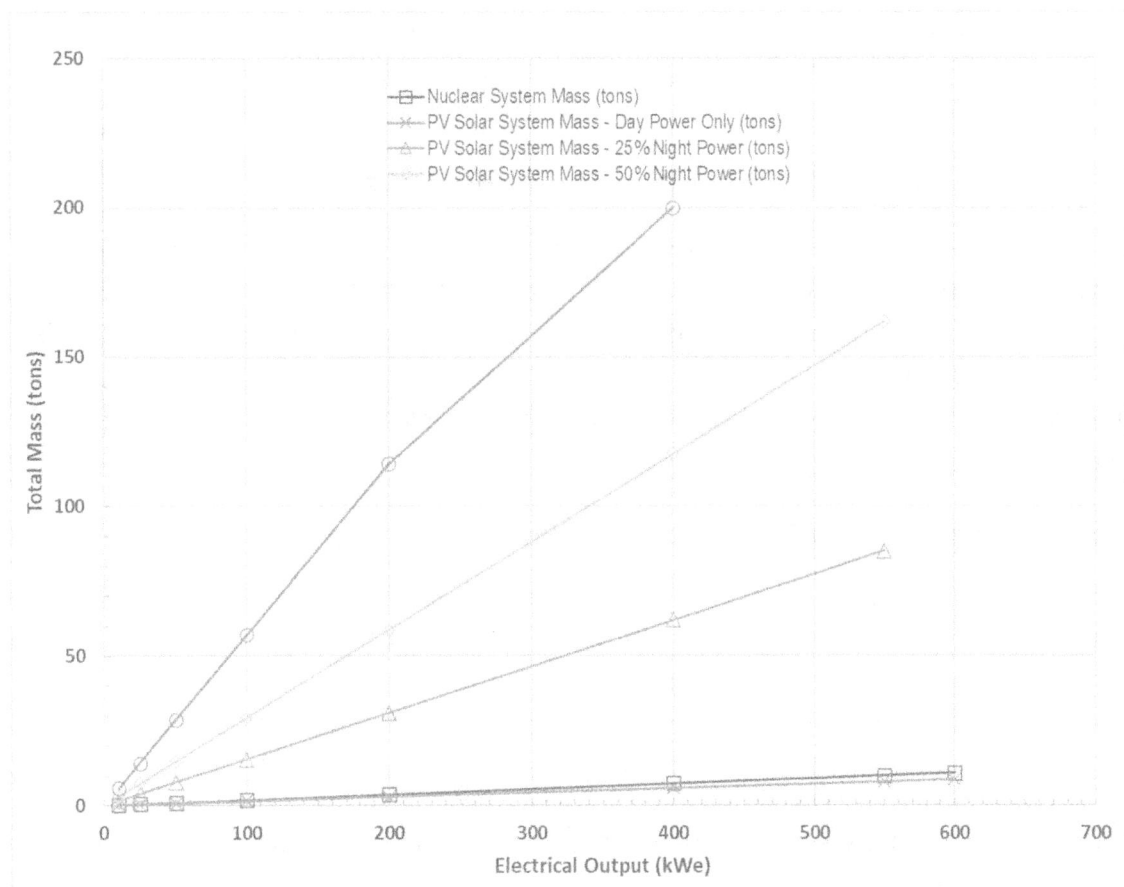

Figure 10-3. Comparison of solar and nuclear fission power for lunar base.

Concurrently, the SNAP-50/SPUR (Space Power Unit Reactor) program, initially part of the Aircraft Nuclear Propulsion Program (ANP), aimed to develop systems ranging from 300 to 1,000 kWe. This program focused on a 1,363 K fast, lithium-cooled, refractory alloy reactor coupled with a potassium Rankine power conversion system. The program's scope included reactor and system conceptual design, nuclear criticality mockups, and fuel materials testing. It was terminated in 1968.

10.4.3 Medium Power Reactor Experiment and 710-Gas Cooled Reactor (1959-1968)

The Medium Power Reactor Experiment (MPRE) began in 1959 to address power needs up to 150 kWe. The reactor design involved boiling potassium in the core and using the vapor in a Rankine cycle turbine. The project included designing and operating various component and system test rigs. Funding ceased in 1966, reallocating resources to other projects.

In 1965, the 710-Gas Cooled Reactor Program commenced, leveraging ANP-developed technologies to design a 200 kWe system capable of 10,000 hours of operation. The design featured a fast reactor using refractory metal-fuel cermets and inert gas cooling, coupled with a Brayton power conversion system. The program was reduced to a fuel element development project in 1968 before its eventual discontinuation.

10.4.4 Thermionic Reactor Development (1959-1973)

The thermionic reactor program, initiated in 1959, explored the potential of reactors with minimal moving parts, high redundancy, high efficiency, small radiator area, low specific mass, and broad power range capability. Although progress was made in high-temperature metallurgy and thermionic components, challenges such as diode lifetimes, emitter cracking, and fuel-clad interactions persisted, leading to the program's end in 1973.

10.4.5 Reactor Safety Testing (1960s-1970s)

The SNAPTRAN series (1, 2, and 3) of reactor safety tests included small transients and destructive excursions of full-scale reactors to simulate scenarios like water flooding and impact. These tests provided valuable insights into reactor disassembly and fuel element burnup during reentry, contributing significantly to the understanding of reactor safety in space missions.

10.4.6 Space Power Advance Reactor (SPAR) and SP-100 (1979-1993)

In 1979, the Space Power Advance Reactor (SPAR) program began modestly, focusing on developing high-temperature heat pipes. In 1983, this effort transitioned to the SP-100 program, a joint initiative between NASA, DOE, and DoD, aiming to provide tens-to-hundreds of kWe of electric power. The program initially considered a wide range of concepts, including liquid-metal, gas-cooled, thermionic, and heat pipe reactors, coupled with various energy conversion systems such as thermoelectric, thermionic, Brayton, Rankine, and Stirling.

By 1985, a high-temperature pin-fuel element reactor with thermoelectric conversion was selected for development. The program also pursued thermionic fuel element verification and high-temperature Stirling engine development. Despite significant progress in component design and testing, the SP-100 program was discontinued in 1993 due to funding constraints and shifting priorities.

10.4.7 Multimegawatt Program (1985-1990)

During the late 1980s, the Multimegawatt Program explored power systems supporting directed energy weapons and electric propulsion. The program investigated open-loop systems for tens-of-megawatts for hundreds of seconds, closed-loop systems for tens-of-megawatts for one year, and open-loop systems for hundreds-of-megawatts for hundreds of seconds. Despite identifying key technology feasibility issues, the program was terminated in 1990 before Phase II design contracts were awarded.

10.4.8 Project Prometheus and Fission Surface Power (2003-2013)

Project Prometheus, established in 2003, aimed to develop nuclear reactor-powered electrical propulsion for deep-space missions, with the Jupiter Icy Moons Orbiter (JIMO) as the initial application. The selected concept featured a gas-cooled reactor coupled with a Brayton energy conversion system. However, the program was canceled in 2006 due to complexity and budget concerns.

In 2007, the NASA Exploration Technology Development Program initiated the Fission Surface Power (FSP) project to develop viable options for lunar outposts and Mars missions. The project focused on fast-spectrum, uranium dioxide (UO_2) fuel pins with stainless steel cladding and structure, exploring both Stirling and Brayton cycle heat engines. The goal was to establish viable options by 2013, culminating in a flight power system.

10.4.9 Russian Space Nuclear Power Programs (1959-Present)

Russia's space nuclear power program, active since 1959, has launched approximately 33 thermoelectric reactor systems, primarily for ocean surveillance radar satellites (RORSAT). These systems, powered by fast reactors using SiGe thermoelectric conversion, provided power levels ranging from several hundred watts to a few kilowatts. Notable incidents include the reentry of Cosmos 954 over Canada in 1978, which resulted in reactor burnup, and the reboost failures of Cosmos 1402 and Cosmos 1900.

The Topaz program marked another significant Russian development, utilizing thermionic power conversion. The Topaz I reactors, launched as Cosmos 1818 and 1867, demonstrated the use of multi-cell thermionic fuel elements. These reactors operated for six months and 346 days, respectively, showcasing advancements in thermionic technology despite some performance degradation.

10.5   Heat Pipes

Heat pipes are passive heat transfer devices that can transport large amounts of thermal energy efficiently over relatively long distances with minimal temperature drop. First developed in the 1960s as part of the space nuclear program, heat pipes have since found wide-ranging applications in thermal management for electronics, spacecraft, and other industries requiring effective heat transfer solutions. The high effective thermal conductivity of heat pipes, which can be orders of magnitude greater than even the most conductive metals, makes them an attractive option for many thermal control needs.

10.5.1 Operating Principles

At its core, a heat pipe consists of a sealed container with a working fluid and wick structure inside. The container is typically made of a highly conductive metal like copper or aluminum. The working fluid is chosen based on the operating temperature range, with options including water, ammonia, methanol, or liquid metals for high-temperature applications. The wick structure lines the inner walls of the container and provides the capillary pumping action to circulate the working fluid.

The heat pipe has three main sections:
1. Evaporator - Where heat is absorbed and the working fluid vaporizes
2. Adiabatic section - Where vapor travels from evaporator to condenser
3. Condenser - Where heat is removed and vapor condenses back to liquid

HEAT PIPE

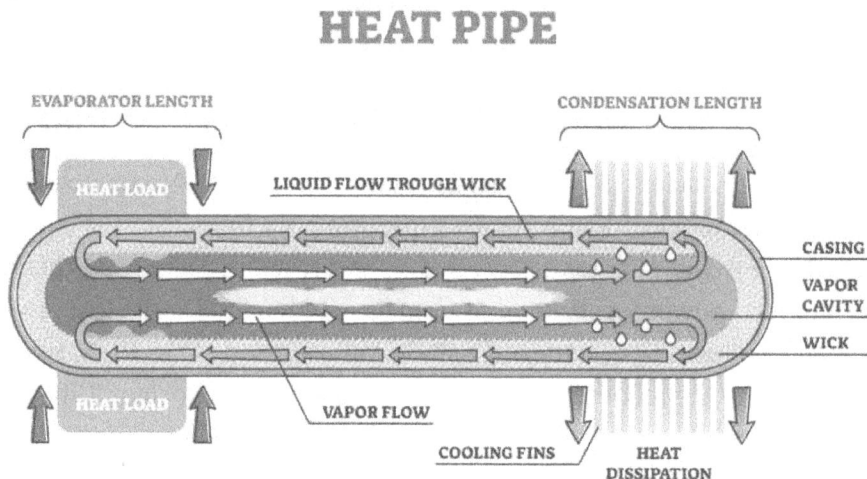

Figure 10-4. Basic heat pipe principal of operation.

As heat is applied to the evaporator, the working fluid vaporizes, absorbing the latent heat of vaporization. The vapor then travels through the adiabatic section to the condenser, where it releases the latent heat as it condenses back to liquid. The wick structure then pumps the liquid back to the evaporator through capillary action, completing the cycle.

This two-phase heat transfer process utilizing the latent heat of vaporization allows heat pipes to transfer large amounts of heat with a very small temperature gradient. The effective thermal conductivity can be 100 to 1000 times greater than solid copper.

10.5.2 Key Components and Design Considerations

Container - Must be leak-proof, maintain structural integrity, and be compatible with the working fluid and wick. Common materials include copper, aluminum, stainless steel, and superalloys.

Working Fluid - Selected based on operating temperature range, compatibility, and thermophysical properties. Water is common for 30-200°C, while liquid metals like sodium are used for high temperatures.

Wick Structure - Provides capillary pumping and fluid distribution. Options include sintered metal powders, wire mesh screens, axial grooves, and composite wicks. The wick must generate sufficient capillary pressure to overcome pressure losses.

Other design factors include:
- Heat pipe size and shape
- Heat flux limits (capillary, sonic, entrainment, boiling)
- Thermal resistance
- Startup behavior
- Gravity effects

10.5.3 Types of Heat Pipes

- Standard Heat Pipe - Simplest design with uniform wick structure. Used for isothermal operations.
- Variable Conductance Heat Pipe (VCHP) - Uses non-condensable gas to provide temperature control.
- Diode Heat Pipe - Allows heat transfer in one direction only.
- Rotating Heat Pipe - Uses centrifugal forces instead of capillary action for fluid return.
- Loop Heat Pipe - Separates vapor and liquid flow paths for reduced flow resistance.
- Pulsating Heat Pipe - Oscillating slug flow provides heat transfer without a wick.

10.5.4 Applications

Electronics Cooling - Heat pipes are widely used to cool CPUs, GPUs, and other high-power electronic components in computers, smartphones, and other devices. Their ability to spread heat and transport it to heat sinks makes them ideal for compact electronics.

Spacecraft Thermal Control - Heat pipes help maintain proper temperature ranges for spacecraft systems and payloads by transferring heat from electronics to radiators. They are lightweight and reliable for long-duration missions.

Industrial Applications - Used in heat exchangers, heat recovery systems, and thermal storage units. Help improve energy efficiency in various processes.

LED Lighting - Manage heat in high-power LED lights to maintain efficiency and lifespan.

Solar Thermal - Transport heat from solar collectors to storage tanks or power generation systems.

HVAC - Used in energy recovery ventilators and for cooling telecommunications equipment.

10.5.5 Advantages and Limitations

10.5.5.1 Advantages:

- Very high effective thermal conductivity
- Passive operation with no moving parts
- Can transport heat over long distances
- Isothermal operation
- Flexible geometry options
- Reliable and long operating life

10.5.5.2 Limitations:

- Temperature range limited by working fluid properties
- Potential for fluid leakage over time
- Orientation sensitivity in gravity environments
- Heat flux limits based on capillary pumping capability
- Startup issues possible in some conditions

10.5.6 Recent Developments and Future Outlook

Recent research has focused on expanding the capabilities and applications of heat pipes:

- Micro heat pipes for electronics cooling
- Flexible/bendable heat pipes
- Novel wick structures like carbon nanotubes
- Improved modeling and simulation tools
- Integration with thermoelectric generators
- Additive manufacturing techniques for heat pipe production

As electronics continue to increase in power density and new thermal management challenges arise in various industries, heat pipes are likely to see expanded use. Their passive operation, high performance, and versatility make them an attractive solution for many heat transfer applications. Continued improvements in materials, manufacturing, and design are expected to further enhance heat pipe capabilities and open up new potential uses.

Heat pipes have come a long way since their initial development for space applications in the 1960s. Their ability to efficiently transfer large amounts of heat with minimal temperature gradients has made them invaluable in electronics cooling, spacecraft thermal control, and various industrial applications. As a passive device with no moving parts, heat pipes offer reliable performance and long operating life. While they do have some limitations in terms of operating conditions and heat transport limits, ongoing research continues to expand their capabilities. Heat pipes are likely to remain an important thermal management technology, helping address heat transfer challenges across many industries for years to come.

10.6   Radiation Attenuation Shielding in Spacecraft

Radiation attenuation shielding is a critical component in the design of spacecraft, especially those equipped with nuclear power systems. The primary function of shielding is to protect the astronaut crew, sensitive spacecraft equipment, and payload from harmful radiation emitted by the nuclear reactor and other sources. The adverse effects of radiation can be immediate or long-term, impacting both biological organisms and material components. We delve into the various aspects of radiation shielding, including the types of radiation, shielding materials, design considerations, and the technical challenges involved in developing effective shielding systems for space missions.

10.6.1 Types of Radiation and Their Effects

Spacecraft operating in space are exposed to various forms of radiation, including galactic cosmic rays, solar particle events, and radiation from onboard nuclear reactors. The primary concerns in shielding design are neutrons and gamma rays, which are emitted from nuclear reactions and radioactive decay.

10.6.1.1 Neutron Radiation

Neutrons are neutral particles that can penetrate deeply into materials, causing nuclear reactions that result in secondary radiation. Neutron radiation is particularly challenging to shield against because it can activate materials, making them radioactive.

10.6.1.2 Gamma Radiation

Gamma rays are high-energy photons that are highly penetrating and can cause ionization and electronic upsets in sensitive equipment. They pose a significant health risk to astronauts and can damage materials and electronic systems.

10.6.2 Radiation Damage Mechanisms

The damage caused by radiation can be broadly categorized into two effects: rate effects and cumulative effects.

10.6.2.1 Rate Effects

Rate effects occur when materials or electronic components are exposed to a high dose rate of radiation over a short period. This can lead to immediate damage, such as noise, electronic upsets, or even burnout of electronic circuits.

10.6.2.2 Cumulative Effects

Cumulative effects result from prolonged exposure to lower levels of radiation, leading to the gradual accumulation of defects in materials. This can degrade their mechanical, thermal, and electrical properties over time, reducing the overall lifespan of the materials and components.

10.6.3 Shielding Requirements and Considerations

Designing effective radiation shielding for space missions involves balancing several factors, including the type of mission, the specific radiation sources, and the required level of protection.

10.6.3.1 Mission-Specific Requirements

The shielding requirements can vary significantly depending on whether the mission is manned or unmanned. Unmanned missions generally have less stringent radiation protection criteria, allowing for lighter and more mass-efficient shielding designs. For manned missions, the protection of human life takes precedence, necessitating more robust shielding solutions.

10.6.3.2 Radiation Exposure Limits

The acceptable radiation exposure levels for crew and sensitive equipment must be defined early in the design process. For instance, the U.S. standard for occupational radiation exposure limits is approximately 5 rem/year for workers, with lower limits for specific organs and tissues.

10.6.4 Shield Configuration and Spacecraft Design

The configuration of the spacecraft plays a crucial role in the design of the shielding system. A common design practice is to place the nuclear reactor and sensitive components at opposite ends of the spacecraft, using an elongated structure to maximize distance and reduce radiation exposure through the inverse square law. Additionally, the "shadow shield" concept is often employed, wherein the shield is designed to protect only the critical areas, allowing radiation to escape in other directions where it poses less risk.

10.6.5 Shielding Materials

The choice of shielding materials is critical for effective radiation protection. Materials must be selected based on their ability to attenuate radiation, mechanical properties, thermal stability, and other factors.

10.6.5.1 Neutron Shielding Materials

Materials rich in light elements, particularly hydrogen, are effective at attenuating neutrons. Hydrogen-containing materials are chosen due to their high neutron cross-section, which makes them efficient at slowing down and absorbing neutrons. Lithium hydride (LiH) is a preferred material for neutron shielding due to its high hydrogen density, low mass, and high melting point. However, natural lithium contains a small fraction of ^6Li, which has a high neutron absorption cross-section, leading to the production of tritium (^3H) and helium (^4He). This can result in pressure buildup within the shielding material, necessitating the use of enriched ^7Li in high neutron flux regions.

10.6.5.2 Gamma Shielding Materials

High-density materials are required to shield against gamma radiation due to their greater mass attenuation coefficients. Materials such as tungsten, lead, and depleted uranium are commonly used for gamma shielding. These materials can absorb gamma rays effectively, reducing the radiation dose to acceptable levels. However, their high density also means they contribute significantly to the overall mass of the spacecraft.

10.6.6 Material Considerations and Challenges

The selection of shielding materials is influenced by several factors, including:
- **Thermal Stability**: Materials must maintain their structural integrity at elevated temperatures, often in the range of 700-900 K.
- **Mechanical Properties**: High compressive strength and resistance to radiation-induced degradation are essential for long-term reliability.
- **Fabrication and Availability**: Materials should be easy to fabricate into the required shapes and available in sufficient quantities.

10.6.7 Shielding Design and Optimization

The design of radiation shields involves iterative processes that balance various factors, including radiation attenuation, mass, thermal management, and mechanical stability.

10.6.7.1 Shadow Shields

Shadow shields are designed to protect specific areas of the spacecraft, typically the crew and sensitive equipment, while allowing radiation to escape in less critical directions. This design minimizes the mass of the shield while providing adequate protection. The shape of the shadow shield is often conical or frustum-shaped, with the base of the cone facing the radiation source. This design takes advantage of the inverse square law, which states that radiation intensity decreases with the square of the distance from the source.

10.6.7.2 Full-Encapsulation Shields

In missions where human presence or sensitive equipment is involved, a full-encapsulation or "four-pi" shield may be used. This design provides uniform protection in all directions, eliminating the need to orient the spacecraft in a specific way to avoid radiation exposure. However, the mass of a four-pi shield is significantly higher than that of a shadow shield, making it less desirable for applications where mass is a critical constraint.

10.6.7.3 Hybrid Shields

In some cases, a combination of shielding configurations may be employed. For example, a spacecraft might use a shadow shield for general protection and additional localized shielding for particularly sensitive components. This approach allows for mass optimization while ensuring adequate protection.

10.6.8 Advanced Shielding Technologies

Recent advancements in material science and engineering have led to the development of novel shielding materials and technologies that offer improved performance and reduced mass.

10.6.8.1 Borated Polymers

Borated polymers are materials that incorporate boron, which has a high neutron absorption cross-section. These materials are used to enhance neutron shielding capabilities while maintaining low mass and good mechanical properties.

10.6.8.2 Metal Matrix Composites

Metal matrix composites (MMCs) combine metals with ceramic or polymer phases to create materials with superior mechanical properties and radiation resistance. MMCs can be engineered to have high thermal conductivity, making them suitable for high-temperature applications.

10.6.8.3 Layered and Graded Shields

Layered shielding involves using multiple layers of different materials to optimize radiation attenuation. For example, a shield might consist of a layer of hydrogen-rich material for neutron attenuation, followed by a layer of high-density material for gamma attenuation. Graded shields use materials with gradually varying compositions to achieve a smoother transition in radiation attenuation, reducing the potential for secondary radiation generation.

Radiation shielding is a critical aspect of spacecraft design, particularly for missions involving nuclear power systems. The choice of shielding materials, configuration, and design strategies must consider the specific mission requirements, radiation sources, and operational conditions. Advanced materials and technologies offer promising solutions for reducing shield mass while maintaining high levels of radiation protection. As space missions become more ambitious, the challenges of radiation shielding will continue to drive innovation and development in this crucial field.

10.7 Space Reactor Design Requirements

Designing a nuclear power system for space applications involves an intricate balance of numerous technical and operational requirements. These systems must cater to diverse mission needs, ensuring adequate power generation, long-term reliability, and safety under a wide range of environmental conditions. Additionally, considerations such as size, mass, and survivability in hostile environments are critical. This section provides an in-depth analysis of the design requirements for space reactors, emphasizing the various factors that influence their development and deployment.

10.7.1 Mission Requirements

The mission requirements for a space nuclear power system are foundational to its design. These requirements typically include power levels, operational lifetime, reliability, and specific environmental tolerances.

10.7.1.1 Power Levels and Lifetime

The power output of a space nuclear reactor must be sufficient to meet the energy demands of the mission, which can vary significantly depending on the spacecraft's payload, propulsion systems, and onboard instruments. Historically, space reactors like SNAP-2 and SNAP-8 were designed to deliver 3 kWe and 35 kWe, respectively, with operational lifetimes ranging from one to several years. More recent projects, such as the Prometheus Program, targeted much higher outputs of up to 200 kWe for durations extending to 20 years.

10.7.1.2 Reliability and Redundancy

Reliability is paramount in space missions, where maintenance opportunities are limited or nonexistent. Design requirements often stipulate high reliability metrics, typically exceeding 99%. To achieve this, systems must be designed to avoid single points of failure. For example, thermoelectric converters are preferred for their inherent redundancy, while systems relying on Brayton cycles may require additional components like valves and sensors to detect and isolate failures. Automated recovery mechanisms are often necessary for missions beyond immediate communication range from Earth, ensuring that the reactor can respond autonomously to unexpected conditions.

10.7.1.3 Radiation Levels and Shielding

Radiation from the reactor poses a risk to both the spacecraft and its occupants. The design must include adequate shielding to protect sensitive electronics and crew. For example, the SP-100 program required neutron fluences not exceeding 1×10^{13} n/cm^2 and gamma doses below 5×10^5 rads Si. Shielding strategies often involve a combination of material shielding and physical separation of the reactor from other components. The mass and configuration of the shielding are critical considerations, as illustrated by the use of shadow shields or full encapsulation shields depending on the mission's specific radiation protection needs.

10.7.2 Space and Surface Environmental Considerations

Space reactors must operate reliably under the harsh conditions of space and, in some cases, planetary surfaces. These environments impose additional design constraints.

10.7.2.1 Space Environment

In space, reactors must operate in a vacuum with temperatures near absolute zero. This requires efficient thermal management systems to dissipate waste heat, typically through radiators. The reactor must also be protected from micrometeoroids and orbital debris, which can damage pressurized components and moving parts. The design must account for radiation from solar, galactic, and planetary sources, ensuring that all critical systems remain within safe operational limits.

10.7.2.2 Planetary Surface Environment

For missions to the Moon or Mars, reactors must contend with unique environmental conditions. The Moon's surface experiences extreme temperature variations and low gravity, while its thin atmosphere offers negligible protection against radiation and micrometeorite impacts. Lunar soil, rich in elements like sulfur and iron, may also be used for in-situ shielding. Mars, with its colder climate and thin CO_2-dominated atmosphere, presents different challenges. The design must account for temperature fluctuations, dust storms, and potential chemical interactions with Martian soil.

10.7.3 Safety and Survivability

Safety considerations are integral to the design of space reactors, encompassing criticality control, heat removal, and structural integrity under various scenarios.

10.7.3.1 Criticality Safety

A fundamental safety requirement is preventing reactor criticality until the spacecraft reaches its operational orbit. This involves design features such as multiple independent shutdown mechanisms, a significant negative reactivity coefficient, and the ability to remain subcritical in the event of water immersion or land impact. These measures ensure that the reactor remains safe during all phases of the mission, including launch, orbit insertion, and in-orbit operations.

10.7.3.2 Heat Removal and Thermal Management

Post-shutdown heat removal is another critical safety concern. The reactor design must include passive or active cooling systems capable of dissipating decay heat. This is particularly challenging in the vacuum of space, where conduction and convection are ineffective, leaving radiation as the primary mode of heat transfer.

10.7.3.3 Structural Integrity and Survivability

The reactor and its associated systems must be robust enough to withstand various environmental stresses, including launch vibrations, space debris impacts, and potential collision scenarios. For military or dual-use missions, additional survivability requirements may be imposed to ensure the system remains operational or safely shuts down in the face of hostile actions or extreme conditions.

10.7.4 Operational Requirements

Operational requirements cover the full lifecycle of the reactor system, from pre-launch preparations to end-of-mission decommissioning.

10.7.4.1 Pre-launch and Launch Considerations

Before launch, the reactor must be safely integrated into the spacecraft and subjected to various tests to ensure it is ready for flight. These tests may include subcriticality checks, system integration tests, and functional verification of all safety and control mechanisms. The reactor design must also comply with transportation and handling requirements, ensuring safe transit from the assembly site to the launch pad.

10.7.4.2 In-orbit Operations

Once in orbit, the reactor must be capable of autonomous startup, power regulation, and shutdown. The power system's control architecture must be capable of managing load variations and responding to commands from mission control or onboard systems. For missions involving long communication delays, the reactor must operate with a high degree of autonomy, handling fault conditions and maintaining safe operations without immediate ground intervention.

10.7.5 Development and Testing

The development of space nuclear power systems involves rigorous testing and validation to meet the stringent requirements of space missions.

10.7.5.1 Testing and Verification

Testing is critical to validate the reactor's performance, reliability, and safety features. This includes ground-based tests to simulate various operational conditions, including thermal, vibrational, and radiation environments. Where practical, component-level and system-level tests are conducted to verify the functionality and integrity of the entire system.

10.7.5.2 Risk Management

Development risk is a key consideration, influencing the choice of technologies and design approaches. The design process must account for potential technical challenges and uncertainties, incorporating redundancy and safety margins to mitigate risks. The selection of mature and proven technologies is often preferred to reduce development time and increase the likelihood of mission success.

10.8   Design Requirements Derived from Launch Vehicles

Nuclear space reactors, unlike their terrestrial counterparts, are bound by the constraints imposed by space travel. The necessity to integrate these reactors into spacecraft and launch them into space presents unique challenges. The design requirements for space nuclear reactors must consider factors such as size, mass, and environmental resilience, all of which are influenced by the capabilities and limitations of launch vehicles. Here, we explore the specific design requirements derived from these launch vehicle constraints, focusing on the implications for reactor size, mass, thermal management, and structural integrity.

10.8.1 Size and Mass Constraints

The most significant difference between space nuclear reactors and terrestrial reactors lies in their size and mass restrictions. Launch vehicles, such as the proposed

Ares I and Ares V, impose strict limits on the payload dimensions and weight. These constraints necessitate the development of compact and lightweight nuclear power systems.

10.8.1.1 Volume and Mass Restrictions

The Ares I and Ares V launch vehicles are designed to carry substantial payloads to low Earth orbit (LEO) and beyond. The Ares I, capable of launching approximately 21,820 kg to the International Space Station and up to 23,640 kg to LEO, is primarily intended for crew and small cargo missions. The Ares V, on the other hand, can carry nearly 188,200 kg to LEO and, in conjunction with Ares I, can send 71,400 kg to the Moon. Despite these impressive capacities, the need for cost-effectiveness in launching nuclear power systems for missions in the tens of kilowatt range may necessitate smaller and less expensive launchers.

To accommodate these restrictions, space nuclear reactors must be designed to be as small and lightweight as possible. This involves using highly enriched uranium to reduce the reactor core size and minimize the amount of shielding required. The use of fast-spectrum reactors, which do not require a moderator, also contributes to reducing the critical radius and overall reactor size, thus lowering the shield mass. The compact design is crucial not only for fitting within the payload fairing but also for minimizing launch costs and maximizing mission flexibility.

10.8.2 Thermal Management and High-Temperature Operation

One of the primary challenges in space nuclear reactor design is the management of waste heat. Unlike terrestrial systems, which can dissipate heat through convection and conduction, space reactors must rely on radiation to dispose of excess thermal energy. This limitation necessitates the use of high-temperature heat rejection systems to maintain the system's thermal balance.

10.8.2.1 High-Temperature Heat Rejection Systems

The efficiency of the power conversion system is a critical factor in determining the size and mass of the heat rejection system. Since heat rejection scales linearly with power output and inversely with temperature to the fourth power, higher operating temperatures significantly reduce the size of the radiators needed. This drives the need for reactors to operate at much higher temperatures than terrestrial reactors, often requiring advanced materials and fuels capable of withstanding these conditions.

Materials such as refractory metals are considered for reactor cores due to their high melting points and excellent thermal properties. However, these materials are often expensive and less well-characterized than those used in conventional reactors. The use of such materials also imposes challenges in terms of fabrication, handling, and operational stability over extended periods.

10.8.3 Structural Integrity and Vibration Resistance

The structural design of space nuclear reactors must ensure they can withstand the mechanical stresses of launch and space operations. The vibrational forces encountered during launch, including those from rocket engines and atmospheric turbulence, can be significant. Additionally, once in space, the power system must endure the rigors of space environments, such as micrometeoroid impacts and thermal cycling.

10.8.3.1 Launch Vibrations and Acoustic Loads

Launch vehicles subject their payloads to a variety of vibrational modes and acoustic loads. These forces can induce mechanical vibrations that may affect the structural integrity of the nuclear reactor and its associated components. The design must include robust structural support systems and damping mechanisms to mitigate these effects. Components such as fuel elements, control mechanisms, and shielding must be secured and supported to prevent damage or displacement during launch.

10.8.3.2 Space Environmental Considerations

Once in space, the reactor must operate in a vacuum, which poses unique challenges for thermal management and material selection. The lack of convective cooling necessitates the use of radiative heat transfer systems, and materials must be chosen for their stability in vacuum conditions. Furthermore, the reactor must be designed to withstand potential impacts from micrometeoroids and debris, which could compromise the system's integrity and safety.

10.8.4 Safety Considerations

Safety is paramount in the design of space nuclear reactors, especially considering the potential consequences of a failure during launch or in orbit. The design must incorporate multiple independent safety systems to prevent criticality accidents and ensure safe shutdown and heat removal.

10.8.4.1 Criticality Control and Safety Mechanisms

To prevent accidental criticality, the reactor design must include features such as negative reactivity coefficients and multiple independent shutdown mechanisms. For instance, control drums or rods can be used to regulate the reactor's reactivity, and the system must ensure subcriticality in all credible accident scenarios, such as water immersion or land impact.

10.8.4.2 Containment and Radiological Protection

Containment of radioactive materials is crucial to prevent environmental contamination. The design must ensure that, in the event of a launch failure or impact, radioactive materials remain securely contained. This involves designing robust containment systems capable of withstanding high impacts and ensuring that any potential releases are minimized.

10.8.5 Development and Testing

Given the unique challenges associated with launching and operating nuclear reactors in space, extensive testing and validation are required. This includes testing for thermal performance, structural integrity, and safety under simulated launch and space conditions.

10.8.5.1 Ground Testing and Validation

Ground testing is essential to verify the reactor's design and ensure it meets all safety and performance requirements. This includes thermal testing to validate the heat rejection systems, vibration testing to simulate launch conditions, and criticality testing to ensure safe operation. These tests must be conducted under controlled conditions to simulate the expected operational environment as closely as possible.

10.8.5.2 In-Flight Testing and Monitoring

In-flight testing and monitoring are also critical for verifying the reactor's performance once deployed. This includes monitoring power output, temperature, radiation levels, and structural integrity. Real-time data collection and analysis allow for immediate response to any anomalies, ensuring the reactor operates safely and effectively throughout its mission.

10.9   Space Reactor Safety

Safety is a paramount concern in the design and operation of space nuclear reactors. Unlike terrestrial reactors, space reactors operate in a unique environment, facing challenges related to launch, space operation, and potential re-entry scenarios. The design philosophy for space reactors incorporates multiple safety strategies, including the principles of "Confine and Contain," "Delay and Decay," and "Dilute and Disperse." These principles are implemented through both hardware design and operational protocols to minimize the risks associated with radiation exposure and criticality accidents. This section provides an in-depth examination of the safety design requirements for space reactors, covering the strategies and methodologies used to ensure safe operation throughout all phases of a mission.

10.9.1 Confine and Contain

The primary safety strategy, "Confine and Contain," involves isolating radioactive materials from the environment and population. This approach is crucial in space applications, where the containment of radioactive materials is primarily achieved through distance and physical barriers.

10.9.1.1 Physical Barriers

In terrestrial reactors, large containment vessels and multiple physical barriers isolate radioactive materials. For space reactors, similar containment principles are applied, albeit with modifications to accommodate the unique challenges of space travel. The reactor core is typically encased in a robust pressure vessel designed to withstand mechanical stresses during launch and operation. This vessel, combined with shielding materials, serves to contain radiation and prevent its escape into the surrounding environment.

10.9.1.2 Distance as a Barrier

Space reactors benefit from the vast distances involved in space missions, which naturally reduce the risk of radiation exposure. The inverse square law of radiation attenuation ensures that radiation levels decrease rapidly with distance from the source. By placing reactors at a considerable distance from other spacecraft components or personnel, designers can significantly reduce radiation exposure. This approach is particularly effective in unmanned missions, where minimal shielding may be used in conjunction with strategic placement to ensure safety.

10.9.2 Delay and Decay

The "Delay and Decay" principle involves designing the reactor system to allow sufficient time for radioactive materials to decay to safe levels before any potential exposure. This strategy is particularly relevant in scenarios involving potential re-entry or disposal of the reactor at the end of its operational life.

10.9.2.1 Long-Life Orbits

One of the most effective implementations of this strategy is placing decommissioned reactors in long-life orbits, where they can remain for extended periods, allowing radioactive decay to reduce the hazard. For instance, placing reactors in orbits with lifetimes of 300 years or more ensures that most short-lived fission products decay to harmless levels before any potential re-entry. This approach minimizes the long-term radiological risk to Earth's population and the environment.

10.9.2.2 Decay Heat Management

After a reactor is shut down, it continues to generate decay heat from the radioactive decay of fission products. Managing this decay heat is critical to preventing overheating and potential structural failure of the reactor system. Space reactors are typically designed with passive or active heat dissipation systems, such as heat pipes or radiators, to safely dissipate decay heat over time. Ensuring adequate heat removal is particularly important during periods when active cooling systems may not be operational, such as during power outages or after reactor shutdown.

10.9.3 Dilute and Disperse

The "Dilute and Disperse" strategy involves spreading out radioactive materials in such a way that radiation levels are diluted below harmful thresholds. This strategy is particularly relevant in scenarios involving accidental re-entry or dispersion of radioactive materials in space.

10.9.3.1 Atmospheric Re-entry and Dispersal

In the event of an unintended re-entry, space reactors are designed to disperse radioactive materials widely to minimize the impact on any single area. This dispersal can be achieved through design features that cause the reactor to disintegrate and burn up upon re-entry, scattering the radioactive materials over a wide area. The aim is to ensure that no single location receives a radiation dose exceeding safety standards. This approach relies on the assumption that the upper atmosphere and vast areas of the Earth's surface can absorb and dilute the radiation to safe levels.

10.9.3.2 Fragmentation Mechanisms

To facilitate dispersal, space reactors may include specific design features such as mechanical or chemical systems that trigger fragmentation upon re-entry. For example, using materials that oxidize rapidly at high temperatures can help disintegrate the reactor structure, promoting dispersal. Alternatively, explosive charges may be used to ensure complete disassembly of the reactor core upon re-entry, further aiding in the dispersal of radioactive materials.

10.9.4 Inadvertent Criticality Prevention

Inadvertent criticality, or an unintentional nuclear chain reaction, poses a significant safety risk. Several design strategies are employed to prevent this scenario, particularly during launch, operation, and post-mission phases.

10.9.4.1 Fluid Immersion Protection

Space reactors must be designed to prevent criticality in the event of fluid immersion, such as in water or liquid hydrogen propellant. In water-moderated scenarios, the introduction of a neutron moderator could potentially cause a fast-spectrum reactor to become critical. To counter this, reactors are designed with features such as control rods or central control plugs that contain neutron-absorbing materials (e.g., boron carbide). These elements ensure the reactor remains subcritical even if submerged in a moderating fluid.

10.9.4.2 Core Compaction and Meltdown Prevention

Core compaction due to mechanical impact or meltdown is another concern. Reactor designs incorporate structural features that prevent core compression, which could otherwise increase the reactor's density and lead to criticality. In the event of a loss of coolant or control system malfunction, redundant safety mechanisms ensure the reactor remains subcritical. These include locking mechanisms for control elements that prevent unauthorized movement, as well as backup power systems to maintain control in the event of primary system failure.

10.9.5 Radiological Hazard Mitigation

The potential release of radioactive materials is a critical concern in space reactor design. Mitigating these hazards involves multiple layers of safety measures, from initial design through end-of-life disposal.

10.9.5.1 Containment and Shielding

Radiological protection begins with robust containment and shielding systems that prevent the release of radioactive materials under normal and accident conditions. The containment system, typically made of high-strength materials, is designed to withstand mechanical stresses and prevent the escape of radioactive gases or particles. Shielding, often using dense materials such as lead or tungsten, attenuates radiation emissions, protecting both onboard equipment and personnel.

10.9.5.2 End-of-Life Disposal

End-of-life disposal strategies ensure that decommissioned reactors do not pose long-term risks. Options include boosting the reactor into a high-altitude orbit, where it remains safely isolated for centuries, or designing the system to burn up and disperse in the atmosphere. For systems that might eventually re-enter the Earth's atmosphere, the goal is to ensure that radiation levels are minimal, either through extensive decay of fission products or by dispersal mechanisms.

10.9.6 Safety Protocols and Testing

Ensuring the safety of space reactors involves rigorous testing and adherence to established protocols throughout the design, launch, operation, and disposal phases.

10.9.6.1 Testing and Validation

Extensive testing is conducted to validate the reactor's design and ensure it meets all safety requirements. This includes simulations of launch and operational environments, criticality testing, and validation of heat dissipation systems. Testing protocols must account for both normal operations and potential accident scenarios, ensuring the system can safely handle a wide range of conditions.

10.9.6.2 Operational Safety Procedures

Operational procedures are established to manage the reactor safely throughout its mission. These procedures include pre-launch checks, operational protocols for startup and shutdown, and emergency response plans. For instance, reactors are typically not allowed to reach criticality until a safe orbit is achieved, preventing the risk of a nuclear accident during launch.

Safety considerations in the design and operation of space nuclear reactors are complex and multifaceted. They require a comprehensive approach that includes robust physical barriers, strategies for managing radioactive decay and dispersal, and stringent measures to prevent inadvertent criticality. Through careful design, rigorous testing, and adherence to established safety protocols, space reactors can be safely deployed and operated, minimizing risks to both human life and the environment. As space exploration continues to advance, these safety measures will remain crucial in ensuring the safe use of nuclear technology in space.

Table 10-3. U-ZrH Reactor Power Plant Summary

Parameter	SNAP-2	SNAP-10A	SNAP-8
Power (kWe)	3	0.5	35
Design lifetime (yr)	1	1	1
Reactor power (kWt)	55	30	600
Efficiency (%)	9	1.6	8
Reactor outlet temp (K)	920	810	975
Reactor type	U-ZrH thermal	U-ZrH thermal	U-ZrH thermal
Primary coolant	NaK-78	NaK-78	NaK-78
Power conversion	Rankine (Hg)	Thermoelectric (SiGe)	Rankine (Hg)
Radiator area (m²)	1.1	5.8	167.2
System unshielded weight (kg)	545	295	4,545
Specific weight (kg/kWe)	182	590	130

10.10   Uranium-Zirconium-Hydride Space Reactors

The development of uranium-zirconium-hydride (U-ZrH) reactor systems began in 1957, motivated by the growing need for reliable electric power sources for space missions. These reactors, designed for use in space, provided a range of power levels suitable for various applications. The U-ZrH reactor technology has been integral in multiple space reactor programs, including SNAP-2, SNAP-10A, and SNAP-8. We explore the technical aspects, operational principles, and challenges associated with U-ZrH reactor power plants.

The inception of U-ZrH reactors was driven by the need to support more ambitious space missions requiring higher power levels. The SNAP (Systems for Nuclear Auxiliary Power) program spearheaded this effort, producing a series of reactor systems that leveraged the unique properties of U-ZrH fuel. The SNAP-2 system, developed starting in 1957, was one of the first to demonstrate the feasibility of U-ZrH reactors, followed by SNAP-10A and SNAP-8. These systems varied in power output, reactor design, and power conversion technologies, as summarized in Table 10-3.

10.10.1   SNAP-2 Reactor System

The SNAP-2 system was a pioneering effort in U-ZrH reactor technology, providing critical insights into the design and operation of space nuclear reactors. The system was designed to deliver 3-5 kWe of electrical power using a mercury Rankine cycle for power conversion. The reactor core was cooled by a liquid metal eutectic, NaK-78, and featured a unique control system based on beryllium reflectors and safety elements.

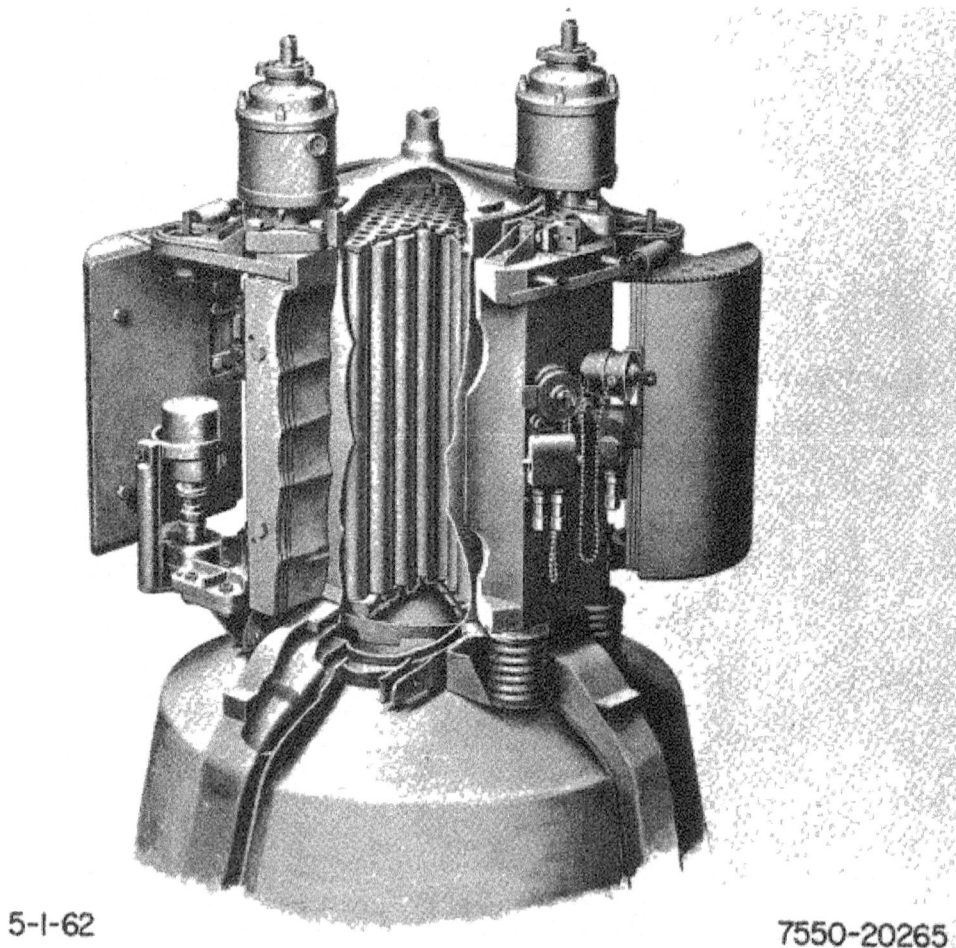

5-1-62   7550-20265

Figure 10-5. SNAP2 Reactor.

10.10.1.1   Reactor Core Design

The SNAP-2 reactor core consisted of 37 fuel elements arranged in a triangular array, forming a hexagonal cylinder. The fuel elements were composed of U-ZrH, with 10 wt% enriched uranium alloy and a hydrogen-to-metal ratio of 1.79. The core was moderated by hydrogen and reflected by beryllium, with a power density that varied radially and axially to optimize the reactor's performance. The fuel elements were encased in Hastelloy N cladding, which included a ceramic barrier to prevent hydrogen diffusion.

The reactor's thermal design aimed for a core outlet temperature of 920 K, with NaK coolant circulating through the core. The primary coolant loop utilized an electromagnetic (EM) pump powered by thermoelectric generators, which also served to preheat the mercury in the power conversion loop.

Table 10-4. SNAP-2 System Boiler Parameters

Parameter	Mercury Side	NaK Side
Flow rate (kg/min)	9.1	32
Outlet temperature (K)	895	920
Pressure drop (kPa)	414	1.7

Table 10-5. Technical Issues Associated with SNAP-2 System Development

Problem Area	Resolution Demonstration	Year Resolved
Criticality of U-ZrH Reactor	Zero Power Critical Experiment	1957
Fabrication of Enriched U-ZrH Fuel	SER Fuel Elements	1958
Mercury-Lubricated Bearings	Prototype Bearing	1958
High-Speed Turboalternator	500-hr CRU-1 Test	1959
Mercury Corrosion Endurance	2,500-hr CRU-V-1A Test	1963

10.10.1.2   Power Conversion Subsystem

The SNAP-2 system employed a mercury Rankine cycle for power conversion. The combined rotating unit (CRU), which included the turbine, alternator, and pump, operated at high speeds (36,000 rpm) and temperatures (895 K). Mercury vapor, generated in the boiler, expanded through the turbine to drive the alternator, which produced electrical power. The vapor then condensed and was recirculated, completing the cycle. The system's design emphasized minimal moving parts and robust containment to ensure reliable operation in the space environment.

10.10.1.3   Challenges and Resolutions

The SNAP-2 program encountered several technical challenges, particularly in managing the materials and components exposed to high temperatures and corrosive mercury. Table 10-4 highlights some of the critical technical issues and their resolutions during the SNAP-2 development.

10.10.2   SNAP-10A Reactor System

The SNAP-10A system represented a continuation and refinement of U-ZrH reactor technology, targeting lower power outputs and leveraging thermoelectric conversion. The SNAP-10A reactor was the only U.S. nuclear reactor to be flown in space, achieving this milestone in 1965. It provided a valuable demonstration of U-ZrH reactor technology's feasibility and reliability in an actual space environment.

10.10.2.1   Design and Operation

The SNAP-10A reactor core was similar to the SNAP-2 design, using U-ZrH fuel with NaK-78 as the primary coolant. However, instead of a Rankine cycle, SNAP-10A utilized thermoelectric converters based on silicon-germanium (SiGe) elements. This choice simplified the power conversion process, as thermoelectric devices have no moving parts and can directly convert heat into electricity.

The reactor operated at a thermal power level of 30 kWt, with a core outlet temperature of 810 K. The generated power was relatively modest, at 500 We, reflecting the mission requirements and the early stage of technology development.

10.10.2.2   Flight Demonstration

The successful launch and operation of SNAP-10A marked a significant achievement. The reactor operated for 43 days before being shut down due to a spacecraft anomaly unrelated to the reactor itself. This mission provided invaluable data on the behavior of U-ZrH reactors in space, including insights into the long-term stability of the fuel and the effectiveness of the thermoelectric power conversion system.

10.10.3   SNAP-8 Reactor System

The SNAP-8 system represented a significant scale-up in power output, targeting 30-60 kWe for more demanding space missions. This reactor continued to use U-ZrH fuel and NaK-78 coolant, but with substantial enhancements in design and performance.

10.10.3.1   Core and Conversion System

The SNAP-8 core featured a larger array of fuel elements and a more robust reflector and control system. The reactor aimed for a core outlet temperature of 975 K, with a corresponding increase in power density and overall thermal power. The power conversion system returned to the mercury Rankine cycle, capitalizing on the lessons learned from SNAP-2 and enhancing the design to support higher power levels.

10.10.3.2 Testing and Challenges

Testing of the SNAP-8 system involved both ground-based experiments and plans for flight tests. Key challenges included managing the higher thermal loads and ensuring the integrity of materials under prolonged exposure to high temperatures and radiation. The program also faced the technical complexities of scaling up the power conversion system while maintaining reliability and efficiency.

10.10.4 Material Considerations and Safety

U-ZrH reactors necessitated careful material selection due to the harsh operating conditions. The use of Hastelloy N for cladding and structural components provided the necessary corrosion resistance and mechanical strength. Additionally, the ceramic barrier coatings in fuel elements prevented hydrogen loss, ensuring fuel integrity over extended operation periods.

Safety considerations were paramount in the design and testing phases. The reactors incorporated multiple safety features, such as control drum systems for reactivity management and redundant safety elements to prevent accidental criticality. The designs also accounted for potential failure scenarios, including coolant loss and system depressurization, ensuring that the reactors could be safely shut down under all conditions.

The development of U-ZrH reactor power plants was a pivotal advancement in space nuclear power technology. These reactors demonstrated the feasibility of using nuclear energy for space applications, providing reliable power for a range of missions. The SNAP program's progression from SNAP-2 to SNAP-8 highlighted the increasing capabilities and technical sophistication of these systems. Despite the challenges faced, the success of these reactors laid the groundwork for future space nuclear power systems, contributing significantly to our understanding of nuclear technology's role in space exploration.

10.11 SNAP-10A Reactor Power Plant

10.11.1 Introduction and Background

The Space Nuclear Auxiliary Power (SNAP) program aimed to develop nuclear reactor systems for space applications. On December 30, 1960, a decision was made to proceed with the design of the SNAP-10A reactor system, a convection-cooled reactor derived from the SNAP-2 technology. The primary objective was to provide 500 watts of electric power at a nominal voltage of 28 volts, with a one-year operational lifespan. The

SNAP-10A reactor became the first and only U.S. nuclear reactor to operate in space, successfully demonstrating nuclear power's feasibility for space applications.

10.11.2   System Design and Configuration

The SNAP-10A reactor system utilized a liquid metal heat transfer mechanism, specifically using a NaK-78 eutectic mixture for thermal energy transport. The system featured a DC electromagnetic pump for fluid circulation and thermoelectric (TE) conversion for electrical power generation. The TE modules were composed of silicon-germanium (SiGe) elements, selected for their stability and efficiency at high temperatures. The waste heat was rejected to space via radiator surfaces integrated with the TE modules, with a total required radiating surface area of approximately 5.8 m².

10.11.2.1   Reactor Core and Fuel Composition

The SNAP-10A reactor core was based on the SNAP-2A design, comprising 37 fuel elements made of uranium-zirconium-hydride (U-ZrH). The uranium-235 fuel was highly enriched, and the zirconium hydride served as a moderator. The fuel elements were clad in Hastelloy N tubing, featuring an inner ceramic barrier to prevent hydrogen diffusion. The core's geometry was a close-packed hexagonal array, with a diameter of 22.6 cm and a height of 41 cm. Beryllium reflectors surrounded the core to enhance neutron economy, with control provided by four movable drums located in the reflector region.

10.11.2.2   Thermal Management and Heat Transfer

The primary coolant, NaK-78, was selected for its favorable thermal properties, including a low melting point and low vapor pressure. The reactor's thermal output was approximately 43.8 kWt, with the NaK exiting the core at around 833 K. The liquid metal was circulated at a rate of 0.8 liters per second by a DC conduction pump, which also integrated a thermoelectric power supply using PbTe elements. The TE modules operated with a temperature differential (ΔT) of 167 K, driving the circulation of NaK through the system.

10.11.2.3   Thermoelectric Conversion and Radiator System

The SNAP-10A utilized SiGe thermoelectric materials for power conversion due to their superior performance at high temperatures compared to alternatives like PbTe. The system employed 40 NaK flow tubes, each equipped with 72 SiGe thermocouple elements, totaling 2,880 individual elements. The TE modules converted thermal energy into electrical power with an efficiency of approximately 1.43%, providing 580 W of electrical power. The waste heat was dissipated through radiators with a mean temperature of 610 K, capable of dissipating about 16.4 Wt/cm².

10.11.3   Control and Safety Mechanisms

The reactor's control system was designed to ensure safe and reliable operation throughout its mission. Control was achieved using four drums embedded in the beryllium reflector, which adjusted the neutron flux by varying the reflector's configuration. The startup sequence was initiated by a ground command, unlocking the control drums and powering the startup controller. The reactor reached criticality approximately seven hours after the drums were positioned, with subsequent power adjustments managed by the reactor's inherent negative temperature coefficient of reactivity.

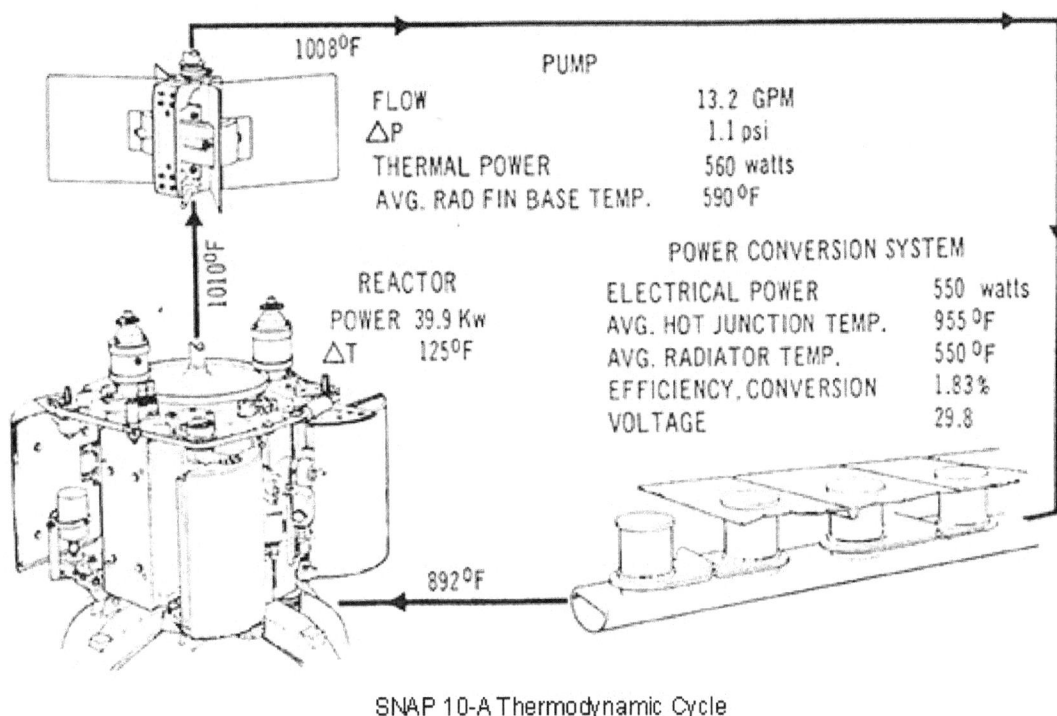

Figure 10-6. SNAP-10A Power conversion system
(source: https://www.energy.gov/etec/system-nuclear-auxiliary-power-snap-overview)

10.11.3.1   Shielding and Radiation Protection

Given the hazardous nature of space radiation, the SNAP-10A featured a shadow shield configuration to protect spacecraft components. The shield was made of cold-pressed lithium hydride (LiH), reinforced with stainless steel honeycomb, and encased in a Type 316 stainless steel casing. The shield weighed 98 kg and was designed to withstand temperatures ranging from 579 K to 712 K. The shield effectively limited radiation exposure, ensuring that fast neutron flux and gamma radiation doses remained within safe limits.

10.11.4   Performance and Operational Characteristics

The SNAP-10A system was rigorously tested and validated through extensive ground testing before its flight demonstration. Key performance metrics included reactor thermal power, electrical power output, system efficiency, and operational stability. The system demonstrated stable thermal management, effective radiation shielding, and reliable power generation capabilities.

Table 10-6. SNAP-10A System Design Parameters

Parameter	Value
Electrical power (W)	533
Reactor thermal power (kWt)	43.8
Reactor type	U-ZrH thermal
Fuel	U-235 enriched
Reflector	Be
Design life (years)	1
Radiator area (m²)	5.8
Overall package length (m)	3.5
Mounting base diameter (m)	1.3
Coolant	NaK-78
Coolant flow rate (gpm)	14.5
Reactor outlet temperature (K)	833
Reactor inlet temperature (K)	829
Power conversion system	SiGe
Voltage	30 V
Mass (kg)	436

10.11.5   Key Challenges and Technological Innovations

The development of the SNAP-10A system involved overcoming several technical challenges, particularly in materials science, thermoelectric conversion, and thermal management. The choice of SiGe over PbTe for thermoelectric conversion marked a significant advancement, driven by SiGe's superior thermal stability and mechanical properties. Additionally, the design and implementation of the shadow shield required precise engineering to balance radiation protection with weight constraints.

10.11.5.1 Material Selection and Compatibility

The selection of Hastelloy N for cladding and structural components was crucial, given its resistance to corrosion by NaK and its mechanical robustness at high temperatures. The integration of a ceramic barrier to prevent hydrogen loss from the U–ZrH fuel was another critical innovation, ensuring long-term fuel integrity and reactor stability.

10.11.5.2 Thermoelectric Module Design

The design and fabrication of the thermoelectric modules required meticulous attention to detail, particularly in the bonding processes and the arrangement of the TE elements. Ensuring uniform thermal contact and minimizing electrical resistance were key factors in achieving the desired conversion efficiency and system reliability.

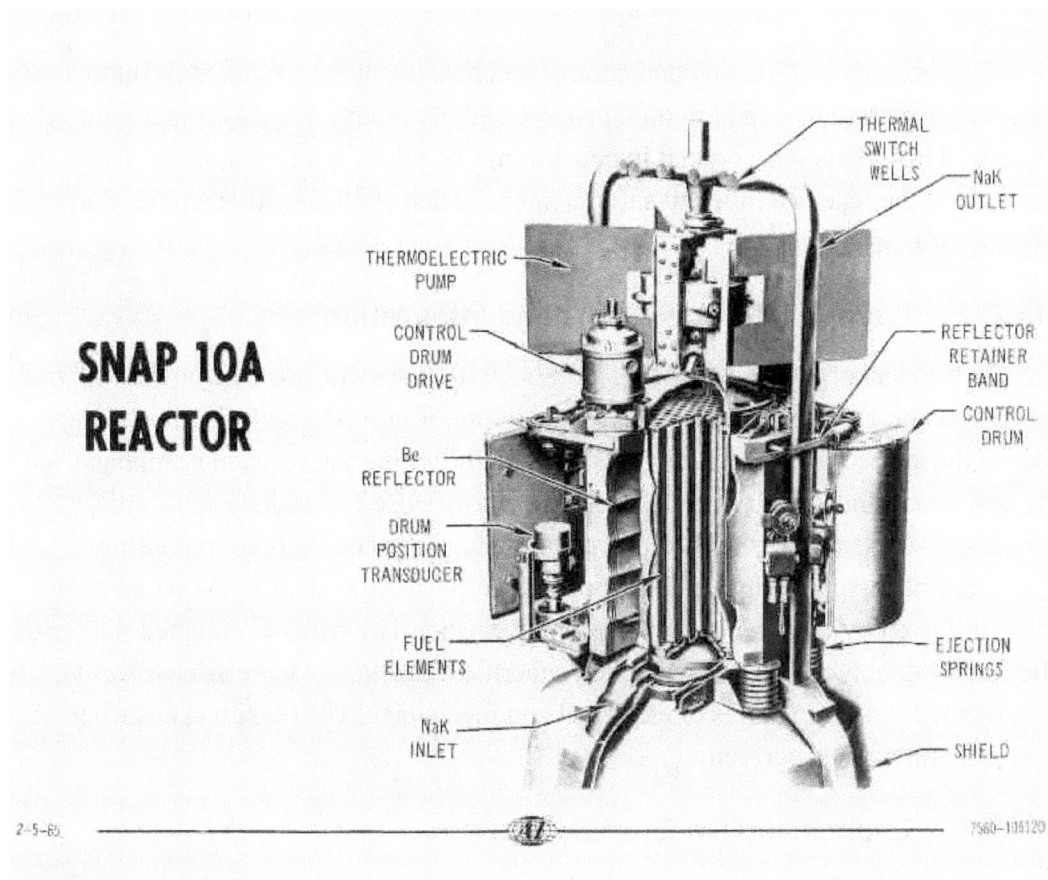

Figure 10-7. Cross section view of the SNAP-10A reactor
(source: https://www.energy.gov/etec/system-nuclear-auxiliary-power-snap-overview).

10.11.6   SNAP-10A Flight Demonstration: The SNAPSHOT Mission

The SNAP-10A reactor, under the SNAPSHOT mission, became the first and only U.S. nuclear reactor to be launched and operated in space. The mission, conducted in 1965, aimed to validate the safety and functionality of space-based nuclear power systems. Here, we provide a technical overview of the SNAP-10A flight demonstration, detailing the pre-launch preparations, mission execution, and post-mission analysis.

10.11.6.1   Pre-Launch Preparations and Testing

Before the launch, the SNAP-10A system underwent rigorous pre-flight testing to ensure its reliability and safety. The system was thermally checked using electric heaters within a simulated space environment to verify thermal management under operational conditions. Criticality tests were performed on the reactor without producing significant fission products, thereby avoiding radiation hazards that could complicate pre-launch activities.

The reactor's fuel loading and control mechanisms were validated through a series of tests, ensuring that the reactor could achieve and maintain criticality under controlled conditions. These tests were crucial in confirming the integrity of the control drums, fuel elements, and the reactor's inherent safety features, such as its negative temperature coefficient of reactivity.

10.11.6.2   Launch Site Operations and Integration

At Vandenberg Air Force Base, the SNAP-10A system was integrated with the Atlas-Agena launch vehicle. The payload integration followed standard procedures, similar to those used for non-nuclear missions, highlighting the mission's emphasis on safety and routine handling. The "cold, clean" reactor posed no radiation hazards, allowing unrestricted access to the payload and launch vehicle during preparation.

The final assembly involved mating the SNAP-10A system to the Atlas-Agena booster, followed by comprehensive subsystem checks. The reactor remained non-critical on the pad, with control drums locked in a subcritical position. The readiness for launch was confirmed through a series of electrical and mechanical verifications, ensuring all components functioned correctly.

10.11.6.3   Launch and On-Orbit Operation

The SNAPSHOT mission launched on April 3, 1965, at 1:24 PM PST. The Atlas-Agena rocket successfully delivered the payload into a near-circular orbit with an apogee of 1,328 km and a perigee of 1,295 km. At 5:05 PM PST, a start command initiated the reactor's startup sequence. The control drums gradually moved to achieve criticality, which was reached at 11:15 PM PST. By 1:45 AM PST on April 4, the reactor had achieved full power.

The initial six days of operation at full power demonstrated the reactor's stability and the efficiency of the thermoelectric power conversion system. The system operated autonomously, with power output regulated by the reactor's negative temperature coefficient. This feature automatically adjusted the reactor's power output in response to temperature variations, maintaining safe and stable operation.

10.11.6.4   Premature Shutdown and Post-Mission Analysis

On May 16, 1965, after 43 days of successful operation, the SNAP-10A reactor system experienced an unexpected shutdown. The shutdown was attributed to a failure in the Agena vehicle's voltage regulator, which interrupted the power supply to the reactor's control system. The design did not include a restart mechanism, as per safety protocols, leaving the reactor in a quiescent state in its orbit.

Despite the premature shutdown, the mission was considered highly successful. The reactor subsystem performed as expected, with criticality achieved at the predicted reactivity levels. The thermoelectric power conversion system proved reliable, converting thermal energy to electrical power with slightly better efficiency than anticipated. The heat rejection system functioned effectively, with no detectable degradation of radiative surfaces or micrometeoroid impacts on the coolant loop tubing.

10.11.6.5   Radiation and Safety Analysis

The mission included comprehensive radiation monitoring to assess the exposure levels and validate shielding effectiveness. Measurements indicated that the fast neutron flux was approximately four times higher along the vehicle centerline than predicted, while gamma radiation levels at the aft end of the Agena vehicle were significantly lower due to attenuation and scattering effects. The design of the shadow shield and the positioning of sensitive components successfully minimized radiation exposure to acceptable levels.

10.11.6.6   SNAPSHOT's Legacy and Ground-Based Testing

As a complement to the on-orbit demonstration, a ground-based duplicate of the SNAP-10A system, designated FS-3, was operated for 10,000 hours in a shielded vacuum chamber. The post-test analysis of FS-3 revealed no signs of wear or incipient failures, further validating the system's design robustness and reliability.

The SNAPSHOT mission provided invaluable data on the behavior of nuclear reactors in space. It demonstrated the feasibility of nuclear power systems for long-duration space missions and established foundational knowledge for future reactor designs. The mission's success confirmed the effectiveness of nuclear safety protocols, radiation shielding, and power management strategies in the harsh space environment.

The SNAP-10A flight demonstration under the SNAPSHOT mission marked a significant milestone in the history of space nuclear power. Despite the unexpected shutdown, the mission achieved its primary objectives, showcasing the safe operation of a nuclear reactor in space. The data obtained from the mission contributed to the broader understanding of space-based nuclear systems, informing the design and safety considerations for future missions.

The SNAPSHOT mission remains a testament to the potential of nuclear technology to support advanced space exploration. It laid the groundwork for the development of more sophisticated and powerful space nuclear reactors, capable of supporting deep-space missions and extended human presence in space.

10.12   SNAP-8

The SNAP-8 program was an ambitious initiative aimed at developing a space-based nuclear power system capable of generating 30 kWe, with the potential to expand to 60 kWe. This system was designed to provide power for extended periods, with an operational lifetime goal of 10,000 hours and a reactor outlet temperature of 975 K. The SNAP-8 system was an evolution of the SNAP-2 reactor, utilizing a uranium-zirconium-hydride (U-ZrH) core cooled by a liquid metal (NaK) and coupled with a mercury Rankine cycle power conversion system.

Figure 10-8. SNAP-10A preflight test.

10.12.1   SNAP-8 System Design and Components

The SNAP-8 system was composed of several key subsystems, each playing a critical role in the overall functionality of the power plant. These subsystems included the reactor core, primary and secondary coolant loops, power conversion system, and auxiliary systems for startup and safety.

10.12.1.1   Reactor Core and Primary Coolant Loop

The SNAP-8 reactor core was based on a U-ZrH fuel design, similar to its predecessors, but with enhancements to support higher power output and operational longevity. The core utilized fully enriched uranium-235 as the fuel, moderated by zirconium hydride. The primary coolant loop used NaK-78, a eutectic mixture of sodium and potassium, chosen for its low melting point and excellent heat transfer properties.

Reactor Characteristics:
- **Design Power Level**: 600 kW_th
- **Core Outlet Temperature**: 975 K
- **Primary Coolant Flow Rate**: 6.7 kg/s
- **Core Mass Flow**: 37 kg/s
- **Control Drums**: 6, providing fine control of the reactor's neutron flux

The reactor core contained 211 fuel elements arranged in a triangular lattice, as shown in Fig. 13. These fuel elements were clad in Hastelloy N, with an internal ceramic coating to prevent hydrogen diffusion. The core's design incorporated a beryllium reflector to enhance neutron economy and control drums for reactivity management.

10.12.1.2   Power Conversion System

The power conversion system in SNAP-8 employed a Rankine cycle using mercury as the working fluid. The choice of mercury was driven by its high thermal conductivity and appropriate boiling point at the reactor's operating temperatures. The system included a boiler, turbine-alternator assembly, condenser, and feed pump.

Key Components:
- **Mercury Boiler**: Single-pass, counterflow design with tantalum containment tubes surrounded by NaK
- **Turbine-Alternator Assembly**: A four-stage axial-flow turbine coupled to a three-phase alternator, designed to operate at 220/208 V, 400 Hz
- **Condenser**: Tube-in-shell design with 73 tapered tubes, using NaK as the coolant
- **Mercury Pump**: A centrifugal pump with a seal-to-space mechanism, crucial for maintaining the mercury loop's integrity

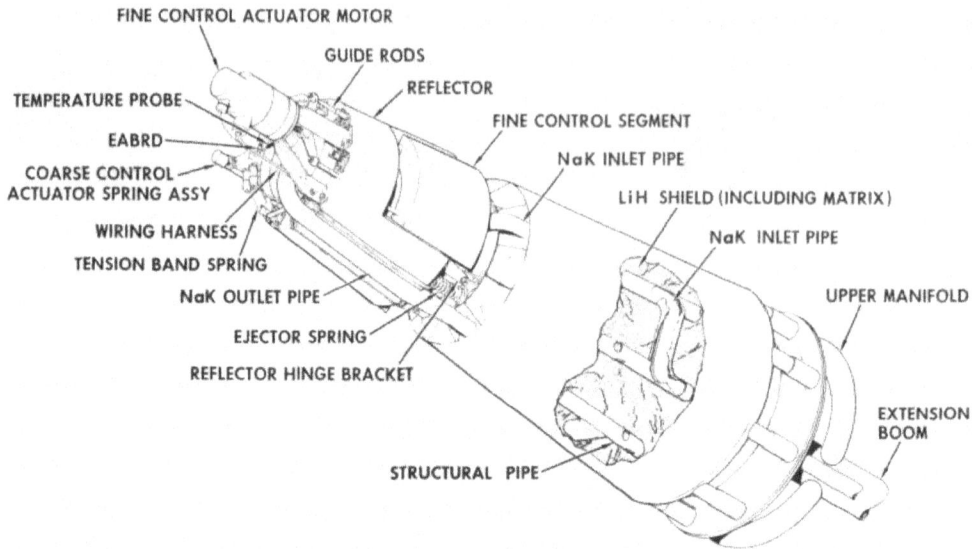

Figure 10-9. SNAP-10A Cutaway.

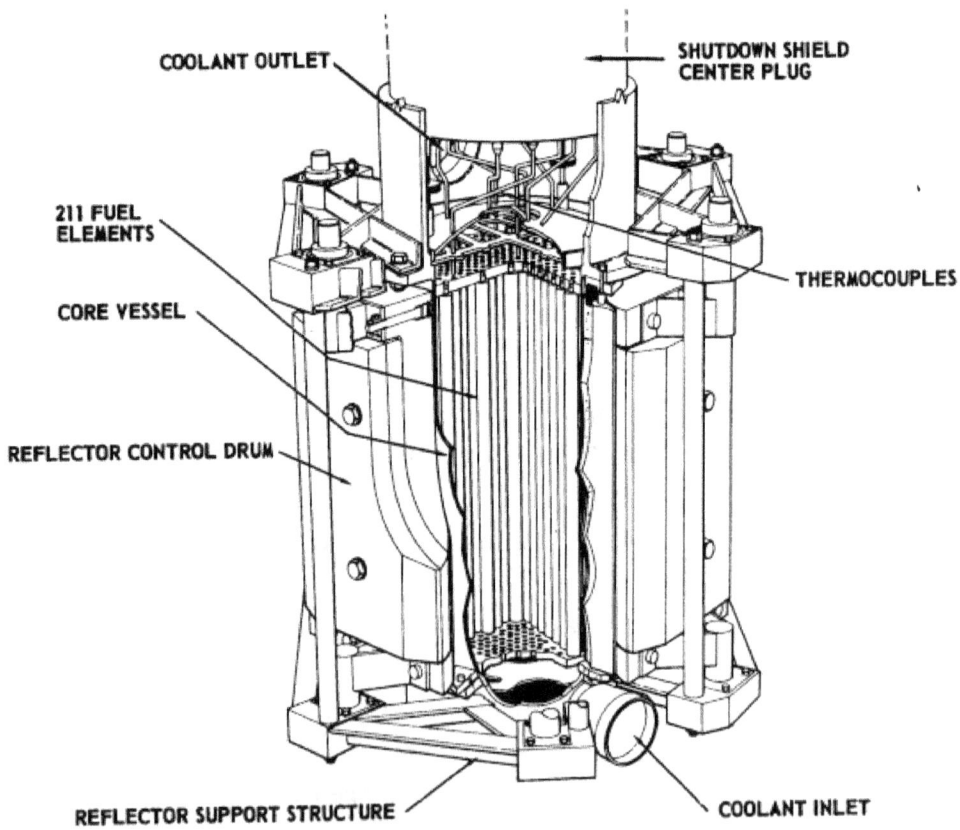

Figure 1.   S8ER Reactor Assembly View

6531-5047

Figure 10-10. SNAP-8 Experimental Reactor.

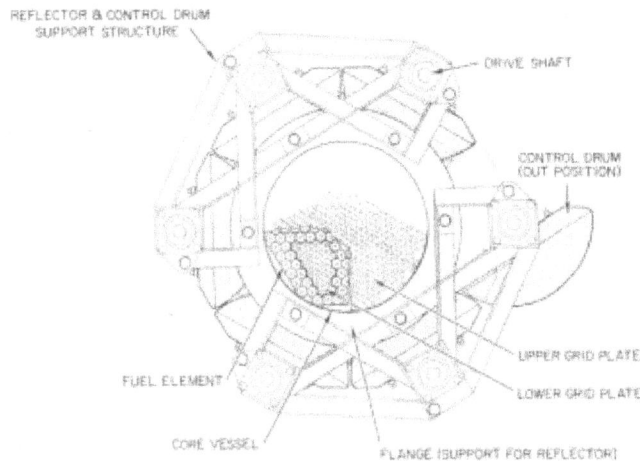

Figure 10-11. SNAP-8 core cross section.

The power conversion system was designed for high efficiency and reliability, with particular attention to minimizing wear and tear on rotating components. The turbine-alternator assembly was lubricated with polyphenyl-ether, an organic fluid suitable for high-temperature applications.

10.12.1.3   Auxiliary Systems

The SNAP-8 system included several auxiliary loops and components to support startup, control, and safety operations. A NaK pump was employed in both the primary and heat rejection loops, providing circulation and cooling. The system also featured a startup loop to manage thermal loads during reactor initiation, as well as various safety mechanisms to ensure reliable shutdown and containment.

10.12.2   Operational Testing and Performance

The SNAP-8 program conducted extensive testing to validate the reactor and power conversion systems. Two main reactor configurations, the SNAP-8 Experimental Reactor (S8ER) and the SNAP-8 Developmental Reactor (S8DR), were developed and tested.

10.12.2.1   SNAP-8 Experimental Reactor (S8ER)

The S8ER was the initial testbed for the SNAP-8 system. It was operated from May 1963 to April 1965, generating over 5.1 million kWh of energy. The reactor demonstrated stable operation at 600 kW$_{th}$ and 975 K, with 100 days of continuous operation at full power. The S8ER also handled 115 rapid power transients, showcasing its dynamic stability.

Key Findings:

- **Fuel Element Cracking**: Post-test analysis revealed cladding cracks in 80% of the fuel elements, attributed to high-temperature swelling and low ductility of Hastelloy N.
- **Thermal Performance**: The reactor maintained stable temperatures, but local overheating due to coolant flow anomalies was observed.

10.12.2.2 SNAP-8 Developmental Reactor (S8DR)

The S8DR, a modified version of the S8ER, incorporated design improvements based on earlier test results. It was tested from January to December 1969, achieving 4.3 million kWh of energy output. The reactor demonstrated improved performance, with modifications to fuel element design, coolant flow management, and control drum mechanisms.

Modifications:

- **Hydrogen Permeation Barrier**: Enhanced ceramic coating for better thermal cycle resistance
- **Fuel Element Design**: Increased clearances and redesigned end caps to reduce stress concentrations
- **Control Drum System**: Improved actuators for more precise control

Despite these improvements, the S8DR experienced similar issues with fuel element cracking, indicating the need for further refinement in cladding material and core temperature management.

10.12.3 Technical Challenges and Solutions

The SNAP-8 program faced several technical challenges, primarily related to fuel element integrity, coolant loop stability, and power conversion efficiency.

10.12.3.1 Fuel Element Integrity

The high operating temperatures of the SNAP-8 reactor led to significant challenges in maintaining the structural integrity of the fuel elements. The primary issue was the swelling of the U-ZrH fuel, which caused stress and eventual cracking in the Hastelloy N cladding.

Solutions Explored:

- **Improved Cladding Materials**: Investigations into alternative alloys with higher ductility and corrosion resistance
- **Enhanced Fuel Element Design**: Optimizing the fuel element geometry and

clearances to accommodate swelling
- **Hydrogen Barrier Technologies**: Developing more effective coatings to prevent hydrogen permeation and maintain fuel integrity

10.12.3.2   Coolant Loop Stability

Maintaining stable coolant flow was critical for preventing localized overheating and ensuring efficient heat transfer. The SNAP-8 system employed a sophisticated NaK pump design, as shown in Fig. 15, to manage coolant circulation in the primary and heat rejection loops.

Challenges:
- **Cavitation Damage**: The NaK pumps experienced cavitation damage, particularly in the impeller area, necessitating design improvements.
- **Coolant Bypass**: Excessive coolant bypass in the core led to uneven temperature distribution and fuel element overheating.

Solutions:
- **Pump Design Optimization**: Enhancing the pump design to minimize cavitation and improve flow characteristics
- **Flow Management**: Implementing measures to control coolant flow distribution and reduce bypass

10.12.3.3   Power Conversion Efficiency

The Rankine cycle power conversion system, using mercury as the working fluid, required careful management of thermal and mechanical stresses. The turbine-alternator assembly, depicted in Fig. 16, was a critical component, requiring precise engineering to ensure reliable operation.

Challenges:
- **Blade Erosion and Corrosion**: Prolonged operation led to erosion and corrosion of turbine blades, impacting efficiency.
- **Lubrication and Cooling**: Maintaining adequate lubrication and cooling for high-speed rotating components was essential.

Solutions:
- **Material Selection**: Utilizing materials with high resistance to erosion and corrosion for critical components
- **Lubrication System Enhancement**: Refining the lubrication and cooling systems to ensure long-term reliability

10.12.4   SNAP-8 Legacy

The SNAP-8 program was a pivotal project in the development of space-based nuclear power systems. Despite encountering technical challenges, the program achieved significant milestones, demonstrating the feasibility of using nuclear reactors for extended space missions.

Key Achievements:
- **Extended Operation**: The S8DR demonstrated the capability for prolonged operation at high power levels, providing valuable data on reactor behavior and system performance.
- **Technical Innovations**: The program pioneered several technical innovations, including advanced fuel element designs, high-efficiency power conversion systems, and robust coolant loop technologies.

The lessons learned from SNAP-8 have informed subsequent nuclear power system designs, contributing to the development of more reliable and efficient space reactors. The program's legacy continues to influence the field of space nuclear power, laying the groundwork for future exploration and scientific missions beyond Earth's orbit.

10.13   The SP-100 Nuclear Power System

The Space Power 100 (SP-100) program, active from 1983 to 1993, aimed to develop a versatile and robust nuclear power system for space applications. This initiative was driven by both Department of Defense (DoD) and NASA requirements for reliable, long-duration power sources in space. The system was designed to deliver tens to hundreds of kilowatts of electric power (kWe) over a 7-year operational period at full power, with an additional 3 years at reduced power, making a total of 10 years. The design incorporated advanced materials, innovative cooling technologies, and robust safety features to ensure performance and reliability. We provide a detailed technical analysis of the SP-100 system, including its design requirements, key components, and operational characteristics.

10.13.1   1. Design Requirements and Objectives

The SP-100 program was conceived to support a wide range of space missions, including surveillance, communication, electric propulsion, and scientific exploration. The primary design requirements included:
- **Operational Life**: 7 years at full power, 10 years total.
- **Power Output**: Up to 100 kWe.
- **Reactor Thermal Power**: 2.45 MWt.
- **Reactor Outlet Temperature**: 1350 K (beginning of mission), 1375 K (end of mission).
- **Radiation Shielding**: Reduce radiation at 25 m to 1×10^{13} n/cm² and 5×10^5 rad (Si).
- **System Mass**: Less than 4600 kg.
- **Launch Compatibility**: Designed for launch on a Titan IV/Centaur rocket with a

2000 km orbit.

These specifications were intended to meet the stringent needs of military and scientific missions, ensuring long-term operation without maintenance or servicing in space.

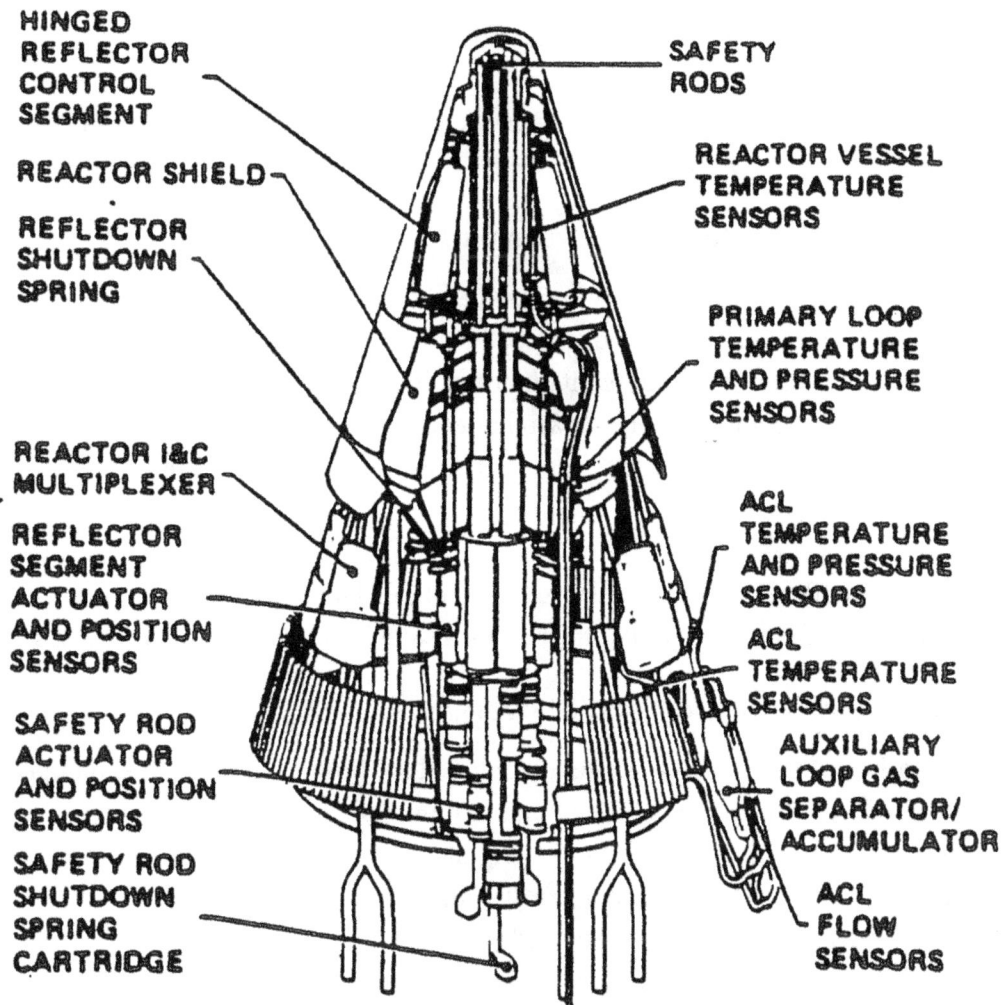

HINGED
REFLECTOR
CONTROL
SEGMENT

REACTOR SHIELD

REFLECTOR
SHUTDOWN
SPRING

REACTOR I&C
MULTIPLEXER

REFLECTOR
SEGMENT
ACTUATOR
AND POSITION
SENSORS

SAFETY ROD
ACTUATOR
AND POSITION
SENSORS

SAFETY ROD
SHUTDOWN
SPRING
CARTRIDGE

SAFETY
RODS

REACTOR VESSEL
TEMPERATURE
SENSORS

PRIMARY LOOP
TEMPERATURE
AND PRESSURE
SENSORS

ACL
TEMPERATURE
AND PRESSURE
SENSORS

ACL
TEMPERATURE
SENSORS

AUXILIARY
LOOP GAS
SEPARATOR/
ACCUMULATOR

ACL
FLOW
SENSORS

Figure 10-12. Sp-100 reactor instrumentation and control systems.

10.13.2   2. System Architecture and Components

The SP-100 system was a complex assembly of multiple subsystems, each designed to perform specific functions. The major components included the reactor core, control and safety systems, heat transport and conversion systems, and the heat rejection system.

10.13.2.1   Reactor Core and Fuel

The SP-100 reactor utilized uranium nitride (UN) fuel, chosen for its high thermal conductivity and high uranium density. The fuel pellets, approximately 50,000 in total, were enriched to 97% U-235 and had a theoretical density of 94.5%. The fuel pellets were contained within cladding made from the niobium-based alloy PWC-11, with a rhenium liner to prevent interaction with the lithium coolant and maintain structural integrity. The fuel pins were arranged in hexagonal assemblies, with 61 pins per full assembly and 50 pins in partial assemblies on the periphery, as shown in Figure 10-13. Sp-100 core cross section.Figure 10-13.

10.13.2.2   Control and Safety Systems

The reactor's neutron economy was managed using sliding neutron reflectors made from beryllium oxide (BeO), housed within Nb-1% Zr shells. Control drums were used to modulate neutron flux and maintain criticality. For safety, boron carbide (B4C) safety rods were incorporated to shut down the reactor in emergencies. These rods could be deployed in response to anomalous conditions, such as loss of coolant or unexpected reactor reactivity changes.

An auxiliary cooling system, featuring 42 U-tubes, was implemented to prevent fuel overheating in the event of primary coolant failure. This system could maintain fuel temperatures below 2000 K, ensuring the reactor's safety under accident conditions.

10.13.2.3   Heat Transport and Conversion

The primary heat transport system employed six liquid lithium loops, pumped by thermoelectric-electromagnetic (TEM) pumps. These pumps utilized the Seebeck effect to generate current and magnetic fields, driving the lithium coolant through the system without mechanical parts, thus enhancing reliability.

Heat from the reactor was transferred to the Power Conversion Assembly (PCA) via these lithium loops. The PCA utilized silicon germanium/gallium phosphide (SiGe/GaP) thermoelectric materials to convert thermal energy to electricity. The system was designed to produce 100 kWe at 200 V DC, with a total of 8640 thermoelectric cells arranged in parallel-series configurations to achieve the desired output voltage.

10.13.2.4   Heat Rejection System

Heat rejection was a critical aspect of the SP-100 design, given the high operating temperatures and power levels. The system used potassium heat pipes within a carbon-carbon matrix structure to transport waste heat to radiators. Each of the twelve radiator panels could be folded for launch and deployed in space, providing a total area of 104 m² for heat dissipation.

10.13.3 Operational Characteristics and Performance

The SP-100 system was designed for autonomous operation with minimal ground intervention. It featured a robust control system capable of managing power output and ensuring safety under varying operational conditions.

10.13.3.1 Power Output and Efficiency

The system's thermoelectric conversion efficiency was approximately 4%, with an expected electrical output of 100 kWe from a reactor thermal power of 2.45 MWt. The PCA's performance was enhanced by the advanced thermoelectric materials used, with a target figure-of-merit (ZT) for SiGe/GaP of 0.85×10^{-3} K^{-1}.

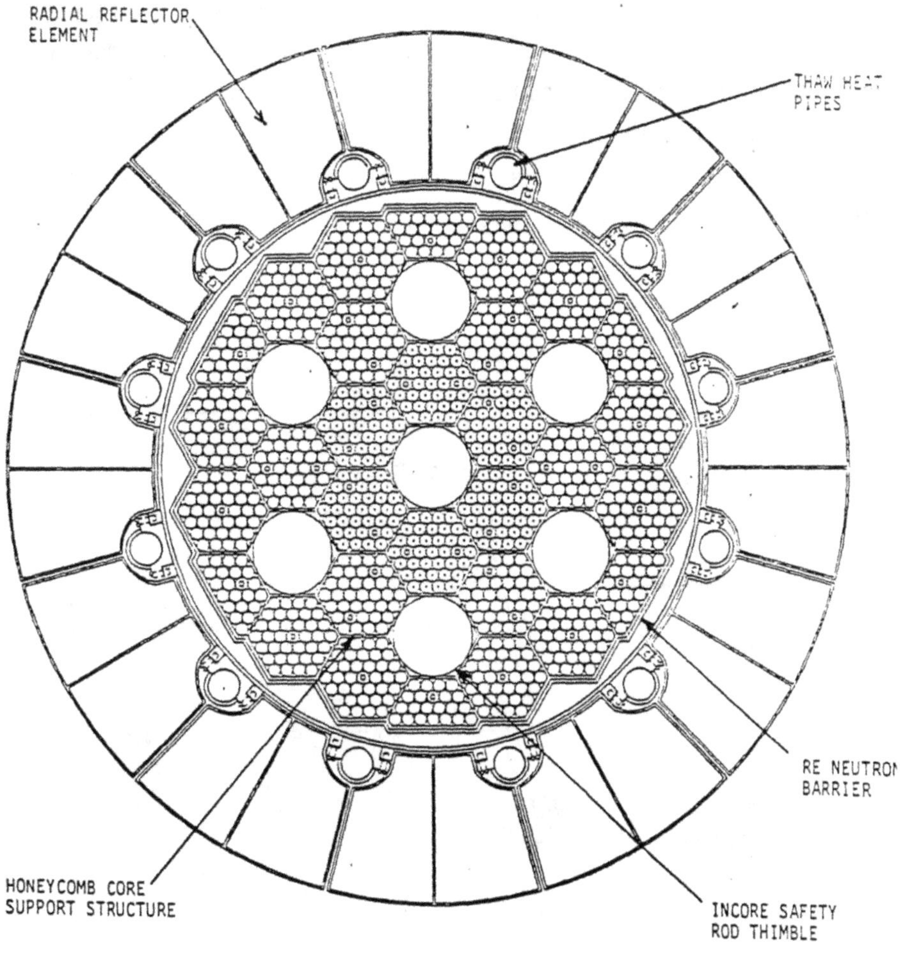

Figure 10-13. Sp-100 core cross section.

10.13.3.2   Safety and Reliability

The SP-100 space nuclear reactor was meticulously designed with safety and reliability as its core principles, incorporating multiple redundant systems and conservative engineering practices to mitigate risks and ensure operational security. Given the high-stakes nature of space missions and the potential hazards of operating a nuclear reactor in space, every aspect of the SP-100's design was crafted to prioritize fail-safe mechanisms, thermal safety, radiation protection, and the ability to withstand reentry.

Reactivity control within the SP-100 reactor was managed by two primary systems: control drums and safety rods. The control drums were equipped with neutron-absorbing materials and could rotate to adjust the reactor's power output dynamically. This system allowed for precise control of the reactor's reactivity during normal operations. In parallel, safety rods were included as a critical backup mechanism, capable of rapidly shutting down the reactor in the event of an unexpected anomaly or emergency. This dual-system approach provided a fail-safe mechanism, ensuring that even if one system failed, the reactor's reactivity could still be safely controlled, thereby reducing the risk of an uncontrolled increase in power.

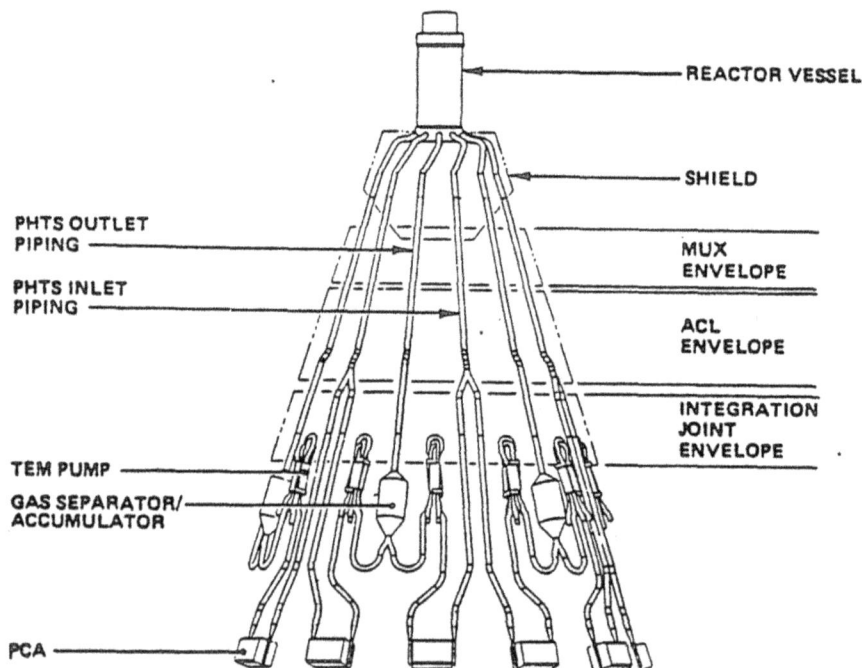

Figure 10-14. Heat transport subsystem.

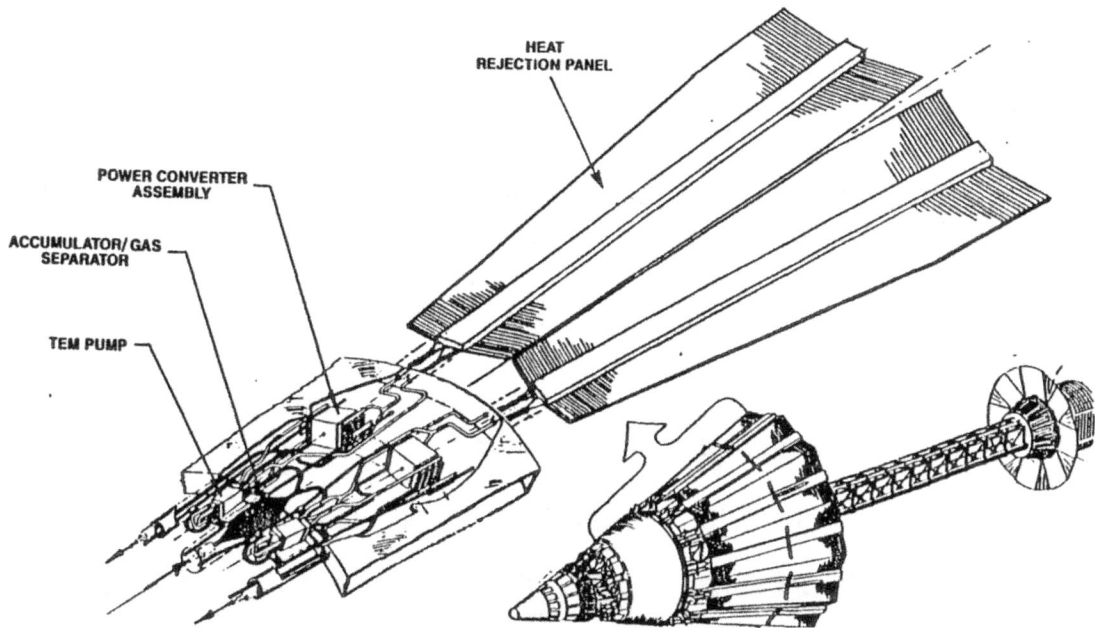

Figure 10-15. Heat rejection radiators.

Thermal safety was another cornerstone of the SP-100 design, particularly given the extreme environments in which the reactor was expected to operate. The system featured an auxiliary cooling system that was specifically designed to maintain safe operating temperatures, even in the event of a failure in the primary cooling loop. This auxiliary system was capable of automatically activating under accident conditions, ensuring continuous heat removal from the reactor core to prevent overheating and potential core damage. This layer of redundancy was critical to preventing thermal runaway and ensuring the reactor's integrity in all scenarios.

Radiation shielding was also a key concern in the design of the SP-100, particularly to protect the spacecraft and any nearby payloads from the reactor's radiation output. The shielding design incorporated multiple layers, including lithium hydride (LiH), depleted uranium, and beryllium oxide (BeO). The lithium hydride served as an effective neutron absorber, significantly reducing neutron radiation levels. Depleted uranium provided additional shielding by attenuating gamma radiation and further reducing neutron flux. The beryllium oxide layer enhanced the shielding by reflecting neutrons, thereby contributing to the overall reduction of radiation exposure from the reactor. This comprehensive shielding system ensured that radiation levels remained within safe limits, protecting both the spacecraft's electronics and any scientific instruments or human passengers aboard.

In addition to operational safety, the SP-100 reactor was engineered to withstand the rigors of reentry into Earth's atmosphere, should such an event occur. The reactor was equipped with a carbon-carbon reentry shield designed to protect the core from the extreme heat and mechanical stresses associated with atmospheric reentry. This shield ensured that the reactor core would not overheat during reentry, thereby preventing the release of radioactive materials and ensuring that the integrity of the reactor was maintained even under these extreme conditions.

The SP-100's design reflects a comprehensive approach to safety and reliability, addressing the unique challenges of operating a nuclear reactor in space. From reactivity control to thermal management and radiation shielding, the SP-100 was equipped with advanced systems to safeguard against potential failures and ensure its safe operation throughout its mission. The inclusion of reentry protection further underscores the system's robustness, ensuring that the SP-100 could meet the stringent safety standards required for space-based nuclear power generation.

Table 10-7. SP-100 development roadmap.

Component	Short-Term	Intermediate Term	Long-Term
System Power	10-40 kWe	10-100 kWe	10-300 kWe
Lifetime (Full Power/Mission)	3/5 Years	5/7 Years	7/10 Years
Reactor			
- Fuel	UN	UN	UN
- Coolant	Lithium	Lithium	Lithium
- Cladding	Nb-1% Zr/Re	PWC-11/Re	PWC-11/Re
- Structure	Nb-1% Zr	Nb-1% Zr	PWC-11
- Outlet Temperature	1350 K	1375 K	1400 K
Reactor Instrumentation & Control (I&C)			
- Mode	Dual	Dual	Separate
- Safety	In-Core	In-Reflector	In-Core
- Control	In-Core	In-Reflector	Reflector
Heat Transport			
- Pump	TEM	TEM	TEM
- Material	Nb-1% Zr/PWC-11	Nb-1% Zr/PWC-11	PWC-11
Converter			
- Thermoelectric (TE) Material	SiGe-0.67 x 10⁻³/K	SiGe(GaP)-0.72 x 10⁻³/K	SiGe(GaP)-0.85 x 10⁻³/K
- Type	Multicell	Multicell	Multicell
- Power per Cell	8.8 We/Cell	10.8 We/Cell	12.8 We/Cell
Radiator			
- Heat Pipe Material	K-Ti (2.5 cm Dia.)	K-Nb-1% Zr (2.5 cm Dia.)	K-Ti (1.3 cm Dia.)
- Fins	Ti	C-C	C-C
- Armor	Ti	C-C	C-C
- Duct	Ti	Nb-1% Zr	Ti

10.13.4 Development and Testing

The SP-100 program included extensive testing of components and subsystems to validate the design and ensure reliability. This testing was critical in identifying and addressing potential issues, such as material compatibility and component durability under space conditions.

10.13.4.1 Fuel and Cladding Testing

The fuel and cladding materials underwent rigorous testing to assess their performance under high temperatures and neutron flux. The rhenium liner and PWC-11 cladding were particularly scrutinized for their ability to contain fission products and withstand the chemical and mechanical stresses of operation.

10.13.4.2 Thermoelectric Module Development

The development of the thermoelectric modules was a key focus, with efforts concentrated on improving material properties and assembly techniques. The use of compliant pads and high-voltage insulators was optimized to ensure reliable electrical insulation and mechanical stability under thermal cycling.

10.13.4.3 Heat Pipe and Radiator Testing

The heat pipes and radiators were tested for their ability to effectively dissipate heat in the vacuum of space. This included testing the durability of the carbon-carbon structure and the performance of the potassium working fluid under various thermal loads.

10.13.5 SP-100 Legacy

The SP-100 program was a pioneering effort in space nuclear power, pushing the boundaries of reactor technology and power conversion systems. Despite the program's cancellation in 1993, it laid the groundwork for future developments in space nuclear power systems. The lessons learned from SP-100 continue to inform the design and development of advanced space power systems, contributing to ongoing efforts in nuclear propulsion and power generation for space exploration.

The SP-100 system's design, characterized by its advanced materials, innovative cooling systems, and robust safety features, represents a significant achievement in space technology. It stands as a testament to the potential of nuclear power in supporting long-duration, high-power missions beyond Earth, paving the way for future explorations and scientific endeavors in the solar system.

10.14 SP-100 Future Development

The table below outlines the progression of SP-100 power plant technology across three development stages: Near-Term, Intermediate Term, and Mature. The evolution is characterized by improvements in system power, operational lifetime, materials, and overall system efficiency.

10.14.1   Evolution in SP-100 Power Plant Technology

10.14.1.1   System Power and Lifetime

The SP-100 system's power capability evolves from an initial range of 10-40 kWe in the Near-Term phase, to a broader range of 10-100 kWe in the Intermediate Term, and finally reaching up to 300 kWe in the Mature phase. The operational lifetime at full power similarly extends from 3-5 years in the Near-Term to 7-10 years in the Mature phase, reflecting improvements in material durability and system reliability.

10.14.1.2   Reactor Design and Materials

- **Fuel**: The reactor consistently uses uranium nitride (UN) fuel across all phases, chosen for its high thermal conductivity and fission product retention.
- **Coolant**: Liquid lithium remains the coolant throughout the development stages due to its excellent thermal properties and low neutron absorption cross-section.
- **Cladding**: Initial designs use Nb-1% Zr with a rhenium (Re) liner for its chemical stability and strength. As technology matures, the transition to PWC-11/Re cladding offers improved resistance to corrosion and radiation damage.
- **Structure**: The use of niobium alloys (Nb-1% Zr) transitions to PWC-11 in the Mature phase, enhancing structural integrity at high temperatures.

10.14.1.3   Reactor Instrumentation and Control (I&C)

The control and safety mechanisms shift from simpler in-core systems to more complex separate control systems in the Mature phase. This shift allows for more precise reactor management and enhanced safety features, such as the use of sliding neutron reflectors and independent safety rods.

10.14.1.4   Heat Transport and Conversion

- **Pump**: The thermoelectric-electromagnetic (TEM) pump technology remains consistent, leveraging the Seebeck effect for reliable coolant circulation.
- **Converter**: The thermoelectric conversion efficiency improves with the transition from SiGe (with a figure-of-merit ZT of 0.67×10^{-3} K^{-1}) to SiGe(GaP) (with a ZT of 0.85×10^{-3} K^{-1}). The power output per cell increases from 8.8 We to 12.8 We, reflecting advancements in TE materials and converter design.
- **5. Radiator and Heat Rejection**

The heat rejection system evolves significantly:

- **Heat Pipes**: The material transitions from K-Ti in the Near-Term to K-Nb-1% Zr in the Intermediate Term, and then to a smaller diameter K-Ti in the Mature phase, improving heat pipe efficiency and system compactness.

- **Fins and Armor**: The fins and armor materials evolve from titanium (Ti) to carbon-carbon composites (C-C), enhancing thermal conductivity and structural robustness.

- **Ducts**: The ducts transition from titanium to Nb-1% Zr, aligning with the improvements in high-temperature materials used throughout the system.

The evolution of the SP-100 power plant technology reflects a concerted effort to increase power output, extend operational lifetime, and enhance system reliability and safety. Each phase introduces material advancements and system optimizations, positioning the SP-100 as a versatile and robust power source for a wide range of space missions. The improvements in thermoelectric materials, structural components, and cooling technologies exemplify the cutting-edge nature of this nuclear power system.

10.15   Thermionic Reactor Power Plants

10.15.1   Introduction

Thermionic reactors represent a class of nuclear power systems that utilize thermionic emission for direct conversion of heat energy into electrical energy. This method exploits the thermionic effect, where electrons are emitted from a heated material, typically a metal or semiconductor, creating a flow of electric current. Thermionic converters can be integrated into the reactor core (in-core) or positioned outside the core (out-of-core), each configuration offering distinct advantages and challenges.

10.15.1.1   Early Developments and In-Core Thermionic Conversion

The development of thermionic reactors began in the mid-1950s, gaining momentum by the 1960s with a focus on in-core thermionic conversion systems. This period saw significant advancements aimed at utilizing these systems for a range of space applications, from powering unmanned satellites with 5-10 kWe systems to larger 120 kWe systems intended for nuclear electric propulsion. The in-core design, often referred to as the "flashlight" configuration, became a primary focus due to its compact nature and potential for high-temperature operation.

The in-core thermionic concept integrates the thermionic converter directly with the nuclear fuel elements, which allows the system to operate at very high temperatures—

up to 2000 K. The thermionic diode, resembling a flashlight battery, is the fundamental building block of this system. The diode comprises a central fuel emitter, surrounded by a cesium reservoir and an outer collector. The fuel emitter, typically a uranium compound, heats up and emits electrons that travel across a gap filled with cesium ions, which help neutralize space charge and reduce the work function, enhancing electron emission efficiency. This setup allows for a compact and efficient direct conversion of thermal energy into electrical energy.

10.15.1.2   Core Configurations and Reactor Designs

In the typical in-core configuration, multiple thermionic fuel elements are arranged in a reactor core, which can be configured for various power outputs. For instance, a 40-60 kWe reactor may include a combination of thermionic fuel elements and additional uranium-zirconium hydride (U-ZrH) fuel elements to achieve criticality and moderate the neutron spectrum. In contrast, a 100-250 kWe system might use a fast reactor design with uranium-235 fuel, optimizing for higher power output without a moderator.

The physical arrangement of these elements can vary significantly, depending on the desired power output and operational parameters. For example, the 40-60 kWe system uses U-ZrH elements alongside thermionic converters, while the 100-250 kWe system employs a fast reactor core configuration without U-ZrH moderation. Control of these reactors typically involves rotating drums containing neutron-absorbing materials, such as boron carbide, to adjust reactivity and power levels.

10.15.1.3   Advantages and Challenges of In-Core Thermionic Systems

In-core thermionic systems offer several advantages, including compact design and the ability to operate at high temperatures, which enhances thermionic efficiency. They are also scalable across a wide range of power levels, making them versatile for various applications. Moreover, by integrating the thermionic converters within the reactor core, these systems minimize the need for complex high-temperature heat transport mechanisms.

However, significant challenges accompany these advantages. The high operating temperatures can lead to fuel swelling and material degradation, particularly in the emitter materials. Additionally, fast neutron damage and issues related to insulator integrity under high radiation environments pose considerable technical hurdles. The need for reliable, long-life insulators and emitters that can withstand the reactor's harsh conditions is critical to the system's overall durability and efficiency.

10.15.1.4   Out-of-Core Thermionic Concepts and Future Directions

To address some of the limitations of in-core systems, out-of-core thermionic concepts have been explored. These designs separate the reactor core from the thermionic converters, typically using heat pipes to transport thermal energy from the core to the converters. This separation can potentially simplify converter development and reduce the thermal and radiation stresses on the conversion elements.

Despite these efforts, out-of-core systems face their own set of challenges, particularly in achieving efficient thermal coupling and managing the structural integrity of the system's components under prolonged high-temperature conditions. As research progresses, the development of advanced materials and innovative designs continues to push the boundaries of what is possible with thermionic reactor power plants.

Thermionic reactor power plants represent a promising yet challenging technology for direct nuclear-to-electric power conversion, with applications ranging from space missions to potential terrestrial uses. The continued advancement in materials science and reactor design will be pivotal in realizing the full potential of this technology.

10.15.2 Thermionic Developments from 1983-1994

From 1983 to 1994, significant advancements were made in thermionic reactor technology, specifically through the SP-100 program, which aimed to develop a 100 kWe thermionic space power system. The design featured a fast spectrum reactor core equipped with 180 six-cell Thermionic Fuel Elements (TFEs). These elements were essential for converting thermal energy from nuclear fission directly into electrical power through the thermionic emission process.

10.15.2.1 Thermionic Fuel Element (TFE) Verification Program

Initiated in 1986, the TFE Verification Program was a critical component of the SP-100 initiative. It aimed to address and resolve several technical challenges identified during the early concept selection phase. The program's primary objectives were to ensure the viability of multi-cell TFEs for use in thermionic reactors with power outputs ranging from 0.5 to 5 MWe and operational lifetimes of 7 to 10 years. The main concerns included managing fuel/clad swelling, ensuring insulator integrity over long irradiation periods, and demonstrating the overall performance and longevity of TFEs and their components.

The program was later refocused to meet the Department of Defense's (DoD) specific requirements for power systems providing 5 to 40 kWe, with operational lifetimes between 1.5 to 5 years. These systems were seen as ideal for space applications due to their compact radiators, which benefited from higher heat rejection temperatures and efficiencies, and the fact that refractory metals were generally not required outside the fuel and emitter zones.

10.15.2.2 TFE Design and Verification

The TFE, the core component of the thermionic power systems, underwent rigorous testing and development. Each TFE is essentially a miniature thermionic converter, designed to withstand the harsh conditions of space, including high temperatures and radiation. The TFE's design includes a central fuel emitter, typically uranium nitride (UN), surrounded by a collector and insulators. Cesium vapor is used within the interelectrode gap to facilitate electron emission and reduce space charge effects, thereby improving conversion efficiency.

Throughout the TFE Verification Program, several key tests were conducted:
1. **1H1 Test:** This initial test aimed to recover earlier thermionic technology, verify the reference cell's performance, and demonstrate the integration of a single-cell TFE within a sheath tube. Despite experiencing a cesium loss issue, the 1H1 test successfully operated for around 12,000 hours, validating the basic TFE design.
2. **1H2 Test:** Focused on verifying an integral graphite cesium reservoir, the 1H2 unit operated for 8,826 hours before encountering a short circuit caused by a top-cap heater failure.
3. **1H3 Test:** This test allowed fission products to mix with cesium in the interelectrode gap, simulating potential operational conditions. It operated for 13,541 hours, demonstrating significant endurance even with potential defects like a cracked emitter or faulty insulator seal.
4. **3H1 Test:** A three-cell TFE that encountered low cesium pressure issues, affecting its power output. Despite these issues, the test provided valuable data on the operational characteristics of multi-cell configurations.
5. **3H5 Test:** This unit functioned for 8,042 hours without significant degradation, showcasing the robustness of the design.
6. **6H1 Test:** Demonstrated the fabricability and performance of a six-cell TFE, running for 4,318 hours and incorporating improved cesium reservoir controls.

10.15.2.3　Key Achievements and Challenges

The TFE Verification Program made substantial progress in developing components capable of enduring the conditions expected in a thermionic space nuclear reactor. Key achievements included demonstrating high burnup levels for fuel emitters and confirming the long-term stability of sheath insulators and insulator seals. These components showed the ability to withstand the intense radiation and thermal environments over extended periods, with some tests running up to 39,000 hours.

Despite these successes, several challenges remained. These included optimizing seal designs to minimize intercell spacing, improving tungsten-to-niobium joints without using tantalum to reduce neutron absorption, and conducting further life testing to ensure long-term reliability.

The developments from 1983 to 1994 laid a solid foundation for future thermionic reactor systems, with the TFE Verification Program providing crucial insights and solutions to key technical challenges. These advancements paved the way for potential space applications, offering a scalable and efficient means of generating electrical power in harsh environments.

10.15.3 Forty Kilowatt Thermionic Power Systems Program

In 1992, a program was initiated to develop a 40 kWe thermionic power plant with an intended design life of 10 years, though the initial operational requirement was set at 1.5 years. Two primary concepts were selected for this program: the S-Prime Thermionic Nuclear Power System and the SPACE-R Thermionic System. These systems aimed to provide scalable and reliable power solutions for space applications, leveraging existing and advanced thermionic technologies.

10.15.3.1 S-Prime Thermionic Nuclear Power System

The S-Prime system drew heavily from previous thermionic and space reactor technologies, particularly the multi-cell Thermionic Fuel Element (TFE) Verification Program and Russian Topaz I technology. This system was designed to be scalable from 5 to 40 kWe, utilizing multi-cell TFEs within a zirconium hydride moderated reactor. The reactor was cooled using a single NaK-78 pumped loop, with waste heat being rejected through a 24 m² heat pipe radiator system. The design incorporated 84 TFEs arranged in a 12-series by 7-parallel configuration, delivering 30.7 V at 1,430 amps. The reactor's core thermal power was less than 500 kWt, with zirconium hydride acting as the moderator, enhancing the system's efficiency and specific mass.

10.15.3.2 Key Design Features and Performance Parameters:

- **Reactor Core:** The reactor core contained 158 Hastelloy-N cladded pins with a zirconium hydride moderator. Internal surfaces were coated with a SCB glass-enamel layer to prevent hydrogen permeation.
- **Control and Safety:** The system included 18 poison back control drums in the radial reflector and nine B4C safety rods to ensure reactor shutdown capability and maintain subcriticality in water immersion scenarios.
- **Structural Components:** The reactor pressure vessel and internal supports were constructed from 316 stainless steel, ensuring durability and structural integrity.
- **System Efficiency:** The S-Prime system achieved a specific power of 15 We/kg, with an end-of-mission (EOM) net power output of 40 kWe. The system's overall length was 6.78 m, with a radiator area of 27 m².

10.15.3.3 Component Testing and Validation:

Critical components required thorough testing to validate their long-term performance and reliability. These included:

- **Sheath Insulators and Insulator Seals:** Tested to withstand significant neutron fluence and operational stress.
- **Graphite Reservoirs and Fueled Emitters:** Demonstrated stability and functionality under high radiation and thermal conditions.
- **TFEs:** Proved capable of maintaining performance for extended periods under operational loads, demonstrating >5% fuel burnup at 1800 K.

10.15.3.4   Program Status and Future Prospects

While the Forty Kilowatt Thermionic Power Systems Program demonstrated promising early concepts and component testing, it did not progress beyond the initial design and validation stages. The insights and technological advancements gained from this program, however, laid the groundwork for future thermionic power systems, offering scalable and efficient power generation options for space missions. The focus on leveraging existing technologies and integrating new advancements highlighted the potential for continued innovation in space power systems.

10.15.4   Space Power Advanced Core-Length Element-Reactor (SPACE-R)

The SPACE-R reactor power system is an advanced in-core thermionic reactor designed to generate 40 kWe at the end-of-life (EOL) over a projected 10-year operational lifespan. The system features a uranium dioxide (UO_2) fuel core with pumped liquid metal cooling (NaK), a zirconium-hydride moderated core, and a single-cell thermionic fuel element (TFE) design similar to the Topaz II reactors. The design facilitates comprehensive full-power operational testing via electric heating before the nuclear fuel is introduced, allowing for system validation and troubleshooting in non-nuclear conditions.

10.15.4.1   Core Design and Configuration

The SPACE-R core has a compact design with a diameter of 46 cm and a length of 35 cm. It comprises 150 TFEs arranged with yttrium hydride moderator elements clad in beryllium, cooled by NaK-78. The TFEs are arranged to optimize reliability and minimize the risk of fission gas contamination in the electrode space. The reflector, made of 10 cm thick beryllium metal, contains 12 control drums that experience a maximum fast neutron fluence of 3×10^{20} nvt. A unique feature of this reactor is its positive temperature coefficient provided by the zirconium-hydride moderator, which enhances the safety and stability of the reactor.

10.15.4.2   Key Components and Materials

- **Fuel and Moderator:** The core utilizes UO_2 fuel with 93% enrichment, yielding a burnup of 3.5% over ten years. The hybrid moderator design consists of beryllium and yttrium hydride.

- **Reactor Vessel and Shielding:** The reactor vessel and internal supports are made of stainless steel and Hastelloy, with a comprehensive radiation shield made from uranium hydride (UH) and zirconium hydride (ZrH_x). The system also includes a reentry shield for added protection during potential atmospheric reentry.

- **Control Mechanisms:** The system is controlled by 12 rotating control drums made of beryllium with boron carbide (B_4C) poison segments. The design includes a single scram drum for emergency shutdown, ensuring robust control and safety.

10.15.4.3 System Performance and Lifetime Concerns

The SPACE-R system has been designed to achieve a high system efficiency of 7.2%, with an end-of-life net electrical output of 44 kWe. The reactor operates at relatively low temperatures, with the primary NaK loop maintained at 900 K and the potassium-cooled radiator at 815 K. This temperature control minimizes material degradation and extends component lifetimes.

Key lifetime issues include managing fuel swelling, preventing fission product leakage into the electrode spaces, ensuring hydrogen containment, mitigating beryllium swelling, preventing corrosion from coolant interaction, and protecting the system from meteorite or debris damage. The low operational temperatures and use of reliable materials like beryllium help maintain the dimensional stability of the components over the system's intended 10-year lifespan.

10.15.4.4 Advantages and Challenges

The SPACE-R design's significant advantages include its simplicity, cost-effectiveness, high reliability, and the ability to conduct comprehensive system tests before nuclear fuel introduction. The single-cell TFE approach simplifies fabrication and assembly, reduces the potential for contamination, and supports a more compact reactor core size. However, challenges include managing the long-term stability of nuclear ceramic fuels, preventing fission product leakage, and ensuring the durability of high-temperature components such as the heat pipes and electrode seals.

The SPACE-R thermionic power system represents a significant advancement in space nuclear power technology, providing a robust, scalable solution for medium-power applications in space environments. The design leverages proven technologies while addressing critical performance and longevity issues, offering a viable option for long-duration space missions.

10.16   Russian Space Nuclear Power Systems

The Soviet Union extensively utilized space nuclear reactors, particularly in its radar ocean reconnaissance satellites (RORSATs), demonstrating a significant commitment to nuclear power in space. The RORSAT program, characterized by its frequent launches and varied operational durations, featured reactors like the Romashka, which primarily used thermoelectric power conversion to supply electricity. These systems were designed for missions of up to 135 days in low Earth orbit, approximately 250 km in altitude. Upon mission completion, the reactors were typically boosted to a long-term storage orbit around 900 to 1,000 km using solid-rocket propulsion. However, not all disposal maneuvers were successful; out of 31 missions between 1970 and 1988, two experienced uncontrolled re-entries, and another had a partial malfunction resulting in a lower-than-intended disposal orbit.

In 1987 and 1988, the Soviet Union tested a new type of reactor power system using thermionic power conversion. The Cosmos 1818 and Cosmos 1867 missions, known as Topaz I, operated for six months and 346 days, respectively. The Topaz I system was notable for its use of multi-cell thermionic fuel elements (TFEs), with one variant using molybdenum and the other tungsten emitters. The tungsten variant demonstrated longer operational longevity, albeit with some degradation in performance over time, necessitating increased thermal power to maintain output. This testing provided valuable data, indicating that swelling and intercell leakage were significant issues that limited the lifespan of these TFEs.

Following the Topaz I program, the Soviets developed Topaz II, featuring single-cell TFEs. Topaz II systems demonstrated the capability of achieving 1.5 years of operational life in nuclear ground testing, with potential for up to three years based on component life data. The design improvements aimed to address issues encountered in Topaz I, particularly those related to hydrogen loss from the zirconium hydride moderator, a concern estimated at about 1% per year. The low excess reactivity of 65 cents posed a risk of life limitation, especially if reactor cooldown and restart were required. Additionally, the integrity of the oxygen getter system and cesium supply needed further refinement, though these were not deemed critical life-limiting factors.

In the early 1990s, a collaborative effort between the United States and Russia sought to further develop thermionic space reactor technology. This cooperation included the delivery of an unfueled Topaz II reactor to the U.S. and subsequent joint testing. Although there were plans for a flight program, these were not realized. The historical context and technical challenges encountered in these programs underscore the complexity and ambition of deploying nuclear power systems in space, highlighting both the advancements made and the hurdles yet to overcome.

Overall, the Soviet Union's space nuclear power systems represent a significant chapter in the history of space exploration, showcasing the technical prowess and strategic importance placed on these technologies during the Cold War era. The transition from thermoelectric to thermionic power conversion in their space reactors marked a notable evolution in their approach, with each step providing valuable lessons for the future of space nuclear power.

10.16.1 The Romashka Thermoelectric Reactor

The Romashka thermoelectric power reactor, utilized in the Soviet RORSAT (Radar Ocean Reconnaissance Satellite) program, was a notable example of early space nuclear power technology. Designed with a fast neutron spectrum, the reactor utilized uranium-molybdenum alloy fuel, with the uranium being enriched to over 90% uranium-235. The fuel core was compact, measuring 0.2 meters in diameter and 0.6 meters in length, and the entire assembly weighed approximately 53 kilograms, with 30 kilograms attributed to the uranium content. The reactor produced around 100 kWth of thermal power, which was converted into approximately 3 kWe of electrical power through silicon-germanium thermoelectric converters operating at high temperatures of up to 1273 K.

The Romashka reactor's design featured 37 steel-clad fuel rods and a control mechanism comprising six beryllium control elements filled with boron carbide (B4C). These elements could move axially to control the reactor's power output and served as a secondary measure to contain radioactivity within the radiation shield, which was composed of lithium hydride (LiH), stainless steel, tungsten, and uranium alloys. The entire reactor, including its radiation shielding, had a mass of around 385 kg.

Safety considerations were a critical aspect of the Romashka reactor's design, particularly given the low Earth orbit (LEO) deployment of the RORSAT satellites. Upon mission completion, a solid-rocket boost system was intended to elevate the reactor to a higher orbit of 900 to 1,000 km to allow for decay of radioactivity before eventual re-entry. In the event of a boost system failure, the reactor was equipped with a core dispersal mechanism designed to minimize radiation exposure to humans by spreading the radioactive materials over a broad area, reducing the potential radiation dose to no more than 0.5 rem (5 mSv) in the first year after re-entry.

The most notable incident involving the Romashka reactor occurred in 1978 when Cosmos 954 re-entered Earth's atmosphere uncontrollably, spreading radioactive debris over Canada's Northwest Territories. The recovery operation, known as Operation Morning Light, covered a vast area of 124,000 square kilometers, and significant radioactive debris was recovered, including small uranium alloy particles and components believed to be part of the reactor's control system. Another incident occurred with Cosmos 1402 in 1983, where the reactor reportedly burned up over the Atlantic Ocean, preventing any recovery of debris. A near-incident with Cosmos 1900 in 1988 was averted when the reactor was successfully boosted to a safe orbit despite initial concerns about its operational state.

Overall, the Romashka reactors played a crucial role in the Soviet space nuclear program, demonstrating both the potential and challenges of using nuclear power in space. The design focused on compactness and safety, with specific measures to manage the reactor's end-of-life and accidental re-entry scenarios. Despite some failures and risks, these reactors marked significant advancements in space-based nuclear technology.

10.16.2   The Topaz I Reactor

The Topaz I reactor power plants were designed and deployed by the Soviet Union as part of their space nuclear power program. The reactors were used in the Cosmos 1818 and Cosmos 1867 satellites, demonstrating the capabilities of thermionic power conversion in space environments. The primary purpose of these missions was to validate the technology rather than achieve long-term operational life.

10.16.2.1   Design Features

Reactor Core and Configuration: The Topaz I reactors utilized a fast neutron spectrum, employing a reactor core that integrated multi-cell thermionic converters within a zirconium hydride moderator. The core's primary control mechanism consisted of 12 rotatable cylindrical control drums made of boron carbide, encased in a beryllium reflector. This setup allowed for fine control over the reactor's thermal power and reactivity compensation.

Figure 10-16. The Topaz-I reactor.

Thermionic Fuel Elements (TFEs): The core contained 79 TFEs, split into two functional sections: 62 TFEs provided power for spacecraft equipment, while 17 TFEs powered the electromagnetic pump (EMP). Each TFE consisted of a uranium diocide (UO2) fuel element, a cathode made from either a tungsten alloy or molybdenum alloy VM-1, and an anode made from niobium alloy VN-2. Beryllia (BeO) insulators and stainless steel outer casings completed the TFE assembly. The TFEs operated at high temperatures, with emitters reaching around 1,865 K, facilitating efficient thermionic conversion.

Cesium Vapor Supply System: A cesium vapor supply system maintained optimal conditions within the TFEs, ensuring the proper function of the thermionic converters. The system included a cesium thermostat, a cesium vapor generator, and a pyrolitic graphite cesium trap to absorb gaseous impurities. The system was designed to allow for ground control of cesium temperature settings, providing flexibility in maintaining optimal operating conditions.

Cooling and Heat Rejection: The cooling system employed a single-loop design with NaK (sodium-potassium eutectic alloy) as the coolant. The heat was rejected via a tube-screen radiator, which had an emissivity of 0.85 or greater, ensuring efficient thermal management. The radiator design was robust, capable of withstanding mechanical loads across various operational conditions.

Control and Safety Systems: The reactor's control system included an automatic controller with a storage battery, capable of regulating start-up, maintaining operational parameters, and controlling output voltage. In case of emergencies or shutdown requirements, the control drums provided a mechanism for rapidly adjusting reactor power output.

10.16.2.2 Technical Specifications

- **Electrical Power Output**: Approximately 6 kWe from the working section, with an output voltage of 32 V and a current of 180 A. The pump section provided a lower voltage of 1.1 V and a current of 1200 A.
- **Thermal Power**: The reactors produced between 110 and 150 kWt of thermal power.
- **Cesium Consumption**: The system had a cesium consumption rate ranging from 6 to 20 g/day, with the cesium vapor pressure maintained between 266 to 730 Pa.
- **Dimensions and Mass**: The SNPS (Space Nuclear Power System) had a total mass of around 1200 kg, with a length of 4.7 m and a diameter of 1.3 m. The radiator had a surface area of approximately 7 m^2.

10.16.2.3   Operational and Flight Performance

The Topaz I reactors were tested in space aboard Cosmos 1818 and Cosmos 1867. Cosmos 1818 operated for 143 days, while Cosmos 1867 operated for 342 days. Both reactors were placed in orbits ensuring a minimum orbital lifetime of 350 years, providing ample time for radioactive decay to safe levels. The primary limiting factor for the operational life of the reactors was the cesium supply in the vapor generator.

Despite encountering short circuits shortly after startup, both reactors successfully transitioned to nominal operating conditions. Differences in operational performance, such as discrepancies in hydrogen leakage rates, were observed between space and ground testing, providing valuable data for future improvements in reactor design and operation.

The Topaz I reactors represented a significant advancement in thermionic space nuclear power systems. They demonstrated the feasibility of using thermionic converters in space, with practical applications for providing power to spacecraft. The design emphasized high-temperature operation, efficient heat rejection, and robust control mechanisms, making them a critical component of the Soviet space program's power systems.

10.16.3   The Topaz II Reactor

The Topaz II reactor is a space nuclear power system designed to provide approximately 6 kWe of electrical power. Unlike its predecessor, Topaz I, which utilized multi-cell thermionic fuel elements (TFEs), Topaz II features single-cell TFEs. This design change allows the system to be fully tested at operational temperatures using electric heating, facilitating comprehensive pre-flight verification and validation.

10.16.3.1   Core Configuration and Components

- **Core Dimensions**: The reactor core of Topaz II is a right circular cylinder with a diameter of 26 cm and a height of 37.5 cm. It houses 37 cylindrical TFEs arranged in a roughly triangular pitch within a zirconium hydride ($ZrH_{1.85}$) moderator block. The moderator is encapsulated in a stainless steel calandria with coolant channels grooved into its surface.
- **Thermionic Fuel Elements (TFEs)**: Each TFE consists of UO_2 fuel pellets, a monocrystalline molybdenum alloy emitter tube coated with tungsten, a collector, and a cesium vapor interelectrode gap. The fuel is enriched to 96% ^{235}U, with pellets approximately 8 mm in height and 17 mm in diameter, some featuring a central hole to flatten the radial power profile. The emitter serves as both the fuel container and the thermionic emitter, operating at temperatures up to 2100 K. The collector is made from monocrystal molybdenum and operates at around 925 K at the outlet.
- **Coolant System**: The reactor is cooled using a eutectic NaK mixture, which

remains liquid throughout the reactor's operational life, except during end-of-mission shutdown. The system includes a single direct current (DC) conduction pump powered by a dedicated set of three TFEs. The pump circulates the NaK coolant, which transfers waste heat from the reactor to a tube-and-fin radiator for dissipation into space.

Table 10-8. Comparison of Topaz I and II reactors

Characteristic	TOPAZ I	TOPAZ II
Electrical Power (kWe)	5 to 6	6
Voltage at Lead (V)	5 to 30	28 to 30
Thermal Power (BOL/EOL) (kWt)		115 / 135
Number of TFEs	79	37
Cells/TFE	5	1
System Mass (kg)		1,061
Emitter Diameter (cm)	1	1.73
Core Length (cm)	30	37.5
Core Diameter (cm)	26	26
Reactor Mass (kg)		290
Moderator	ZrH$_{1.8}$	ZrH$_{1.85}$
Emitter Material	Mo/W	Mo/W
Emitter Temperature (K)	1,773	1,800 to 2,100
Coolant	Pumped NaK	Pumped NaK
Number of Pumps	1	1
Type of Pump	Conduction	Conduction
Pump Power	19 TFEs	3 TFEs
Reactor Outlet Temperature (BOL/EOL) (K)	773 / 873	560 / 600
Coolant Flow Rate (kg/s)		1.5
Cesium Supply	Flow Through	Flow Through
Axial Reflector	Be Metal	Be Metal
Radial Reflector	Be Metal	Be Metal
Number of Control Drums	12	12
Shield Mass (kg)		390
Radiator Type		Tube and Fin
Radiator Area (m²)		7.2
Radiator Mass (kg)		50

10.16.3.2 Key Systems and Features

- **Cesium Supply and Management**: A still supplies cesium vapor to the interelectrode gap, venting impurities directly into space. The system maintains a cesium consumption rate of about 0.5 g/day.

- **Reflectors and Control Drums**: The reactor includes a beryllium metal axial reflector and a radial reflector made of beryllium. Twelve rotating control drums, containing boron carbide for neutron absorption, manage reactor power and safety, with nine used for regulation and three for safety shutdown.
- **Radiator and Shielding**: The waste heat is rejected through a tube-and-fin radiator, constructed from steel tubes and copper fins, coated with glass to enhance emissivity. The radiation shield, a truncated cone filled with lithium hydride (LiH), provides neutron and gamma radiation attenuation. The shield's design minimizes radiation streaming, with coolant pipes passing through at specific angles.
- **Gas Systems**: The reactor contains various gas systems, including a $CO2\_22$/He mix within the moderator region to inhibit hydrogen release, helium in the TFE gas gap for heat transfer, and helium for the radiation shield. An argon/helium mix is used in the volume accumulator.

10.16.3.3   Testing and Performance

- **System Testing**: The Topaz II underwent extensive testing, including component and system-level evaluations. Testing phases involved both preliminary functional sufficiency checks and formal tests to meet defined acceptance criteria. The system demonstrated stable operation under various conditions, with the longest test duration reaching 14,000 hours.
- **Operational Considerations**: The reactor operates with a maximum coolant temperature rise of 40 K from inlet to outlet. It has a calculated operational life of 4 to 5 years, based on fuel behavior, including swelling and fission gas release, as well as the durability of TFE materials under irradiation.

The Topaz II reactor's design advances over Topaz I, particularly with the single-cell TFE configuration, allow for a more streamlined and reliable power system for space applications. This system has demonstrated the potential for extended operation in space environments, with comprehensive pre-flight testing ensuring reliability and safety.

10.17   Project Prometheus: Gas-Cooled Reactor

Project Prometheus was initiated by NASA in 2003 with the goal of developing nuclear-powered systems to enable long-duration space missions utilizing electric propulsion. The project aimed to overcome the limitations of solar energy for powering spacecraft in the outer reaches of the solar system, where solar irradiance is significantly reduced. The primary mission envisioned under Project Prometheus was the Jupiter Icy Moons Orbiter (JIMO), which was intended to explore the Jovian moons Callisto, Ganymede, and Europa. Additionally, other potential missions considered under this initiative included explorations of Saturn and its moons, Neptune and its moons, comet and multi-asteroid sample return missions, and Kuiper Belt rendezvous missions.

The missions were designed to span durations ranging from 10 to 12 years, as in the case of the JIMO mission, up to a maximum of 20 years for missions targeting the Kuiper Belt. To support these extended mission durations and the significant power requirements—estimated at approximately 200 kWe—Project Prometheus focused on developing a nuclear reactor system that could reliably provide the necessary energy. The payload accommodation envelope of the proposed Deep Space Vehicle was designed to support a mass capability of no less than 1,500 kg.

A key technological component of Project Prometheus was the selection of a gas-cooled reactor system, coupled directly with a Brayton energy conversion system. This combination was chosen for its potential to provide efficient and reliable power in the harsh conditions of space. The gas-cooled reactor design offered advantages in thermal efficiency and power density, making it suitable for long-duration missions with significant energy demands.

The projected timeline for the JIMO mission included a planned launch date in 2015. However, the project's success hinged on several critical developments, including the fabrication and testing of reactor module materials capable of withstanding the harsh environmental conditions and prolonged operational periods associated with deep space missions. Another critical aspect was the establishment of nuclear testing facilities to support the rigorous testing and validation of the reactor system design.

Despite the ambitious goals and the initial progress made, Project Prometheus was terminated in 2005. The decision to halt the project was influenced by various factors, including budget constraints and the technological challenges associated with developing such an advanced nuclear power system. Nonetheless, the research and development efforts undertaken during the project provided valuable insights and laid the groundwork for future endeavors in nuclear-powered space exploration.

In addition to deep space missions, the reactor module technologies developed under Project Prometheus were also considered for potential use in Lunar and Mars surface power reactors. These applications would require similar advancements in nuclear fuel, reactor core materials, coolant systems, and instrumentation and control technologies to support sustainable human and robotic exploration on extraterrestrial surfaces.

10.17.1   Project Prometheus Design Concept

Project Prometheus aimed to develop advanced space nuclear propulsion and power systems, particularly focusing on a gas-cooled reactor with a Brayton cycle energy conversion system. This design was chosen to support long-duration space missions, such as the Jupiter Icy Moons Orbiter (JIMO) and other deep space exploration missions. The following sections detail the key requirements, impacts, and implementation strategies of the Project Prometheus design concept, alongside technical specifics of the chosen gas-cooled reactor system.

10.17.1.1   Key Requirements and Implementation

The key requirements for the reactor module development are outlined below. These requirements address various aspects, including extensibility to surface operations (e.g., Lunar and Mars missions), electrical power output, operating lifetime, fault tolerance, and radiation shielding. These criteria influenced the design and selection of materials and technologies, ensuring the reactor module could meet the demanding conditions of deep space missions.

10.17.1.2   Design and Implementation Details:

1. **Extensibility and Compatibility:** The reactor's materials and design must be suitable for potential future surface missions, necessitating compatibility with environments like the Moon and Mars. This includes considering temperature, radiation, and pressure variations.
2. **Power Output and Efficiency:** The reactor was designed to deliver a nominal 200 kWe to the spacecraft, with a thermal power output of approximately 1 MWt. The electrical power output is essential for sustaining spacecraft systems and electric propulsion during mission durations of up to 20 years.
3. **Lifetime and Redundancy:** While the JIMO mission specifically required a 12-year operational life, the reactor design aimed for a 20-year lifespan to accommodate potential extended missions. To ensure reliability, particularly in deep space where repair is not feasible, the design includes redundant systems where practical, such as dual Brayton cycle loops.
4. **Fault Tolerance and Autonomy:** The system needed to autonomously detect and correct faults, particularly those preventing thrust, within one hour. This capability was critical to maintaining mission trajectory and safety.
5. **Radiation Shielding:** The design incorporated shielding to protect the spacecraft's payload from radiation emitted by the reactor. The shielding was designed to reduce neutron and gamma fluxes to acceptable levels, ensuring the safety of sensitive instrumentation and potential human presence.

10.17.1.3   Reactor and Power Conversion System

The gas-cooled reactor design selected for Project Prometheus utilizes a helium-xenon (He-Xe) gas mixture as the coolant. This choice was based on several advantages, including chemical inertness, compatibility with high temperatures, and favorable thermal conductivity. The Brayton cycle was selected for energy conversion due to its higher efficiency at the expected operating temperatures compared to other cycles, such as Stirling or Rankine.

10.17.1.4   Core and Fuel Element Design

The reactor core consisted of cylindrical fuel pin elements arranged within a structural matrix. These fuel elements are critical to the reactor's operation and safety, as they contain the nuclear fuel and facilitate the transfer of thermal energy to the coolant. The core and fuel element design options included various configurations to optimize for mass, thermal performance, and structural integrity under irradiation and thermal stresses.

- **Fuel Composition and Cladding:** The core utilized UO_2 as fuel, with potential cladding materials including refractory metals like molybdenum, niobium alloys, and advanced ceramics such as silicon carbide (SiC). The choice of cladding material significantly impacts the reactor's ability to withstand high temperatures and corrosive environments, as well as its neutron economy.
- **Core Geometry Options:** Several core geometries were considered, including open lattice designs, pin-in-block configurations, and modular cermet designs. Each design had unique advantages and challenges, such as mechanical support, thermal management, and neutron moderation.
- **Reactor Control and Safety:** The core included fixed and movable reflector segments, alongside safety rods. These components were essential for controlling the reactor's reactivity, managing power output, and ensuring safe shutdowns during emergencies or at the end of the mission.

10.17.1.5   Brayton Cycle Energy Conversion

The thermal energy generated in the reactor core was converted to electrical energy using a closed Brayton cycle. The system included:

- **Turbine-Alternator-Compressor Unit:** This unit converted thermal energy into electrical energy. The hot gas exiting the reactor core expanded through a turbine, which was mechanically coupled to both a compressor and an alternator. The alternator generated electrical power for spacecraft systems, while the compressor recycled the working fluid.
- **Heat Exchangers and Radiator:** After expanding through the turbine, the working fluid passed through a recuperator and a cooler. The recuperator preheated the gas entering the reactor, improving system efficiency. The cooler transferred residual heat to a pumped liquid loop, which then radiated the heat to space via a large radiator panel system.
- **Radiator Design and Heat Rejection:** The design aimed to maximize heat rejection efficiency while minimizing mass. A tube-and-fin radiator structure, possibly with glass-coated fins for enhanced emissivity, was used to radiate waste heat into space, ensuring the system's thermal balance.

Project Prometheus's gas-cooled reactor and Brayton cycle energy conversion system represent a sophisticated approach to providing reliable and long-lasting power for deep space missions. The project's termination in 2005 halted further development, but the design and technological advancements achieved laid a solid foundation for future space nuclear power systems. The technical considerations encompassed advanced materials, intricate thermal and structural management systems, and rigorous safety and fault tolerance mechanisms, showcasing the complexity and innovation required for nuclear-powered space exploration.

10.17.2 Prometheus Program Status at Termination

At the termination of the Prometheus Project, significant advancements had been made in the development of a space nuclear reactor system designed to generate 200 kWe for deep space exploration. This project was ambitious in its scope, aiming for longer mission lifetimes of up to 20 years, greater autonomy for deep-space operations, and flexibility to support a variety of missions, including those with propulsion and scientific exploration components. The program faced numerous challenges, both technical and infrastructural, which ultimately influenced its conclusion.

10.17.2.1 Key Achievements

10.17.2.1.1 System Suitability and Technology Readiness:

- The direct gas Brayton cycle was identified as a highly suitable system for deep space and planetary surface missions. This system's compatibility with the operational environment and power requirements of such missions was validated, highlighting its potential for future space reactor applications.
- The program determined that the necessary technologies were within reach for a first space reactor mission. However, further work was required to resolve uncertainties, particularly regarding material performance under irradiation conditions.

10.17.2.1.2 Material Performance Challenges:

- A significant limitation encountered was the uncertainty in material performance, particularly concerning the fuel and cladding materials in an irradiation environment. The fast neutron spectrum of the proposed reactor necessitated extensive testing of candidate materials under similar conditions to ensure long-term reliability and safety.
- There was a pressing need for additional research to quantify the behavior of these materials, especially under the expected high-radiation, high-temperature conditions. Specific concerns included swelling, creep, embrittlement, and chemical compatibility of materials over the extended mission durations.

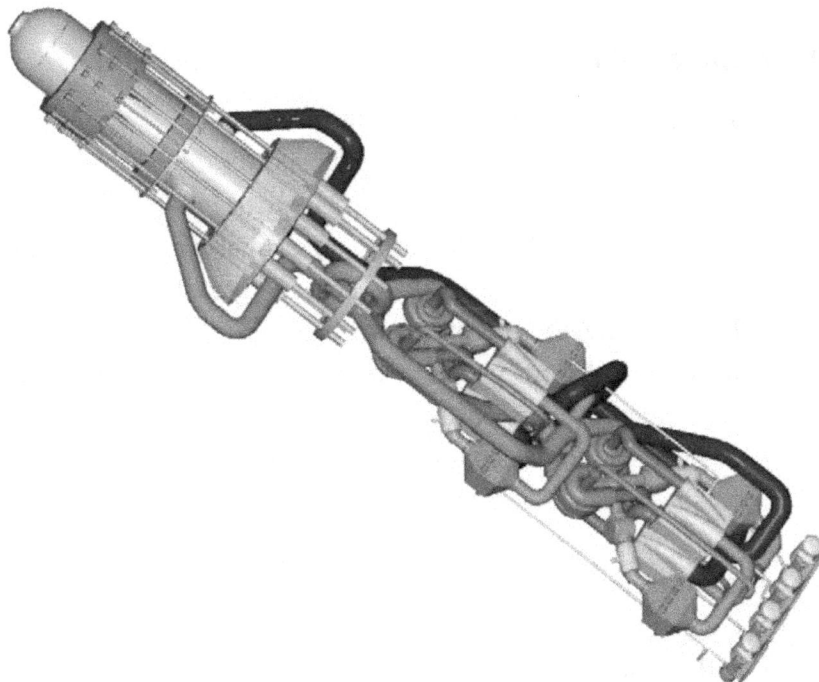

Figure 10-17. Prometheus reactor module

10.17.2.1.3 Infrastructural and Facility Limitations:

- **Nuclear Facilities for Fast Spectrum Testing**:
  - The U.S. faced a critical shortage of facilities capable of conducting necessary fast spectrum irradiation tests. The decommissioning of key facilities such as the Experimental Breeder Reactor-II (EBR-II) and the Fast Flux Test Facility (FFTF) left a significant gap in the infrastructure needed for testing and validating materials and systems under fast neutron flux conditions.
- **Critical Experiment Facilities**:
  - The absence of operational critical experiment facilities posed another significant challenge. These facilities are crucial for performing benchmark critical experiments, which are essential for verifying reactor physics calculations and ensuring reactor safety.
- **Mock-Up and Full-Scale Testing Facilities**:
  - There were also deficiencies in facilities capable of conducting

comprehensive reactor mock-up tests and full-scale reactor testing of prototype and flight systems. Such facilities are vital for validating the integrated system performance and reliability before deployment in space.

10.17.2.2   Developmental Roadblocks and Recommendations:

- **Material Characterization and Testing**:
  - The program identified an urgent need for the establishment or reactivation of facilities capable of performing fast spectrum irradiation tests. Such facilities are critical for providing the data needed to finalize material selection and reactor design.
- **Critical Facility Development**:
  - Recommendations were made for the development of new critical experiment facilities and test reactors to support the continued development of space nuclear systems. These facilities would enable the necessary safety and performance testing required to meet NASA's rigorous standards for space missions.
- **International Collaboration**:
  - Given the infrastructural limitations within the U.S., international collaboration was suggested as a potential avenue to access necessary testing facilities and expertise, leveraging existing global infrastructure to advance the program's objectives.

The termination of the Prometheus Project marked the end of a significant effort to advance space nuclear reactor technology for deep space missions. While considerable progress was made, particularly in system design and the identification of suitable technologies, critical gaps remained in material testing and infrastructural support. Addressing these gaps would require significant investment in new facilities and international partnerships. The project's findings and the challenges identified provide a valuable foundation for future initiatives in space nuclear power systems, highlighting the technological and infrastructural needs to realize such ambitious goals.

10.18   Fission Surface Power (FSP)

The advancement of human space exploration necessitates the development of reliable and efficient power systems capable of sustaining long-duration missions in challenging extraterrestrial environments. As missions to the Moon and Mars become increasingly ambitious, the demand for robust power systems has intensified, especially given the unique conditions faced on these celestial bodies. Both the Moon and Mars present significant challenges for traditional solar power systems due to their environmental conditions. On the Moon, a 29.5-day rotational period results in prolonged lunar nights of approximately 354 hours, during which solar power is unavailable. Similarly, Mars experiences reduced sunlight due to its distance from the Sun, frequent dust storms, and higher latitudes, which further diminish the effectiveness of solar power. Consequently, there is a pressing need for alternative power solutions that can reliably provide the necessary energy for sustained human presence and operations.

Fission Surface Power (FSP) systems have emerged as a leading candidate to meet these power demands. Nuclear fission, the process of splitting atomic nuclei to release energy, offers significant advantages over solar power systems, particularly in terms of mass and volume efficiency. Unlike solar arrays, which require extensive surface area and energy storage systems to cope with periods of darkness, FSP systems can provide a continuous and dependable power output. Additionally, recent studies by NASA suggest that FSP systems are competitive with solar power in terms of cost, making them a viable option for future missions.

Under the NASA Exploration Technology Development Program, in collaboration with the Department of Energy, a comprehensive program has been initiated to develop and validate FSP technologies. The objective is to produce viable FSP designs by 2013, with a view towards supporting lunar outpost deployments and subsequent Mars missions. The key goals of this program include the development of FSP concepts that are cost-effective, technically sound, and capable of meeting the stringent power requirements of surface missions. Moreover, the initiative aims to establish a solid technical foundation for FSP designs, mitigate development risks, and provide credible cost estimates for future flight systems.

FSP systems must adhere to rigorous safety and operational standards to ensure their feasibility for space missions. Safety protocols mandate that the reactor remains in a subcritical state during launch and transit phases, preventing the production of fission products and associated radiation risks. Upon deployment, the reactor will be activated only when necessary, with shielding mechanisms in place to protect crew members and equipment from radiation. Shielding can be provided using either in-situ resources or materials transported from Earth. Post-operation, the reactor's radiation levels are designed to diminish to relatively low residual levels, ensuring minimal long-term environmental impact.

A key feature of FSP systems is their operational robustness, which is crucial for the success of long-duration space missions. The design philosophy emphasizes conservatism, simplicity, and reliability. The system is engineered to be multi-fault tolerant, with components designed for long life and minimal maintenance requirements. This includes a self-regulating reactor design with negative temperature reactivity feedback, ensuring safe responses to perturbations and maintaining consistent power output despite variations in electrical loads. The use of established terrestrial nuclear technologies and extensive databases of prior nuclear power system experiences further enhance the reliability and cost-effectiveness of FSP systems.

In terms of design specifications, FSP systems are required to deliver at least 40 kWe net power output, with an operational life of no less than eight years. They must be capable of functioning under various environmental conditions, including the cold temperatures and sun angles of the lunar surface. Additionally, the systems must be resilient to credible component failures, maintaining at least 50% power output even in the event of a failure. Safety considerations include limiting radiation exposure to less than 5 rem/year for unshielded crew members at a distance of 100 meters, assuming appropriate shielding measures are implemented.

As NASA and other space agencies progress towards establishing a sustained human presence on the Moon and Mars, the development of reliable power systems becomes increasingly critical. FSP offers a promising solution to the challenges posed by these harsh environments, providing a steady and reliable power source essential for life support, scientific exploration, and industrial activities. The ongoing efforts to advance FSP technology represent a significant step towards achieving long-term human exploration and habitation of extraterrestrial surfaces.

10.18.1   Description of the Fission Surface Power (FSP) Conceptual Design

The Fission Surface Power (FSP) system is being developed as a reliable power source for lunar and Martian missions. This system is designed to provide continuous power in environments where traditional solar power systems are less effective due to extended periods of darkness or reduced sunlight. The architecture of the FSP system is illustrated in Fig. 1, which outlines the major components and their interactions. The system consists of three primary modules: the Reactor Module, the Power Conversion Module, and the Heat Rejection Module.

10.18.1.1   Reactor Module

The Reactor Module encompasses the reactor core, the primary coolant loop, and shielding elements. The chosen reactor design is fueled by Uranium Dioxide (UO_2), a decision driven by its favorable characteristics in fast-spectrum nuclear reactions. The coolant used in this system is a liquid metal alloy of sodium-potassium (NaK), selected for its excellent thermal conductivity and low melting point, which makes it suitable for the low-temperature operational regime of the reactor. The system avoids the use of refractory metals by limiting the reactor outlet temperature to 900 K. This limitation simplifies material selection, allowing for the use of stainless steel and other established materials that are well-characterized under these conditions. The reactor design also includes substantial radiation shielding to protect nearby equipment and personnel, with the option to use in-situ resources such as lunar regolith for additional protection.

Figure 10-18. Mars FSP Concept.

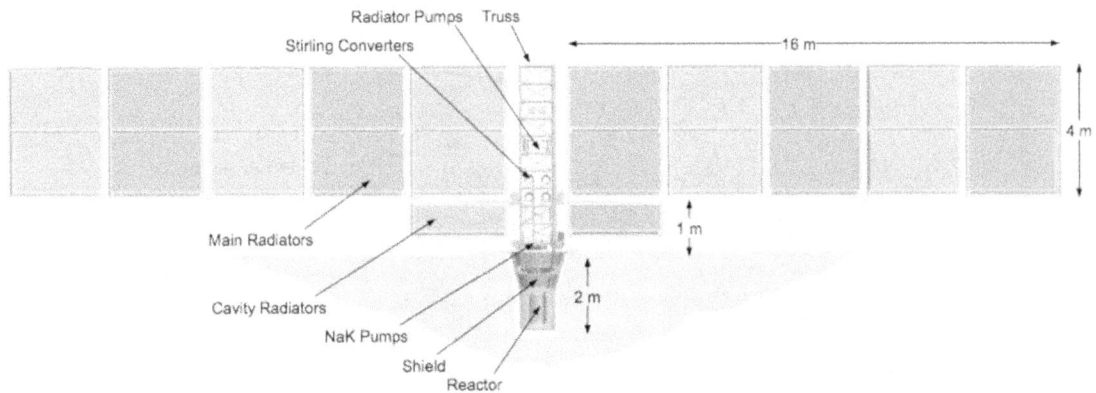

Figure 10-19. FSP Concept layout.

10.18.1.2 Power Conversion Module

The Power Conversion Module is responsible for converting the thermal energy generated by the reactor into electrical power. The primary conversion technology selected is the Free-Piston Stirling (FPS) engine, known for its high efficiency and reliability in converting heat to electricity. The FPS engines operate in a closed cycle, where a working fluid is heated by the reactor, expands to do work, and is then cooled before being reheated. Each Stirling converter in the module includes two axially opposed Stirling engines coupled with linear alternators, producing electrical power at 400 Vac, 100 Hz. This design choice provides redundancy and ensures partial power production even if one unit fails. The Stirling system operates with a hot-end temperature of 830 K and a cold-end temperature of 400 K, which is maintained by a water cooling loop.

10.18.1.3   Heat Rejection Module

The Heat Rejection Module is critical for maintaining the thermal balance of the system. It dissipates excess heat from the power conversion process to space. The module includes a set of radiators that operate at an effective temperature of 380 K. The design incorporates vertical radiators deployed autonomously or with crew assistance, depending on the configuration. The radiators are mounted above the reactor and power conversion equipment to minimize contamination from lunar or Martian dust and to maximize thermal efficiency. The radiators are crucial for the system's operation, as they remove waste heat from the Stirling converters and other components, ensuring stable and efficient operation.

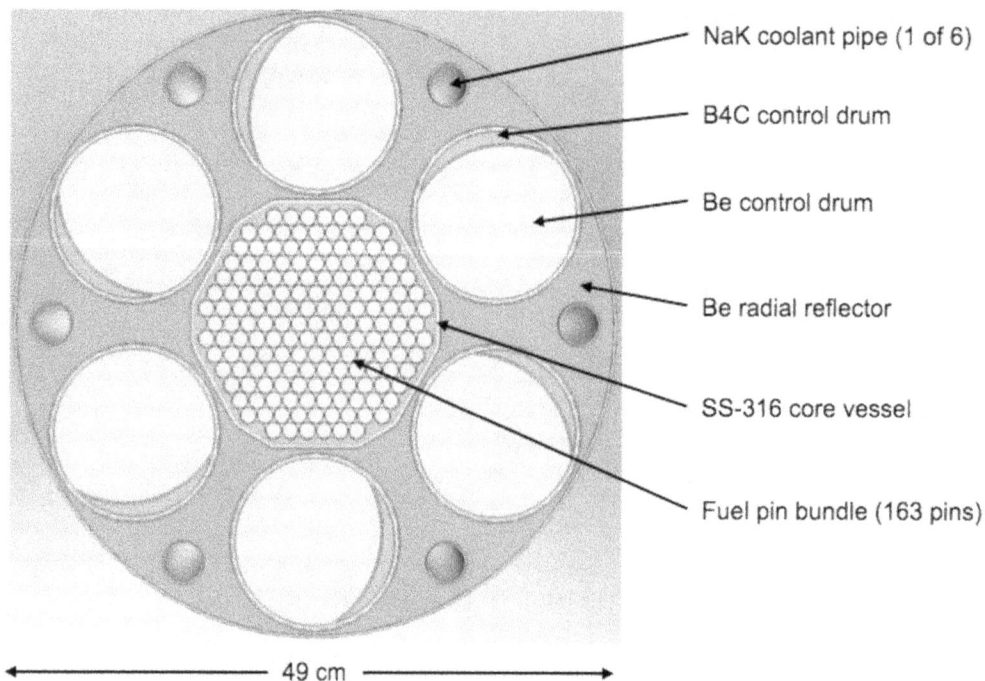

NaK coolant pipe (1 of 6)

B4C control drum

Be control drum

Be radial reflector

SS-316 core vessel

Fuel pin bundle (163 pins)

49 cm

Figure 10-20. FSP reactor module radial cross-section.

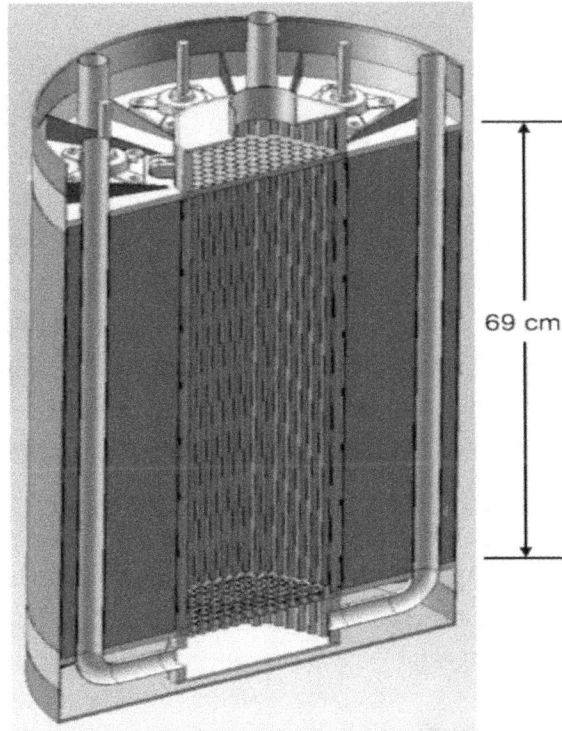

Figure 10-21. FSP reactor module axial cut-away view.

10.18.1.4 System Configurations

The FSP system can be deployed in two primary configurations: *landed* and *emplaced*.

- **Landed Configuration**: In this setup, the entire FSP system remains on a dedicated lander platform. This configuration is advantageous for its ease of deployment, as it does not require additional construction equipment or significant crew involvement. The lander includes pre-installed shielding to protect the outpost from radiation, with a designed separation distance of approximately 1 km. This separation helps to minimize radiation exposure, reducing it to less than 5 rem/year at the outpost location. The system also features a 4π-shaped shield to contain radiation, ensuring safety across the operational area.

- **Emplaced Configuration**: This configuration involves off-loading the FSP system from a cargo lander and installing it with the assistance of crew members and construction equipment. The reactor and associated systems may be buried or shielded with local regolith, offering superior radiation protection and allowing for closer proximity to the outpost—typically around 100 meters. This setup facilitates easier maintenance and potentially reduces system mass, as it can utilize locally available materials for shielding. The emplaced configuration also supports direct power transmission at 400 Vdc, which is then converted to 120 Vdc for distribution.

10.18.1.5   Notational Concept Layout and Stowed Configuration

The FSP system's notational layout involves a deployed span of approximately 34 meters from tip to tip, with radiators positioned about 1 meter above the surface to reduce dust accumulation. The reactor is situated at the bottom of a 2-meter excavation, with a plug shield protecting the equipment above. In its stowed configuration, the system measures approximately 3 by 3 by 7 meters, compact enough for transport and initial deployment. The deployment mechanism involves a truss structure with a scissor mechanism similar to that used on the International Space Station radiators.

The FSP system's conceptual design incorporates proven nuclear technologies with innovative adaptations for space applications. It balances safety, efficiency, and reliability, making it a promising candidate for providing power to lunar and Martian missions. The design leverages terrestrial nuclear experience and aims for high operational reliability, ensuring a stable power supply for critical mission operations.

10.18.2   Power Conversion Subsystem

The Power Conversion Module of the Fission Surface Power (FSP) system employs advanced Stirling converters to efficiently transform thermal energy into electrical power. The system comprises four Stirling converters, each capable of generating approximately 12 kWe, resulting in a total power output of 48 kWe. This design leverages technology from the Advanced Radioisotope Power System program, scaling it for the FSP's requirements.

Each Stirling converter features two axially opposed free-piston engines connected to linear alternators. This configuration minimizes vibrations and enhances the system's efficiency. The converters are approximately 1.2 meters long and are heated by NaK (sodium-potassium) coolant circulating through intermediate heat exchangers. These heat exchangers distribute the thermal energy evenly across the Stirling heater heads, which are constructed from Inconel 718 and operate at temperatures of 830 K on the hot end and 415 K on the cold end. The overall efficiency of the Stirling converters is around 28%, and they generate electricity at 400 Vac and 100 Hz, which is then conditioned by the Power Management and Distribution (PMAD) system to meet the mission's power needs at 120 Vdc.

The PMAD system manages the electrical output, ensuring a consistent power supply despite fluctuations in user demand. It includes features such as a shunt to manage user load variations and a parasitic load radiator to dissipate excess power. The system is designed to handle auxiliary loads, including pumps and motors, and to maintain a net output of 40 kWe for the outpost. The system's resilience is enhanced by redundant components and parallel fluid loops, allowing for continued operation even in the event of individual component failures.

The Stirling converters' design and construction are informed by extensive testing and experience from the Advanced Stirling Radioisotope Generator (ASRG) program. This includes innovations like the use of MarM-247 superalloy for higher temperature tolerances, enabling potential future enhancements in efficiency. The design also includes measures for autonomous operation, fault isolation, and the capability to continue operating in reduced capacity if one engine fails. Overall, the Power Conversion Module's sophisticated engineering ensures reliable and efficient power generation for the FSP system.

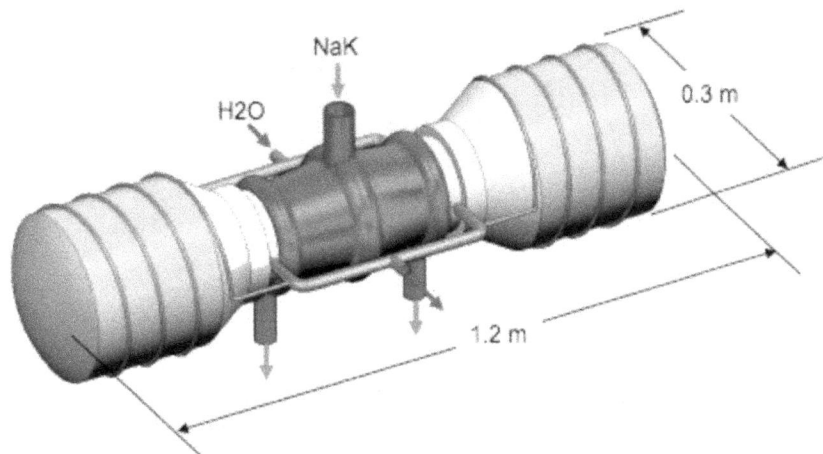

Figure 10-22. FSP 12 kWe dual-opposed Stirling convertor.

10.18.3   Heat Rejection Subsystem

The Heat Rejection Module of the Fission Surface Power (FSP) system is designed to dissipate excess thermal energy generated during power conversion, ensuring the system's stability and efficiency. The module features a water-based heat transport system coupled with two-sided composite radiator panels. The radiators utilize titanium-water heat pipes, which are integrated into four heat transport loops connected to two main radiator wings. Each wing comprises 10 sub-panels, each approximately 3.5 meters wide and 1.75 meters tall, collectively providing a two-sided radiator area of 175 square meters. This design includes a 10% area margin to account for operational uncertainties.

The system operates by circulating water heated to 400 K from the Stirling converters through the radiator panels, where the water releases heat before returning at 370 K. The total heat rejected by the radiators is approximately 120 kW. Each loop, equipped with a dedicated pump and volume accumulator, handles about 30 kW of heat rejection. The radiator panels are constructed using a sandwich design with aluminum honeycomb cores and polymer-matrix composite face sheets, incorporating titanium-water heat pipes for efficient heat transfer. These panels are connected via flexible interconnects, allowing them to be folded compactly for transport.

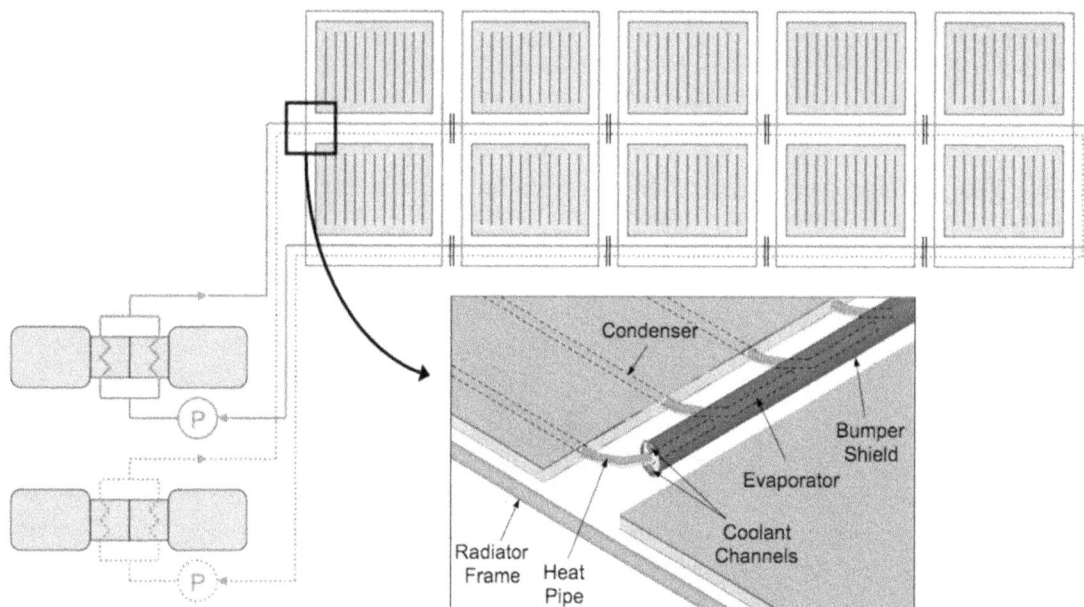

Figure 10-23. FSP heat transport.

To protect against micrometeoroids, the manifold transferring heat from the water loops to the heat pipes is shielded by a bumper. The radiator's design also includes low-absorptivity and low-emissivity mylar surface aprons, reducing thermal contributions from the surroundings and ensuring effective cooling even in worst-case sun angle conditions at the equator. The effective radiator sink temperature is maintained at around 250 K, optimizing the heat rejection process.

10.19   The Kilopower Project

The Kilopower Project was initiated by NASA in 2015 to demonstrate subsystem-level technology readiness of small space fission power systems (FPSs) in the 1-10 kWe range. The centerpiece was the Kilowatt Reactor Using Stirling TechnologY (KRUSTY) test, which involved developing and testing a 1 kWe-class FPS ground technology demonstrator.

Key objectives were to:
1. Demonstrate technologies extensible to 1-10 kWe space FPSs
2. Enable modular surface power for human exploration and future deep space science missions
3. Achieve Technology Readiness Level 5 (component validation in relevant

environment)

The project was co-funded by NASA and the DOE National Nuclear Security Administration. Key participants included:

- NASA Glenn Research Center: Led test design, built balance of plant components
- NASA Marshall Space Flight Center: Developed electrical reactor simulator, shielding
- Los Alamos National Laboratory: Led reactor design, performed nuclear testing
- Y-12 National Security Complex: Fabricated HEU reactor core
- Nevada National Security Site: Supported nuclear testing

10.19.1   KRUSTY Development Plan

The KRUSTY demonstration proceeded through several phases:

1. Flight Concept Development
   - Examined 1 kWe deep space and 10 kWe planetary surface configurations
   - Selected 1 kWe as baseline for affordability and testing feasibility
   - Developed preliminary 10 kWe Mars surface concept to identify future challenges
2. Thermal Prototype Development and Test
   - Validated heat transfer from core simulator to sodium heat pipes
   - Tested heat pipe clamping methods
   - Enabled progression to depleted uranium (DU) testing
3. Depleted Uranium Test
   - Demonstrated Y-12 readiness to produce HEU core
   - Refined experiment preparation and assembly procedures
   - Electrically heated testing through anticipated test conditions
   - Evaluated Stirling engine interface configurations
4. Preparation for Nuclear Test
   - Y-12 cast HEU core sections
   - Performed assembly rehearsals
   - Shipped hardware to Nevada test site
   - Began criticality testing of nuclear components
5. HEU Testing The nuclear testing proceeded through four phases:

a) Component Critical Testing
   - Characterized reactivity of core, reflector, shielding components
   - Verified analytical models

b) Cold Criticals
   - Tested integrated core, heat pipes, and power conversion

- Further refined analytical models

c) Warm Criticals
- Determined temperature feedback with increasing reactivity insertions
- Validated models for high-temperature run

d) Full-Power Test
- 28-hour test simulating space mission profile
- Demonstrated start-up, steady-state, and transient operations

10.19.2   KRUSTY Experiment Design

Key features of the 1 kWe KRUSTY experiment design included:

Reactor:
- Highly enriched uranium-molybdenum core (3 sections)
- Beryllium oxide neutron reflector
- Boron carbide control rod
- ~4 kWth at 800°C nominal operating point

Heat Transfer:
- 8 sodium-filled Haynes 230 heat pipes
- Clamped to core with heated stainless steel rings

Power Conversion:
- 2 Advanced Stirling Converters (~70 We each)
- 6 thermal simulators to represent additional engines

The core, heat pipes, and power conversion were integrated into a vacuum chamber on a criticality test stand at the Nevada site. Beryllium oxide reflector segments on a lift table below the core-controlled reactivity.

10.19.3   Test Results

The KRUSTY experiment successfully demonstrated the Kilopower reactor concept:
- Achieved criticality and validated physics models
- Start-up to 800°C in 1.5 hours (goal <3 hours)
- Steady-state operation >4 kWth at 800°C
- Stable response to simulated failures and transients
- <15°C temperature change on loss of cooling (goal <50°C)
- 30% thermal-to-electric conversion efficiency (goal >25%)
- Power turn-down ratio >16:1 (goal >2:1)

The test marked the first nuclear steady-state and transient testing of a U.S. space reactor prototype since the 1960s, establishing technology readiness for further development.

10.19.4   Potential Missions

The Kilopower technology enables a range of potential planetary surface and deep space missions, including:

Planetary Surface Power:
- Mars: 4-5 x 10 kWe units to support human exploration
- Moon: 1 kWe demonstration for rover or ISRU applications

Deep Space Science (with Nuclear Electric Propulsion):
- Neptune-Triton Orbiter: 3700 kg spacecraft, 13-year flight
- Chiron Orbiter: Potential double Centaur object mission
- Titan-Enceladus Orbiter: 4000 kg, <14 year dual moon mission
- Pluto Orbiter: 2600 kg spacecraft, 14 year flight
- Titan Saturn System Mission: Orbiter, lander, balloon, 15+ year mission
- Kuiper Belt Object Orbiter: 16-year flight to orbit 2001 XH255

Figure 10-24. Fully assembled KRUSTY core.
(source: https://discover.lanl.gov/news/0502-game-changing-space-mission/)

These missions leverage the high power, long life, and constant operation of fission systems to enable increased payload mass, higher data rates, and shorter flight times compared to radioisotope power.

The Kilopower Project successfully demonstrated the first end-to-end testing of a new U.S. space reactor system in over 50 years. The innovative use of a monolithic fuel form, heat pipe heat transfer, and Stirling power conversion was shown to enable simple, low-cost fission power for space applications. The KRUSTY experiment validated the reactor physics, thermal performance, and operation of a prototypic system, establishing the technology foundation for affordable kilowatt-class fission power for challenging deep space science and human exploration missions.

With a successful technology demonstration completed, further work is needed to develop optimized flight systems, perform additional testing, and mature the technology for specific mission applications. The Kilopower concept provides NASA a scalable, versatile option to provide abundant power for exploring the most demanding destinations in the solar system.

11 Nuclear Thermal Propulsion

11.1 Introduction

Nuclear Thermal Propulsion (NTP) technology represents a significant advancement in space propulsion systems, leveraging nuclear reactions to achieve superior performance metrics compared to conventional chemical propulsion. This chapter provides a detailed examination of the historical development, technical milestones, and future prospects of NTP, focusing on the scientific principles, engineering challenges, and recent innovations driving this technology.

11.1.1 Theoretical Foundations and Early Developments

The conceptual basis for NTP emerged from early 20th-century explorations into nuclear energy. The discovery of nuclear fission by Otto Hahn and Fritz Strassmann in 1938, followed by the subsequent identification of the chain reaction process by Lise Meitner and Otto Frisch, established the foundational principles necessary for harnessing nuclear reactions for propulsion. The Manhattan Project during World War II demonstrated the practical feasibility of controlled nuclear reactions, further stimulating interest in their application to rocketry.

In 1944, Stanislaw Ulam proposed using nuclear explosions for propulsion, a concept that evolved into Project Orion in the late 1950s under the direction of Theodore Taylor and Freeman Dyson. While Orion focused on nuclear pulse propulsion, it validated the potential of nuclear energy in space travel and laid the groundwork for future NTP initiatives.

11.1.2 The NERVA Program

The Nuclear Engine for Rocket Vehicle Application (NERVA) program, initiated in 1959, marked the first major effort to develop a viable NTP system. Managed jointly by the U.S. Atomic Energy Commission and NASA, NERVA aimed to design, develop, and test nuclear thermal rocket engines for potential use in crewed space missions. The program built upon the foundational work of Project Rover, which had begun exploring nuclear reactor designs for propulsion in 1955.

Table 11-1. Timeline of nuclear thermal propulsion tests.

Reactor	Date	Maximum Power (MW$_h$)	Maximum Runtime (s)	Net I$_{sp}$ (s)	Exit T (°C)
KIW-A	7/1/1959	70	300		
KIW-A'	7/8/1960	88	307		
KIW-A3	10/19/1960	112.5	259		
KIW-B1A	12/7/1961	225	36		
KIW-B1B	9/1/1962	880	Multiple		
KIW-B4A	12/30/1962	450	Multiple		
KIW-B4D	5/1/3/1964	990	40		
KIW-B4E	8/28/1964	937	480		
KIW-B4E	9/10/1964	882	150		
NRX-A2	9/24/1964	1096	40		
NRX-A3	4/23/19656	1093	210		
NRX-A3	5/20/1965	1072	792		
PHOEBUS-1A	6/25/1965	1090	630		
NRX-A4/EST	March 1966	1055	1740	820	2127
NRX-A5	June 1966	1120	1776		
PHOEBUS-1B	2/23/1967	1450	1800	835	2171
NRX-A6	10/15/1967	1125	3720	847	2282
PHOEBUS-2A	6/25/1968	4082	750	805	2037
PEEWEE-1	Fall 1968	514	2400	845	2266
XE-PRIME	7/1/1969	1140	210	710	2127
NF-1	July 1972	44	6528	830	2277

Key technical achievements of the NERVA program included the development and testing of several nuclear rocket engines, such as the NRX-A6, NRX-A5, and XE Prime. These engines utilized high-temperature, graphite-moderated reactors to heat hydrogen propellant to temperatures exceeding 2500 K, achieving specific impulses in the range of 850-900 seconds, significantly higher than those of chemical rockets (300-450 seconds). The NRX and XE engines demonstrated the feasibility of using nuclear reactors to achieve efficient and high-thrust propulsion, validating the core principles of NTP.

Despite these successes, the NERVA program was terminated in 1973 due to budgetary constraints and shifting national priorities. However, the program's technical achievements and accumulated knowledge provided a crucial foundation for future NTP research and development.

11.1.3 Post-NERVA Developments

The discontinuation of the NERVA program led to a period of diminished focus on NTP during the 1970s and 1980s, as the United States concentrated on the Space Shuttle program and robotic space missions. Nonetheless, the underlying potential of NTP for deep space exploration remained compelling.

Renewed interest in NTP emerged in the 1990s, driven by initiatives such as President George H.W. Bush's Space Exploration Initiative, which proposed human missions to Mars. During this period, studies conducted by NASA and other organizations highlighted the advantages of NTP, including reduced transit times and increased payload capacity for interplanetary missions.

Table 11-2. Key design parameters for post-NERVA conceptual NTP reactors

Design Parameter/ Concept	I_{sp} (s)	Thrust (kN)	Core Thermal Power (MW)	Core Average Temperature (°C)
ENABLER	825	333	1586	2427
SMALL ENGINE	780	73	367	2361
SNRE	1000	44	210	2227
710	873	444	2010	2371
CERMET	930	445	2000	2234
PBR1	971	334	1945	2927
PBR2	780	33	150	2477
PeBR	1000	315	1500	2727
LPNTR1	1075	111	525	2927
LPNTR2	1050	48	260	2927
WIRE CORE	930	914	4400	2727

Key challenges during this era included:

1. **Radiation Shielding**: Developing effective shielding to protect astronauts from the intense radiation generated by the reactor was a significant technical hurdle. Advances in materials science and innovative shielding concepts were essential to address this issue.
2. **High-Temperature Materials**: The extreme operating conditions of NTP reactors necessitated the development of materials capable of withstanding high temperatures and radiation. Research focused on high-temperature ceramics, refractory metals, and carbon-carbon composites.
3. **Reactor Design and Control**: Achieving precise control over the reactor's thermal output and ensuring safe and reliable operation were critical for the

viability of NTP systems. Innovations in reactor design, control mechanisms, and safety systems were required to address these challenges.

11.1.4 Contemporary Advances

The past two decades have seen significant advancements in NTP technology, driven by both government and private sector initiatives. Notable developments include:

1. **Technological Innovations**: Modern research has benefited from advancements in reactor design, fuel fabrication, and computational modeling. High-assay low-enriched uranium (HALEU) has emerged as a promising alternative to highly enriched uranium (HEU), offering enhanced safety and regulatory compliance.

2. **Government Programs**: NASA's Nuclear Thermal Propulsion Project, part of the Space Technology Mission Directorate, aims to develop a flight-ready NTP system. The program focuses on advancing reactor technology, developing high-temperature materials, and conducting ground tests to validate system performance.

3. **Private Sector Contributions**: Companies such as Blue Origin and Lockheed Martin have shown interest in NTP technology, exploring its potential for commercial space exploration missions. Collaborative efforts between government agencies and private industry are accelerating the development and deployment of NTP systems.

4. **International Collaboration**: Global interest in NTP has led to collaborative research efforts involving organizations such as the European Space Agency (ESA) and Russia's Roscosmos. These partnerships are fostering the exchange of knowledge and resources, contributing to the advancement of NTP technology on a global scale.

11.1.5 Future Prospects

NTP technology holds significant promise for future space exploration endeavors, particularly for crewed missions to Mars and beyond. Key advantages of NTP include higher specific impulse, which translates to greater efficiency and reduced transit times, and the ability to carry larger payloads compared to chemical propulsion systems. As humanity sets its sights on ambitious interplanetary missions, NTP offers a viable solution to the challenges of deep space travel.

Ongoing research and development efforts are focused on addressing the remaining technical and regulatory challenges associated with NTP. Innovations in materials science, reactor design, and radiation shielding continue to enhance the feasibility and safety of NTP systems. Additionally, international collaboration and commercial partnerships are expected to play a crucial role in the future development and deployment of NTP technology.

Nuclear Thermal Propulsion stands as a transformative technology with the potential to revolutionize space exploration. From its early theoretical foundations and the pioneering NERVA program to contemporary advancements and future prospects, NTP has demonstrated its capability to achieve superior performance metrics compared to traditional chemical propulsion. Continued research, technological innovation, and international cooperation will be essential to realizing the full potential of NTP, enabling humanity to embark on ambitious missions to Mars and beyond.

11.2   Fundamentals of Nuclear Propulsion

Nuclear thermal propulsion (NTP) offers a promising alternative to chemical rockets, particularly for deep space missions such as manned missions to Mars. Unlike chemical rockets that rely on the combustion of chemical propellants, nuclear thermal rockets utilize a nuclear reactor to heat a propellant, typically hydrogen, to high temperatures, producing thrust through the expulsion of the heated gas.

11.2.1 Specific Impulse and Thrust Generation

The performance of a rocket engine is often measured by its specific impulse (I_{sp}), which indicates the efficiency of the propulsion system. Specific impulse is defined as the thrust produced per unit weight flow rate of the propellant. It can be expressed as:

$$I_{sp} = \frac{F}{\dot{m} g_0} \tag{11-1}$$

where:
- F is the thrust (N),
- $\dot{m}$ is the mass flow rate of the propellant (kg/s),
- g_0 is the standard acceleration due to gravity (9.81 m/s^2).

In SI units, specific impulse is given in seconds (s). This relationship shows that higher specific impulse values indicate more efficient engines.

For nuclear thermal rockets, the specific impulse can be further detailed using the properties of the propellant and the conditions within the reactor:

$$I_{sp} = \frac{C_f \sqrt{\dfrac{2\gamma}{\gamma - 1} \dfrac{R T_c}{M}}}{g_0} \tag{11-2}$$

where:
- C_f is the thrust coefficient, a function of nozzle design,

- γ is the specific heat ratio of the propellant gas,
- R is the specific gas constant for the propellant ($J/kg \cdot K$),
- T_c is the chamber temperature (K),
- M is the molar mass of the exhaust gases (kg/mol).

The above equation illustrates that a higher chamber temperature T_c and a lower molar mass M result in a higher specific impulse, which is why hydrogen, with its low molar mass, is typically used as the propellant in nuclear thermal rockets.

11.2.2 Thrust Generation Mechanism

In a nuclear thermal rocket, the reactor core heats the hydrogen propellant to temperatures typically between 2400 K and 3100 K . The propellant is stored cryogenically and then pumped through the reactor core, where it absorbs thermal energy. The heated hydrogen gas expands and is expelled through a rocket nozzle, producing thrust. The thrust F can be calculated using:

$$F = \dot{m}v_e \tag{11-3}$$

where v_e is the exhaust velocity, which can be expressed as:

$$v_e = C_f \sqrt{\frac{2\gamma}{\gamma - 1}\frac{RT_c}{M}} \tag{11-4}$$

The exhaust velocity v_e is directly related to the specific impulse and the efficiency of the propulsion system.

11.2.3 Thermodynamic Considerations

The efficiency and performance of nuclear thermal rockets depend heavily on thermodynamic processes. The reactor core must operate at high temperatures to maximize the thermal energy transferred to the propellant. However, materials limitations pose challenges, as the reactor must withstand extreme conditions without degradation. High-temperature materials, such as refractory metals or graphite, are typically used, though they must be protected from erosion by hot hydrogen.

The overall energy conversion efficiency can be analyzed using the concept of thermal efficiency, which for a nuclear thermal rocket can be approximated as:

$$\eta_{th} = \frac{\text{Kinetic energy of exhaust}}{\text{Thermal energy supplied by reactor}} \tag{11-5}$$

Given that the kinetic energy of the exhaust is $\frac{1}{2}\dot{m}v_e^2$ and the thermal energy supplied is $\dot{Q}$, where $\dot{Q} = \dot{m}C_p(T_c - T_{in})$ with C_p being the specific heat capacity at constant pressure and T_{in} the inlet temperature, the thermal efficiency can be written as:

$$\eta_{th} = \frac{\frac{1}{2}\dot{m}v_e^2}{\dot{m}C_p(T_c - T_{in})} \tag{11-6}$$

This formula emphasizes the importance of high reactor outlet temperatures and efficient thermal management in achieving high specific impulses and overall propulsion efficiency.

11.2.4 System Configuration

A typical nuclear thermal rocket system consists of:
1. Reactor Core: Generates heat through nuclear fission.
2. Cryogenic Hydrogen Tank: Stores the propellant at low temperatures.
3. Turbopumps: Circulate the hydrogen through the reactor.
4. Nozzle: Expands the heated propellant to generate thrust.

The design of the reactor core, including fuel arrangement and material selection, is crucial for optimizing performance and ensuring safety. In some designs, a portion of the hydrogen propellant is bled off to drive the turbopumps, known as a "hot bleed cycle."

Nuclear thermal propulsion offers significant advantages over chemical propulsion, particularly for deep space missions where high specific impulse is crucial for reducing transit times and overall mission mass. The development and optimization of nuclear thermal rockets involve complex thermodynamic and material challenges, requiring careful consideration of reactor design, propellant properties, and thermal management systems.

11.3 Radiation Shielding in Nuclear Thermal Propulsion Systems

Radiation shielding is crucial for nuclear thermal propulsion systems to protect astronauts, spacecraft components, and sensitive instruments from harmful radiation. This section delves into the fundamental principles of radiation shielding, explores various materials and configurations, and presents formulas for calculating shielding effectiveness.

11.3.1 Fundamentals of Radiation Shielding

Radiation emitted by nuclear reactors can be broadly classified into two categories: neutrons and gamma rays. Effective shielding requires materials that can attenuate these types of radiation through various mechanisms:

- Neutron Attenuation: Neutrons can be slowed down or absorbed by materials with high neutron cross-sections. Materials rich in hydrogen, such as water or metal hydrides, are particularly effective for neutron shielding due to hydrogen's ability to slow down neutrons through elastic scattering.
- Gamma Ray Attenuation: Gamma rays are high-energy photons that require dense materials for effective attenuation. High atomic number materials, such as lead or tungsten, are commonly used to absorb gamma rays through photoelectric absorption, Compton scattering, and pair production.

11.3.2 Shielding Materials and Properties

The choice of shielding material depends on several factors, including the type of radiation, thermal and mechanical stability, ease of fabrication, and mass. The attenuation of radiation can be described using the attenuation coefficient μ, which depends on the material and the energy of the radiation.

11.3.2.1 Neutron Shielding Materials

Materials with high hydrogen content are ideal for neutron shielding. Lithium hydride (LH) is particularly favored due to its high hydrogen atom density and relatively low mass density. The attenuation of neutrons in a material can be expressed as:

$$I(x) = I_0 e^{-\Sigma x} \tag{11-7}$$

where:

- $I(x)$ is the neutron intensity after passing through a thickness x of the material,
- I_0 is the initial neutron intensity,
- Σ is the macroscopic cross-section of the material.

The macroscopic cross-section Σ is given by:

$$\Sigma = N\sigma \tag{11-8}$$

where:

- N is the number density of atoms in the material (atoms $/cm^3$),
- σ is the microscopic cross-section (cm^2).

For lithium hydride, with a hydrogen atom density N_H of

5.85×10^{22} atoms /cm^3, the effectiveness of neutron attenuation can be significant.

11.3.2.2 Gamma Ray Shielding Materials

Gamma ray shielding requires materials with high atomic numbers and densities. Lead and tungsten are commonly used. The attenuation of gamma rays follows the exponential law:

$$I(x) = I_0 e^{-\mu x} \tag{11-9}$$

where:

- μ is the linear attenuation coefficient (cm^{-1}), dependent on the material and photon energy.

The mass attenuation coefficient μ/ρ(cm^2/g) is often used, where ρ is the material density (g/cm^3). The mass attenuation coefficient helps in comparing materials' effectiveness independent of their density.

11.3.3 Shield Configuration and Optimization

Shielding design must consider not only material selection but also the configuration to optimize mass and protect all critical components. Two common configurations are:

- Shadow Shields: These are used to minimize mass by blocking radiation in specific directions, typically towards crewed sections. The shadow shield must cover the solid angle subtended by the reactor and any scattering sources.
- Distributed Shields: These surround the reactor and other sensitive components, providing uniform protection. They often combine layers of different materials to handle both neutron and gamma radiation

The total shielding effectiveness can be represented as:

$$I(x) = I_0 e^{-(\Sigma_n + \mu)x} \tag{11-10}$$

where:

- Σ_n accounts for neutron attenuation,
- μ accounts for gamma attenuation.

The thickness x is chosen based on the acceptable radiation dose levels, which are significantly lower for crewed missions compared to uncrewed missions. The dose D received can be calculated by integrating the intensity over time and accounting for the material's exposure:

$$D = \int_0^t I(t')dt' \tag{11-11}$$

11.3.4 Practical Considerations in Spacecraft Shielding

Spacecraft shielding design also accounts for operational scenarios, such as reactor startup and shutdown, and the need for safe zones, such as a "storm cellar" for solar flare protection. The total mass of the shielding is a critical factor, influencing the mission's feasibility and cost. The balance between adequate protection and minimizing mass requires careful analysis and optimization.

Materials like lithium hydride (LiH) are preferred for neutron shielding due to their low density and high neutron attenuation properties. For gamma radiation, uranium alloys and tungsten are effective but come with high mass penalties. A combination of materials in a laminated shield can provide comprehensive protection, with neutron moderators like LiH and gamma absorbers like tungsten.

The thickness of the shield x required to reduce the radiation to a safe level can be estimated using:

$$x = \frac{\ln\left(\frac{I_0}{I(x)}\right)}{\Sigma_n + \mu} \tag{11-12}$$

11.3.5 Shielding Challenges and Advanced Materials

Challenges in shield design include handling the thermal load, especially in space environments where heat dissipation is limited. Advanced materials, such as borated polyethylene for neutron absorption and high-Z compounds for gamma shielding, are being explored to enhance effectiveness while reducing mass.

Boron is particularly useful for neutron shielding as it has a high neutron capture cross-section and produces minimal secondary radiation. Borated materials are often used in combination with hydrogen-rich materials for improved neutron shielding.

Radiation shielding is a critical component in the design of nuclear thermal propulsion systems. The selection of materials and the configuration of the shield must balance protection, mass, and cost. The use of advanced materials and composite structures allows for effective shielding against both neutron and gamma radiation, ensuring the safety of crew and equipment during space missions.

11.4   High-Temperature Materials

The core of a high-temperature nuclear reactor for nuclear thermal propulsion is where the fission reactions occur. It must be made of materials that can withstand the extremely high temperatures and radiation levels generated by these reactions. The melting point of the materials used in the reactor core is an essential factor in determining the safety and performance of the reactor. We will examine the melting points of several materials commonly used in high-temperature nuclear reactor cores. High-performance and temperature liquid-fueled systems are often proposed that use propellant flow to keep all core structural and moderator materials at reasonable temperatures (< 800 K) while still allowing molten fuel to heat the propellant to a very high temperature before expansion through a nozzle. One potential concept is the Centrifugal Nuclear Thermal Rocket (CNTR).

For traditional solid-fuel NTP engines and the structural and moderator components of liquid-fueled engines, the first material we will consider is graphite, which has been used as a moderator in some high-temperature reactors. Graphite has a high melting point of around 3600°C and is an excellent thermal conductor, making it an ideal material for high-temperature reactors. However, graphite is also highly flammable and can become highly reactive in the presence of oxygen and high temperatures, making it less attractive for air-breathing high-temperature reactors.

Another commonly used material in high-temperature reactor cores is beryllium, which has a melting point of around 1278°C. Beryllium is a good thermal conductor with a high thermal expansion coefficient, making it well-suited for high-temperature reactors.

Tungsten is another material that is sometimes used in high-temperature reactors. Tungsten has a melting point of around 3410°C, making it an ideal choice for high-temperature reactors. Tungsten is also a good thermal conductor and is highly resistant to thermal shock, making it a good choice for high-temperature reactors. However, tungsten is also a very dense material, which can make it challenging to handle and can increase the weight of the reactor. Tungsten is also a strong neutron absorber, making it difficult to use in moderated systems fueled by high assay low enriched uranium (HALEU).

Hafnium is another material that is sometimes used in high-temperature reactors. Hafnium has a high melting point of around 2227°C and is highly resistant to thermal shock, making it well-suited for high-temperature reactors. However, hafnium is a highly reactive material that can be difficult to work with and has a high neutron absorption cross-section. Hafnium is also expensive, making it a less attractive choice for high-temperature reactors.

Molybdenum is another material that is sometimes used in high-temperature reactors. Molybdenum has a high melting point of around 2620°C and is a good thermal conductor, making it well-suited for use in high-temperature reactors. Molybdenum is also highly resistant to thermal shock, making it a good choice for use in high-temperature reactors. However, molybdenum is also a very dense material, which can make it difficult to handle and can increase the weight of the reactor.

The thermal properties of materials are critical in determining their suitability for use in high-temperature nuclear reactors. In this essay, we will compare the thermal properties of five materials commonly used in high-temperature reactors: uranium dioxide (UO2), uranium carbide (UC), carbon, niobium carbide (NbC), and tungsten.

Uranium dioxide (UO2) is a commonly used fuel in nuclear reactors and has a melting point of around 2800°C. UO2 has a low thermal conductivity, meaning it does not conduct heat well and can lead to overheating in high-temperature reactors. Additionally, UO2 is highly reactive and can become unstable at high temperatures, which can pose a safety risk in high-temperature reactors.

Uranium carbide (UC) is a relatively new material that is being investigated for use in high-temperature reactors. UC has a high melting point of around 2900°C and higher thermal conductivity than UO2. However, UC is also highly reactive and can become unstable at high temperatures, which can pose a safety risk in high-temperature reactors.

Carbon is a common material used in high-temperature reactors as a moderator and reflector. Carbon has a high melting point of around 3600°C and is a good thermal conductor, making it well-suited for use in high-temperature reactors. However, carbon is also highly flammable and can become highly reactive in the presence of high temperatures, which can pose a safety risk in high-temperature reactors.

Niobium carbide (NbC) is a refractory material that is being investigated for use in high-temperature reactors. NbC has a high melting point of around 3300°C and is highly resistant to thermal shock, making it well-suited for use in high-temperature reactors. However, NbC is also a highly reactive material that can be difficult to work with, which can pose a challenge in high-temperature reactors.

Tungsten is another material that is commonly used in high-temperature reactors. Tungsten has a high melting point of around 3410°C and is a good thermal conductor, making it well-suited for use in high-temperature reactors. Tungsten is also highly resistant to thermal shock, making it a good choice for use in high-temperature reactors. However, tungsten is also a very dense material, which can make it difficult to handle and can increase the weight of the reactor.

In conclusion, the thermal properties of the materials used in high-temperature reactors are critical in determining their suitability for use in these reactors. Uranium dioxide, uranium carbide, carbon, niobium carbide, and tungsten all have unique thermal properties that make them suitable for different applications in high-temperature reactors. The choice of material for use in a high-temperature reactor will depend on the specific requirements of the reactor, including the desired thermal conductivity, resistance to thermal shock, and ease of handling.

The choice of material for use in a high-temperature reactor will depend on the specific requirements of the reactor, including the desired thermal conductivity, resistance to thermal shock, and ease of handling.

Table 11-3 shows the melting points of some high-temperature reactor materials of interest.

Table 11-3. Melting points of some nuclear reactor materials of interest in nuclear thermal propulsion

Application	Material	Melting Temperature (°C)
Fuel	Uranium	1132
Fuel	Uranium Dioxide	2865
Fuel	Uranium Carbide	3900
Fuel	Uranium Ntride	2000
Structures/Refractory Metal	Tungsten	3422
Structures/Refractory Metal	Molybdenum	2610
Structures/Refractory Metal	Rhenium	3180
Structures/Refractory Metal	Tantalum	2996
Structures/Cladding Refractory Non-Metal	Carbon	3600
Structures/Cladding Refractory Non-Metal	Nobium Carbide	4500
Structures/Cladding Refractory Non-Metal	Zirconium Carbide	3300
Structures/Cladding Refractory Non-Metal	Hafnium Carbide	3900
Structures/Cladding Refractory Non-Metal	Tantalum Carbide	4215

11.5   Nuclear Thermal Propulsion Requirements

Nuclear Thermal Propulsion (NTP) systems are integral to the future of deep-space exploration, offering distinct advantages over conventional chemical propulsion systems. The requirements for NTP systems are shaped by mission-specific parameters such as specific impulse, thrust, system mass, reliability, safety, and cost. This section outlines the critical factors influencing the design and operation of NTP systems.

11.5.1 Performance Requirements

11.5.1.1 Specific Impulse (Isp)

Specific impulse is a key measure of propulsion efficiency, indicating how effectively a propulsion system uses propellant. NTP systems achieve significantly higher specific impulses compared to chemical rockets, generally in the range of 800 to 1000 seconds. This is due to the ability of NTP systems to heat propellants, such as hydrogen, to extremely high temperatures, resulting in high exhaust velocities. The choice of hydrogen as a propellant is particularly advantageous due to its low molecular weight, which enhances specific impulse.

11.5.1.2 Thrust

Thrust is a critical parameter for determining the acceleration and maneuverability of a spacecraft. For missions to Mars and other deep-space destinations, NTP systems must generate sufficient thrust to meet mission requirements. Typical thrust levels for NTP systems range from 65.8 kN to 439 kN. The thrust requirement must be balanced with the engine's burn time and the total mass of the propulsion system, as higher thrust generally necessitates more powerful reactors and larger propellant reserves.

11.5.2 Safety and Reliability Requirements

11.5.2.1 Safety

Safety is a paramount concern in the design and operation of NTP systems, especially for manned missions. The reactor core must be designed to prevent unintended criticality during all phases of the mission, including launch, transit, and operation. This is typically achieved using control mechanisms like control drums or rods to maintain the reactor in a subcritical state until deliberate activation. Additionally, effective radiation shielding is crucial to protect the crew and sensitive equipment from neutron and gamma radiation.

11.5.2.2 Redundancy and Reliability

To ensure mission success, NTP systems must be highly reliable. This reliability is often achieved through redundancy, including multiple independent control systems and backup power supplies. The design must account for potential system failures, providing fail-safe mechanisms and contingency plans. Given the long-duration nature of deep-space missions, the reliability of all components, from the reactor to the propulsion system, is essential.

11.5.3 System Mass and Size Constraints

NTP systems must optimize performance while adhering to mass and size constraints. The total system mass includes the reactor, shielding, propellant tanks, and associated systems. Minimizing mass is critical to maximize payload capacity and reduce launch costs. This can be achieved through the use of advanced materials and innovative engineering solutions, such as compact reactor designs and efficient thermal management systems. The amount of propellant required depends on the mission's delta-v requirements, highlighting the importance of specific impulse in reducing propellant mass.

11.5.4 Cost, Schedule, and Space Qualification

11.5.4.1 Cost Considerations

The development and deployment of NTP systems involve substantial costs, driven by the complexity and stringent safety standards of nuclear technologies. Cost considerations include research and development, manufacturing, testing, and integration with spacecraft. Efforts to reduce costs may include using modular components, leveraging existing technologies, and optimizing design for manufacturability.

11.5.4.2 Schedule

The timeline for developing NTP systems is influenced by extensive testing and qualification requirements. These include reactor design, fuel production, system integration, and thorough ground and flight testing. The development schedule must align with specific mission timelines, especially for missions with narrow launch windows, such as Mars missions.

11.5.4.3 Space Qualification

Space qualification ensures that NTP systems can withstand the harsh conditions of space, including vacuum, radiation, and temperature extremes. This involves a series of tests, including simulations and potentially in-space demonstrations, to validate the system's performance and reliability over the mission's duration.

11.5.5 Mission Flexibility and Multi-Mission Capability

While Mars exploration is a primary focus for NTP development, these systems should be adaptable to various mission profiles, including lunar missions, asteroid missions, and broader deep-space exploration. Flexibility can be achieved through modular designs that allow for adjustments in propellant load, thrust levels, and reactor configurations. This versatility ensures that NTP systems can meet a wide range of mission requirements, from short-term scientific missions to long-term human exploration.

Nuclear Thermal Propulsion systems are vital for the next generation of space exploration, offering unparalleled efficiency and thrust capabilities. The design and implementation of NTP systems must address a comprehensive set of requirements, encompassing performance, safety, reliability, mass, cost, and qualification. These systems enable more ambitious missions, reducing travel time and enhancing mission flexibility. As development continues, NTP systems will play a crucial role in enabling human exploration beyond Earth, from the Moon to Mars and beyond.

11.6   Safety Issues Relevant to Nuclear Thermal Propulsion

Nuclear Thermal Propulsion (NTP) systems present unique safety challenges that are critical for the success of human space exploration missions, particularly those aiming to reach Mars. The safety considerations encompass protecting the crew from radiation, ensuring the structural integrity of the propulsion system, and preventing environmental contamination. This section discusses the essential safety requirements and methodologies employed to address the inherent risks of NTP systems.

Table 11-4. Nuclear thermal rocket parameters.

Parameter	NERVA	Small NERVA Engine	RD-0410	Timberwind	Timberwind 45
Country	USA	USA	Russia	USA	USA
Fuel Composition	(UC-ZrC)C composite	(UC-ZrC)C composite	(U,Nb,Zr)C	UC-TaC core with layers of PyC/TaC and ZrC	UC-TaC core with layers of PyC/TaC and ZrC
Fuel Form	Prismatic	Prismatic	Twisted ribbon	Particle bed	Particle bed
Propellant	LH2	LH2	LH2+Hexane	LH2	LH2
Vacuum Thrust [kN]	337	72	35.3	178	441.3
Thrust [lbf]	75.1	16.4	7.9	39.7	98.3
Specific Impulse (s)	825	875	910	869	1000
Speed (m/s)	8085	8575	8918	8516	9800
Nozzle Expansion Ratio	100	100			
Chamber pressure (MPa)	3.1	3.1	7		
Core outlet temperature (K)	2360	2695	3000	2750	
Power Density (MWth/liter)		2.3-5.2	35-40	40	40
Burn time (s)	36,000		3600	100	449
Engine diameter (m)			1.2		
Engine length (m)		4.4	3.7		
Mass with shield (kg)	12300	2550	2000		1500
Thermal power (MWth)	1556	365	196		

11.6.1 Importance of Safety in NTP Systems

The primary safety concern in NTP systems is to protect the crew from harmful radiation. Radiation risks include exposure to galactic cosmic radiation and solar flares, both of which can be mitigated by reducing travel times. NTP systems, with their higher specific impulse, can significantly shorten mission durations compared to chemical rockets, thus reducing radiation exposure and associated risks. Additionally, NTP systems eliminate the need for chemical rocket oxidizers, reducing the complexity and potential hazards related to chemical propulsion systems.

The reliability of NTP systems is also enhanced by the simplicity of their design, which generally includes fewer moving parts than chemical rockets. This can lead to increased reliability and fewer opportunities for critical failures. Moreover, NTP systems offer greater flexibility in mission scheduling and can potentially reduce the overall cost and complexity of missions by minimizing the need for in-orbit assembly.

11.6.2 Key Safety Requirements and Approaches

Safety must be integrated from the initial design stages of NTP systems. The following are the key safety requirements and approaches used to achieve these goals:

1. **Prevent Unplanned Criticality**: The reactor must be designed to remain subcritical until intentional startup. This includes implementing independent systems to reduce reactivity to a subcritical state in case of anomalies.
2. **Maintain Thrust for Safe Return**: The propulsion system must ensure sufficient thrust capability to return the crew safely in case of mission abort or other contingencies.
3. **Core Integrity**: The reactor core must maintain integrity throughout the mission, except in controlled reentry scenarios. This includes robust containment to prevent the release of radioactive materials.
4. **Radiological Safety in Launch Abort Scenarios**: The system must be designed to contain radioactive materials in the event of a launch abort, minimizing environmental contamination.
5. **Safe Reactor Disposal**: Procedures must be in place for the safe disposal of the reactor, either through reentry and burn-up or disposal in deep space.
6. **Radiation Shielding**: Adequate shielding must protect the crew and sensitive equipment from radiation, both during normal operations and in case of reactor malfunctions.
7. **Redundant Systems**: The design should incorporate independent and redundant control and safety systems to handle potential failures and ensure mission continuation or safe abort.
8. **Minimize Environmental Impact**: Both terrestrial and non-terrestrial environments must be protected from contamination. This includes the design

and operation of the NTP system, as well as post-mission disposal.

Figure 11-1. Nerva XE engine.

Figure 11-2. Russian RD-0410 engine.

11.6.3 Radiation Exposure and Protection

Exposure to radiation is one of the most significant risks for crewed missions. The primary sources of radiation include:

- **Galactic Cosmic Radiation (GCR)**: A continuous source of high-energy particles that is difficult to shield against, making short mission durations preferable.
- **Solar Particle Events (SPEs)**: Solar flares can deliver significant radiation doses in short bursts. Crews can be protected by moving to shielded areas, such as a "storm cellar."

Radiation exposure limits for crew members are based on established standards, such as those from the National Council on Radiation Protection and Measurements (NCRP). These standards dictate maximum allowable doses to various organs and tissues, ensuring long-term health and safety.

11.6.4 Risk and Safety Analysis Methodology

To ensure the safety of NTP systems, comprehensive risk and safety analyses are conducted using various methodologies:

- **Fault Tree Analysis (FTA)**: A top-down approach that identifies potential causes of system failures. It helps in understanding the interdependencies and failure probabilities of different components.
- **Failure Modes and Effects Analysis (FMEA)**: A bottom-up approach that examines the potential failure modes of components and their effects on the overall system. It is useful for identifying critical components and necessary redundancies.
- **Event Tree Analysis (ETA)**: An approach that evaluates the outcomes of initiating events and their potential consequences, helping in the assessment of system success and failure scenarios.

These methodologies are applied throughout the design, testing, and operational phases of NTP systems to identify, evaluate, and mitigate potential risks.

Safety is the cornerstone of Nuclear Thermal Propulsion system design and operation. By addressing key safety requirements, such as preventing unplanned criticality, ensuring radiological protection, and incorporating redundant systems, NTP systems can be developed to safely support human exploration missions to Mars and beyond. Comprehensive risk analysis methodologies and adherence to strict radiation exposure limits are essential in protecting the crew and environment from the inherent risks of nuclear propulsion.

11.7   ROVER and NERVA Programs

11.7.1 Introduction

The concept of utilizing nuclear fission energy for rocket propulsion emerged as early as 1946, during the nascent stages of nuclear reactor development. The compelling advantages of nuclear thermal propulsion, characterized by high energy density and efficiency, spurred interest in this technology for space exploration. In 1955, the United States embarked on an ambitious program to develop a solid-core, hydrogen-cooled nuclear reactor for rocket propulsion. The Rover/NERVA (Nuclear Engine for Rocket Vehicle Application) programs aimed to leverage the high energy potential of nuclear reactions to achieve superior propulsion capabilities. Despite technical successes, the programs were ultimately terminated in 1973 due to shifting national priorities following the Apollo missions.

The Rover/NERVA programs were foundational in nuclear propulsion research, comprising two primary elements:

1. **The Rover Program**: Conducted at Los Alamos National Laboratory, this program focused on advanced technology development, including reactor design, fuel development, and establishing testing facilities.
2. **The NERVA Program**: This program involved industrial partners Aerojet and Westinghouse, working towards the realization of a flight-ready nuclear propulsion system. It included the demonstration of prototype flight reactors and rocket engines.

The NERVA engine development program prioritized reliability as the primary design criterion, with the specific impulse being the second most important factor, followed by minimizing overall weight. The stringent reliability requirement was set at 0.997 with a 90% confidence level. Additional requirements included a minimum specific impulse of 760 seconds, the ability to operate across a wide range of power conditions, and a thrust capability of 337 kN.

11.7.2 Development and Achievements

The Rover/NERVA programs saw the development of various reactor test series, each contributing to the advancement of nuclear rocket technology. These included:

1. **Kiwi Series**: These early research reactors were critical in establishing the foundational technology for nuclear rockets. The Kiwi reactors were instrumental in developing high-temperature fuels and were the first to operate with stored liquid hydrogen. The program successfully addressed thermal-hydraulic vibrational issues, culminating in the Kiwi-B4E reactor achieving operation at over 1,890 K.

2. **NRX Series**: Under the NERVA program, the NRX reactors were developed to demonstrate the specific impulse and power output necessary for a practical nuclear thermal rocket. The NRX-A6 test was a significant milestone, achieving a specific impulse of 7,450 m/s and operating for 62 minutes at 2,220 K and 1,100 MW.

3. **Phoebus Series**: These reactors aimed to push the boundaries of nuclear propulsion by increasing specific impulse and power density. The Phoebus-2A reactor, for example, operated above 4,000 MW and reached peak power of 4,100 MW, demonstrating the feasibility of high-power nuclear thermal propulsion.

4. **Pewee and Nuclear Furnace (NF) Series**: These reactors focused on testing higher temperature fuel elements and achieving longer operational lifetimes. The Pewee-I reactor operated at 2,555 K and 514 MW, while the NF-I reactor achieved 109 minutes of operation at high power densities and temperatures.

11.7.3 Key Milestones and Testing

The NERVA engine development program was notable for its groundbreaking achievements, including the first and only tests of a nuclear rocket engine in a downward firing configuration. This experimental setup aimed to simulate the conditions of space and to validate engine startup, full power, and shutdown behaviors. The program successfully demonstrated multiple starts and restarts, solidifying the operational feasibility of nuclear thermal rockets.

The culmination of the Rover/NERVA programs was marked by the XE test, which involved a prototype flight engine system. The XE engine underwent extensive testing, including 28 starts and restarts in a simulated space environment. The test incorporated non-nuclear flight components alongside a flight-type reactor, demonstrating the integrated functionality required for actual space missions.

11.7.4 Challenges and Termination

Despite the technical successes, the Rover/NERVA programs faced significant challenges. The development of high-temperature reactors with stable and long-lasting fuel elements was a primary focus, as these parameters directly influenced the specific impulse and operational duration of the engines. The programs also grappled with political and budgetary constraints. The decision to suspend the Saturn V launch vehicle program in 1969, which was to serve as the primary launch vehicle for nuclear rockets, effectively curtailed further development.

The final chapter of the Rover/NERVA programs was marked by the NF-I reactor test in 1972, which demonstrated the capability of nuclear reactors to operate at high temperatures and power densities for extended periods. However, in 1973, the programs were officially terminated, not due to technical failure but due to changing national priorities and the cessation of the human exploration of Mars.

11.7.5 Legacy and Impact

The Rover/NERVA programs represent a critical chapter in the history of space propulsion. They established the technical groundwork for nuclear thermal propulsion, which remains a vital consideration for future deep-space missions. The programs' achievements in reactor design, fuel technology, and system integration continue to inform current and future efforts in nuclear propulsion.

As the interest in human exploration of Mars and other deep-space destinations revives, the lessons learned from the Rover/NERVA programs provide invaluable insights. The potential of nuclear thermal propulsion to reduce mission durations and increase payload capacities positions it as a key technology for the next era of space exploration.

11.8   NERVA Engine Development

The NERVA (Nuclear Engine for Rocket Vehicle Application) engine development program focused on maximizing overall engine performance, particularly specific impulse, thrust levels, mass minimization, and operational longevity. The ultimate goal was to extend the engine's operational lifespan from an initial one hour to ten hours. Achieving high reliability was paramount, with a target reliability of 0.997 at a 90% confidence level.

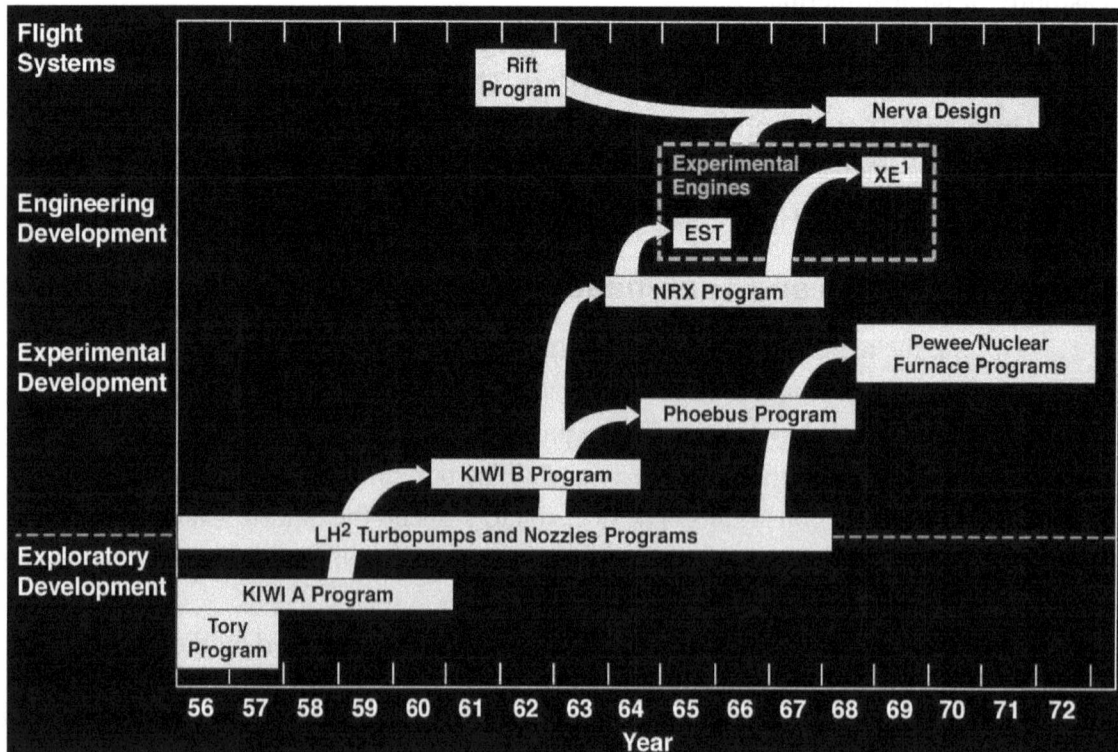

Figure 11-3. NERVA/ROVER rocket development timeline..

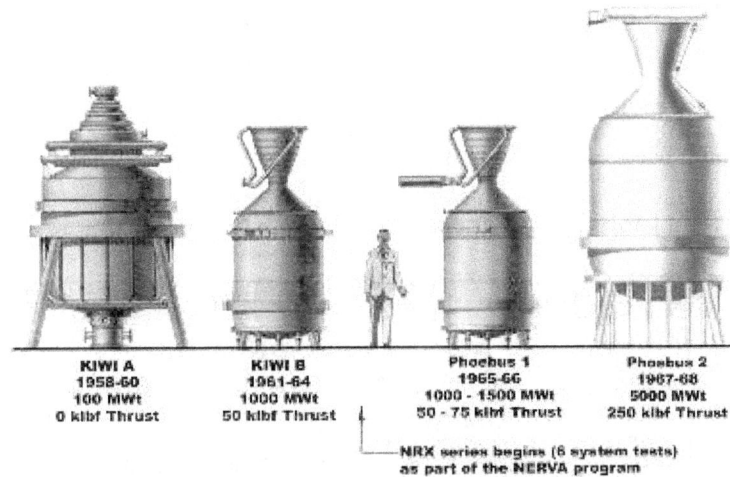

Figure 11-4.Size comparison of KIWI and Phoebus series.

11.8.1 Core Design and Configuration

One of the primary challenges in NERVA engine development was to increase the operating temperatures of the reactor fuel while enhancing the overall reactor design. Initial designs incorporated tie-rods attached to the core's cold end support plate, cooled by hydrogen discharged into the nozzle chamber. However, this configuration resulted in a reduction of specific impulse due to the relatively low temperature of the tie-rod coolant compared to the hydrogen passing through the main fuel elements.

To address this, later reactor designs replaced tie-rods with tie-tubes, which functioned as regeneratively cooled heat exchangers. The coolant from the tie-tubes was directed back into the core inlet, thereby eliminating the loss in specific impulse. Moreover, the peripheral flow rate around the core, initially set high to protect the outer core sections, was gradually reduced. This reduction was aimed at increasing the hydrogen temperature discharged into the nozzle chamber. Additionally, the incorporation of a regenerative heat exchanger-cooled peripheral system further optimized the design.

11.8.2 Engine Cycle Evolution

The NERVA engine's specific impulse was enhanced by transitioning from a hot-bleed cycle to a full-flow "topping cycle." In the hot-bleed cycle, some heated hydrogen from the reactor exit was extracted to drive the turbopump, with the turbine exhaust discharged into space. This configuration resulted in a lower overall specific impulse due to the relatively low temperature of the exhaust gases.

In contrast, the topping cycle used fluid from the tie-tube outlet to drive the turbine, with the turbine discharge directed into the core inlet. This allowed for a significant increase in specific impulse, as the turbine inlet temperature was considerably lower, necessitating a higher turbine flow rate and consequently a higher pump discharge pressure. The full-flow topping cycle thus represented a major advancement in nuclear rocket propulsion technology.

Table 11-5. Timeline of the ROVER program.

Year	Event
1955	Initiation of the nuclear rocket program at Los Alamos National Laboratory as Project Rover. The focus was on developing a solid-core, hydrogen-cooled reactor for rocket propulsion.
Jul-59	First reactor test (Kiwi-A) at 70 MW for 5 minutes.
Oct-60	Completion of proof-of-principle tests with the Kiwi-A series reactors.
Jul-61	Selection of Aerojet-General and Westinghouse Electric Corporation as industrial contractors for the rocket engine and reactor, respectively. Initiation of the Reactor In-Flight Tests (RIFT) program
1961-1964	Kiwi-B series of reactor tests, including resolving vibration problems and demonstrating design power with a series of 1,000 MW reactors and several cold-flow unfueled reactors.
May-64	First full power test (Kiwi-B4D) at design power with no core vibration issues; demonstrated restart capability.
Sep-64	NRX-A2 reactor test, reaching full power of 1,100 MW for about 5 minutes, marking the first tests of the NERVA (Nuclear Engine for Rocket Vehicle Application) reactor.
Jan-65	Kiwi-B-type reactor deliberately destroyed on fast transient as part of a safety program
Jun-65	Phoebus-1A reactor operated at full power for 10.5 minutes, marking the prototype of a new class of reactors.
Mar-66	NRX/EST, the first rocket engine breadboard powerplant, operated at full power (1,100 MW) for 13.5 minutes.
Dec-67	The fifth fueled NRX reactor in the NERVA engine series exceeded the design goal of 60 minutes at 1,100 MW
Jun-68	Phoebus-2A, the most powerful nuclear rocket reactor ever built, operated above 4,000 MW for 12 minutes.
Dec-68	Pewee reactor set records in power density and temperature, operating at 503 MW for 40 minutes at 2,550 K, with a core power density of 2,340 MW/m³.
Mar-69	Successful operation of the first down-firing prototype nuclear rocket engine, XE-prime, at 1,100 MW
1969	Suspension of Saturn V production, the prime launch vehicle for NERVA
Jun-72	Nuclear Furnace (NF-I) demonstrated fuel at peak power densities of approximately 4,500 MW/m³ and temperatures up to 2,500 K for 109 minutes.
Jan-73	Termination of the nuclear rocket program, deemed a technical success but canceled due to shifting national priorities.

11.8.3 Reliability and Safety

Achieving the high reliability target was a critical aspect of the NERVA program. A probabilistic design methodology was employed, involving the assignment of probabilities and confidence levels to each component. This process entailed identifying potential failure modes, operational environments, and synergistic failures, and evaluating all operational modes, including standby, startup, full power, and shutdown. This rigorous approach highlighted the need for additional testing and, in some cases, redundancy in components such as valves, controls, and instrumentation.

11.8.4 Fuel and Thrust Developments

The evolution of engine designs throughout the NERVA program saw the development of fuel capable of sustaining higher chamber temperatures. The experimental XE engine, which utilized the hot-bleed cycle, achieved a specific impulse of approximately 6,960 m/s (710 seconds). Later designs, incorporating the full-flow topping cycle and improved fuel, projected specific impulses as high as 8,575 m/s (875 seconds) for the small engine variant. This improvement was closely linked to advancements in fuel technology, which allowed for reactor outlet temperatures up to 2,695 K.

The development trajectory also included increasing thrust capabilities. The XE engine was designed for a thrust of 245 kN (55,000 lbf), while the NERVA flight engine aimed for 337 kN (75,000 lbf), and the small engine was targeted at 72 kN (16,000 lbf). These thrust levels required corresponding increases in reactor power, from 1,140 MW for the XE engine to 1,556 MW for the NERVA flight engine and 367 MW for the small engine.

11.8.5 Turbopump and Nozzle Design

The turbopump played a crucial role in the propulsion system, responsible for pressurizing the hydrogen propellant. The design of the turbopump evolved through the program, with the XE turbopump featuring a single-stage radial exit flow centrifugal pump and a two-stage turbine. Operational experience from the NRX/EST and XE engines informed further improvements, including resolving issues with shaft system binding at bearings.

The nozzle assembly was designed to maximize thrust by efficiently expanding the heated gas from the reactor. For the NERVA flight engine, the nozzle had an area ratio of 100:1 and incorporated a regeneratively cooled section made of ARMCO 22-13-5 alloy. The extension to the nozzle, made of graphite, did not require cooling. The major engineering challenges involved ensuring structural integrity and fabricability, which were largely resolved by the end of the program.

11.8.6 Control and Instrumentation

Extensive development work was conducted on controls and instrumentation, essential for managing the complex operations of a nuclear rocket engine. Control drum actuators, thermocouples, displacement sensors, pressure sensors, and vibration sensors were all developed and tested for durability and accuracy. A key focus was on ensuring reliable performance across the entire operating range, from startup through full power and shutdown.

11.8.7 Operational Considerations: Startup, Throttling, and Shutdown

The startup procedure for a nuclear rocket engine involves a careful sequence of steps to condition the engine components before reaching high power. This includes chilling down the turbopump, nozzle, reflector, and core inlet to cryogenic temperatures. The reactor is then gradually brought up to power, transitioning to closed-loop temperature control as chamber temperatures rise.

One unique capability of the NERVA engine was throttling, allowing for precise control over thrust levels. This capability was crucial for minimizing decay heat removal requirements and for managing safe retreat operations in case of component malfunctions. Decay heat, generated by fission product radioisotopes, required careful management to prevent overheating post-shutdown. The shutdown process involved a controlled retreat from power, followed by a cooldown phase where residual heat was dissipated using the remaining propellant.

The NERVA program, despite its eventual termination, represented a significant leap in the development of nuclear thermal propulsion technology. The advancements in fuel technology, reactor design, and engine cycle configurations laid a robust foundation for future nuclear rocket engines. The meticulous focus on reliability, safety, and performance optimization ensured that the NERVA engines were not only powerful but also safe and reliable for potential future space missions.

11.9   Russian Nuclear Rocket Development

The Soviet Union began its pioneering work on Nuclear Thermal Propulsion (NTP) in the late 1950s, exploring innovative concepts and designs. The primary goal was to develop a reliable and efficient nuclear rocket engine capable of specific impulses between 900 and 1,000 seconds, using hydrogen as the propellant heated by solid carbide fuel elements. This effort was significant given the high energy density and efficiency potential of nuclear fission, which offered a promising alternative to chemical propulsion.

11.9.1 Design and Experimental Test Results

11.9.1.1 Core and Fuel Element Design

The USSR's Nuclear Rocket Engine (NRE) design was distinct from the homogeneous core designs prevalent in U.S. programs like NERVA. The Soviet approach utilized a heterogeneous reactor core, characterized by the modular arrangement of fuel elements and separate moderator materials. The fuel elements, composed of high-temperature, corrosion-resistant ternary carbides (UC-ZrC-NbC and UC-ZrC-C), were configured as twisted ribbons, providing a surface-to-volume ratio 2.6 times greater than that of U.S. counterparts. This configuration significantly enhanced the heat transfer efficiency between the fuel and the hydrogen propellant.

Fuel elements were assembled into bundles, with each rod-type fuel assembly containing six to eight twisted ribbon elements. These assemblies were then placed into channels within a zirconium hydride (ZrH) moderator that filled the reactor core. The modular design allowed for precise control over the fuel and moderator arrangement, enabling optimization of the reactor's neutron economy and thermal performance. The reactor core could achieve power densities up to 40 MWt per liter, with a minimum core mass characteristic of approximately 0.3 MWt per kilogram.

One notable feature of this design was the flexibility in axial and radial profiling. By adjusting the placement and composition of the fuel elements (e.g., UC-ZrC-NbC upstream and UC-ZrC-C downstream), the reactor could localize high temperatures to specific areas, allowing the rest of the core to operate at significantly lower temperatures (<1,073 K). This differential allowed for the use of common structural materials in the core and facilitated the use of neutron moderators like zirconium and lithium hydrides.

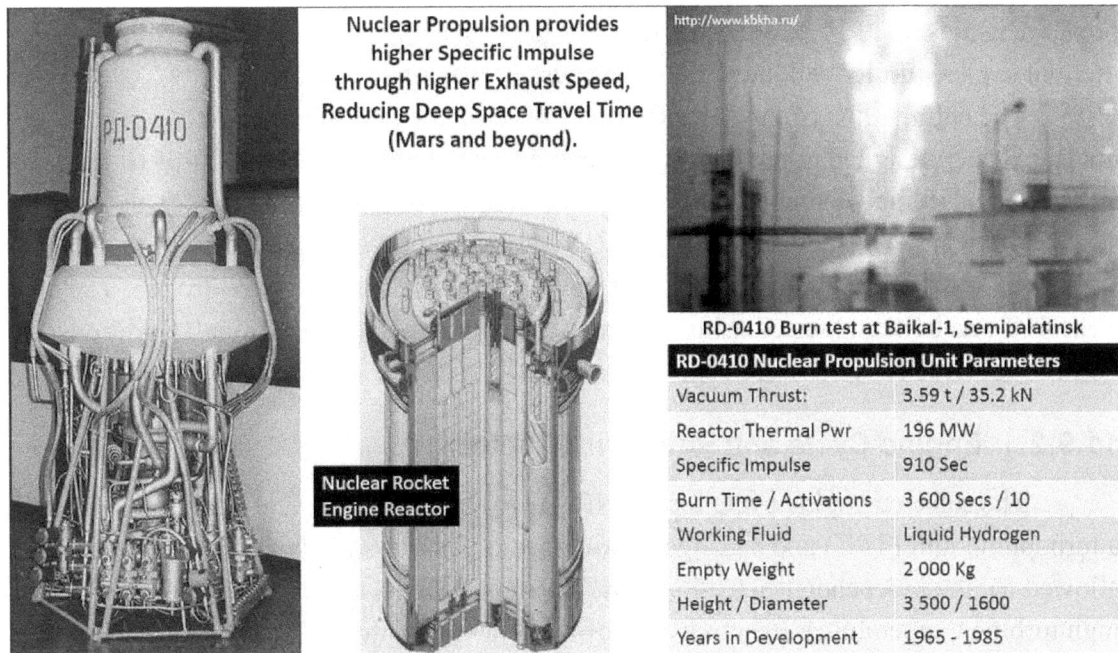

Nuclear Propulsion provides higher Specific Impulse through higher Exhaust Speed, Reducing Deep Space Travel Time (Mars and beyond).

http://www.kbkha.ru/

RD-0410 Burn test at Baikal-1, Semipalatinsk

RD-0410 Nuclear Propulsion Unit Parameters	
Vacuum Thrust:	3.59 t / 35.2 kN
Reactor Thermal Pwr	196 MW
Specific Impulse	910 Sec
Burn Time / Activations	3 600 Secs / 10
Working Fluid	Liquid Hydrogen
Empty Weight	2 000 Kg
Height / Diameter	3 500 / 1600
Years in Development	1965 - 1985

Figure 11-5. Russian RD-0410 nuclear propulsion reactor.

11.9.1.2 Experimental Facilities and Testing

Multiple test facilities were constructed to support the development and testing of individual NRE components, such as fuel elements and moderator modules. A key facility was the IVG-I experimental reactor, designed specifically for testing fuel elements under realistic operational conditions. This facility enabled the Soviet program to conduct approximately 15 tests in the EWG-1 research reactor, where they achieved hydrogen outlet temperatures as high as 3,100 K. These tests corresponded to a specific impulse of 925 seconds and demonstrated a total operational lifetime of 4,000 seconds with up to 10 consecutive restarts. The fuel elements also showed robust performance at lower temperatures, with 4,000-hour operation at 2,000 K, highlighting their durability and stability under extended operation.

From 1970 to 1988, the Soviet program conducted around 30 firings in research reactors, exploring thermal power levels up to 230 MW and propellant mass flow rates up to 16.5 kg/s. The power density of the fuel elements reached 25 MWt/L, with uranium enrichments of 90% and varied fuel loadings between 6.7 and 15.9 kg. Remarkably, the release of radioactive products from the reactor core through the exhaust was kept below 1% by mass.

11.9.2 Key Developments and Engine Prototypes

By the mid-1970s, the reliability of the Soviet carbide fuel elements had been well established, with successful testing of 330 fuel element bundles. These tests achieved power densities up to 35 MWt/L and temperature transients up to 1,000 K/s with 12 thermal cycles, further validating the robustness of the design.

The culmination of this development effort was the RD-0410 engine, a prototype nuclear rocket designed to produce approximately 35 kN of thrust. The RD-0410 featured a twisted ribbon fuel element design, which allowed for efficient heat transfer and high specific impulse. A notable aspect of the RD-0410 was its use of hexane as a propellant additive, aimed at reducing the erosion of fuel elements caused by hot hydrogen. The RD-0410 was built and tested using electric heaters to simulate operational conditions, as no full-scale nuclear tests were conducted.

11.9.2.1 Engine Cycle and Reactivity Control

The selected engine cycle for the NRE involved a closed-cycle configuration with a turbopump unit (TPU) and a gas-generating heat release assembly (HRA). This design allowed for the independent selection of turbine components and enabled operation at high turbine temperatures. Reactivity control was achieved through the use of twelve control drums situated around the reactor periphery, which could modulate the reactor's neutron economy.

To maximize the mass-averaged working fluid temperature at the reactor exit, the Soviet engineers designed the core with a gradient in uranium concentration, decreasing towards the reactor exit. This design created a thermal neutron flux distribution that facilitated efficient reactor operation. The core's "cold" end utilized a beryllium reflector moderator, while the "hot" end allowed for enhanced thermal neutron leakage, thus optimizing the reactor's thermal profile.

11.9.2.2 Brayton Cycle Power Loop and NPPS Concept

An advanced concept explored during the Soviet program was the Nuclear Propulsion and Power System (NPPS), which integrated a Brayton cycle power loop. This system aimed to provide continuous electrical power for spacecraft systems, using a mixture of xenon and helium as the working fluid. The NPPS concept included a closed-cycle power loop capable of generating 50 kW of electrical power, with a maximum working fluid temperature of 1,500 K and a maximum pressure of 9 MPa.

The addition of the Brayton cycle provided several operational benefits, including continuous reactor operation, minimized thermal cycling stresses, rapid reactor restart capability in emergencies, efficient decay heat removal, and the potential for high specific impulse attitude control and orbital maneuvering.

11.9.3 Summary and Challenges

The Soviet nuclear rocket development program featured several unique design elements, including the use of twisted ribbon fuel elements, a heterogeneous reactor core, and the incorporation of a closed-cycle turbopump arrangement. Despite the significant advancements and the construction of a full-scale RD-0410 prototype, the program did not proceed to full-scale nuclear testing. Challenges such as the fabrication complexity of twisted ribbon fuel, susceptibility to brittle failure, and high fission gas release rates remained unresolved at the time of the program's termination in 1988, largely influenced by the Chernobyl disaster's impact on nuclear research.

The Soviet efforts in NTP laid important groundwork for future research and provided valuable insights into the potential of nuclear propulsion systems for space exploration. The unique approaches and innovations developed during this period continue to inform contemporary NTP research and development.

11.10   Particle-Bed Reactor Nuclear Rockets

Particle-Bed Reactor (PBR) nuclear rockets represent a significant evolution in nuclear thermal propulsion technology. The concept draws heavily from advancements made in the development of High-Temperature Gas-Cooled Reactors (HTGRs), particularly the innovation in particle-sized fuel. The core idea revolves around utilizing fuel particles, typically 700 micrometers in diameter, each containing a zirconium carbide kernel infused with several percent of uranium carbide. This innovative design offers a superior heat transfer capability, enabling the reactor to achieve higher outlet temperatures—potentially up to 3,000 K—and consequently, higher specific impulse values approaching 1,000 seconds.

The inception of PBR technology dates back to the mid-1980s, during which time the particle-bed concept was explored for various applications, including orbital transfer rockets. The idea gained further traction in 1987 when the Strategic Defense Initiative (SDI) launched the Timberwind program. The program aimed to develop a PBR engine for long-range anti-missile interceptors, promising significant reductions in system mass compared to solid-core reactors. The PBR design proposed for Timberwind was distinguished by its remarkable potential in achieving a high-power density of 40 MWt per liter, a thrust-to-weight ratio of 30:1, and a specific impulse of 869 seconds. This would enable the system to deliver burn-out velocities exceeding 7 km/s, sufficient for intercepting ballistic missile trajectories over 5,000 km away within approximately 18 minutes of launch. However, despite the promising design and developmental efforts, the Timberwind program was ultimately terminated in 1992.

11.10.1   Particle-Bed Reactor Description

The particle-bed reactor's design is centered around the use of small, coated solid carbide fuel particles. The selection of this fuel type is crucial for achieving high-temperature operation and optimal specific impulse performance. The particles are designed with a mixed uranium-refractory metal carbide kernel coated with an additional layer of refractory metal carbide. These coatings serve a dual purpose: they enhance the fuel's thermal stability at high temperatures and improve the retention of fission products, thereby ensuring operational safety and efficiency.

In a typical PBR design, the fuel particles are confined within a cylindrical annulus formed by porous frits—referred to as the cold and hot frits. This assembly constitutes the basic fuel element, wherein the packed fuel particles are held within a frit structure. The configuration allows for the coolant, typically hydrogen, to flow axially through the system, first entering a moderator block and then moving through the cold frit, fuel bed, and finally the hot frit. The coolant flow is radial within the fuel element, ensuring efficient heat transfer and minimal temperature gradients, which are critical for maintaining fuel integrity and maximizing thermal efficiency.

The unique architecture of PBRs offers several advantages. One of the key benefits is the extensive heat transfer area provided by the small particles, which allows for high power densities—up to 10 MW/L—and results in a compact and lightweight reactor design. The small size of the fuel particles also minimizes the local temperature differential between the particles and the coolant, typically around 50 K, enabling rapid reactor startup and shutdown. This feature is particularly advantageous for space missions requiring quick response times and high maneuverability.

Another notable aspect of the PBR design is that only a small portion of the core operates at extremely high temperatures. While the hot frit and the inner regions of the fuel bed reach temperatures above 2,000 K, the majority of the reactor components, including the pressure vessel, control drums, core moderator, and cold frit, remain at much lower temperatures (300 K or less). This characteristic not only simplifies the reactor's thermal management but also enhances its structural integrity and operational reliability.

11.10.2 Structural and Operational Details

The PBR reactor is typically configured as a thermal reactor, utilizing enriched uranium fuel (93.5%) and moderated by beryllium (Be). The Be moderator block contains channels for multiple fuel elements, with additional channels reserved for launch poison safety chains, which are crucial for reactor safety during launch and ascent phases. These safety chains are designed to remain in place until the reactor reaches orbit, at which point they are withdrawn to allow for reactor startup.

Coolant entry into the core is facilitated through guide tubes in the Be moderator, flowing through distribution holes to an annular plenum surrounding each fuel element. From there, the coolant flows radially inward, passing through the outer cool frit, the packed fuel bed, and the hot frit. The heated coolant then exits through central channels in each fuel element, converging into a common outlet plenum and ultimately the nozzle.

The design ensures that the high-temperature portions of the reactor, such as the hot frit and the outlet plenum, are effectively isolated from the cooler sections, thereby protecting the structural components from thermal degradation. The pressure vessel, typically made of aluminum, is located between the core and the reflector, with rotating control drums situated outside the pressure vessel. These drums are cooled by the cold propellant before it enters the pressure vessel, further enhancing the system's thermal management.

11.10.3 Performance and Applications

The PBR design offers considerable performance advantages, including a specific impulse potentially reaching up to 1,000 seconds with hydrogen propellant. The system's total power output can be substantial, with designs like the 200 MWt model capable of producing thrust levels around 50,000 N with hydrogen and 80,000 N with ammonia coolant. The reactor's assembly mass is relatively low, around 600 kg, making it an attractive option for space applications requiring high efficiency and minimal weight.

The flexibility in using different coolants, such as hydrogen or ammonia, provides additional versatility, allowing the PBR to adapt to various mission profiles and environmental conditions. This adaptability, combined with the reactor's high efficiency and rapid startup capabilities, positions PBRs as a promising technology for future space exploration missions, particularly those requiring high-thrust, high-efficiency propulsion systems.

The Particle-Bed Reactor represents a sophisticated and highly efficient approach to nuclear thermal propulsion. Its development, highlighted by the Timberwind program and subsequent research, underscores the potential of this technology to revolutionize space propulsion. By leveraging the unique properties of particle-sized fuel and advanced thermal management techniques, PBRs offer a viable path toward achieving high specific impulses and efficient space travel. Despite the termination of active development programs, the foundational work on PBRs continues to inform and inspire ongoing research in nuclear propulsion, holding promise for future advancements in space exploration capabilities.

11.10.4   PBR Technology Status

The development and testing of Particle-Bed Reactor (PBR) nuclear rocket technology made significant strides during the Timberwind program. Several critical experiments and analyses were conducted to validate the theoretical models and practical viability of PBR designs, focusing on fuel element behavior, system stability, and operational efficiency.

11.10.4.1   Experimental Testing and Fuel Development

The experimental foundation for PBR technology was laid through a series of tests, notably at the Sandia Pulsed Reactor. These tests included zero-power PBR reactor core experiments crucial for benchmarking neutronics simulation codes. The baseline particle fuel design, leveraging technology from High Temperature Gas-Cooled Reactors (HTGR), was rigorously tested under the program "Pulsed Irradiation of PBR Fuel Element" (PIPE). The PIPE experiments aimed to evaluate the performance of fuel particles comprising a uranium carbide (UCx) kernel surrounded by multiple protective layers. The kernels, fabricated using an internal-gelation technique, typically had a uranium carbide composition with a stoichiometric value (x) of around 1.65.

The PIPE fuel particle tests, conducted at 5% of the scale of a full operational system in Sandia National Laboratories' Annular Core Research Reactor (ACRR), revealed critical insights into the PBR fuel element's thermo-mechanical behavior. These tests indicated that the PBR fuel element is a complex system, sensitive to variations in power densities and hydrogen flow rates. The first test, PIPE-1, was conducted in October 1988 and involved five cycles at increasing power levels, achieving average and peak outlet temperatures of 1,600 K and 1,900 K, respectively. The power density during these tests reached 1.5 to 2.0 MW/L, and the results were largely normal, with no significant anomalies observed.

However, the subsequent test, PIPE-2, conducted in July 1989, encountered significant issues. The experiment had to be terminated prematurely after just 24 seconds of operation due to over-temperature conditions in the fuel element. The primary cause was attributed to blockages in the cold frit, likely caused by foreign material, which led to extensive fuel particle fractures. This resulted in frit clogging by particle fragments and the release of fission products. These issues underscored the need for a major redesign of the fuel element configuration to enhance its robustness and reliability.

11.10.4.2   Challenges and Technical Issues

The PIPE tests highlighted several critical challenges in PBR technology. One of the primary issues was the clogging of small, micron-sized holes in the frits supporting the fuel particles. This clogging can cause localized disruptions in propellant flow, leading to thermal anomalies that could damage the fuel particles and other structural components. The control of flow distribution within the reactor remains another significant challenge. Anomalies in propellant flow and thermal profiles can create local hotspots in the fuel particle bed, potentially causing fuel particle and structural failure.

The low heat capacity of the particle bed, while beneficial for efficient heat transfer, poses risks during reactor startup and shutdown. The low thermal inertia requires precise and responsive reactivity control to prevent power overshoots, which could lead to excessive temperatures and potential melting of the fuel particles or other core components. This stringent reactivity control is crucial to maintain the integrity and safety of the reactor during operation.

The tests conducted thus far have shown that the power densities achieved were relatively low compared to those anticipated in an operational system. Therefore, extensive testing at more representative power densities is essential to validate the technology's scalability and reliability.

11.10.4.3   Advanced Fuel Development

In response to the challenges identified during the PIPE experiments, an advanced fuel particle development program was initiated. The goal was to achieve a mixed mean outlet propellant temperature of 3,000 K while ensuring that the fuel particles remain solid and free of free carbon, which can cause eutectic melting and significant material loss. The advanced fuel particles were designed with a uranium-containing refractory metal carbide kernel, either bare or coated with one or more layers of refractory-carbide material. Near-term developments included using a (U, Zr)C kernel with a NbC (or ZrC) coating, aiming to provide high solidus temperatures and robust fission product retention capabilities. Longer-term goals focused on developing (U, Zr, Nb)C particle fuel to maximize operational temperatures.

11.10.4.4   Analytical Models and Flow Stability

Several analytical models have been developed to understand and predict the flow stability within PBR systems. The system's multiple heated flow channels, cooled by a flowing gas, pose a risk of temperature and flow instabilities. Such instabilities can arise from perturbations in one channel, causing temperature increases, viscosity changes, and density reductions, leading to decreased mass flow rates and further temperature rises. This positive feedback loop could potentially lead to system failure if not adequately controlled. Analytical studies suggest that maintaining a sufficiently high flow resistance in the cold frit is crucial to creating a stable pressure drop across the fuel region, which is essential for maintaining overall system stability. Nonetheless, full-scale experimental engine tests are necessary to verify the flow stability and refine the reactor design.

11.10.4.5   Summary

The Particle-Bed Reactor offers potential advantages, including specific impulses approaching 1,000 seconds and a significantly lower reactor mass compared to traditional designs like NERVA. However, the technology remains in the experimental phase, with substantial advancements needed in fuel particle development, fuel element design, flow stability, and system controls. The experimental work conducted from 1987 to 1991 has laid the groundwork for understanding the technology's current state and identifying key areas for future research and development. The ongoing efforts aim to address these challenges and unlock the full potential of PBR technology for future space exploration missions.

11.11   Pellet Bed Reactor Concept

The Pellet Bed Reactor (PeBR) is a novel nuclear thermal propulsion concept characterized by its fast neutron flux (with peak neutron energies ranging from 1 to 50 keV) and hydrogen coolant. The PeBR utilizes spherical fuel pellets, each approximately 1.0 cm in diameter, composed of graphite-ZrC coated fuel. These pellets are self-supported within the reactor core, eliminating the need for internal structural components. This design simplicity not only reduces the core's size and weight but also facilitates reactor fueling operations either at the launch site or potentially in orbit.

11.11.1   Fuel Composition and Structure

The fuel pellets in the PeBR consist of TRISO-coated microspheres embedded in a graphite matrix. Each microsphere features a fuel kernel made from either uranium carbide (UC-NbC) or uranium nitride-tantalum carbide (UN-TaC), with a diameter of about 500 micrometers. Surrounding this kernel are three protective layers:

1. **Inner Layer**: A low-density pyrolytic graphite (PyC) layer, 15-20 micrometers thick, which accommodates fission product recoil and partially absorbs fission gases.
2. **Intermediate Layer**: A high-density graphite layer, 5-10 micrometers thick, providing structural reinforcement and preventing acid absorption during the chemical vapor deposition process of the outer coating.
3. **Outer Layer**: A tantalum carbide (TaC) or niobium carbide (NbC) coating, 10-20 micrometers thick, offering protection against chemical interactions with the propellant.

This multi-layered design is essential for maintaining the integrity of the fuel under high temperatures, potentially reaching up to 3,100 K for a hydrogen exit temperature of 3,000 K. Notably, NbC is often preferred over TaC due to its lower neutron absorption cross-section, despite its slightly lower melting temperature.

11.11.2   Reactor Design and Operation

The PeBR core has a compact diameter of approximately 80 cm, with the spherical fuel pellets arranged in a manner that allows efficient heat transfer. The lack of internal core structures simplifies the design, enabling passive cooling and reducing overall reactor mass. The core's thermal management is augmented by a dual shielding system: a hot shield within the pressure vessel and a shadow shield outside the reactor. Both shields are composed of layered lithium hydride (LiH) and tungsten (W), effectively managing the thermal and radiation environment.

Hydrogen Flow Path: Liquid hydrogen is pumped from the tank, first cooling the nozzle structure and then flowing through the Be_2C radial reflector. By the time it exits the reflector, the hydrogen has transformed into gas, which then drives the turbopumps. Subsequently, the hydrogen is directed back to the reactor dome, where it flows downward through the hot shield, maintaining the LiH portion's temperature between 600 to 680 K. The flow continues to cool the upper axial Be_2C reflector before moving radially through the core to remove the heat generated in the fuel pellets. Finally, the heated hydrogen exits through a central channel with a diameter of approximately 19 cm.

Core Configuration: The PeBR core is surrounded by a set of 26 control drums, each 11 cm in diameter, integrated into the radial reflector structure. These control drums, along with eight 4-cm diameter safety rods, provide the necessary reactivity control and ensure the reactor remains subcritical during potential accidents, such as water immersion scenarios. The core's design is optimized to achieve a height-to-diameter ratio greater than 1.85, which helps reduce pressure losses and supports efficient passive cooling.

11.11.3   Technical Challenges and Potential Advantages

The PeBR concept promises several advantages, including a potential specific impulse of 1,000 seconds, which is higher than traditional designs like NERVA. The absence of internal core structures simplifies the design and allows for flexible fueling and disposal options. The reactor's ability to passively remove decay heat enhances safety, particularly in post-operation scenarios.

However, several critical challenges remain. These include ensuring the nuclear fuel's stability at peak operating temperatures of 3,100 K, addressing materials compatibility issues, particularly with hydrogen at high temperatures, and developing robust fuel coatings to prevent erosion. Additionally, the fabrication of the spherical fuel pellets requires precise control to meet stringent specifications.

The PeBR design represents a significant step forward in nuclear thermal propulsion technology, offering a blend of efficiency, simplicity, and safety. However, further experimental research and development are necessary to overcome the remaining technical challenges and fully realize the potential of this innovative reactor concept.

11.12   Wire Core Reactor Concept

The Wire Core Reactor (WCR) is an advanced nuclear thermal propulsion concept that utilizes tungsten alloy wire fuel elements. This reactor design is derived from work initially conducted under the aircraft nuclear propulsion program, specifically the 710 Program. The WCR features a compact, efficient design, characterized by a high surface area-to-volume ratio, with approximately 19 cm² of heat transfer area per cubic centimeter of core. This is significantly greater compared to about 5 cm² for traditional smooth passage reactors.

11.12.1   Core and Fuel Element Design

The active core of the WCR is annular, allowing for radial coolant flow. Hydrogen enters the core through ducts at each end, flows through the active core, and exits via the exhaust nozzle after making a series of 90-degree turns. The core structure is built from multiple layers of tungsten wires, with a diameter of 0.89 mm, interspersed with spacer wires averaging 0.34 mm in diameter. The tungsten wires serve as fuel elements, each containing a cermet fuel composed of 70 vol% uranium nitride (UN) and 30 vol% tungsten.

Fuel Cladding: The fuel elements are clad in a tungsten-rhenium alloy, containing 3-5% rhenium. The cladding thickness varies from 0.076 mm at the inner core radius to 0.178 mm at the outer core radius. This cladding provides structural integrity and thermal conductivity, ensuring effective heat transfer and containment of the nuclear fuel.

Fabrication Process: The fabrication of the tungsten wire fuel elements begins with braided tungsten tubing, consisting of 0.2 mm tungsten wires braided to form a 3.2 mm diameter tube. This tubing is filled with 0.1 mm UN particles, which are coated with tungsten via chemical vapor deposition (CVD). The tubing is then swaged down to a diameter of 1.9 mm and drawn further to achieve the final wire fuel element diameter of 0.89 mm.

Core Assembly: The core is assembled in ring-shaped segments, each approximately 10 cm in the axial direction and 2.5 cm thick radially. The segments are built up layer by layer, alternating between fueled and spacer wires. Once assembled, the core segments are fused into a solid, continuous structure using tungsten CVD, ensuring structural cohesion and durability.

11.12.2   Reactor and Engine Configuration

The WCR is designed to provide a thrust of 914 kN with a specific impulse of 1,000 seconds, employing an expander cycle for its operation. In this cycle, heat generated within the nozzle and internal components drives the hydrogen turbopump, as illustrated in the reactor engine schematic. The reactor and shield assembly's total mass is approximately 2,304 kg.

Engine Cycle and Heat Management: The WCR utilizes hydrogen as the primary coolant, which flows through the reactor core, absorbing heat from the fuel elements. The high thermal conductivity of the tungsten wire and the large surface area facilitate efficient heat transfer, allowing for rapid thermal response and high-power densities.

11.12.3   Advantages and Challenges

The WCR offers several key advantages:

1. **High Heat Transfer Efficiency**: The large surface area per unit volume enhances heat transfer efficiency, critical for achieving high power densities.
2. **Separation of Fuel and Structural Components**: This design feature mitigates potential issues with structural integrity and thermal expansion.
3. **Material Compatibility**: The use of tungsten and uranium nitride ensures compatibility with hydrogen coolant, even at elevated temperatures.

However, the WCR concept faces several technical challenges that require further research and development:

- **Fuel Fabrication**: The process of fabricating the complex wire fuel elements, including swaging, drawing, and CVD coating, needs optimization and validation.
- **Thermal and Mechanical Stability**: Ensuring the stability of the fueled wires under high temperature and neutron flux conditions is critical. This includes addressing potential issues such as nitrogen overpressure in the UN matrix and the mechanical properties of the fueled wires.
- **Experimental Validation**: Extensive testing is necessary to verify the reactor's performance, particularly the compatibility of the W-Re-UN system with hydrogen at high temperatures and the behavior of the fuel elements under reactor conditions.

The Wire Core Reactor concept presents a promising approach to nuclear thermal propulsion, offering high efficiency and compact design. However, significant experimental work and material science advancements are needed to realize its full potential.

11.13   Conical Fuel Core Concept

The Conical Fuel Core, employed in the IMPULSE propulsion concept, is a distinctive nuclear thermal propulsion design that utilizes conical fuel elements to achieve efficient heat transfer and high specific impulse. This design is characterized by the use of thin, truncated cone-shaped fuel plates, which are stacked together to form a fuel assembly. Each conical fuel element features a perimeter lip that serves multiple purposes, including structural support, spacing between elements, and regulation of coolant flow.

11.13.1   Fuel Element Design

The conical fuel elements are designed to optimize the surface area for heat transfer while minimizing the structural mass. The thin, conic shell of each element is complemented by radial ribs that provide structural integrity and enhance thermal conduction. The perimeter lip, being unfueled, serves as a mechanical support and defines the inter-element spacing. This lip can be precisely orificed to control the flow of the propellant, allowing for individual adjustment of coolant flow rates to match the power distribution requirements of the core.

The versatility of the conical fuel element design allows for the use of various fuel types, each with distinct performance characteristics. The specific impulse and maximum operating temperatures for different fuels is as follows:

- **Carbide Fuel**: Capable of operating at temperatures up to 3,100 K, carbide fuel is highly resistant to hydrogen corrosion, making it suitable for high-temperature applications. It has a fuel loading fraction capability of approximately 0.2 and can achieve a specific impulse of up to 985 seconds.
- **Composite Fuel**: Operable at temperatures up to 2,700 K, with a specific impulse potential of around 900 seconds, and a fuel fraction of 0.2.
- **Moly-Cermet**: This fuel type can operate at temperatures up to 2,200 K, with a fuel fraction of 0.5 and a specific impulse of 805 seconds.
- **Tungsten-Cermet**: Suitable for temperatures up to 2,800 K, with a fuel loading fraction of 0.5 and a specific impulse of 920 seconds.

11.13.2   Core Configuration

The reference configuration for the IMPULSE system is a thermal reactor moderated by zirconium hydride (ZrH), utilizing a three-pass core layout. This design allows for a heterogeneous core, where different fuel types can be employed in various regions of the core to optimize performance. For instance, lower temperature regions can utilize fuels with lower maximum temperatures, while the high-temperature regions can employ fuels like carbide that can withstand higher operating temperatures.

In the IMPULSE design, a core with 19 fuel assemblies is utilized, designed to produce a thrust of 334 kN (75,000 lbf). With carbide fuel cones, the system can achieve a specific impulse of 970 seconds and a thrust-to-weight ratio of 30. The core layout, as shown in Fig. 6, includes positions for fuel and control drums, as well as provisions for turbine gas heaters, enhancing the reactor's operational flexibility and safety.

11.13.3   Advantages and Challenges

The conical fuel core design offers several significant advantages:

1. **Enhanced Heat Transfer**: The conical shape and thin design maximize the surface area available for heat transfer, facilitating efficient thermal energy extraction.
2. **Structural Efficiency**: The perimeter lip provides necessary structural support without adding significant mass, thus improving the thrust-to-weight ratio.
3. **Fuel Versatility**: The ability to use various fuel types allows for optimization based on specific mission requirements, whether for high thrust or high specific impulse.

However, there are also challenges associated with this design:

- **Thermal and Structural Stress**: The thin structure of the conical elements must withstand high thermal and mechanical stresses, especially at high power densities.
- **Coolant Flow Control**: Precise control of coolant flow is critical to ensure uniform temperature distribution and prevent hot spots, which could damage the fuel elements.
- **Material Compatibility**: Ensuring the long-term stability and compatibility of materials, particularly at the high temperatures encountered, remains a key technical challenge.

The Conical Fuel Core concept offers a promising approach for advanced nuclear thermal propulsion systems, capable of delivering high performance and efficiency. However, further research and development are required to address the technical challenges and optimize the design for practical application.

11.14   Conical Fuel Core Concept for IMPULSE Reactor

The Conical Fuel Core is an innovative nuclear thermal propulsion concept employed in the IMPULSE propulsion system, characterized by the use of conically shaped fuel elements. These elements are designed to optimize the thermal and structural performance of the reactor while ensuring efficient heat transfer and coolant management. This concept is notable for its unique geometry and potential to achieve high specific impulse and thrust-to-weight ratios, making it an attractive option for advanced space propulsion.

11.14.1   Design and Geometry of Conical Fuel Elements

The core of the IMPULSE reactor consists of thin fueled plates shaped as truncated cones, known as conical fuel elements. Each element features a thin conic shell with a perimeter lip that serves multiple functions:

1. **Structural Support**: The perimeter lip provides mechanical support and maintains the spacing between adjacent fuel elements.
2. **Coolant Flow Regulation**: Orifices in the perimeter lip regulate the flow of coolant through each fuel element, ensuring that the flow is matched to the local power distribution within the core.
3. **Heat Conduction Pathway**: The geometry of the elements, with radial ribs and a short heat conduction path, enhances the thermal conductivity and stability of the fuel, allowing for efficient heat transfer from the fuel to the coolant.

The conical shape of the elements is designed to maximize surface area, thereby increasing the contact area with the coolant and enhancing heat transfer. This is crucial for maintaining the integrity of the fuel under high-temperature conditions and optimizing the reactor's thermal efficiency.

11.14.2   Fuel Options and Performance Characteristics

The conical fuel elements can be fabricated from a variety of fuel materials, each offering different thermal and structural properties. Table 3 outlines the specific impulse and maximum operating temperatures for several fuel types:

*   **Carbide Fuels**: Capable of operating at temperatures up to 3,000 K, carbide fuels are highly resistant to corrosion and hydrogen interactions, making them suitable for high-temperature applications. These fuels have a maximum fuel fraction of 0.2, providing a specific impulse of up to 985 seconds.
*   **Composite Fuels**: With a maximum operating temperature of 2,700 K and a fuel fraction of 0.2, composite fuels offer a balance between performance and structural integrity, achieving specific impulses around 900 seconds.
*   **Moly-Cermet Fuels**: These fuels can operate at temperatures up to 2,200 K, with a fuel loading fraction of 0.5. They are particularly effective for small cores and provide a specific impulse of 805 seconds.
*   **Tungsten-Cermet Fuels**: Capable of withstanding temperatures up to 2,800 K, tungsten-cermet fuels offer a fuel fraction of 0.5 and a specific impulse of 920 seconds. They are suitable for applications requiring robust structural performance at high temperatures.

11.14.3   Core Configuration and Reactor Design

The IMPULSE reactor's core configuration includes 19 fuel assemblies moderated by zirconium hydride (ZrH), arranged in a three-pass core layout. This design allows for a heterogeneous core, enabling the use of different fuels in distinct regions of the core to optimize thermal and neutronic performance. The three-pass core design facilitates efficient moderation and ensures even temperature distribution throughout the core.

In the reference design for the IMPULSE reactor, the system is configured to deliver a thrust of 334 kN (75,000 lbf) with a specific impulse of 970 seconds. The use of carbide fuel cones enables the reactor to achieve a high thrust-to-weight ratio of 30, making it suitable for a wide range of space missions, from low Earth orbit maneuvers to deep space exploration.

11.14.4 Advantages and Technical Challenges

The Conical Fuel Core concept offers several key advantages:

- **Enhanced Heat Transfer Efficiency**: The conical geometry and thin design of the fuel elements maximize the surface area for heat transfer, improving the efficiency of thermal energy conversion.
- **Structural Integrity**: The perimeter lip and radial ribs provide necessary structural support, maintaining the stability of the fuel elements under high thermal and mechanical loads.
- **Versatility in Fuel Choice**: The ability to use various fuel types allows for optimization based on specific mission requirements, whether for high thrust or high specific impulse.

However, the concept also presents several technical challenges:

- **Thermal Stress Management**: The thin structure of the fuel elements must withstand significant thermal stresses, particularly at high power densities. Ensuring material integrity under these conditions is critical.
- **Coolant Flow Control**: Precise control of coolant flow through the orifices in the perimeter lip is essential to prevent local overheating and ensure uniform temperature distribution.
- **Material Compatibility**: The choice of materials for the fuel and structural components must ensure long-term compatibility and resistance to corrosion, especially under high-temperature hydrogen exposure.

The Conical Fuel Core concept represents a promising approach to nuclear thermal propulsion, offering the potential for high performance and efficiency. The design's unique geometry and use of advanced fuel materials provide a versatile and robust solution for future space missions, though further research and development are needed to address the associated technical challenges and fully realize its potential.

11.15 Pressure Fed Nuclear Thermal Rocket (PFNTR) Concept

The Pressure Fed Nuclear Thermal Rocket (PFNTR) represents an innovative propulsion technology that simplifies the nuclear thermal propulsion system by eliminating the need for a turbopump. Instead, the PFNTR relies on conventional propellant tank pressure to expel liquid hydrogen through the flow control system into the reactor. This simplification reduces mechanical complexity, enhances system reliability, and potentially increases specific impulse due to hydrogen dissociation effects at high temperatures and low pressures.

11.15.1   Design and Operational Principles

The PFNTR operates at significantly lower pressures compared to traditional nuclear thermal rockets, typically around 0.1 MPa (15 psia). This low-pressure operation allows for the exploitation of hydrogen's unique dissociation-recombination properties at elevated temperatures. When hydrogen is heated above 3,000 K, it begins to dissociate into atomic hydrogen. The recombination of these atoms upon expansion in the nozzle releases additional energy, thereby increasing the exhaust velocity and specific impulse of the rocket.

The core design of the PFNTR can vary, but a common configuration features a spherical core structure. This geometry maximizes the surface area for heat transfer, thereby improving the thermal efficiency of the reactor. The fuel within the core can be arranged in various forms, such as particles, pellets, or platelets, depending on the specific design requirements and desired performance characteristics.

11.15.2   Core and Fuel Element Design

In the PFNTR, the radial outward flow of hydrogen is a key design feature, allowing the coolant to pass through the core and absorb heat from the fuel elements. The fuel elements themselves are typically constructed from high-temperature-resistant materials capable of withstanding the thermal and mechanical stresses associated with nuclear fission and high-temperature hydrogen flow. Reactivity control within the core is uniquely managed by adjusting the hydrogen content rather than relying on traditional control drums. This approach simplifies the reactor design and potentially enhances the system's safety and responsiveness.

11.15.3   Performance Characteristics and Theoretical Considerations

One of the primary advantages of the PFNTR system is its potential for achieving high specific impulses due to hydrogen dissociation. The theoretical specific impulse can reach up to 15,000 N·s/kg at hydrogen temperatures of 3,500 K. However, practical limitations such as kinetic losses, nozzle divergence, and boundary-layer effects often reduce the delivered performance. The delivered specific impulse at 3,500 K is typically around 12,000 N·s/kg, significantly lower than the ideal values but still highly competitive compared to traditional chemical propulsion systems.

11.15.4   Reactor Control and Safety Considerations

Reactor control in a PFNTR system is inherently different due to the absence of control drums. Instead, the reactivity is adjusted by manipulating the amount of hydrogen present in the core, which can be challenging but offers a streamlined approach to reactor management. This method requires precise monitoring and control systems to ensure safe and stable reactor operation, especially given the potential for rapid power changes due to hydrogen's thermal properties.

Shielding is another critical consideration in PFNTR design. The low-pressure operation and the resulting high thermal fluxes necessitate robust radiation shielding to protect both the reactor components and the surrounding spacecraft. Advanced materials and structural designs are required to provide adequate protection while minimizing the overall weight and complexity of the system.

11.15.5   Challenges and Future Development

Despite its potential advantages, the PFNTR concept faces several technical challenges:

1. **Hydrogen Dissociation Management**: Accurately predicting and managing the dissociation and recombination of hydrogen is complex and requires precise thermal management and flow control.
2. **Material Selection and Durability**: The materials used in the reactor core and fuel elements must withstand high temperatures and corrosive hydrogen environments, which can limit material choices and impact the reactor's operational lifespan.
3. **Reactor Control and Stability**: The novel reactivity control method requires development of new control systems and safety protocols to ensure reliable operation.

To fully realize the potential of PFNTR technology, further experimental validation is necessary. This includes testing at high temperatures and low pressures to confirm theoretical specific impulse values and assess the practicality of reactor control methods. Additionally, advancements in material science and thermal management will be crucial to overcoming the current limitations and ensuring the system's viability for future space missions.

11.16   Liquid Core Nuclear Thermal Propulsion Concepts

Liquid Core Nuclear Thermal Propulsion (NTP) concepts, such as the **Liquid Annulus Reactor System (LARS)** and the **Droplet Core Reactor**, represent advanced designs aiming to achieve extremely high operating temperatures and specific impulses. These concepts involve maintaining the nuclear fuel in a liquid state, allowing the fuel temperature to reach between 3,000 and 9,000 K. This high-temperature operation is expected to significantly enhance the thermal efficiency and propulsion capabilities of NTP systems.

11.16.1   Liquid Annulus Reactor System (LARS)

The LARS concept utilizes a rotating cylindrical fuel element that contains a liquid core of high-temperature refractory material, typically comprising uranium mixed with diluents such as uranium carbide (UC2) and zirconium carbide (ZrC). The design is characterized by a solid outer layer adjacent to the reactor vessel walls, while the inner portion, in contact with flowing hydrogen, remains in a liquid state. The centrifugal force generated by the rotation of the fuel element stabilizes the liquid fuel layer, preventing it from dispersing or mixing with the coolant.

11.16.1.1   Design and Operation

In the LARS design, the fuel element's outer layer is solid, providing structural integrity and containment, while the molten inner layer facilitates efficient heat transfer to the hydrogen propellant. The high rotation speed required to stabilize the molten fuel introduces significant engineering challenges, particularly concerning material durability and stability. Key components of the system include:

1. **Rotating Fuel Element**: The rotating canister stabilizes the liquid fuel layer through centrifugal force, with rotational speeds calibrated to maintain the desired layer thickness and stability.
2. **Coolant Gap**: A gap exists between the solid outer fuel layer and the reactor's internal structure, allowing for effective cooling and thermal isolation.
3. **Moderator and Reflector**: Stationary moderators and reflectors surround the fuel elements, providing neutron moderation and reflection to sustain the nuclear reaction.

11.16.1.2   Technical Challenges

Several critical technical challenges need to be addressed for the successful implementation of the LARS concept:

- **Liquid Layer Stability**: Maintaining a stable and uniform liquid fuel layer is crucial, requiring precise control over rotational speed and reactor geometry.
- **Material Compatibility**: The materials used must withstand high temperatures, radiation, and corrosive hydrogen environments. This includes ensuring that the solid and liquid phases of the fuel do not interact detrimentally.
- **Evaporative Losses**: At the high operating temperatures, there is a risk of evaporative loss of fuel components, potentially leading to contamination or degradation of the system.
- **Radiative Heat Transfer**: Efficient radiative heat transfer within the molten layer

and to the hydrogen propellant is essential for achieving the desired specific impulse and thermal efficiency.

11.16.2 Droplet Core Reactor Concept

The Droplet Core Reactor concept eliminates the need for structural support for solid fuel materials at high temperatures by suspending fuel particles, such as tungsten-coated uranium dioxide (UO2), in a flowing hydrogen stream. The reactor's design relies on a body force—such as centrifugal force from reactor rotation or vortex flow—to retain the fuel particles within the reactor core, preventing them from being carried away with the coolant flow.

11.16.2.1 Design and Operation

In this concept, the hydrogen coolant flows through a mixture of suspended fuel particles, directly contacting and heating the hydrogen. The hot hydrogen then expands through a nozzle to produce thrust. The system can utilize several methods to achieve the necessary body force:

1. **Rotational Force**: The reactor can be rotated, creating a centrifugal force that pushes the fuel particles outward, preventing them from escaping with the coolant flow.
2. **Vortex Flow**: A vortex flow pattern can be established within the reactor, using the centrifugal force of the vortex to retain the fuel particles.
3. **Technical Challenges**

The Droplet Core Reactor also faces numerous challenges:

- **Fuel Particle Retention**: Ensuring that the fuel particles remain within the reactor core and do not escape with the exhaust flow is a significant challenge. This requires precise control of flow dynamics and particle behavior.
- **Fission Product Management**: The release of fission products during operation must be controlled to prevent contamination and ensure safety.
- **Agglomeration**: Fuel particles may tend to clump together or sinter at high temperatures, leading to potential blockages or uneven heating.

The Liquid Core and Droplet Core reactor concepts represent innovative approaches to achieving high specific impulses in nuclear thermal propulsion. By utilizing liquid or suspended particulate fuels, these designs aim to operate at temperatures significantly higher than traditional solid-core reactors, thereby enhancing thermal efficiency and propulsion performance. However, substantial technical challenges remain, particularly in materials science, thermal management, and reactor control, which must be overcome before these concepts can be realized in practical applications. The successful development of these systems could revolutionize space propulsion, enabling faster and more efficient travel beyond Earth's orbit.

11.17   Gaseous Core Nuclear Thermal Propulsion Concepts

Gaseous Core Nuclear Thermal Propulsion (GCNTP) represents an advanced category of nuclear propulsion technology where the nuclear fuel exists in a gaseous or plasma state. This configuration allows for extremely high temperatures, which can lead to significant increases in specific impulse compared to solid-core designs. Specific impulses as high as 50,000 m/s have been postulated, with nuclear fuel temperatures potentially exceeding 10,000 K. The gaseous core approach leverages the potential benefits of operating at temperatures where materials limitations are mitigated by separating the fuel from structural components through various innovative designs.

11.17.1   Coaxial Flow Gaseous Core Reactor

The Coaxial Flow Gaseous Core Reactor concept involves a configuration where a low-velocity inner stream of fissioning uranium metal plasma is surrounded by a high-velocity propellant gas, typically hydrogen. The primary mechanism for energy transfer from the hot plasma to the hydrogen propellant is through both convection and radiation. The hydrogen gas acts as both a propellant and a coolant, absorbing thermal energy from the uranium plasma. This setup allows for efficient heat transfer and rapid acceleration of the propellant.

11.17.1.1   Key Challenges:

- **Fuel Loss:** A significant issue with the coaxial flow design is the potential mixing of uranium plasma with the hydrogen propellant, which can lead to fuel losses.
- **Thermal Management:** Managing the extreme temperatures and ensuring efficient heat transfer while avoiding damage to structural materials are critical challenges.

11.17.1.2   Nuclear Lightbulb Reactor

The Nuclear Lightbulb Reactor concept confines the fissioning uranium plasma within a transparent quartz or similar material cell, termed the "lightbulb." The design uses a buffer gas, such as argon, injected tangentially to create a swirling flow that prevents the plasma from contacting the cell walls. The primary mode of heat transfer from the plasma to the propellant in this concept is radiation, which occurs without direct contact between the plasma and the hydrogen propellant.

Advantages:
- **Isolation of Plasma:** The transparent cell effectively isolates the fissioning plasma, preventing fuel losses and reducing the risk of

contamination.

- **Radiative Heat Transfer:** Although radiative transfer is limited to certain wavelengths due to the transparency and resistance of the cell material to radiation-induced darkening, this method allows for efficient energy transfer.

Challenges:

- **Material Transparency and Durability:** The transparent cell must remain resistant to nuclear radiation-induced darkening and withstand the thermal and mechanical stresses at high temperatures.
- **Efficient Radiative Transfer:** Ensuring that sufficient energy is transferred from the plasma to the hydrogen propellant through radiation is a significant challenge.

11.17.2   Minicavity Plasma Core Reactor

The Minicavity Plasma Core Reactor employs a small cavity design surrounded by a moderator region and a nuclear-fueled "driver" region. The driver region provides the necessary neutrons to sustain the chain reaction, thus reducing the criticality dimensions of the gaseous cavity. The design aims to minimize the size of the gaseous core while maximizing the neutron economy and thermal efficiency.

11.17.2.1   Concept Operation:

- **Fuel Injection and Containment:** The reactor utilizes vortex injection of uranium hexafluoride gas (UF_6) to create a confined plasma region. The swirling motion helps maintain a stable plasma core while ensuring efficient heat transfer to the hydrogen propellant.

11.17.2.2   Technical Challenges and Considerations

The primary challenges across gaseous core reactor concepts include:

- **Fuel Confinement:** Effective containment of the gaseous or plasma nuclear fuel to prevent losses and maintain criticality.
- **Thermal and Structural Management:** The extreme operating temperatures necessitate advanced materials capable of withstanding thermal stress, radiation damage, and chemical compatibility with the fuel and propellant.
- **Radiation Shielding and Safety:** Managing the intense radiation emitted from the fissioning plasma and ensuring safety for both the spacecraft and any nearby structures or personnel.
- **Fuel Handling and Reprocessing:** Efficient handling, replenishment, and potentially reprocessing of nuclear fuel in a gaseous state present unique challenges not found in solid-core reactors.

Gaseous Core Nuclear Thermal Propulsion concepts offer the potential for unprecedented specific impulses and efficiency in space propulsion. However, they also introduce significant technical and engineering challenges that require extensive research and development. Advances in material science, nuclear physics, and thermal management are essential to overcoming these challenges and realizing the potential of GCNTP systems for deep-space exploration and other advanced space missions.

11.18   The Centrifugal Nuclear Thermal Rocket (CNTR)

The concept of liquid-fuel nuclear rocket engines has been a focal point in propulsion research since at least 1954, driven by the need for advanced systems that could enable rapid and efficient travel within the solar system. Throughout the 1960s, a variety of nuclear thermal propulsion (NTP) design concepts were proposed, each aiming to leverage the immense energy produced by nuclear reactions to heat a propellant—typically hydrogen—to generate thrust. Among the various designs, three primary approaches have emerged: (1) the bubble-through reactor, (2) the radiation reactor, and (3) the particle or droplet reactor. These designs are depicted in Figures 1, 2, and 3, respectively, and each offers unique advantages and challenges.

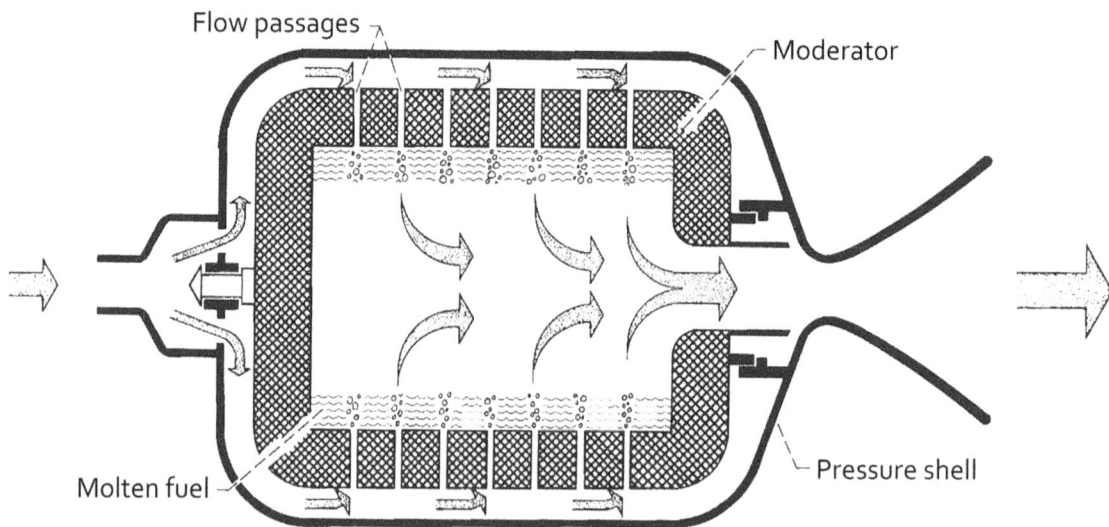

Figure 11-6. Bubble-Through Liquid Fuel Reactor.

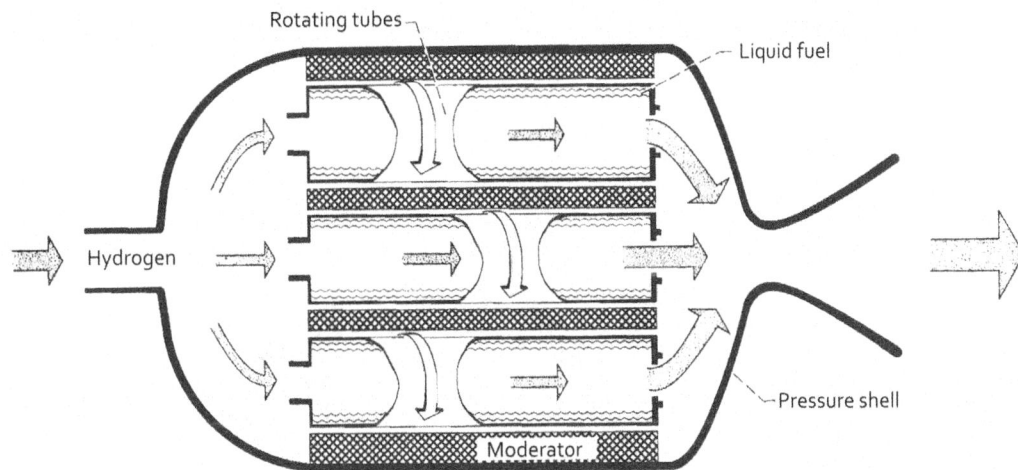

Figure 11-7. Radiation Liquid Fuel Reactor.

11.18.1   The Bubble-Through Reactor

The bubble-through reactor, illustrated in Figure 1, represents one of the earliest and most straightforward concepts for liquid-fuel nuclear rocket engines. In this design, the nuclear reactor fuel is maintained in a liquid state and is spun at high speeds to form a stable annular layer along the inner cylindrical surface of the reactor. This high-speed rotation is essential for containing the liquid fuel within the reactor and preventing it from dispersing under the microgravity conditions encountered in space.

The heating process in the bubble-through reactor involves bubbling hydrogen propellant through the liquid nuclear fuel. As the hydrogen gas passes through the liquid, it absorbs thermal energy from the surrounding fuel, rapidly heating to the same temperature as the liquid fuel. The now-hot hydrogen is then expelled through the rocket nozzle, generating thrust.

The bubble-through reactor design is attractive due to its relative simplicity and its ability to achieve high fuel temperatures, which are crucial for maximizing the specific impulse—a measure of propulsion efficiency. However, several technical challenges must be addressed, including the stability of the liquid fuel and the efficiency of heat transfer to the hydrogen propellant. Additionally, maintaining the liquid state of the fuel without it vaporizing or escaping the containment area is critical for the engine's safe and effective operation.

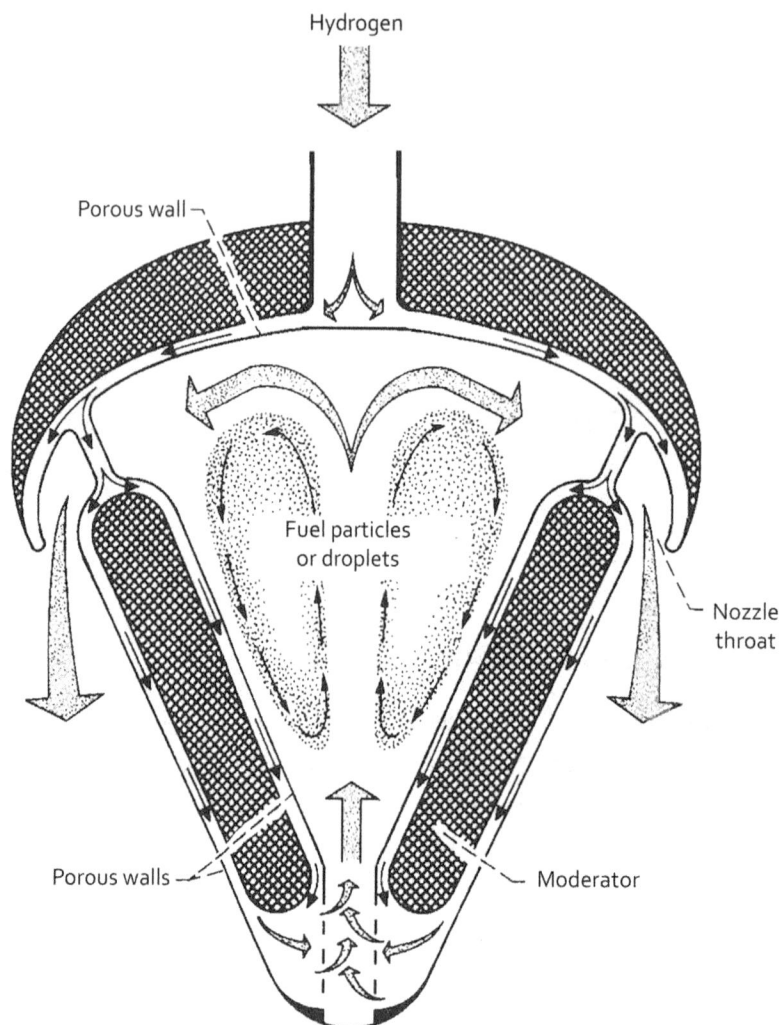

Figure 11-8. Particle or Droplet Liquid Fuel Reactor.

11.18.2   The Radiation Reactor

The radiation reactor, shown in Figure 2, builds on the concept of liquid fuel but introduces a more sophisticated method for heating the hydrogen propellant. In this design, the liquid fuel is housed within rotating cylindrical tubes, which are designed to maintain a uniform layer of liquid along their inner surfaces through rapid rotation.

Hydrogen is introduced axially, flowing down the center of these rotating tubes. As it passes through the reactor, the hydrogen is heated primarily by radiation emitted from the liquid fuel, as well as through surface convection from the tube walls. The rotating action of the tubes enhances the heat transfer by ensuring a consistent distribution of liquid fuel and maintaining the necessary temperature gradients.

The radiation reactor design is particularly effective at achieving very high propellant temperatures, leading to greater engine efficiency and higher specific impulse. The reliance on radiation as the primary heating mechanism also allows for precise control over the heating process, potentially improving both engine performance and fuel efficiency. Despite these advantages, the design is more complex than the bubble-through reactor and requires advanced materials capable of withstanding the intense radiation and high temperatures involved. The rotating components must be engineered to operate reliably in space, where mechanical failures could compromise the mission.

11.18.3   The Particle or Droplet Reactor

The particle or droplet reactor, depicted in Figure 3, represents the most innovative and complex of the three liquid-fuel nuclear rocket engine designs. This approach differs fundamentally from the previous concepts by dispersing the nuclear fuel into the propellant stream in the form of fine particles or liquid droplets. These fuel particles or droplets are continuously recirculated through the propellant stream within the reactor's active zone, where the nuclear reactions occur.

As the hydrogen propellant flows through the reactor, it directly contacts the highly energetic fuel particles or droplets, absorbing heat through conduction, radiation, and convection. This direct interaction enables exceptionally efficient heat transfer, allowing the hydrogen to reach much higher temperatures than those possible with other reactor designs. The superheated hydrogen is then expelled through the rocket nozzle, producing the desired thrust.

The particle or droplet reactor design offers significant theoretical potential for unmatched efficiency and specific impulse, making it an attractive option for deep-space missions where propulsion efficiency is critical. The continuous circulation of fuel particles or droplets allows for sustained operation without the need for frequent refueling or complex maintenance procedures.

However, the particle or droplet reactor also presents significant engineering challenges. Ensuring the continuous and stable recirculation of fuel particles or droplets requires sophisticated control systems and precise engineering to prevent issues such as clumping or settling of the fuel. Moreover, the reactor must be designed to contain and shield the extremely high levels of radiation produced in the active zone, as well as to manage potential fuel loss or degradation over time.

Despite the simplicity of the radiation reactor engine concept, it faces significant challenges due to the location of maximum heat generation. This heat is concentrated at the outer boundary of the liquid uranium, specifically along the inner wall of the rotating cylinder. Unfortunately, there are no known containment materials that can maintain the required structural integrity at these extremely high temperatures. In contrast, the droplet reactor, which disperses fuel as fine particles or droplets within the propellant stream, encounters insurmountable difficulties with current neutronics modeling tools and techniques. The complexity of accurately simulating neutron interactions within such a dynamic environment renders the droplet reactor design infeasible with today's technology.

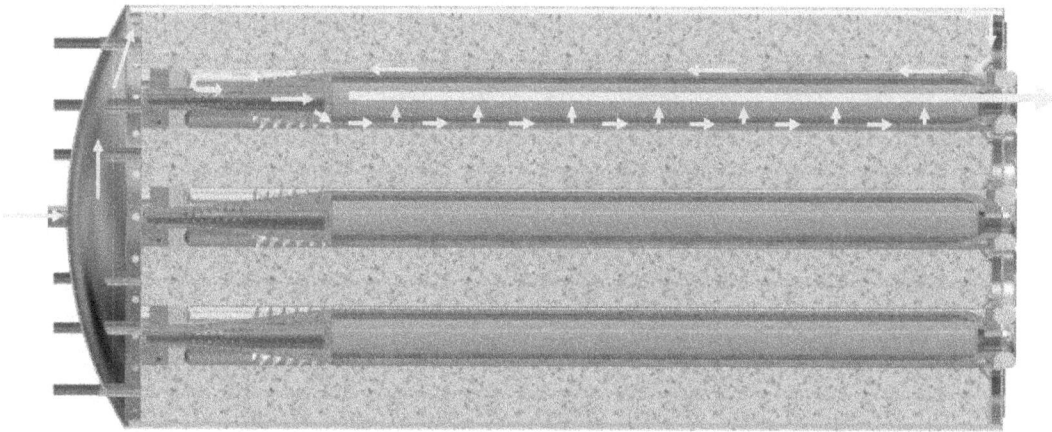

Figure 11-9. Propellant Flow Path in the CNTR (not to scale).

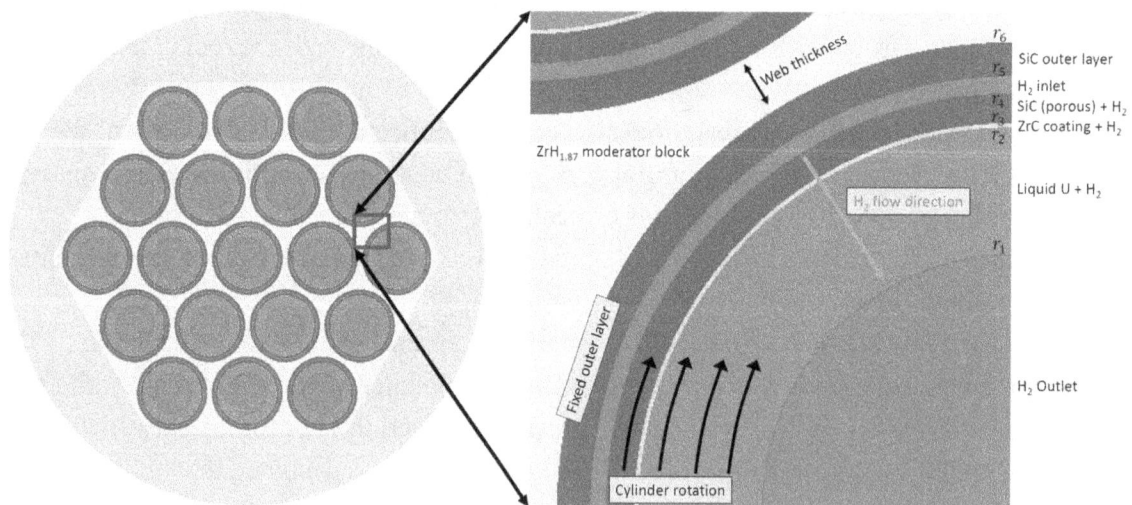

Figure 11-10. Full-core slice of the engine shown in Figure 4 at the axial midplane (top) and a close-up of a single CFE (bottom).

As a result, research efforts have increasingly shifted towards the bubble-through reactor concept, which presents a more manageable approach in terms of both thermodynamics and neutronics. A promising liquid fuel reactor design, currently under study by a NASA-sponsored university team, leverages the bubble-through reactor approach while incorporating multiple elongated Centrifugal Fuel Elements (CFEs). This design, illustrated in Figure 11-9 and Figure 11-10, features a configuration of nineteen CFEs and shares some operational similarities with solid fuel NTP systems.

In this setup, the propellant—originating from a tank (not shown)—is directed through a series of components, including the neutron reflector, a regeneratively cooled section of the nozzle, and the neutron moderator, before reaching the fueled region. This flow configuration ensures that all moderators and structural materials within the engine are maintained at relatively low temperatures, below 800 K. As depicted in Figure 11-9, the propellant enters through the porous wall of the rotating cylinder at approximately 800 K. It then moves radially through the molten uranium fuel annulus, absorbing heat as it passes. The heated propellant exits axially through the bore into a common plenum, before being accelerated through a converging/diverging nozzle to produce thrust.

A significant advantage of this design is its ability to maintain liquid uranium at a temperature of around 1500 K near the inner cylinder wall of each CFE. Importantly, the temperature near the center of the rotating cylinder can reach up to 5500 K, but this extreme heat only affects the propellant, never coming into contact with any structural materials. This design feature effectively mitigates the issues associated with material degradation under high temperatures. Additionally, the system operates at high pressures, exceeding 3.5 MPa, to prevent the bulk boiling of the uranium metal, ensuring the reactor's stability and efficiency throughout its operation.

While ongoing modeling and analysis continue to support the viability of this concept, numerous engineering challenges must be overcome before its feasibility can be fully established. These challenges include:

1. **Heat Transfer Efficiency:** It is essential to demonstrate effective heat transfer between the metallic liquid uranium and the propellant. This is crucial for achieving the desired thermal efficiency and ensuring the propellant reaches the required high temperatures.

2. **Porous Rotating Cylinder Wall Development:** A key engineering hurdle is the development of a porous rotating cylinder wall. This wall must allow the propellant to flow into the Centrifugal Fuel Element (CFE) while simultaneously preventing molten uranium from being expelled by centrifugal force through the propellant flow passages. The porous structure should finely distribute the incoming propellant and align the flow with the axial power profile within the rotating cylinder, ensuring adequate mixing between the uranium and propellant.

3. **High-Temperature Coating:** A suitable coating must be formulated for the inner surface of the rotating cylinder wall. This coating needs to be compatible with both liquid uranium and all potential propellants at temperatures around 1500 K,

maintaining structural integrity and preventing chemical reactions that could compromise the system.

4. **Rotating Cylinder Design and Cooling:** The design and fabrication of the rotating cylinder itself pose significant challenges. The cylinder must be engineered to handle high rotational speeds and extreme temperatures, with the addition of transpiration and film cooling techniques to prevent the formation of hot spots that could lead to material failure.

5. **High-Speed Rotation and Redundancy:** Reliable mechanisms must be developed to rotate the cylinders at several thousand RPM. Additionally, systems must be devised to manage the failure of individual cylinders without compromising the overall reactor performance, ensuring the engine can continue to operate safely and efficiently.

6. **Transient Accommodation:** Methods for handling operational transients, including startup and shutdown phases, are critical. These methods must minimize the loss of uranium fuel and prevent vibrational instabilities that could jeopardize the reactor's stability and safety.

7. **Uranium Loss Control:** The reactor and cylinder exit must be designed to minimize uranium loss, with a target of losing less than 0.01% of the propellant mass as High Assay Low Enriched Uranium (HALEU). Achieving this goal is essential for maintaining the reactor's long-term efficiency and sustainability.

8. **Reactivity Control:** Effective reactivity control mechanisms must be developed to manage the reactor's operation during startup, shutdown, and fuel burnup. These mechanisms must also address any potential entrainment of uranium in the propellant stream, ensuring stable and controlled reactor behavior.

9. **Neutronic Optimization:** The neutronic design of the fuel must be carefully optimized. Lessons learned from previous liquid reactor development programs, particularly those operating at lower temperatures, should be applied to ensure stable operation throughout the reactor's lifecycle, including during startup, steady-state operation, and shutdown.

10. **Integration with NTP Engine:** Finally, methods for integrating the Centrifugal Nuclear Thermal Reactor (CNTR) into an NTP engine must be developed. The CNTR design relies on a moderator block approach, and existing techniques for incorporating traditional NTP reactors with moderator blocks may be applicable to this advanced system.

The extensive list of engineering challenges highlights why a liquid-fuel nuclear thermal rocket engine has not yet been developed or prototyped. However, these challenges are inherently tied to the high operating temperatures, which are also the key to unlocking the engine's high-performance potential—specifically, achieving a specific impulse of 1800 seconds with a thrust-to-weight ratio comparable to that of solid-fuel

NTP engines. Overcoming these obstacles could revolutionize space propulsion, enabling more ambitious and far-reaching missions in the future.

Current scientific missions to the outer planets of the Solar System often rely on planetary flyby trajectories to gain the necessary velocity from gravitational assists. While effective, these trajectories come with significant limitations: they create infrequent and narrow launch windows and result in transit times from Earth to the destination that are typically twice as long as those achievable with direct trajectories. Unfortunately, chemical propulsion systems lack the necessary performance to support direct trajectories to these distant targets.

Mission analyses have demonstrated that solid-fuel nuclear thermal propulsion (NTP) systems can enable direct trajectory rendezvous missions to planets like Jupiter and Saturn. Such missions could benefit from launch windows that open approximately once a year, using commercially available heavy-lift rockets. Preliminary studies have further suggested that liquid-fuel NTP systems could enable direct trajectories to even more distant targets, such as selected Kuiper Belt objects, including Pluto and Quaoar.

Beyond expanding the boundaries of scientific exploration within the Solar System, liquid-fuel NTP has the potential to drastically reduce travel times for human missions. By enabling the use of trajectories that deviate from minimum energy paths, this technology could make rapid transit to many planetary destinations feasible, opening new possibilities for human exploration. It is this transformative potential that justifies ongoing research into liquid-fuel NTP systems.

11.19 The Demonstration Rocket for Agile Cislunar Operations (DRACO)

The Demonstration Rocket for Agile Cislunar Operations (DRACO) is a pioneering nuclear thermal propulsion (NTP) project, set to become the world's first in-orbit demonstration of an NTR engine. Developed under a collaboration between the Defense Advanced Research Projects Agency (DARPA), NASA, Lockheed Martin, and BWX Technologies, DRACO aims to revolutionize space travel efficiency and capabilities. The program is scheduled for a 2027 demonstration, with the spacecraft expected to be launched aboard a Vulcan Centaur provided by the U.S. Space Force.

11.19.1 Project Overview and Objectives

The DRACO initiative focuses on developing a reusable spacecraft powered by a next-generation nuclear thermal rocket engine. This engine will use low-enriched uranium (LEU) as fuel, with an enrichment level of approximately 20% U-235, significantly lower than weapons-grade material but higher than that used in commercial nuclear power plants. The primary objective of DRACO is to demonstrate the viability and advantages of NTP technology in space, specifically targeting missions beyond Earth's orbit, such as crewed missions to Mars.

11.19.2   Reactor and Propulsion System

DRACO's NTP engine design includes a fission reactor that heats a liquid hydrogen propellant. The reactor operates by transferring heat to the hydrogen, converting it into gas at extremely high temperatures (around 2,700 K), which then expands through a nozzle to generate thrust. The propulsion system employs an expander cycle, wherein liquid hydrogen first cools various components before being superheated by the reactor and expelled to produce thrust.

The core design emphasizes safety and efficiency. The use of LEU reduces regulatory complexities and safety risks associated with highly enriched uranium. The propellant management system, including cryogenic storage and handling of liquid hydrogen, is a critical component, ensuring the propellant remains in a liquid state over extended periods, a significant challenge for long-duration missions.

11.19.3   Specific Impulse and Thrust

One of the significant advantages of NTP over traditional chemical propulsion is its higher specific impulse, a measure of propulsion efficiency. DRACO aims for a specific impulse exceeding 800 seconds, nearly doubling the efficiency of conventional chemical rockets. This efficiency translates to a significant reduction in travel time for interplanetary missions, potentially halving the time required for a journey to Mars.

11.19.4   Safety and Regulation

The DRACO program follows stringent safety protocols, particularly concerning nuclear material handling and launch safety. According to a 2019 U.S. presidential memorandum, the launch of spacecraft using uranium enriched below 20% only requires approval from the sponsoring agency's head, streamlining the regulatory process. The spacecraft will only activate its reactor once in a stable orbit, minimizing risks associated with nuclear materials.

11.19.5   Development Phases and Testing

The DRACO program is divided into three primary phases:
1. **Phase 1** focused on initial design concepts and was completed with contracts awarded to General Atomics, Blue Origin, and Lockheed Martin for reactor and spacecraft designs.
2. **Phase 2** involves non-nuclear testing, including cold-flow tests with a non-nuclear engine mock-up to evaluate mechanical integrity and design robustness under operational conditions.
3. **Phase 3** encompasses the integration of the reactor and engine, environmental testing, and the in-orbit demonstration. This phase includes rigorous testing to validate the reactor's performance under space conditions, including thermal

management and radiation shielding.

The demonstration mission will place the DRACO spacecraft into a high orbit between 700 and 2,000 kilometers above Earth. The reactor will not be activated until a safe orbit is confirmed, ensuring that even in the unlikely event of a failure, the radioactive materials would not pose a risk to Earth's environment.

11.19.6   Potential Impacts and Future Applications

DRACO represents a significant step forward in space propulsion technology. The higher efficiency and specific impulse of NTP systems promise more agile and flexible space operations, particularly in cislunar space and beyond. With potential applications ranging from crewed missions to Mars to extended lunar operations and deep-space exploration, DRACO's successful demonstration could pave the way for a new era of space travel.

NASA and DARPA's collaboration highlights the growing interest in nuclear technologies for space applications, recognizing their potential to extend human presence deeper into the solar system. The DRACO program not only aims to demonstrate technical feasibility but also to lay the groundwork for future NTP systems that could drastically reduce mission costs, increase payload capacities, and shorten travel times in space exploration.

Bibliography

Nassersharif, Bahram, "Nuclear Reactor Engineering: Volume 1: Nuclear Engineering Fundamentals," 2024, ISBN- 979-8325597350.

Nassersharif, B., & Thomas, D. (2023). Nuclear Propulsion. IntechOpen. doi: 10.5772/intechopen.110616

National Research Council, New Frontiers in the Solar System: An Integrated Exploration Strategy, The National Academies Press, Washington, D.C., 2003.

National Research Council, Vision and Voyages for Planetary Science in the Decade 2013–2022, The National Academies Press, Washington, D.C., 2011.

L. S. Mason et al., "Kilowatt-Class Fission Power Systems for Science and Human Precursor Missions," presented at American Nuclear Society Topl. Mtg. Nuclear and Emerging Technologies for Space (NETS-2013), Albuquerque, New Mexico, February 25–28, 2013.

L. S. Mason et al., "Versatile Stirling Technology for Radioisotope and Fission Power Systems," Proc. Topl. Mtg. Nuclear and Emerging Technologies for Space, John C. Stennis Space Center, Mississippi, February 24–26, 2014, p. 1024, American Nuclear Society (2014).

J. T. Creasy et al., "Fuel Selection and Development for Small Fission Power System," Proc. Topl. Mtg. Nuclear and Emerging Technologies for Space, John C. Stennis Space Center, Mississippi, February 24–26, 2014, p. 2006, American Nuclear Society (2014).

D. I. Poston et al., "Notional Design of the Kilopower Space Reactor," Proc. Topl. Mtg. Nuclear and Emerging Technologies for Space, John C. Stennis Space Center, Mississippi, February 24–26, 2014, p. 2007, American Nuclear Society (2014).

M. A. Gibson et al., "Development of NASA's Small Fission Power System for Science and Human Exploration," Proc. Topl. Mtg. Nuclear and Emerging Technologies for Space, John C. Stennis Space Center, Mississippi, February 24–26, 2014, p. 2015, American Nuclear Society (2014).

P. R. McClure et al., "The Use of the Nevada National Security Site as a Reactor Test Center," Proc. Topl. Mtg. Nuclear and Emerging Technologies for Space, John C. Stennis Space Center, Mississippi, February 24–26, 2014, p. 2016, American Nuclear Society (2014).

M. A. Gibson et al., "Development of NASA's Small Fission Power System for Science and Human Exploration," Proc. Propulsion Energy Forum, Cleveland, Ohio, July 28–30, 2014, AIAA-2014-3458, American Institute of Aeronautics and Astronautics (2014); https://arc.aiaa.org/doi/abs/10.2514/6.2014-3458 (current as of Feb. 2020).

M. A. Rucker, "Integrated Surface Power Strategy for Mars," Proc. Topl. Mtg. Nuclear and Emerging Technologies for Space, February 23–26, 2015, Albuquerque, New Mexico, p. 5074, American Nuclear Society (2015).

D. I. Poston et al., "Reactor Design of the Kilowatt Reactor Using Stirling Technology (KRUSTY)," Proc. Topl. Mtg. Nuclear and Emerging Technologies for Space, Huntsville, Alabama, February 22–25, 2016, p. 6090, American Nuclear Society (2016).

D. I. Poston et al., "Shielding Options for Kilopower Mars Surface Reactors," Proc. Topl. Mtg. Nuclear and Emerging Technologies for Space, Huntsville, Alabama, February 22–25, 2016, p. 6089, American Nuclear Society (2016).

D. T. Palac et al., "Nuclear Systems Kilopower Update," Proc. Topl. Mtg. Nuclear and Emerging Technologies for Space, Huntsville, Alabama, February 22–25, 2016, p. 6057, American Nuclear Society (2016).

M. Briggs and M. A. Gibson, "Status of the Kilowatt Reactor Using Stirling Technology (KRUSTY)," Proc. Topl. Mtg. Nuclear and Emerging Technologies for Space, Orlando, Florida, February 27–March 2, 2017, p. 21151, American Nuclear Society (2017).

M. A. Gibson et al., "NASA's Kilopower Reactor Development and the Path to Higher Power Missions," Proc. IEEE Aerospace Conf. 2017, Big Sky, Montana, March 6–10, 2017, Institute of Electrical and Electronic Engineers (2017).

J. R. Casani, "Space Fission Power: NASA's Best Bet to Continue to Explore the Outer Solar System," Proc. Topl. Mtg. Nuclear and Emerging Technologies for Space, Richland, Washington, February 25–28, 2019, American Nuclear Society (2019).

R. Sanchez et al., "Kilowatt Reactor Using Stirling Technology (KRUSTY) Component-Critical Experiments," Nucl. Technol., vol. 206, p. S56, 2020; https://doi.org/10.1080/00295450.2020.1722553.

T. Grove et al., "Kilowatt Reactor Using Stirling Technology (KRUSTY) Cold Critical Measurements," Nucl. Technol., vol. 206, p. S68, 2020; https://doi.org/10.1080/00295450.2020.1712950.

D. I. Poston et al., "Results of the KRUSTY Nuclear System Test," Nucl. Technol., vol. 206, p. S89, 2020; https://doi.org/10.1080/00295450.2020.1730673.

H. M. Dieckamp, *Nuclear Space Power Systems*, Atomics International, Canoga Park, CA, Sep. 1967.

Atomic Industrial Forum, *Guidebook for the Application of Space Nuclear Power Systems*, New York, Jan. 1969.

U.S. Department of Energy, "Environmental Development Plan (EDP)-Space Applications," DOE/EDP-0026, Apr. 1978.

D. Buden et al., "Selection of Power Plant Elements for Future Reactor Space Electric Power Systems," LA-7858, Los Alamos National Laboratory, NM, Sep. 1979.

G. L. Bennett, J. J. Lombardo, and B. J. Rock, "Space Nuclear Electric Power Systems," *Advances in Astronautical Sciences*, vol. 44, American Astronautical Society, 1980.

Robert G. Lange and Edward F. Mastal, "A Tutorial Review of Radioisotope

Power Systems," *Critical Review of Space Nuclear Power and Propulsion 1984-1993*, Edited by M. S. El-Genk, American Institute of Physics Press, New York, ISBN 1-56396-317-5, pp. 1-20, 1994.

John C. Mankins, "Technology Readiness Levels," White Paper, Apr. 6, 1995, Office of Space Access and Technology, NASA.

Lisa Herrera and Beverly A. Cook, "Support Facilities For Radioisotope Applications In Space Environments," *NTSE-92 Nuclear Technologies For Space Exploration*, American Nuclear Society Meeting, Jackson, WY, Aug. 16-19, 1992, pp. 114-122.

Jerry L. Wert, Dennis L. Oberg, and Tommy L. Criswell, "Effect of Radiation from an RTG on the Installation, Personnel, and Electronics of a Launch System," *Space Nuclear Power Systems*, Edited by M. El-Genk and M. D. Hoover, Orbit Book Co., Melbourne, FL, 1992, pp. 151-157.

Gary L. Bennett, et al., "Development and Implementation of a Space Nuclear Safety Program," *Space Nuclear Power Systems 1987*, Edited by M. El-Genk and M. D. Hoover, Orbit Book Co., Malabar, FL, 1988, pp. 59-92.

Gary L. Bennett, "A Look At The Soviet Space Nuclear Power Program," *The 24th Intersociety Energy Conversion Engineering Conference*, Washington, D.C., Aug. 6-11, 1989.

Gary L. Bennett, "Lessons Learned From The Galileo and Ulysses Flight Safety Review Experience," CP420, *Space Technology and Applications International Forum-1998*, Edited by M. S. El-Genk, 1998, pp. 1269-1274.

"NASA Document, Final Environmental Impact Statement for the New Horizons Mission," July 2005.

G. Bennett, et al., "Mission of Daring: The General-Purpose Heat Source Radioisotope Thermoelectric Generator," *4th International Energy Conversion Engineering Conference*, AIAA 2006-4096, June 2006.

B. Alan Harmon and David B. Lavery, "NASA Radioisotope Power Systems Program Update," CP969, *Space Technology and Applications International Forum—STAIF 2008*, Edited by M. S. El-Genk, 2008 American Institute of Physics, 978-0-7354-0486-1/08, pp. 396-402.

Dennis Miotla, "Assessment of Plutonium-238 Production Alternatives," Briefing for Nuclear Energy Advisory Committee, Apr. 21, 2008.

Brian Berger, "Disagreement Arises About Plutonium-238 Storage," *Space News*, Apr. 7, 2008, p. 6.

H. Bateman, Cambridge Philosophical Society Proceedings, vol. 15, pp. 423-427, 1910.

A. A. Pitrolo, B. J. Rock, W. Remini, and J. A. Leonard, "SNAP-27 Program Review," Intersociety Energy Conversion Engineering Conference Proceedings, pp. 153-170, 1969.

W. C. Remini and J. H. Grayson, "SNAP-27 /ALSEP Power Subsystem Used in the Apollo Program," Intersociety Energy Conversion Engineering Conference, 1970 Proceedings, pp. 13-10 to 13-15.

H. Lurie and S. Rocklin, "Transit RTG-A Status Report," Intersociety Energy Conversion Engineering Conference, 1970 Proceedings, pp. 15-111 to 15-116.

G. Stapfer and V. Truscello, "The Long-Term Behavior of SNAP-19 Generators Which Use the TAGS Thermoelectric Material," Intersociety Energy Conversion Engineering Conference Proceedings, pp. 1184-1189, 1971.

W. M. Brittain, "SNAP-19 Radioisotope Thermoelectric Generator Thermal/Electrical Integration With the Viking Mars Lander," Intersociety Energy Conversion Engineering Conference Proceedings, pp. 1189-1199, 1971.

P. Ronklove and V. Truscello, "Long-Term Tests of a SNAP-19 Thermoelectric Generator," Intersociety Energy Conversion Engineering Conference Proceedings, pp. 186-193, 1972.

"TRANSIT RTG Final Safety Analysis Report, Vol. I," TRW Report TRW(A)-11464-0491, Mar. 1971.

Multi-Hundred Watt Radioisotope Thermoelectric Generator Program, "Final Safety Analysis Report," General Electric Document No. GEMS-419, Mar. 1975.

C. E. Kelly, "The MHW Converter (RTG)," Intersociety Energy Conversion Engineering Conference, 1975 Proceedings, pp. 880-886.

D. Buden, et al., "Selection of Power Plant Elements for Future Reactor Space Electric Power Systems," LA-7858, Los Alamos National Laboratory, Sep. 1979.

C. J. Goebel and L. R. Putnam, "SNAP-19 Performance Update for Pioneer and Viking Missions," Nth Intersociety Energy Conversion Engineering Conference, pp. 1476-1479, Aug. 1979.

L. Garvey and G. Stapfer, "Performance Testing of Thermoelectric Generators Including Voyager and LES 8/9 Flight Results," Intersociety Energy Conversion Engineering Conference, pp. 1470-1475, 1979.

F. Kreith, Radiation Heat Transfer for Spacecraft and Solar Power Plant Design, International Textbook Co., Scranton, 1962.

Multi-Hundred Watt Radioisotope Thermoelectric Generator Program, "Final Safety Analysis Report for the MJS-77 Mission," General Electric Document No. 77SDS4206, Jan. 1977.

R. V. Anderson, et al., "Space Reactor Electric Systems," ESG-DOE-13398, Rockwell International, Mar. 29, 1983.

NASA, "Space Shuttle Program-Space Shuttle System Payload Accommodations, Level 1 Program Definition and Requirements, Vol. XIV," JSC-07700, Rev H, May 16, 1983.

NASA, Space and Planetary Environment Criteria Guidelines for Use in Space Vehicle Development, 1982 Revision (Volume I), NASA TM-82478, Jan. 1983.

G. L. Bennett, J. J. Lombardo, and B. J. Rock, "U.S. Radioisotope Thermoelectric Generator Space Operating Experience (June 1961-December 1982)," 18th Intersociety Energy Conversion Engineering Conference, Orlando, FL, Aug. 21-26, 1983.

C. Wood, "Optimization of Thermoelectric Materials," Proceedings of 1982 Working Group on Thermoelectrics, Jet Propulsion Laboratory, JPL-D-497, Jan. 1983.

Joseph A. Angelo, Jr. and David Buden, Space Nuclear Power, Orbit Book Company, 1985.

G. Stapfer and V. Truscello, "The Long-Term Behavior of SNAP-19 Generators Which Use the TAGS Thermoelectric Material," Intersociety Energy Conversion Engineering Conference Proceedings, pp. 1184-1189, 1981.

B. M. French and S. P. Maran, "A Meeting With the Universe," AT&4, EP-177, 1981.

J. A. Van Allen, "Pioneer's Unfunded Reach for the Stars," Aviation Week & Space Technology, Apr. 12, 1982, p. II.

Robert G. Lange and Edward F. Mastal, "A Tutorial Review of Radioisotope Power Systems," in Critical Review of Space Nuclear Power and Propulsion 1984-1993, Edited by M. S. El-Genk, American Institute of Physics, pp. 1-20, 1994.

John Dassoulas and Ralph L. McNutt, Jr., "RTGs on Transit," CP880, Space Technology and Applications International Forum—STAIF 2007, Edited by M. S. El-Genk, 2007, pp. 195-204.

H. Bateman, "Cambridge Philosophical Society Proceedings," vol. 15, pp. 423-427, 1910.

G. R. Grove, J. A. Powers, N. Goldenberg, D. P. Kelly, and D. L. Prosser, "Plutonium-238 Isotopic Heat Sources: A Summary Report," Mound Laboratory Report MEM-1270, Miamisburg, Ohio, June 7, 1965.

A. A. Pitrolo, B. J. Rock, W. Remini, and J. A. Leonard, "SNAP-27 Program Review," Intersociety Energy Conversion Engineering Conference Proceedings, pp. 153-170, 1969.

W. C. Remini and J. H. Grayson, "SNAP-27 /ALSEP Power Subsystem Used in the Apollo Program," Intersociety Energy Conversion Engineering Conference, 1970 Proceedings, pp. 13-10 to 13-15.

"TRANSIT RTG Final Safety Analysis Report, Vol. I," TRW report TRW(A)-11464-0491, Mar. 1971.

G. Stapfer and V. Truscello, "The Long-Term Behavior of SNAP-19 Generators Which Use the TAGS Thermoelectric Material," Intersociety Energy Conversion Engineering Conference Proceedings, pp. 1184-1189, 1971.

W. M. Brittain, "SNAP-19 Radioisotope Thermoelectric Generator Thermal/Electrical Integration With the Viking Mars Lander," Intersociety Energy Conversion Engineering Conference Proceedings, pp. 1189-1199, 1971.

P. Ronklove and V. Truscello, "Long-Term Tests of a SNAP-19 Thermoelectric

Generator," Intersociety Energy Conversion Engineering Conference Proceedings, pp. 186-193, 1972.

W. Beale et al., "Free Piston Stirling Engines - A Progress Report," SAE Paper 730647, June 1973.

Multi-Hundred Watt Radioisotope Thermoelectric Generator Program, "Final Safety Analysis Report," General Electric Document No. GEMS-419, Mar. 1975.

C. E. Kelly, "The MHW Converter (RTG)," Intersociety Energy Conversion Engineering Conference, 1975 Proceedings, pp. 880-886.

B. Goldwater, "Study of a Free Piston Stirling Engine-Linear Power Conversion System for a 10.5 kWe and 51.0 kWe Output Power," Mechanical Technology Inc., Nov. 1975.

E. C. Snow, R. W. Zocher, I. M. Grinberg, and L. E. Hulbert, "General-Purpose Heat Source Development Phase II-Conceptual Designs," Los Alamos National Laboratory Report LA-7546-SR, Nov. 1978.

A. Schock, "Design, Evolution and Verification of the General Purpose Heat Source," Intersociety Energy Conversion Engineering Conference, 1980 Proceedings, pp. 1032-1042.

G. L. Bennett, J. L. Lombardo, and B. J. Rock, "Nuclear Electric Power for Space Systems: Technology Background and Flight Systems Program," Intersociety Energy Conversion Engineering Conference, 1981 Proceedings, pp. 362-368.

J. A. Van Allen, "Pioneer's Unfunded Reach for the Stars," Aviation Week & Space Technology, Apr. 12, 1982, p. II.

Sundstrand Corporation, "Final Report, Technology Verification Phase, Dynamic Isotope Power System," Report No. 2032, Sunstrand Energy Systems, Rockford, IL, 1982.

NASA, "Space and Planetary Environment Criteria Guidelines for Use in Space Vehicle Development, 1982 Revision (Volume I)," NASA TM-82478, Jan. 1983.

R. V. Anderson, et al., "Space Reactor Electric Systems," ESG-DOE-13398, Rockwell International, Mar. 29, 1983.

C. Wood, "Optimization of Thermoelectric Materials," Proceedings of 1982 Working Group on Thermoelectrics, Jet Propulsion Laboratory, JPL-D-497, Jan. 1983.

NASA, "Space Shuttle Program-Space Shuttle System Payload Accommodations, Level 1 Program Definition and Requirements, Vol. XIV," JSC-07700, Rev H, May 16, 1983.

G. L. Bennett, J. J. Lombardo, and B. J. Rock, "U.S. Radioisotope Thermoelectric Generator Space Operating Experience (June 1961-December 1982)," 18th Intersociety Energy Conversion Engineering Conference, Orlando, FL, Aug. 21-26, 1983.

Joseph A. Angelo, Jr. and David Buden, Space Nuclear Power, Orbit Book Company, 1985.

Gary L. Bennett, et al., "Development and Implementation of a Space Nuclear

Safety Program," Space Nuclear Power Systems, 1987, Edited by M. S. El-Genk and M. D. Hoover, Orbit Book Co., Malabar, FL, 1988, pp. 59-92.

Gary L. Bennett and James J. Lombardo, "The Dynamic Isotope Power System: Technology Status and Demonstration Program," Chapter 20, Space Nuclear Power Systems 1988, Edited by M. S. El-Genk and M. D. Hoover, Orbit Book Company, Malabar, FL, 1989, pp. 101-115.

Richard J. Pearson, "Dynamic Isotope Power System Critical Component Verification," Space Nuclear Power Systems 1988, Editors Mohamed S. El-Genk and Mark D. Hoover, Orbit Book Co., ISSN 1-41-2824, 1989, pp. 117-125.

Marshall B. Eck and Meera Mukunda, "On the Nature of the Response of the General Purpose Heat Source to the Impact of Large Solid Rocket Motor Casing Fragments," Space Nuclear Power Systems 1989, Edited by M. S. El-Genk and M. D. Hoover, Orbit Book Co., Malabar, FL, 1992, pp. 223-238.

Richard B. Harty, "Comparison Of DIPS And RFCs For Lunar Mobile And Remote Power Systems," Space Nuclear Power Systems Ninth Symposium, Editors Mohamed S. El-Genk and Mark D. Hoover, AIP Conference Proceedings 246, L.C. Catalog Card No. 91-58793, 1992, pp. 202-207.

Gary L. Bennett, "Update to the Safety Program for the General Purpose Heat Source Radioisotope Thermoelectric Generators for the Galileo and Ulysses Missions," Space Nuclear Power Systems 1989, Edited by M. S. El-Genk and M. D. Hoover, Orbit Book Co. Malabar, FL., 1992, pp. 199-222.

William D. Otting, et al., "Dynamic Isotope Power System, Integrated System Test," CONF 940101, 1994 American Institute of Physics, 1994, pp. 365-369.

Robert G. Lange and Edward F. Mastal, "A Tutorial Review of Radioisotope Power Systems," in Critical Review of Space Nuclear Power and Propulsion 1984-1993, Edited by Mohamed S. El-Genk, American Institute of Physics, L.C. Catalog Card No. 94-70780, 1994, pp. 1-20.

Gary L. Bennett, "Safety Aspects of Thermoelectrics in Space," published in CRC Handbook of Thermoelectrics, Edited by D. M. Rowe, CRC Press, New York, 1995, pp. 551-572.

John C. Mankins, "Technology Readiness Levels," NASA Advanced Concepts Office, A White Paper, Apr. 6, 1995.

Christopher J. Crowley, et al., "Thermophotovoltaic Converter Performance for Radioisotope Power Systems," CP746, Space Technology and Applications International Forum-STAIF 2005, edited by M. S. El-Genk, 2005, pp. 601-614.

B. Xu, et al., "Integrated Bandpass Filter Contacts for GaSb Thermophotovoltaic Cells," CP746, Space Technology and Applications International Forum-STAIF 2005, edited by M. S. El-Genk, 2005, pp. 615-622.

Alan V. Von Arx, "MMRTG Heat Rejection Summary," CP813, Space Technology and Applications International Forum-STAIF 2006, edited by M. S. El-Genk,

2006, pp. 743-750.

National Aeronautics and Space Administration, "Final Environmental Impact Statement for the Mars Science Laboratory Mission," Volume 1, Nov. 2006.

Appendix

Physical Constants

Category	Constant	Value	Unit
Mathematical Constants	Pi (π)	3.141592653589793	-
	Euler's number (e)	2.718281828459045	-
Physical Constants	Speed of light (c)	299,792,458	m/s
	Gravitational constant (G)	6.67259×10^{-11}	Nm^2/kg^2
	Universal gas constant (R)	8,314.4621	$J/kmol - K$
	Stefan-Boltzmann constant (σ)	5.670373×10^8	$W/m^2 - K^4$
	Acceleration of gravity (g)	9.80665	m/s^2
	Standard atmosphere, sea level	101,325	Pa
Astronomical Constants	Astronomical unit (AU)	149,597,870	km
	Light year (ly)	9.460530×10^{12}	km
	Parsec (pc)	3.261633	ly
	Sidereal year	365.256366	days
	Mass of Sun	1.9891×10^{30}	kg
	Radius of Sun	696,000	km
	Mass of Earth	5.9737×10^{24}	kg
	Equatorial radius of Earth	6,378.137	km
	Earth oblateness	1/298.257	-
	Obliquity of the ecliptic (epoch 2000)	23.4392911	degrees
	Mean lunar distance	384,403	km
	Radius of Moon	1,738	km
	Mass of Moon	7.348×10^{22}	kg
	Luminosity of Sun	3.839×10^{26}	W
	Solar constant, at 1 AU	1,366	W/m^2
	Solar maxima	$1990 + 11n$	(date)
Spaceflight Constants	GM (Sun)	$1.32712438 \times 10^{20}$	m^3/s^2
	GM (Earth)	3.986005×10^{14}	m^3/s^2
	GM (Moon)	4.902794×10^{12}	m^3/s^2
	GM (Mars)	4.282831×10^{13}	m^3/s^2
	J_2 (Earth)	0.00108263	-
	J_2 (Moon)	0.0002027	-
	J_2 (Mars)	0.00196045	-

Planetary Data

Planet	Equatorial Radius (km)	Mass (10^{24} kg)	Surface Gravity (m/s²)	GM (10^{15} m³/s²)	Semimajor Axis (10^6 km)	Eccentricity	Orbit Inclination (°)	Sidereal Period (days)	Sidereal Rotation Period (hours)
Mercury	2,439.7	0.3302	3.70	0.02203	57.91	0.2056	7.00	87.960	1,407.6
Venus	6,051.8	4.8685	8.87	0.3249	108.21	0.0067	3.39	224.701	-5,832.5
Earth	6,378.1	5.9742	9.80	0.3986	149.60	0.0167	0.000	365.256	23.9345
Mars	3,397.0	0.64185	3.71	0.04283	227.92	0.0935	1.850	686.980	24.6229
Jupiter	71,492	1,898.6	24.79	126.686	778.57	0.0489	1.304	4,332.59	9.9250
Saturn	60,268	568.46	10.44	37.931	1,433.53	0.0565	2.485	10,759.2	10.656
Uranus	25,559	86.832	8.87	5.794	2,872.46	0.0457	0.772	30,685.4	-17.24
Neptune	24,764	102.43	11.15	6.835	4,495.06	0.0113	1.769	60,189	16.11
Pluto	1,195	0.0125	0.58	0.00083	5,869.66	0.2444	17.16	90,465	-153.29

Alpha Emitters

Isotope	Atomic Mass (u)	Half-Life	Decay Constant (λ) (s^{-1})	Alpha Decay Energy (MeV)
Uranium-238	238.05078	4.468 billion years	4.916×10^{-18}	4.27
Radium-226	226.02541	1,600 years	1.373×10^{-11}	4.78
Thorium-232	232.03805	14.05 billion years	1.564×10^{-18}	4.01
Polonium-210	209.98287	138.376 days	5.008×10^{-7}	5.30
Radon-222	222.01758	3.8235 days	2.083×10^{-6}	5.49
Plutonium-238	238.04956	87.7 years	2.505×10^{-11}	5.59
Plutonium-239	239.05216	24,100 years	9.131×10^{-13}	5.16
Americium-241	241.05682	432.2 years	5.020×10^{-11}	5.49
Curium-244	244.06275	18.1 years	1.214×10^{-9}	5.80
Actinium-227	227.02775	21.77 years	1.010×10^{-11}	4.95
Bismuth-212	211.98885	60.55 minutes	1.908×10^{-3}	6.09

Index

www.ingramcontent.com/pod-product-compliance
Lightning Source LLC
Chambersburg PA
CBHW060947210326
41598CB00031B/4747